COLLEGE MATHEMATICS FOR STUDENTS OF BUSINESS, LIFE SCIENCES, AND SOCIAL SCIENCES

COLLEGE MATHEMATICS FOR STUDENTS OF BUSINESS, LIFE SCIENCES, AND SOCIAL SCIENCES

Second Edition

DENNIS G. ZILL
Loyola Marymount University

EDWIN F. BECKENBACH
University of California, Los Angeles

WILLIAM WOOTON
Los Angeles Pierce College

IRVING DROOYAN
Los Angeles Pierce College

Wadsworth Publishing Company
Belmont, California
A division of Wadsworth, Inc.

Mathematics Editor: Richard Jones
Editorial Production Services: Cobb/Dunlop Publisher Services, Inc.
Cover painting: Janet Wood
Signing representative: Kay Chamberlain

© 1982 by Wadsworth, Inc.

© 1977, 1973, 1968, 1964 by Wadsworth Publishing Company, Inc. All rights reserved. No part of this book may be reproduced, stored in a retrieval system, or transcribed, in any form or by any means, electronic, mechanical, photocopying, recording, or otherwise, without the prior written permission of the publisher, Wadsworth Publishing Company, Belmont, California 94002, a division of Wadsworth, Inc.

Printed in the United States of America

1 2 3 4 5 6 7 8 9 10————86 85 84 83 82

Library of Congress Cataloging in Publication Data

Main entry under title:

College mathematics for students of business, life sciences, and social sciences.

   First ed. published in 1977 as: College mathematics for students of business and the social sciences.
   Includes index.
   1. Mathematics — 1961–   . I. Zill, Dennis G., 1940–   . II. College mathematics for students of business and the social sciences.
QA37.2.C65   1982   510   81-2211
ISBN 0-534-00886-0   AACR2

CONTENTS

PREFACE

PART I

1
REVIEW: SETS — 3

1.1	Definitions and Symbols	3
1.2	The Real Numbers	10
	Chapter Test	15

2
REVIEW: POLYNOMIALS AND THE LAWS OF EXPONENTS — 16

2.1	Sums of Polynomials	16
2.2	Products of Polynomials	21
2.3	Factoring Polynomials	25
2.4	Quotients of Polynomials	28
2.5	Equivalent Fractions	33
2.6	Sums of Rational Expressions	36
2.7	Products and Quotients of Rational Expressions	39

2.8	Powers with Integral Exponents	43
2.9	Powers with Rational Exponents	47
2.10	Radical Expressions	52
	Chapter Test	59
	Cumulative Review of Part I	63

PART II

3
EQUATIONS AND INEQUALITIES IN ONE VARIABLE 67

3.1	Equivalent Equations; First-Degree Equations	67
3.2	Second-Degree Equations	73
3.3	Equations Involving Radicals	80
3.4	Substitution in Solving Equations	84
3.5	Solution of Linear Inequalities	85
3.6	Solution of Quadratic Inequalities	90
3.7	Equations and Inequalities Involving Absolute Values	95
3.8	Word Problems	99
	Chapter Test	106

4
RELATIONS AND FUNCTIONS 109

4.1	Ordered Pairs of Real Numbers	109
4.2	Linear Functions	117
4.3	Forms of Linear Equations	127
4.4	Quadratic Functions	134
4.5	Polynomial Functions	142
4.6	Special Functions and Variation	145
4.7	Inverse Relations and Functions	151
	Chapter Test	156

5
EXPONENTIAL AND LOGARITHMIC FUNCTIONS — 158

5.1	The Exponential Function	158
5.2	The Logarithmic Function	162
5.3	Logarithms to the Base 10	167
5.4	Applications of Common Logarithms	175
5.5	Computations with Logarithms	179
5.6	Exponential and Logarithmic Equations	184
5.7	Logarithms to the Base e	188
	Chapter Test	194

6
SYSTEMS OF EQUATIONS — 197

6.1	Systems of Linear Equations in Two Variables	197
6.2	Systems of Linear Equations in Three Variables	203
6.3	Systems of Nonlinear Equations	209
6.4	Break-Even Point and Equilibrium Point	216
	Chapter Test	221

7
MATRICES AND DETERMINANTS — 222

7.1	Definitions; Matrix Addition	222
7.2	Matrix Multiplication	229
7.3	Solution of Linear Systems Using Row-Equivalent Matrices	241
7.4	Determinants	246
7.5	The Inverse of a Square Matrix	251
7.6	Solution of Linear Systems Using Inverses of Matrices	259
7.7	Cramer's Rule	265
	Chapter Test	271

PART III

8
INEQUALITIES AND LINEAR PROGRAMMING — 275

8.1	Graphs of First-Degree Relations	275
8.2	Systems of Linear Inequalities	278
8.3	Convex Sets; Polygonal Regions	282
8.4	Linear Programming	285
8.5	The Simplex Method	291
8.6	The Dual Problem	298
	Chapter Test	304

9
SEQUENCES AND SERIES — 306

9.1	Sequences	306
9.2	Series	314
9.3	Limits of Sequences and Series	321
9.4	The Binomial Theorem	328
	Chapter Test	336

10
PROBABILITY — 338

10.1	Basic Counting Principles; Permutations	338
10.2	Combinations	345
10.3	Probability Functions	348
10.4	Probability of the Union of Events	353

10.5	Probability of the Intersection of Events	358
10.6	Binomial Probability	363
10.7	Conditional Probability	368
	Chapter Test	373

11
MATHEMATICS OF FINANCE — 375

11.1	Computations of Interest	375
11.2	Annuities	380
11.3	Present Value of an Annuity	384
11.4	Continuous Compounding of Interest	388
	Chapter Test	394

PART IV

12
DIFFERENTIAL CALCULUS — 397

12.1	Tangent Line to a Graph	397
12.2	The Derivative	404
12.3	Derivatives of the Power Function, Constants, and Sums	408
12.4	Derivatives of Products and Quotients	413
12.5	The Power Rule for Functions	418
12.6	Implicit Differentiation	421
12.7	Higher-Order Derivatives	425
12.8	Derivative of the Logarithmic Function	429
12.9	Derivative of the Exponential Function	434
	Chapter Test	438

13
APPLICATIONS OF THE DERIVATIVE 439

13.1	The First and Second Derivatives and the Shape of a Graph	439
13.2	Maximum and Minimum Values of a Function	444
13.3	Absolute Extrema — Applications	450
13.4	The Derivative as a Rate of Change	456
13.5	Related Rates	461
13.6	Further Applications of the Derivative	464
	Chapter Test	471

14
INTEGRAL CALCULUS 473

14.1	Antiderivatives	473
14.2	Integration by Substitution	478
14.3	Integration by Parts	485
14.4	The Definite Integral	490
	Chapter Test	495

15
APPLICATIONS OF THE INTEGRAL 496

15.1	Area as a Definite Integral	496
15.2	Area Between Two Graphs	503
15.3	The Average Value of a Function	507
15.4	Further Applications of the Definite Integral	510
	Chapter Test	518

16
MULTIVARIATE DIFFERENTIAL CALCULUS 519

16.1	Functions of Several Variables	519
16.2	Partial Differentiation	524
16.3	Higher-Order Derivatives	529
16.4	Maxima and Minima	533
16.5	Constrained Extrema	538
	Chapter Test	544

APPENDIX I
THE FUNDAMENTAL PROPERTIES OF THE REAL NUMBER SYSTEM 545

APPENDIX II
TABLES 549

I	Common Logarithms	550
II	Exponential Functions	552
III	Natural Logarithms of Numbers	553
IV	Compound Amount of $1	554
V	Present Value of $1	555
VI	Amount of an Annuity of $1 per Period	556
VII	Present Value of an Annuity of $1 per Period	557

ODD-NUMBERED ANSWERS 559

INDEX 605

PREFACE

This second edition of *College Mathematics*, like the first, is designed as an introductory course in mathematics for those students pursuing concentrations in business and the social sciences. However, in this edition we have expanded the applications so that the material is equally suitable for students in biological, pre-medical or pre-dental curricula. Although many applications have been added to the text, we have adhered to the philosophy of introducing these applications gradually to parallel the increasing utility of the mathematics. In this manner we hope the applications reflect some measure of "reality."

College Mathematics is written with flexibility strongly in mind. The division of the book into four parts should provide suitable and ample material for either a two semester or three quarter course sequence in beginning college-level mathematics.

Part I is a review of the basic properties of sets, real numbers, the laws of exponents, and polynomial expressions. Since it is likely that many students are familiar with some or all of these topics, this part can be covered quickly, skipped entirely, or assigned as an independent study. For the convenience of the student, as well as the instructor who may choose to use it as a diagnostic test, Part I ends with a cumulative review.

Part II could be the starting point of the course. It contains equations, inequalities, functions, logarithms, systems of equations, matrices, and determinants.

Part III contains linear programming, sequences, series probability, and the mathematics of finance.

Part IV is an introduction to the fundamental concepts of differential and integral calculus. The amount of material in this section should be sufficient for a one-quarter course in calculus.

This second edition also contains some new features:

Chapter Tests have replaced the Chapter Reviews.

A section on applications of the logarithm has been added. The pH of a solution, the intensity of sound, and the Richter scale are some of the topics considered.

A section on Bernoulli trials and binomial probability has been added.

The material on the calculus has been entirely rewritten and expanded. The tangent line and area problems are considered while keeping the notion of a limit on an intuitive level. The differential calculus of several variables is now placed in a separate chapter. The concept of Lagrange multipliers has been introduced.

The number of problems and examples has been increased considerably. We feel that abundant examples and problems are the primary strength of any text in mathematics.

The metric system of units has received increased emphasis.

Definitions and theorems are encased in a box for easy recognition and reference.

In conclusion we would like to express our appreciation to the many users of the first edition who have taken the time to communicate their ideas for improvement of the text. A special word of thanks goes to the following reviewers for their helpful comments: Edith W. Ainsworth, University of Alabama; Richard Black, Cleveland State University; Leonard Bruening, Cleveland State University; Evi Nemeth, State University of New York at Utica/Rome; Paul Pontius, Pan American University.

Dennis G. Zill
Edwin F. Beckenbach
William Wooton
Irving Drooyan

PART I

1 REVIEW: SETS
2 REVIEW: POLYNOMIALS AND THE LAWS OF EXPONENTS

1

REVIEW: SETS

1.1

DEFINITIONS AND SYMBOLS

A **set** is a collection of some kind. It may be a collection of people, colors, numbers, or anything else. In algebra, we are interested in sets of numbers of various sorts and in their relations to sets of points or lines in a plane or in space. Any one of the collection of things in a set is called a **member** or **element** of the set, and is said to be **contained** or **included in** the set. For example, the counting numbers 1, 2, 3, ... (where the dots indicate that the sequence continues indefinitely) are the elements of the set we call the set of **positive integers** or **natural numbers**.

SET NOTATION Sets are usually designated by capital letters, A, B, C, etc. They are identified by means of **braces**, $\{\ \}$, and the members are either listed or described within the braces. For example, the elements might be listed as in $\{1, 2, 3\}$, or described as in {first three positive integers}. The expression "$\{1, 2, 3\}$" is read "the set whose elements are one, two, and three"; "{first three positive integers}" is read "the set whose elements are the first three positive integers."

SUBSETS If every member of a set A is also a member of a set B, then we say that A is a **subset** of B. For example, if $A = \{1, 2, 3\}$ and $B = \{1, 2, 3, 4\}$, then A is a subset of B. The symbol $\subset$ (read "is a subset of" or "is contained in") will be used to denote the subset relationship. Thus,

$$\{1, 2, 3\} \subset \{1, 2, 3, 4\} \qquad \{2, 3\} \subset \{1, 2, 3\} \qquad \{1, 2, 3\} \subset \{1, 2, 3\}.$$

Notice that every set is a subset of itself.

The set that contains no elements is called the **empty set** or **null set,** and is denoted by the symbol $\emptyset$ (read "the empty set" or "the null set"); $\emptyset$ is a subset of every set.

**SET-
MEMBER-
SHIP
NOTATION**

The symbol ∈ (read "is a member of" or "is an element of") is used to denote membership in a set. Thus,
$$2 \in \{1, 2, 3\}.$$
Note that we write
$$\{2\} \subset \{1, 2, 3\} \qquad \text{and} \qquad 2 \in \{1, 2, 3\},$$
since $\{2\}$ is a *subset* of $\{1, 2, 3\}$, whereas 2 is an *element* of $\{1, 2, 3\}$.

An individual element of a set that contains more than one element is usually denoted by a lowercase italic letter (for example, a, d, s, x), or sometimes by a letter from the Greek alphabet such as α (alpha), β (beta), γ (gamma), etc. Symbols used in this way are called **variables**.

**NEGATION
SYMBOL**

The slant bar, /, drawn through certain symbols indicates negation. Thus, $\neq$ is read "is not equal to," $\not\subset$ is read "is not a subset of," and $\notin$ is read "is not an element of." For example,
$$\{1, 2\} \neq \{1, 2, 3\} \qquad \{1, 2, 3\} \not\subset \{1, 2\} \qquad 3 \notin \{1, 2\}.$$

**SET-
BUILDER
NOTATION**

Another symbolism useful in discussing sets is illustrated by
$$\{x \mid x \in A \quad \text{and} \quad x \notin B\}$$
(read "the set of all x such that x is a member of A and x is not a member of B"). This symbolism, called **set-builder notation**, specifies a variable (in this case, x) and, at the same time, states a condition on the variable (in this case, that x is contained in the set A and not in the set B.)

**THE
UNIVERSE**

A general set containing all the objects under discussion is called the **universe**. Ideas involving universal sets, their subsets, and certain operations on sets can be depicted using plane geometric figures called **Venn diagrams**. Figure 1.1 shows such a diagram representing a universe U which has as its elements all points of the rectangle and its interior. It also shows subsets A, B, C, D, and E of the universe, each denoted by a circle and its interior. In this figure, sets A, B, and C have no elements in common, D is a subset of C, and E has some common elements with C.

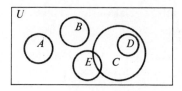

Figure 1.1

1.1 Definitions and Symbols

OPERATIONS ON SETS There are several mathematically important operations on the subsets of a given universe. One such operation is defined as follows.

> The **union** of two subsets A and B of a universe U is the set of all elements of U that belong either to A or to B or to both.

SET UNION SYMBOL The symbol $\cup$ is used to denote the union of sets. Thus, $A \cup B$ can be written as
$$A \cup B = \{x \mid x \in A \quad \text{or} \quad x \in B\}.$$

Example If
$$A = \{1, 2, 3, 4, 5\}$$
and
$$B = \{2, 3, 4, 5, 6\},$$
then
$$A \cup B = \{1, 2, 3, 4, 5, 6\}.$$

Notice that each element in $A \cup B$ is listed only once in this example, since repetition would be redundant. Figure 1.2 is a Venn diagram in which the shaded region depicts $A \cup B$.

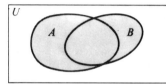

Figure 1.2

A second set operation is defined as follows.

> The **intersection** of two subsets A and B of a universe U is the set of all elements of U that belong to both A and B (the common elements).

SET INTERSECTION SYMBOL The symbol $\cap$ is used to denote intersection. Thus, $A \cap B$ can be written as
$$A \cap B = \{x \mid x \in A \quad \text{and} \quad x \in B\}.$$

Example If
$$A = \{1, 2, 3, 4, 5\}$$
and
$$B = \{2, 3, 4, 5, 6\},$$
then
$$A \cap B = \{2, 3, 4, 5\}.$$

In the Venn diagram of Figure 1.3, the shaded region depicts $A \cap B$.

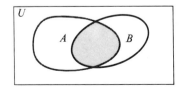

Figure 1.3

Example A company C_1 makes

$$A = \{\text{television sets, refrigerators, stoves, dishwashers}\},$$

and company C_2 makes

$$B = \{\text{refrigerators, toasters, electronic ovens, stoves}\}.$$

If company C_1 buys company C_2, then together they will manufacture

$$A \cup B = \{\text{television sets, refrigerators, stoves, dishwashers, toasters, electronic ovens}\}.$$

The new company can decrease costs by eliminating the duplication in the manufacturing of

$$A \cap B = \{\text{refrigerators, stoves}\}.$$

Example In a survey of 150 families that viewed TV on a particular evening it was found that
  86 families watched ABC
  76 families watched CBS
  65 families watched NBC
and
  37 families watched both ABC and CBS
  30 families watched both ABC and NBC
  16 families watched both NBC and CBS
and
  6 families watched all three networks.
How many families watched only ABC on the particular evening? Only CBS? Only NBC?

Solution The six families that watch all three networks have already been counted in the intersections; namely, in the given number of families that watch two networks. Thus, in the three shaded areas in the first figure there are $37 - 6 = 31$, $30 - 6 = 24$, and $16 - 6 = 10$ families. It follows that the number of families that watch only

$$\text{ABC are } 86 - (31 + 6 + 24) = 25,$$
$$\text{CBS are } 76 - (31 + 6 + 10) = 29,$$
$$\text{NBC are } 65 - (24 + 6 + 10) = 25.$$

These numbers are indicated in the shaded areas in the second figure.

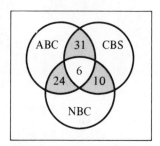
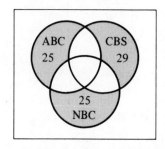

1.1 Definitions and Symbols

The set operation of intersection can be used to describe sets that have no elements in common.

> Two sets A and B are **disjoint** if and only if $A \cap B = \emptyset$.

For example, in Figure 1.1, $A \cap B = \emptyset$, $B \cap E = \emptyset$, $B \cap C = \emptyset$, $D \cap E = \emptyset$, and so on. Also, $A = \{1, 2, 3\}$ and $B = \{5, 6, 7\}$ are disjoint, since they have no common elements.

COMPLEMENT OF A SET Both union and intersection are applied to sets in relation to each other. The following operation on sets, however, applies to only one set in relation to the universe U.

> The **complement** of a set A in U is the set of all elements of U that do not belong to A.

The symbol A' is used to denote the complement of A in U.

Example If

$$U = \{1, 2, 3, 4, 5\} \quad \text{and} \quad A = \{2, 4\},$$

then

$A' = \{\text{elements of } U \text{ which are not in } A\}$

$\quad = \{1, 3, 5\}.$

The shaded part of the Venn diagram in Figure 1.4 represents A'. Notice that $A \cap A' = \emptyset$ (that is, a set and its complement are disjoint), and that $A \cup A' = U$.

Figure 1.4

EXERCISE 1.1

Replace the question mark with either $\in$ or $\notin$ to make a true statement.

1. 3 ? $\{2, 3, 4\}$
2. 15 ? $\{2, 4, 6, \ldots\}$
3. $\{2\}$? $\{2, 3, 4\}$
4. $\emptyset$? $\{2, 3, 4\}$

Replace the question mark with either $\subset$ or $\not\subset$ to make a true statement.

5. $5 \;?\; \{4, 5, 6\}$
7. $\emptyset \;?\; \{4, 5, 6\}$
6. $\{5, 4, 6\} \;?\; \{4, 5, 6\}$
8. $\{4\} \;?\; \{4, 5, 6\}$

9. Let $U = \{5, 6, 7\}$. List the subsets of U that contain
  a. three members **b.** two members **c.** one member **d.** no members

10. Let $U = \{1, 2, 3, 4\}$. List the subsets of U that contain
  a. four members **b.** three members **c.** two members
  d. one member **e.** no members **f.** 2 and one other member

Designate each of the sets listed in Problems 11–16 by using set-builder notation.

Example {even positive integers}

Solution $\{x \mid x = 2n, \; n \text{ a positive integer}\}$

11. {odd positive integers}
12. {positive integers}
13. $\{1, \frac{1}{2}, \frac{1}{3}, \frac{1}{4}, \dots\}$
14. $\{\frac{1}{2}, \frac{1}{4}, \frac{1}{6}, \frac{1}{8}, \dots\}$
15. {elements not in set A}
16. {elements in set B}

In Problems 17–28, list the elements of the given sets if $U = \{1, 2, 3, 4, 5, 6, 7, 8, 9, 10\}$, $A = \{2, 4, 6, 8, 10\}$, $B = \{1, 2, 3, 4, 5\}$, and $C = \{1, 3, 5, 7, 9\}$.

17. A' **18.** B' **19.** C' **20.** $A \cap B$
21. $A \cup B$ **22.** $A \cup C$ **23.** $A \cap C$ **24.** $A' \cap B'$
25. $A' \cup C'$ **26.** $(A \cap B)'$ **27.** $A' \cup C$ **28.** $C' \cap B$

For each of Problems 29–40, copy the Venn diagram shown here on a sheet of paper. Shade the part of the diagram corresponding to each set.

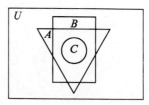

29. $A \cap B$ **30.** $C \cap A$ **31.** $C \cap B$ **32.** $A \cap C'$
33. $B \cap C'$ **34.** $B' \cup A'$ **35.** $B' \cap A'$ **36.** $B' \cap A$
37. $A' \cap B$ **38.** $(C \cap B)'$ **39.** $(C' \cap A)'$ **40.** $(A' \cap B)'$

1.1 Definitions and Symbols

Using Venn diagrams if necessary, complete each equation.

41. $(A')' =$ **42.** $A \cap A' =$ **43.** $A \cap U =$ **44.** $A \cup A' =$

45. $A \cup U =$ **46.** $A \cap \emptyset =$ **47.** $A \cup \emptyset =$ **48.** $A \cap A =$

49. $\emptyset' \cap \emptyset =$ **50.** $A \cup A =$ **51.** $\emptyset \cup \emptyset =$ **52.** $U \cap U =$

State under what conditions each of the following statements is true.

53. $A \cup B = \emptyset$ **54.** $A \cup \emptyset = \emptyset$ **55.** $A \cap U = U$

56. $A \cup B = A$ **57.** $A \cup \emptyset = U$ **58.** $A' \cap U = U$

59. $A \cap B = A$ **60.** $A' \cup \emptyset = \emptyset$ **61.** $A \cup B = A \cap B$

62. $A \cap U = B \cap U$ **63.** $A \cap \emptyset = \emptyset$ **64.** $A' \cup \emptyset = A$

Applications

65. *Biology* Of 200 laboratory rats made to inhale cigarette smoke over a period of time, it was found that 80 developed hypertension, 140 developed emphysema, and 99 developed cancer. Also, 39 rats developed both hypertension and cancer, 50 developed both hypertension and emphysema, and 60 developed both emphysema and cancer. Twenty rats developed all three of the diseases. How many rats developed only hypertension? Only emphysema? Only cancer? How many rats developed none of the diseases?

66. There are eight possible blood classifications according to whether blood contains antigens A, B, or Rh. For example, type O− blood contains none of the antigens, O+ contains Rh but neither A nor B, AB+ contains all three antigens. The rarest type of blood is AB− and the most common is O+. The possible sets are indicated in the following Venn diagram. In 1000 people it was found that 400 have antigen A, 150 have antigen B, 800 have antigen Rh,

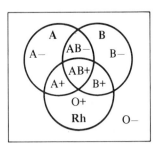

340 have both antigens A and Rh, 120 have both antigens B and Rh, 38 have both antigens A and B, and 32 have all three antigens. Determine how many persons have the blood types:

    a. A+   **b.** B+   **c.** AB−
    d. A−   **e.** B−   **f.** O+
    g. O−.

1.2

THE REAL NUMBERS

You should already be somewhat familiar with the five sets of numbers described below.

1. The set N of **positive integers** or **natural numbers**, whose elements are the counting numbers:
$$N = \{1, 2, 3, \ldots\}.$$

2. The set J of **integers**, whose elements are the counting numbers, their negatives, and zero:
$$J = \{\ldots, -2, -1, 0, 1, 2, \ldots\}.$$

3. The set Q of **rational numbers**, whose elements are all those numbers that can be represented as the quotient of two integers $\frac{a}{b}$ (or a/b, or $a \div b$), where b is not zero. Among the elements of Q are such numbers as $-3/4$, $18/27$, $3/1$, and $-6/1$. In symbols:
$$Q = \left\{x \mid x = \frac{a}{b}, \quad a, b \in J, \quad b \neq 0\right\}.$$

The members of the set of rational numbers are the numbers with terminating or repeating decimal representations. For example, $1/4 = 0.25$ and $1/9 = 0.1111\ldots$ are elements of Q.

4. The set H of **irrational numbers**, whose elements are the numbers that cannot be represented in the form a/b, where a and b are integers. Irrational numbers are the numbers with decimal representations that are nonterminating and nonrepeating. Among the elements of H are such numbers as $\sqrt{2}$, π, $-\sqrt{7}$, and $\sqrt{3} = 1.73205\ldots$.

5. The set R of **real numbers**, which is the union of the set of all rational numbers and the set of all irrational numbers:
$$R = \{x \mid x \in Q \cup H\}.$$

THE NUMBER LINE

You are probably familiar with the fact that there is a correspondence between the real numbers and the points on a geometric line (to each real number there corresponds one and only one point on the line, and vice versa). To illustrate this, we imagine the line scaled in convenient units, with the positive direction (from 0 to 1) denoted by an arrowhead. The line is then called a **number line**. The real number corresponding to a point on the line is called the **coordinate** of the point and the point is called the **graph** of the number. For example, a number-line representation of the set {1, 3, 5} is shown in Figure 1.5.

Figure 1.5

A horizontal number-line graph directed to the right can be used to illustrate the separation of the real numbers into three disjoint subsets: {negative real numbers}, {0}, {positive real numbers}. The point associated with 0 is called the **origin**. The set of numbers whose elements are associated with the points on the right-hand side of the origin belong to the set of **positive real numbers**, and the set whose elements are associated with the points on the left-hand side belong to the set of **negative real numbers**. The union: {0} ∪ {positive real numbers} is called the set of **nonnegative real numbers**.

The word "negative" is used in two ways. In one case, we refer to the negative of a number, whereas, in the other, we refer to a negative number, which is the negative of a positive number. For example, -2 is a negative number and is the negative of the positive number 2, whereas $-(-2)$ is the negative of the negative number -2 and is equal to 2. In algebra we frequently encounter symbolism such as $-x$; the presence of the negative sign does not necessarily mean that this is a negative number. The symbol $-x$ could represent a negative or a nonnegative real number, depending on whether x has a positive or nonpositive value.

ORDER RELATIONS If the graph of a real number a lies to the left of the graph of a real number b on the number line, we say that a is **less than** b or b is **greater than** a. Symbolically we write $a < b$ (read "a is less than b") and $b > a$ (read "b is greater than a"). Figure 1.6 illustrates the relations $a > c$, $a < b$, $b > c$, and so on.

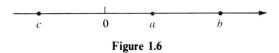

Figure 1.6

The following is a summary of the basic properties of the "less than" relation.

For any real numbers a, b, and c:

   I   *If $a < b$ and $b < c$, then $a < c$.*
   II *If $a < b$, then $a + c < b + c$.*
  III *If $a < b$ and $c > 0$, then $ac < bc$.*
  IV *If $a < b$ and $c < 0$, then $ac > bc$.*

Property I, states, for example, that if $-1 < 3$ and $3 < 8$, then $-1 < 8$. To illustrate property II we note that if $2 < 5$, then

$$2 + (-3) < 5 + (-3)$$
$$-1 < 2.$$

Properties III and IV state that if a "less than" relation is multiplied by a positive number, then the **sense** or **direction** of the inequality symbol is retained, whereas multiplying by a negative number reverses the sense of the symbol. For example, if $-4 < 9$, then

$$-4(3) < 9(3)$$
$$-12 < 27;$$

but

$$-4(-3) > 9(-3)$$
$$12 > -27.$$

Often we use the symbols $\leq$ and $\geq$ (read "is less than or equal to" and "is greater than or equal to," respectively) as contractions for two symbols, one of equality and one of inequality, connected by the word "or." For example, $x \leq 7$ means that $x < 7$ *or* $x = 7$.

ABSOLUTE VALUE The graphs of the numbers $-a$ and a on a number line are the same **distance** from the origin, but on opposite sides of it. If we wish to refer to this distance on either side of the origin, then we use the term **absolute value.** The absolute value of $-a$ and the absolute value of a are the same nonnegative number. The symbol $|a|$ is used to denote the absolute value of a.

> If a is a real number, then the **absolute value** of a is given by
> $$|a| = \begin{cases} a & \text{if } a \geq 0, \\ -a & \text{if } a < 0. \end{cases}$$

For example, $|-3| = -(-3) = 3$, $|7| = 7$, and $|0| = 0$.

EXERCISE 1.2

Let $N = \{positive\ integers\}$, $J = \{integers\}$, $Q = \{rational\ numbers\}$, $H = \{irrational\ numbers\}$, and $R = \{real\ numbers\}$. State whether each statement is true or false.

1.2 The Real Numbers

1. $-5 \in N$
2. $0 \in Q$
3. $\sqrt{3} \in R$
4. $\pi \in Q$
5. $-3 \in R$
6. $-7 \in H$
7. $\{-1, 1\} \subset N$
8. $\{-1, 1\} \subset J$
9. $\{-1, 1\} \subset Q$
10. $\{0.121212\ldots, 0.33\} \subset Q$
11. $\{0.1789340243\ldots\} \subset Q$
12. $\left\{\frac{1}{4}, \frac{1}{2}, \sqrt{9}, \sqrt{4}\right\} \subset H$

For real numbers x and y, justify each statement by citing one of the properties I–IV.

Example If $2 < 3x$, then $-4 > -6x$.

Solution Property IV. Each member of $2 < 3x$ is multiplied by -2.

13. If $x < 3$ and $y < x$, then $y < 3$.
14. If $x + 1 < 0$, then $x < -1$.
15. If $y < 8$, then $3y < 24$.
16. If $y < 4$, then $y + 2 < 6$.
17. If $x < 7$, then $x - 2 < 5$.
18. If $x < 9$, then $-2x > -18$.
19. If $-6x < 12$, then $x > -2$.
20. If $x - 3 < 5$, then $x < 8$.

Express each statement by means of symbols.

Examples
 a. 5 is not greater than 7
 b. x is between 5 and 8

Solutions
 a. $5 \not> 7$, or $5 \leq 7$
 b. $5 < x < 8$

21. 7 is greater than 3
22. 2 is less than 5
23. -4 is less than -3
24. -4 is greater than -7
25. x is between -1 and 1, inclusive
26. x is negative
27. x is positive
28. x is nonpositive
29. x is nonnegative
30. $2x$ is less than or equal to 8

Write each expression without using absolute-value notation.

Examples
 a. $|-11|$
 b. $|x - 6|$

Solutions
 a. $-(-11) = 11$
 b. $x - 6$ if $x - 6 \geq 0$
    $-(x - 6)$ if $x - 6 < 0$

31. $|-3|$
32. $|-5|$
33. $|7|$
34. $|4|$
35. $-|-2|$
36. $-|-5|$
37. $|3x|$
38. $|2y|$
39. $|x+1|$
40. $|x+2|$
41. $|y-3|$
42. $|y-4|$

Replace the question mark with an appropriate order symbol to form a true statement.

43. $|-2| \,?\, |-5|$
44. $|3| \,?\, |-4|$
45. $-7 \,?\, |-1|$
46. $5 \,?\, |-2|$
47. $|-3| \,?\, 0$
48. $-|-4| \,?\, 0$

Write an equivalent relation without using the negation symbol, /.

49. $2 \not= 5$
50. $-1 \not\leq -2$
51. $7 \not\geq 8$
52. $|x| \not< 3$
53. $|x| \not\geq 3$
54. $x \not> |y|$

For real numbers a and b, $a \leq b$ if and only if there exists a nonnegative number d such that $a + d = b$. In Problems 55–58, find the number d that converts the inequality into an equality.

Example $-3 \leq 9$

Solution   We want $-3 + d = 9$, so $d = 9 - (-3) = 12$.

55. $2 \leq 11$
56. $-7 \leq -3$
57. $24 \geq -5$
58. $3 \geq -3$

CHAPTER TEST

Replace the question mark with either $\in$ or $\subset$ to make a true statement.

1. $6 \underline{?} \{2, 4, 6, 8\}$
2. $\{2\} \underline{?} \{1, 2, 3, 4\}$
3. $\emptyset \underline{?} \{2, 4\}$
4. $25 \underline{?} \{5, 10, 15, ...\}$
5. List the subsets of $\{1, 2, 3, 4\}$ that have $\{1, 3\}$ as a subset.
6. Designate {multiples of 5} by using set-builder notation.

Let $U = \{1, 2, 3, 4, 5, 6, 7, 8, 9\}$, $A = \{2, 4, 6, 8\}$, $B = \{5, 6, 7, 8, 9\}$, and $C = \{3, 5, 7, 9\}$. List the members of the sets.

7. $A \cup B$
8. $A \cap B$
9. $A \cap C$
10. $A' \cap B$
11. $(A \cap C)'$
12. $(B \cup C)'$

Let $A = \left\{-8, -\dfrac{15}{7}, -\sqrt{5}, -1, 0, 3, \dfrac{13}{2}, \sqrt{50}, 21\right\}$. List the members of the sets.

13. {positive integers in A}
14. {integers in A}
15. {rational numbers in A}
16. {irrational numbers in A}

Justify each statement by citing one of the properties of the less-than relation.

17. If $x < 3$, then $-2x > -6$.
18. If $x + 7 < 3$, then $x < -4$.

Express each statement by means of symbols.

19. -6 is greater than -9
20. 4 is less than or equal to y
21. y is nonpositive
22. $y + 2$ is nonnegative

Rewrite each expression without using absolute-value notation.

23. $|-7|$
24. $|x - 5|$

Write each statement without using the negation symbol.

25. $x \not\geq 4$
26. $|x| \not< 6$

2
REVIEW: POLYNOMIALS AND THE LAWS OF EXPONENTS

In the following discussion we shall use the commutative, associative, and distributive laws of the real number system. If the reader needs to review these topics, a summary of all the basic laws of algebra is given in Appendix I. Also, in each case we shall assume that letters such as a, b, and c represent members of the set of real numbers.

2.1

SUMS OF POLYNOMIALS

Any grouping of constants and variables generated by applying a finite number of the elementary operations—addition, subtraction, multiplication, division, or the extraction of roots—is called an **algebraic expression**. For example,

$$\frac{3x^2 + \sqrt{2x-1}}{3x^3 + 7} \quad \text{and} \quad xy + 3x^2z - \sqrt[5]{z}$$

are algebraic expressions. If two expressions have equal values for all values of the variables for which both expressions are defined, then we say that on the set of these values the expressions are **equivalent**. For example,

$$4x - x + 3x = 4x + (-x) + 3x$$
$$= (4 - 1 + 3)x$$
$$= 6x.$$

POSITIVE-INTEGER POWERS You should recall that an expression of the form a^n is called a **power** of a, where a is the **base** of the power and n is the **exponent** of the power.

2.1 Sums of Polynomials

> If n is a positive integer, then
> $$a^n = \underbrace{a \cdot a \cdot a \cdot \cdots \cdot a.}_{n \text{ factors}}$$

POLYNOMIAL EXPRESSIONS

In any algebraic expression of the form $A + B + C + \cdots$, where $A, B, C, \ldots$ are algebraic expressions, $A, B, C, \ldots$ are called **terms** of the expression. For example, in $x + (y + 3)$ the terms are x and $(y + 3)$, but in $x + y + 3$ the terms are $x, y,$ and 3. An algebraic expression that contains only nonnegative integral powers of a variable and contains no variable in a denominator is a **polynomial**. For example,

$$5x, \qquad \frac{3x^2}{2} - \frac{7x}{2}, \qquad 0, \qquad 2x^2 - 3x + 4, \qquad \frac{y}{4} - \frac{\sqrt{7}}{4}$$

are polynomials, whereas

$$\frac{3}{x}, \qquad 3 + \sqrt{x}, \qquad \frac{2\sqrt{x-1}}{\sqrt{x+1}}$$

are algebraic expressions, but not polynomials in the variable x.

Polynomials consisting of one, two, or three terms are also called **monomials**, **binomials**, and **trinomials**, respectively. Thus, $3x^2y$ is a monomial, $x + 4x^2$ is a binomial, and $x + y + z$ is a trinomial.

The **degree** of a monomial is given by the exponent of the variable in the monomial. Thus, 5 is of degree zero, $2x$ is of first degree, and $3x^4$ is of fourth degree; but no degree is assigned to the special monomial 0. If a monomial contains more than one variable, its degree is given by the sum of the exponents on the variables; $3x^2y^3z$ is of sixth degree in $x, y,$ and z. It can also be described as being of second degree in x, third degree in y, fifth degree in x and y, and so on. The constant factor in the monomial, 3, is called the **coefficient** of the monomial. The **degree of a polynomial** is the same as the degree of its term of highest degree. Since no degree is assigned to the monomial 0, no degree is assigned to the polynomial 0.

Because $a - b$ is defined to be $a + (-b)$, we shall view the signs in any polynomial as signs denoting numbers or their negatives, and the operation involved to be addition. Thus,

$$3x - 5y + 4z = (3x) + (-5y) + (4z),$$

and its terms are $3x, -5y,$ and $4z$. Again, an expression such as

$$a - (bx + cx^2),$$

in which a set of parentheses is preceded by a negative sign, can be written

$$a + [-(bx + cx^2)],$$

or

$$a + [-bx - cx^2],$$

or, finally,

$$a + (-bx) + (-cx^2).$$

Also, since

$$\frac{a}{b} = a\left(\frac{1}{b}\right),$$

we can view division by a constant as multiplication by its **reciprocal (multiplicative inverse)**, and, for example, write

$$\frac{3x^2}{4} + \frac{x}{2} \quad \text{as} \quad \frac{3}{4}x^2 + \frac{1}{2}x.$$

SIMPLIFI-CATION OF POLY-NOMIALS By applying the commutative, associative, and distributive laws in various ways, we can frequently rewrite polynomials and sums of polynomials in what might be termed "simpler" forms.

Example
$$(2x^2 + 3x + 5) + 2x + (6x^2 + 7) = 2x^2 + 6x^2 + 3x + 2x + 5 + 7$$
$$= (2 + 6)x^2 + (3 + 2)x + 5 + 7$$
$$= 8x^2 + 5x + 12$$

Here we have reduced the number of terms from six to three, and the original expression is said to have been "simplified."

We shall often be concerned with polynomials in one variable. A polynomial of degree n, $n \geq 0$, in x can be represented—when its terms are rearranged, if need be—by an expression of the form

$$a_0 x^n + a_1 x^{n-1} + a_2 x^{n-2} + \cdots + a_{n-1} x + a_n \qquad (a_0 \neq 0),$$

where it is understood that the a's are the (constant) coefficients of the powers of x in the polynomial.

The term $a_0 x^n$ is called the **leading term**, and the coefficient a_0 is called the **leading coefficient** in the polynomial.

2.1 Sums of Polynomials

SYMBOLS FOR POLYNOMIALS   Polynomials are frequently represented by symbols such as

$$P(x), \qquad D(y), \qquad Q(z),$$

where the symbol in parentheses designates the variable. Thus, we might write

$$P(x) = 2x^3 - 3x + 2,$$
$$D(y) = y^6 - 2y^2 + 3y - 2,$$
$$Q(z) = 8z^4 + 3z^3 - 2z^2 + z - 1.$$

VALUE OF A POLYNOMIAL   The notation $P(x)$ can be used to denote values of the polynomial for specific values of x. Thus, $P(2)$ means the value of the polynomial $P(x)$ when x is replaced by 2. For example, if

$$P(x) = x^2 - 2x + 1,$$

then

$$P(2) = 2^2 - 2(2) + 1 \qquad = 1,$$
$$P(3) = 3^2 - 2(3) + 1 \qquad = 4,$$
$$P(-4) = (-4)^2 - 2(-4) + 1 = 25.$$

In some applications (see Chapter 13), the notation $P(x)|_a^b$ denotes the difference $P(b) - P(a)$. Thus, if

$$P(x) = \frac{x^2}{2} - 4x,$$

then

$$P(x)|_2^3 = P(3) - P(2) = \left[\frac{3^2}{2} - 4(3)\right] - \left[\frac{2^2}{2} - 4(2)\right] = -\frac{3}{2}.$$

EXERCISE 2.1

Simplify each expression.

Example $(y^2 - 3y + 2) + (4y^2 - 7) - (2y^2 + 6y - 5)$

Solution $y^2 - 3y + 2 + 4y^2 - 7 - 2y^2 - 6y + 5$
$= (y^2 + 4y^2 - 2y^2) + (-3y - 6y) + (2 - 7 + 5)$
$= 3y^2 - 9y$

1. $(z^3 - 3z^2 + 2) + (z^3 + 5z - 3) - (2z^3 + 2z^2 - 3z + 1)$
2. $(x^2 - 3x + 5) - (x^2 + 3x - 5) - (2x^2 + 3x - 1)$
3. $(2y^4 - 3y^2 + 2) - (y^4 + 3y^3 - 2) + (-y^4 + 2y^3 - 3y^2 + y - 1)$
4. $(3z^4 + 2z^3 - 3z) + (5z^3 - 2z^2 + z) - (z^4 + 5z^3 - 2z^2 + 4z - 1)$
5. $(3x^2 - xy + 4y^2) - (x^2 + 2xy + y^2) + (-x^2 + 3xy - y^2)$
6. $(2y^2z - 3yz^2 + 5yz) + (y^2z + yz^2 - 4yz) - (3y^2z + yz^2 - yz)$
7. $(x^2 - 2x + 3) - [(2x^2 + x - 5) - (x^2 + 3x + 1)]$
8. $(2y^2 + 3y - 2) - [(y^2 + 5y - 3) - (-y^2 + 3y + 2)]$

Express $P(x) + Q(x)$ and $P(x) - Q(x)$ as polynomials.

9. $P(x) = 3x - 2;\quad Q(x) = 3 - x$
10. $P(x) = x^2 + 3x - 2;\quad Q(x) = 2x^2 + x - 2$
11. $P(x) = x^2 - 2x + 3;\quad Q(x) = 2x^2 - 2x - 1$
12. $P(x) = 2x^3 - 3x^2 + x - 1;\quad Q(x) = x^3 + 3x - 2$
13. $P(x) = x^3 + 4x^2 - 2x + 1;\quad Q(x) = x^2 - 2x + 3$
14. $P(x) = 2x^2 - 3x + 4;\quad Q(x) = 2x^3 + x^2 - x - 1$
15. $P(x) = 3x^4 - 2x^2 + 1;\quad Q(x) = x^5 - 3x^3 + 4x$
16. $P(x) = 4x^5 - x^3 + 3x;\quad Q(x) = 2x^4 + x^2 - 5$

For the polynomials

$$P(x) = 2x^2 - 3x + 2, \quad Q(x) = 3 - 2x + x^2, \quad S(x) = -2x^2 + 3x - 5,$$

write each given expression as an equivalent polynomial.

17. $P(x) + Q(x)$
18. $P(x) + [Q(x) - S(x)]$
19. $P(x) - [Q(x) + S(x)]$
20. $P(x) - [Q(x) - S(x)]$
21. $[Q(x) - P(x)] - S(x)$
22. $S(x) - [-P(x) - Q(x)]$

Example Given $P(x) = 2x^2 - x + 3$, find $P(3)$, $P(-3)$, $P(0)$, $P(a)$, and $P(x)|_0^3$.

Solution
$P(3) = 2(3)^2 - (3) + 3 = 18$
$P(-3) = 2(-3)^2 - (-3) + 3 = 24$
$P(0) = 2(0)^2 - (0) + 3 = 3$
$P(a) = 2a^2 - a + 3$
$P(x)|_0^3 = P(3) - P(0) = 18 - 3 = 15$

23. Given $P(x) = x^3 - 3x^2 + x + 1$, find $P(2)$, $P(-2)$, $P(0)$, $P(x)|_0^2$, and $P(x)|_{-2}^2$.
24. Given $P(x) = 2x^3 + x^2 - 3x + 4$, find $P(3)$, $P(-3)$, $P(0)$, $P(x)|_0^3$, and $P(x)|_{-3}^0$.

25. Given $P(x) = x^{12}$, find $P(1)$, $P(-1)$, $P(0)$, $P(x)|_0^1$, and $P(x)|_{-1}^1$.
26. Given $P(x) = x^{13}$, find $P(1)$, $P(-1)$, $P(0)$, $P(x)|_0^1$, and $P(x)|_{-1}^1$.
27. Given $Q(x) = 2x^4 - x^2 + 3$, find $Q(0)$, $Q(h)$, and $Q(-h)$.
28. Given $R(x) = x^4 + 2x^3 - 5$, find $R(0)$, $R(t)$, and $R(-t)$.
29. Given $P(x) = x^2 + 2x - 3$, find $P(x + h)$, $P(x - h)$, and $P(x^2)$.
30. Given $N(x) = 2x^2 - x + 5$, find $N(x - h)$, $N(x + h)$, and $N(x^3)$.

Example Given $P(x) = x - 4$ and $Q(x) = x + 2$, find $P(Q(2))$.

Solution $Q(2) = 2 + 2 = 4$; hence,
$$P(Q(2)) = P(4) = 4 - 4 = 0.$$

31. Given $P(x) = x + 2$ and $Q(x) = x - 3$, find $P(Q(2))$ and $Q(P(2))$.
32. Given $P(x) = 2x + 1$ and $Q(x) = \frac{1}{2}(x - 1)$, find $P(Q(-2))$ and $Q(P(-2))$.
33. Given $P(x) = x^2 + 3$ and $Q(x) = 6$, find $Q(P(2))$ and $P(Q(0))$.
34. Given $P(x) = 2x^3 - 3x$ and $Q(x) = x^2 + 1$, find $P(Q(0))$ and $Q(P(0))$.
35. If $P(x)$ is of degree n and $Q(x)$ is of degree $n - 2$, what is the degree of $P(x) + Q(x)$? Of $P(x) - Q(x)$?
36. If $P(x)$ and $Q(x)$ are polynomials, with $P(0) = 4$ and $Q(0) = 3$, what is the value of $P(x) + Q(x)$ for $x = 0$? Of $P(x) - Q(x)$ for $x = 0$?

2.2

PRODUCTS OF POLYNOMIALS

LAWS OF EXPONENTS FOR POSITIVE-INTEGER EXPONENTS

When n is a positive integer, we have
$$a^n = \underbrace{a \cdot a \cdot a \cdots a}_{n \text{ factors}}.$$

The product $a^m \cdot a^n$, where m and n are positive integers, is then given by
$$a^m \cdot a^n = \underbrace{(a \cdot a \cdot a \cdots a)}_{m \text{ factors}} \underbrace{(a \cdot a \cdot a \cdots a)}_{n \text{ factors}}$$
$$= \underbrace{a \cdot a \cdot a \cdots a}_{(m + n) \text{ factors}}.$$

We state the result formally, together with two related properties.

> If m and n are positive integers, then
>
> I $a^m \cdot a^n = a^{m+n}$,
>
> II $(a^m)^n = a^{mn}$,
>
> III $(ab)^n = a^n b^n$.

The justifications of properties II and III are similar to the one above for property I and are therefore omitted.

Examples a. $x^2 x^3 = x^{2+3} = x^5$

b. $(x^2)^3 = x^{2 \cdot 3} = x^6$

c. $(xy)^4 = x^4 y^4$

Example $(3x^2 y)(2xy^2) = 3 \cdot 2 x^{2+1} y^{1+2} = 6 x^3 y^3$

USING THE DISTRIBUTIVE LAW In the real number system the sum of two (or more) terms can always be replaced by one term representing the sum. For example, the three terms of $x + y + z$ can be replaced by the two terms $x + (y + z)$. Thus, by the distributive law,

$$3x(x + y + z) = 3x[x + (y + z)]$$
$$= 3x(x) + 3x(y + z)$$
$$= 3x^2 + 3xy + 3xz.$$

It is not necessary in practice to group terms; we could have proceeded as follows:

$$3x(x + y + z) = 3xx + 3xy + 3xz = 3x^2 + 3xy + 3xz.$$

The distributive law can be applied successively to the products of polynomials containing more than one term.

Example $(3x + 2y)(x - y) = 3x(x - y) + 2y(x - y)$
$$= 3x^2 - 3xy + 2xy - 2y^2$$
$$= 3x^2 - xy - 2y^2$$

Of course, products of polynomials should ordinarily be simplified mentally if it is convenient to do so. The following binomial products are types so frequently

2.2 Products of Polynomials

encountered that you should learn to recognize them on sight:
$$(x + a)(x + b) = x^2 + (a + b)x + ab,$$
$$(x + a)^2 = x^2 + 2ax + a^2,$$
$$(x + a)(x - a) = x^2 - a^2.$$

EXERCISE 2.2

Simplify by writing each product in equivalent polynomial form. In each term combine all constants and all powers of each variable.

Examples   a. $(-2x^3)(3xy)(y^2)$ b. $a^n \cdot a^{n+1}$

Solutions   a. $-6x^4 y^3$ b. $a^{n+(n+1)} = a^{2n+1}$

1. $(-3x^2)(-2xy)(-y^3)$
2. $(a^3)(-2ab^2)(-b^3)$
3. $(4x^2 y)(2xy^3)(-xy)$
4. $(2ab^3)(-3a^3 b^2)(-a^2 b^2)$
5. $x^{n+2} \cdot x^{2n-2}$
6. $z^{n-3} \cdot z^3$
7. $3^{n+1} \cdot 3^{1-n}$
8. $4^{2n+1} \cdot 4^{2-2n}$

Examples   a. $2(x^2 - x - 1)$ b. $(x - 3)(x + 5)$

Solutions   a. $2x^2 - 2x - 2$ b. $x^2 + 5x - 3x - 15 = x^2 + 2x - 15$

9. $abc(a - b + 2c)$
10. $-ab(2a - b + 3c)$
11. $(x + 2)(x + 5)$
12. $(x - 3)(x + 2)$
13. $(x + 2y)^2$
14. $(2x + y)^2$
15. $(5x + 1)(2x + 3)$
16. $(2x + 3)(x - 5)$
17. $2(3a + 2b)(3a - 2b)$
18. $4(5x - y)(5x + y)$
19. $-(2a - b)(c - 3d)$
20. $-(3a - b)(c + 3d)$
21. $(4a - 3b)^2$
22. $(5a - 2b)^2$

Example $(y - 3)(y^2 + 3y - 1)$

Solution
$$(y-3)(y^2+3y-1) = y(y^2+3y-1)-3(y^2+3y-1)$$
$$= y^3+3y^2-y-3y^2-9y+3$$
$$= y^3-10y+3$$

Alternative format
$$\begin{array}{r} y^2+3y-1 \\ \underline{y-3} \\ y^3+3y^2-y \\ \underline{-3y^2-9y+3} \\ y^3-10y+3 \end{array}$$

23. $(x+4)(x^2+2x-1)$
24. $(x-2)(x^2-x+3)$
25. $(3x-1)(2x^2+3x-1)$
26. $(2x-3)(3x^2-2x+5)$
27. $a(a-b)(a^2+ab+b^2)$
28. $b(a+b)(a^2-ab+b^2)$

Example $5\{2x-[x-3(x+2)+3]-2\}$

Solution Eliminate the inner grouping symbols first and work toward the outside grouping symbols, removing them last:
$$5\{2x-[x-3(x+2)+3]-2\} = 5\{2x-[x-3x-6+3]-2\}$$
$$= 5\{2x-[-2x-3]-2\}$$
$$= 5\{2x+2x+3-2\}$$
$$= 5\{4x+1\}$$
$$= 20x+5.$$

29. $2\{a-[a-2(a+1)+1]+1\}$
30. $-\{4-[3-2(a-1)+a]+a\}$
31. $2x\{x+3[2(2x-1)-(x-1)]+5\}$
32. $-x\{4-2[(x+1)-3(x+2)]-x\}$
33. $5y^2-3[y^2-(y-2)^2+1]-4$
34. $2x^2+5[x-(x+3)^2-x^2]+2x^2$
35. $2\{z^2+2[2z-(5+z)+2]^2-3\}+2$
36. $-1\{3z^2-2[3-(3z+1)+2]^2-z^2\}+2z$
37. Given $P(x) = x^2-3x+7$, find $P(x-1)$ and $P(2-x)$.
38. Given $P(x) = x^2+2x+1$, find $P(x+h)$ and $P(x-h)$.
39. Given $P(x) = 3-x^2$, find $[P(3)]^2$ and $P(3^2)$.
40. Given $P(x) = x^2-3x$, find $[P(-x)]^2$ and $P(-x^2)$.
41. Simplify the expression $(a+b)^2-(a^2+b^2)$. What are the conditions on a and b for a^2+b^2 to be greater than $(a+b)^2$? For a^2+b^2 to be less than $(a+b)^2$?
42. If $P(x)$ and $Q(x)$ are polynomials of degree m and n, respectively, what is the degree of $P(x) \cdot Q(x)$?

2.3

FACTORING POLYNOMIALS

What do we mean when we say that we have **factored** an integer or a polynomial? It is true, for example, that

$$2 = 4\left(\frac{1}{2}\right),$$

but we would not ordinarily say that 4 and 1/2 are factors of 2. On the other hand, since

$$10 = (2)(5),$$

we do say that 2 and 5 are integer factors of 10.

Now consider the polynomial

$$2x^2 - 10,$$

which is completely factored as

$$2x^2 - 10 = 2(x^2 - 5)$$

if we are limited to **integral** coefficients—that is, coefficients that are integers. But if we consider polynomials in x having real numbers as coefficients, its complete factorization is

$$2x^2 - 10 = 2(x - \sqrt{5})(x + \sqrt{5}).$$

Thus, the result depends in part on the coefficients we consider permissible. In this book, we are primarily concerned with polynomials whose coefficients are integers.

In factoring polynomials having integer coefficients, we consider as factors only polynomials having integer coefficients with no common integer factor other than 1 or -1. We say that such a polynomial is **prime** if it is not the product of two polynomials of this sort and its leading coefficient is positive. We say that a polynomial other than 0 or 1 is **completely factored** if it is written equivalently as a product of prime polynomials or as such a product times -1.

FACTORS OF QUADRATICS

One very common type of factoring involves quadratic (second-degree) binomials or trinomials with integer coefficients. From Section 2.2, we recall that

(1) $\quad (x + a)(x + b) = x^2 + (a + b)x + ab,$

(2) $\quad (x + a)^2 = x^2 + 2ax + a^2,$

(3) $\quad (x + a)(x - a) = x^2 - a^2.$

These three forms involve prime polynomial factors. In this section, we are interested in viewing these relationships from right to left—that is, from polynomial to factored form.

Example
$$6x^2 - 96 = 6(x^2 - 16)$$
$$= 2 \cdot 3(x - 4)(x + 4)$$

A few other polynómials occur frequently enough to justify a study of their factorization. In particular, the forms

(4) $\quad (a + b)(x + y) = ax + ay + bx + by,$

(5) $\quad (x + a)(x^2 - ax + a^2) = x^3 + a^3,$

(6) $\quad (x - a)(x^2 + ax + a^2) = x^3 - a^3$

are often encountered in mathematics. We are again interested in viewing these relationships from right to left. Expressions such as the right-hand member of form (4) are factorable by grouping. For example, to factor

$$3x^2y + 2y + 3xy^2 + 2x,$$

we write it in the form

$$3x^2y + 2x + 3xy^2 + 2y$$

and factor the common monomial x from the first group of two terms and y from the second group of two terms, obtaining

$$x(3xy + 2) + y(3xy + 2).$$

If we now factor the common binomial $(3xy + 2)$ from each term, we have

$$(3xy + 2)(x + y),$$

in which both factors are prime.

The application of forms (5) and (6) is direct.

Example $\quad 8a^3 + b^3 = (2a)^3 + b^3$
$$= (2a + b)[(2a)^2 - 2ab + b^2]$$
$$= (2a + b)(4a^2 - 2ab + b^2)$$

Example $\quad a^3 - 1 = (a - 1)(a^2 + a + 1)$

2.3 Factoring Polynomials

EXERCISE 2.3

Factor completely into products of polynomials with integer coefficients. (Assume that all variables in exponents represent positive integers.)

Examples   **a.** $18x^2y - 24xy^2 + 6xy$ **b.** $x^{2n} + x^n$ **c.** $4a^3 - 5a^2 + a$

Solutions   **a.** $(2)(3)xy(3x - 4y + 1)$ **b.** $x^n(x^n + 1)$ **c.** $a(4a^2 - 5a + 1)$
$\phantom{\textbf{Solutions}\quad\textbf{a.}\ (2)(3)xy(3x - 4y + 1)\quad\textbf{b.}\ x^n(x^n + 1)\quad\textbf{c.}\ }= a(4a - 1)(a - 1)$

1. $9x^5y - 3x^4y + 6x^3y$
2. $x^2y^2z^2 + 2xyz - xz$
3. $x^2 - 3x - 4$
4. $y^2 + 5y - 6$
5. $x^2 - 8x + 12$
6. $z^2 + 4z - 12$
7. $y^2 + 4y + 4$
8. $t^2 - 6t + 9$
9. $x^2 - 25$
10. $z^2 - 36$
11. $2x^2 + 5x + 3$
12. $2n^2 - n - 3$
13. $1 + 5a + 6a^2$
14. $3 - 7z + 2z^2$
15. $6z^2 + 8z + 2$
16. $6n^3 + 21n^2 + 9n$
17. $x^4y^2 - x^2y^2$
18. $3xy^2 - 12xy^2$
19. $x^{2n} - 1$
20. $x^{4n} - x^{2n}y^{2n}$
21. $x^{n+2} - x^{n+1} + 2x^n$
22. $x^{n-2} - 3x^{n-1} + x^n$

Examples   **a.** $by - ay + bx - ax$ **b.** $8x^3 - y^3$

Solutions   **a.** $y(b - a) + x(b - a)$ **b.** $(2x)^3 - y^3$
$\phantom{\textbf{Solutions}\quad\textbf{a.}\ }= (b - a)(y + x)$ $\phantom{\textbf{b.}\ }= (2x - y)(4x^2 + 2xy + y^2)$

23. $y^4 + 3y^2 + 2$
24. $x^4 - 5x^2 + 4$
25. $2a^4 - a^2 - 1$
26. $3z^4 - 11z^2 - 4$
27. $x^4 - (y - 2x)^4$
28. $ax^2 + x + ax + 1$
29. $x^2 + ax + xy + ay$
30. $3x + y - 6x^2 - 2xy$
31. $a^3 + 2ab^2 - 4b^3 - 2a^2b$
32. $6x^3 - 4x^2 + 3x - 2$
33. $y^3 - 27x^3$
34. $8 + x^3y^3$
35. $x^3 + (x - y)^3$
36. $(x + y)^3 - z^3$

Examples   **a.** $a^{2n} - 9$ **b.** $x^{4n} - 3x^{2n} - 4$

Solutions a. $(a^n - 3)(a^n + 3)$ b. $(x^{2n} - 4)(x^{2n} + 1)$
$\qquad\qquad\qquad\qquad\qquad\qquad\qquad = (x^n - 2)(x^n + 2)(x^{2n} + 1)$

37. $a^{2n} - 4$
38. $x^{2n} - y^{2n}$
39. $x^{4n} - y^{4n}$
40. $x^{4n} - 2x^{2n} + 1$
41. $3x^{4n} - 10x^{2n} + 3$
42. $6y^{2n} + 30y^n - 900$
43. $2y^{2n} - 12y^n - 1440$
44. $2x^{2n} - 23x^n y^n - 39y^{2n}$
45. Show that $ac - ad + bd - bc$ can be factored both as $(a - b)(c - d)$ and as $(b - a)(d - c)$.
46. Show that $a^2 - b^2 - c^2 + 2bc$ can be factored as $(a - b + c)(a + b - c)$.
47. Consider the polynomial $x^4 + x^2y^2 + 25y^4$. If $9x^2y^2$ is both added to and subtracted from this expression, we have

$$x^4 + x^2y^2 + 25y^4 + 9x^2y^2 - 9x^2y^2 = (x^4 + 10x^2y^2 + 25y^4) - 9x^2y^2$$
$$= (x^2 + 5y^2)^2 - (3xy)^2$$
$$= [(x^2 + 5y^2) - 3xy][(x^2 + 5y^2) + 3xy]$$
$$= (x^2 - 3xy + 5y^2)(x^2 + 3xy + 5y^2).$$

By adding and subtracting an appropriate monomial, factor $x^4 + x^2y^2 + y^4$.

48. Use the method of Problem 47 to factor $x^4 - 3x^2y^2 + y^4$.

2.4

QUOTIENTS OF POLYNOMIALS

It is easily shown that the quotient of two polynomials need not be another polynomial. For example, $1/x$ is the quotient of the polynomials 1 and x but is not, itself, a polynomial.

If a polynomial Q exists such that $P = DQ$, then the polynomial P is said to be **exactly divisible** by the polynomial D, and we write $P/D = Q$. If P is not exactly divisible by D, then the quotient P/D cannot be written as a polynomial. For example, $P = x^5$ is exactly divisible by $D = x^3$ since if $Q = x^2$, it follows that $DQ = x^3 \cdot x^2 = x^{3+2} = x^5 = P$. On the other hand, $P = x^5$ is not exactly divisible by $D = x^6$ since no polynomial Q exists for which $x^6 Q = x^5$.

Let us examine some ways in which we can rewrite quotients of polynomials, even if the resulting expressions are *not* always polynomials. We begin with the simplest case, where P and D are monomials and their quotient *is* a polynomial (m and n are positive integers, and $m > n$):

$$\frac{a^m}{a^n} \qquad (a \neq 0).$$

2.4 Quotients of Polynomials

We have

$$\frac{a^m}{a^n} = a^m \cdot \frac{1}{a^n}$$

$$= (a^{m-n} \cdot a^n) \cdot \frac{1}{a^n}$$

$$= a^{m-n} \cdot \left(a^n \cdot \frac{1}{a^n}\right)$$

$$= a^{m-n} \cdot 1$$

$$= a^{m-n}.$$

LAWS OF EXPONENTS FOR QUOTIENTS

This establishes part I of the properties stated below. Justification of part II is similar and is therefore omitted.

> *If m and n are positive integers, then*
>
> $$\text{I} \quad \frac{a^m}{a^n} = a^{m-n} \quad (m > n),$$
>
> $$\text{II} \quad \left(\frac{a}{b}\right)^m = \frac{a^m}{b^m},$$
>
> *where $a, b \neq 0$.*

Example

$$\frac{12a^5 b^3}{4a^2 b^2} = \frac{12}{4} \cdot a^{5-2} b^{3-2} = 3a^3 b \quad (a, b \neq 0).$$

Note in the preceding example that a and b are not permitted to take the value 0, because if they were, $3a^3 b$ would represent a real number, 0, whereas $12a^5 b^3 / 4a^2 b^2$ would not be defined.

QUOTIENTS OF POLYNOMIALS

If a, b, and c are real numbers and $c \neq 0$, then recall that

$$\frac{a}{c} + \frac{b}{c} = \frac{a+b}{c}.$$

This demonstrates the familiar practice of adding fractions with a common denominator. Conversely, we can read the above statement from right to left. For example,

$$\frac{3+5}{2} = \frac{3}{2} + \frac{5}{2}.$$

This process is called **termwise division**. Similarly, we can rewrite quotients of polynomials of the form $(A + B)/C$, $C \neq 0$, in the form $A/C + B/C$, and then rewrite the expression further to obtain the simplest form. For example,

$$\frac{2x^3 + 4x^2 + 8x}{2x} = \frac{2x^3}{2x} + \frac{4x^2}{2x} + \frac{8x}{2x},$$

for all $x \neq 0$. Then,

$$\frac{2x^3}{2x} + \frac{4x^2}{2x} + \frac{8x}{2x} = x^2 + 2x + 4 \qquad (x \neq 0).$$

Note that $8x/2x = 8/2 = 4$ since $x/x = 1$.

As mentioned at the start of this section, quotients of polynomials cannot always be represented by polynomials. For example, we have

$$\frac{2x^3 + 4x + 1}{x} = \frac{2x^3}{x} + \frac{4x}{x} + \frac{1}{x}$$

$$= 2x^2 + 4 + \frac{1}{x} \qquad (x \neq 0),$$

where the resulting expression is not a polynomial.

If the divisor of a quotient contains more than one term, **long division** involving successive subtractions can be used to rewrite the quotient. For example, the computation

$$\begin{array}{r}
x - 3 \\
x^2 + 2x - 1 \overline{\smash{\big)}\, x^3 - x^2 - 7x + 3} \\
\underline{x^3 + 2x^2 - x } \\
-3x^2 - 6x + 3 \\
\underline{-3x^2 - 6x + 3} \\
0
\end{array}$$

shows that, for $x^2 + 2x - 1 \neq 0$,

$$\frac{x^3 - x^2 - 7x + 3}{x^2 + 2x - 1} = x - 3.$$

It is most convenient to arrange the dividend and the divisor in descending powers of the variable before dividing, and to leave an appropriate space for any missing terms (terms with coefficient 0) in the dividend.

2.4 Quotients of Polynomials

When the divisor is not a factor of the dividend, the division process will produce a nonzero remainder. For example, from

$$
\begin{array}{r}
x^3 - 3x^2 + 10x - 28 \\
x+3 \overline{\smash{\big)}\, x^4 + x^2 + 2x - 1} \\
\underline{x^4 + 3x^3 } \\
-3x^3 + x^2 \\
\underline{-3x^3 - 9x^2 } \\
10x^2 + 2x \\
\underline{10x^2 + 30x } \\
-28x - 1 \\
\underline{-28x - 84} \\
83 \text{ (remainder)}
\end{array}
$$

we see that

$$\frac{x^4 + x^2 + 2x - 1}{x+3} = x^3 - 3x^2 + 10x - 28 + \frac{83}{x+3} \qquad (x \ne -3).$$

Observe that a space was left in the dividend for a term involving x^3, even though the dividend contains no such term.

EXERCISE 2.4

Write each quotient as a polynomial. (Assume that all variables in exponents represent positive integers.)

Examples

a. $\dfrac{6x^2 y^3}{2xy}$

b. $\dfrac{x^{2n+3}}{x^{n+1}}$

c. $\dfrac{2y^3 - 6y^2 + y}{y}$

Solutions

a. $\dfrac{6}{2} \cdot x^{2-1} y^{3-1}$

$= 3xy^2 \quad (x, y \ne 0)$

b. $x^{2n+3-(n+1)}$

$= x^{n+2} \quad (x \ne 0)$

c. $\dfrac{2y^3}{y} - \dfrac{6y^2}{y} + \dfrac{y}{y}$

$= 2y^2 - 6y + 1 \quad (y \ne 0)$

1. $\dfrac{8a^3 y^5}{2a^2 y^3}$

2. $\dfrac{27x^4 b^2}{3x^2 b}$

3. $\dfrac{38a^3 b^5}{19ab^3}$

4. $\dfrac{121 y^2 x^5 z}{11 y^2 xz}$

5. $\dfrac{x^{2n}}{x^n}$ 6. $\dfrac{a^{2n+3}}{a^{n-4}}$ 7. $\dfrac{x^{2n}y^{n+1}}{x^n y}$ 8. $\dfrac{r^{2n}s^{n+5}}{r^{n-1}s^{n+3}}$

9. $\dfrac{8a^2 + 4a + 4}{2}$ 10. $\dfrac{12x^3 - 8x^2 + 36x}{4x}$

11. $\dfrac{x^3 - 4x^2 - 3x}{x}$ 12. $\dfrac{8a^2x^2 - 4ax^2 + ax}{ax}$

Write each quotient, P/D, either in the form Q or in the form $Q + R/D$, where Q and R are polynomials and the degree of R is less than that of D.

Examples a. $\dfrac{2y^3 - 6y^2 + 2y - 4}{y}$ b. $\dfrac{y^3 + y^2 - 5y + 2}{y^2 - 2y}$

Solutions a. $\dfrac{2y^3}{y} - \dfrac{6y^2}{y} + \dfrac{2y}{y} - \dfrac{4}{y}$

$= 2y^2 - 6y + 2 - \dfrac{4}{y}$ $(y \neq 0)$

b.
$$y^2 - 2y \overline{\smash{\big)}\, \begin{aligned} &y + 3 \\ &y^3 + y^2 - 5y + 2 \\ &\underline{y^3 - 2y^2} \\ & 3y^2 - 5y \\ & \underline{3y^2 - 6y} \\ & y + 2 \end{aligned}}$$

$y + 3 + \dfrac{y+2}{y^2 - 2y}$ $(y \neq 0, y \neq 2)$

13. $\dfrac{15x^3 - 10x^2 + 3}{5x}$ 14. $\dfrac{38a^2 - 19a + 3}{19a}$

15. $\dfrac{2x^2 + x - 15}{x + 3}$ 16. $\dfrac{4y^2 - 4y - 5}{2y + 1}$

17. $\dfrac{4y^3 + 12y + 5}{2y + 1}$ 18. $\dfrac{3x^3 + 11x^2 + 11x + 15}{x + 3}$

19. $\dfrac{2x^3 - 5x^2 + 8x + 3}{2x - 1}$ 20. $\dfrac{2x^4 + 13x^3 - 7}{2x - 1}$

21. $\dfrac{4y^5 - 4y^2 - 5y + 1}{2y^2 + y + 1}$ 22. $\dfrac{2x^3 - 3x^2 - 15x - 1}{x^2 + 5}$

23. $\dfrac{x^4 - 4x^3 + 10x^2 - 12x + 9}{x^2 - 2x + 3}$ 24. $\dfrac{z^4 + 4z^3 - 2z^2 - 4z + 1}{z^2 + 2z + 1}$

2.5 EQUIVALENT FRACTIONS

RATIONAL EXPRESSIONS   A fraction is an expression denoting a quotient. If the numerator (dividend) and the denominator (divisor) are polynomials, then the fraction is said to be a **rational expression**. Trivially, any polynomial can be considered a rational expression, since it is the quotient of itself and 1. For each replacement of the variable(s) for which the numerator and denominator of a fraction represent real numbers and for which the denominator is not zero, the fraction represents a real number. Of course, for any value of the variable(s) for which the denominator vanishes (is equal to zero), the fraction does not represent a real number and its value is undefined.

If A, B, and C represent polynomials, then for values of the variables for which the denominators do not vanish expressions such as

$$-\frac{A}{B} = \frac{-A}{B} = \frac{A}{-B} = -\frac{-A}{-B}$$

and

$$\frac{AC}{BC} = \frac{A}{B}$$

are said to be **equivalent**.

Examples a. $\dfrac{x-1}{x-2},\ \dfrac{1-x}{2-x},\ -\dfrac{1-x}{x-2},\ -\dfrac{x-1}{2-x}$ b. $\dfrac{x^2(x+1)}{x(x^2-1)}$ is equivalent to $\dfrac{x^2}{x(x-1)}$

These are all equivalent provided $x \neq 2$.   provided $x \neq 0$, $x \neq -1$, and $x \neq 1$.

REDUCING FRACTIONS   We note that in the last example the numerator and the denominator still contain a common factor of x. A fraction is said to be in **lowest terms** only when the numerator and the denominator do not contain prescribed types of factors in common. The arithmetic fraction a/b, where a and b are integers and $b \neq 0$, is in lowest terms provided a and b are relatively prime—that is, provided they contain no common positive integral factors other than 1. If the numerator and the denominator of a fraction are polynomials with integral coefficients, then the fraction is said to be in lowest terms if the numerator and the denominator cannot be expressed as products of polynomials with integral coefficients having a common factor other than ± 1. Expressing a given fraction in lowest terms is called **reducing** the fraction.

Example $\dfrac{y}{y^2} = \dfrac{1 \cdot y}{y \cdot y} = \dfrac{1}{y}$ $(y \neq 0)$

Diagonal cancellation lines are sometimes used to abbreviate the procedure of reducing a fraction. Thus, in the above example, we may write

$$\frac{y}{y^2} = \frac{\cancel{y}^{1}}{\cancel{y^2}_{y}} = \frac{1}{y} \qquad (y \neq 0).$$

Reducing a fraction to lowest terms should be accomplished mentally whenever convenient.

To reduce fractions with polynomial numerators and denominators, you should, when possible, write them in factored form. Common factors are then evident by inspection.

Example
$$\frac{2x^2 + x - 15}{2x + 6} = \frac{(2x - 5)(x + 3)}{2(x + 3)} = \frac{2x - 5}{2} \qquad (x \neq -3)$$

BUILDING FRACTIONS

We can also change fractions of the form A/B to equivalent fractions in higher terms by multiplying the numerator and the denominator by the same polynomial C. That is,

$$\frac{A}{B} = \frac{AC}{BC} \qquad (B, C \neq 0).$$

We might want to do this, for instance, in order to express two given fractions A/B and D/C as equivalent fractions with the same denominator BC.

EXERCISE 2.5

Reduce fractions to lowest terms where possible. Specify restrictions on the variable(s) for which the reduction is not valid.

Examples a. $\dfrac{6x^2 y^3}{2xy^4}$ b. $\dfrac{a - b}{b^2 - a^2}$ c. $\dfrac{a + b}{b}$

Solutions a. $\dfrac{3x(2xy^3)}{y(2xy^3)}$ b. $\dfrac{-(b - a)}{(b - a)(b + a)}$ c. Expression is in lowest terms $(b \neq 0)$.

$= \dfrac{3x}{y} \quad (x, y \neq 0)$ $= \dfrac{-1}{b + a} \cdot (b \neq a, b \neq -a)$

2.5 Equivalent Fractions

1. $\dfrac{a^2bc}{ab^2c}$
2. $\dfrac{24x^4y^2z}{16x^3y^2z}$
3. $\dfrac{2x+2y}{x+y}$
4. $\dfrac{x^2+x}{x+1}$
5. $\dfrac{a-b}{b-a}$
6. $\dfrac{x^2-xy}{y-x}$
7. $\dfrac{x^2-1}{1-x}$
8. $\dfrac{x^2-16}{4-x}$
9. $\dfrac{2x^3-4x^2-3x}{2x}$
10. $\dfrac{8a^2x^2-4ax^2+ax}{2ax}$
11. $\dfrac{y^2+5y-14}{y-2}$
12. $\dfrac{x^2+5x+6}{x+3}$
13. $\dfrac{4y^2+8y-5}{1-2y}$
14. $\dfrac{2x^2+13x-7}{1-2x}$
15. $\dfrac{x^2+6x+9}{x^2+2x-3}$
16. $\dfrac{x^2+5x+6}{x^2+6x+9}$
17. $\dfrac{y^2-2y+1}{y^2-1}$
18. $\dfrac{4-9z^2}{9z^2+12z+4}$
19. $\dfrac{n^2-8n+15}{n^2+n-12}$
20. $\dfrac{t^2-2t-8}{t^2-t-6}$
21. $\dfrac{3x^2-27}{x^2-11x+24}$
22. $\dfrac{5s^2-45s+90}{180-5s^2}$
23. $\dfrac{x^3-y^3}{x^2-y^2}$
24. $\dfrac{x^4-y^4}{x^2+y^2}$
25. $\dfrac{y^4-16}{y^4-y^2-12}$
26. $\dfrac{a^3-a^2+b^2+b^3}{3a^2+6ab+3b^2}$

Express each given fraction as an equivalent fraction with the given denominator. Specify values for the variable(s) for which the fractions are equivalent.

27. $\dfrac{3}{4}; \dfrac{?}{12}$
28. $\dfrac{1}{5}; \dfrac{?}{10}$
29. $\dfrac{b}{a}; \dfrac{?}{a^2b}$
30. $\dfrac{b}{2a}; \dfrac{?}{6a^3b^2}$
31. $\dfrac{3}{y+2}; \dfrac{?}{y^2-y-6}$
32. $\dfrac{2}{x+3}; \dfrac{?}{x^2+x-6}$
33. $\dfrac{3}{a+3}; \dfrac{?}{a^3+27}$
34. $\dfrac{-2}{x^2+y^2}; \dfrac{?}{x^4-y^4}$

35. Is the value of the fraction $\dfrac{x(1-x)}{x^2-3x+2}$ equal to that of $\dfrac{x}{2-x}$ for all values of x? If not, for what value(s) of x does the equality fail to hold?

36. Write three equivalent forms of the fraction $\dfrac{1}{a-b}$ $(a \neq b)$ by changing the sign or signs of the numerator, denominator, or fraction itself.

37. What is the condition on a and b for the fraction $N = \dfrac{-1}{a-b}$ to represent a positive number? A negative number?

38. Is the fraction $\dfrac{x-2}{1+x^2}$ defined for all real values of x? For what value(s) of x does the fraction equal zero?

2.6 SUMS OF RATIONAL EXPRESSIONS

If the fractions in a sum have unlike denominators, we can replace the fractions with equivalent fractions having common denominators and then write the sum as a single fraction. Thus,

$$\frac{A}{B} + \frac{C}{D} = \frac{AD}{BD} + \frac{CB}{DB}$$

$$= \frac{AD + CB}{BD} \qquad (B, D \neq 0).$$

The difference

$$\frac{A}{B} - \frac{C}{D}$$

may be viewed as the sum

$$\frac{A}{B} + \left(-\frac{C}{D}\right),$$

and then we can write it as

$$\frac{AD}{BD} + \left(\frac{-CB}{DB}\right) = \frac{AD - CB}{BD} \qquad (B, D \neq 0).$$

LEAST COMMON MULTIPLE In rewriting fractions in a sum so that they share a common denominator, any such denominator may be used. If the **least common multiple** of the denominators (called the **least common denominator**) is used, however, the resulting fraction will be in simpler form than if any other common denominator is employed. The least common multiple of two or more positive integers is the least positive integer that is exactly divisible by each of the given numbers (each quotient is a positive integer).

The notion of a least common multiple among several polynomial expressions is, in general, meaningless without further specification of what is desired. We can, however, define the least common multiple of a set of polynomials with integer coefficients to be the polynomial of lowest degree with integer coefficients yielding a polynomial quotient upon division by each of the given polynomials, and to be, among all such polynomials, the one having the least possible positive leading coefficient.

2.6 Sums of Rational Expressions

Very often, the least common multiple of a set of positive integers or polynomials can be determined by inspection. When inspection fails us, however, we can find the least common multiple as follows:

1. Express each polynomial in completely factored form.
2. Set up as factors of a product each *different* factor that occurs in any of the polynomials, including each factor the greatest number of times it occurs in any one of the given polynomials.

Example Find the least common multiple of 12, 15, and 18.

Solution

$$12 \qquad 15 \qquad 18$$
$$2 \cdot 2 \cdot 3 \qquad 3 \cdot 5 \qquad 3 \cdot 3 \cdot 2$$

The least common multiple is $2^2 \cdot 3^2 \cdot 5$, or 180.

Example Find the least common multiple of x^2, $x^2 - 9$, and $x^3 - x^2 - 6x$.

Solution

$$x^2 \qquad x^2 - 9 \qquad x^3 - x^2 - 6x$$
$$x \cdot x \qquad (x-3)(x+3) \qquad x(x-3)(x+2)$$

The least common multiple is $x^2(x+2)(x-3)(x+3)$.

To simplify sums of fractions with different denominators, we can ascertain the least common denominator of the fractions, determine the factor necessary to express each fraction as a fraction with this common denominator, write the fractions accordingly, and then express the sum as a single fraction.

Example Write $\dfrac{3}{x} + \dfrac{2}{x^2} + \dfrac{3}{xy}$ as a single fraction.

Solution The least common denominator of the fraction is $x^2 y$. We have

$$\frac{3}{x} + \frac{2}{x^2} + \frac{3}{xy} = \frac{3(xy)}{x(xy)} + \frac{2(y)}{x^2(y)} + \frac{3(x)}{xy(x)}$$

$$= \frac{3xy}{x^2 y} + \frac{2y}{x^2 y} + \frac{3x}{x^2 y}$$

$$= \frac{3xy + 2y + 3x}{x^2 y} \qquad (x, y \neq 0).$$

EXERCISE 2.6

Write each sum or difference as a single fraction in lowest terms. Assume that no variable in a denominator takes a value for which the denominator vanishes.

1. $\dfrac{x-1}{2y} + \dfrac{x}{2y}$

2. $\dfrac{y+1}{x} + \dfrac{y-1}{x}$

3. $\dfrac{2a-b}{a} - \dfrac{a-b}{a}$

4. $\dfrac{3a-1}{b} - \dfrac{2-a}{b}$

5. $\dfrac{a+2}{3} - \dfrac{a-3}{9}$

6. $\dfrac{a-2}{9} - \dfrac{a+1}{3}$

7. $\dfrac{2}{a+b} + \dfrac{1}{2a+2b}$

8. $\dfrac{7}{5x-10} + \dfrac{5}{3x-6}$

9. $\dfrac{2}{3-x} - \dfrac{1}{x-3}$

10. $\dfrac{7}{y-3} + \dfrac{3}{3-y}$

11. $\dfrac{a+1}{a+2} - \dfrac{a+2}{a+3}$

12. $\dfrac{5x-y}{3x+y} - \dfrac{6x-5y}{2x-y}$

13. $\dfrac{x+2y}{2x-y} - \dfrac{2x+y}{x-2y}$

14. $\dfrac{x-2y}{x+y} - \dfrac{2x-y}{x-y}$

15. $\dfrac{1}{y^2-y-2} + \dfrac{1}{y^2+2y+1}$

16. $\dfrac{2}{z^2-z-6} + \dfrac{3}{z^2-9}$

17. $\dfrac{3n}{n^2+3n-10} - \dfrac{2n}{n^2+n-6}$

18. $\dfrac{5u}{u^2+3u+2} - \dfrac{3u-6}{u^2+4u+4}$

19. $\dfrac{y}{y^2-16} - \dfrac{y+1}{y^2-5y+4} + \dfrac{1}{y+4}$

20. $\dfrac{1}{b^2-1} - \dfrac{1}{b^2+2b+1} + \dfrac{1}{b+1}$

21. $x + \dfrac{1}{x-1} - \dfrac{1}{(x-1)^2}$

22. $y - \dfrac{2y}{y^2-1} + \dfrac{3}{y+1}$

23. $x - 1 + \dfrac{3}{2x-1} - \dfrac{x}{4x^2-1}$

24. $2y - 3 - \dfrac{1}{y^2+2y+1} + \dfrac{3}{y+1}$

25. $\dfrac{y+3}{3y^2+7y+4} - \dfrac{y-7}{3y^2+13y+12}$

26. $\dfrac{z+4}{2z^2-5z-3} + \dfrac{2z-1}{2z^2+3z+1}$

27. $\dfrac{xy}{(z-x)(x-y)} + \dfrac{yz}{(z-y)(x-z)} + \dfrac{xz}{(y-x)(y-z)}$

28. $\dfrac{a+b}{a^2+2ab-3b^2} - \dfrac{a-2b}{a^2-b^2} + \dfrac{2a+b}{a^2+4ab+3b^2}$

29. $\dfrac{1}{(a-b)(b-c)} + \dfrac{1}{(b-c)(c-a)} + \dfrac{1}{(c-a)(a-b)}$

30. Any set of fractions has an infinite number of common denominators. Why is it convenient to use the *least* common denominator in finding sums or differences of fractions?

2.7

PRODUCTS AND QUOTIENTS OF RATIONAL EXPRESSIONS

Recall from arithmetic that in order to multiply fractions we multiply corresponding numerators and denominators; to divide fractions, we invert the divisor and multiply. For example,

$$\frac{3}{4} \cdot \frac{5}{7} = \frac{15}{28} \quad \text{and} \quad \frac{3}{4} \div \frac{5}{7} = \frac{3}{4} \cdot \frac{7}{5} = \frac{21}{20}.$$

Similarly for polynomials A, B, C, and D for which the denominators do not vanish we have,

$$\frac{A}{B} \cdot \frac{C}{D} = \frac{AC}{BD},$$

$$\frac{A}{B} \div \frac{C}{D} = \frac{AD}{BC}.$$

The quotient $\frac{A}{B} \div \frac{C}{D}$ can also be written

$$\frac{\frac{A}{B}}{\frac{C}{D}}$$

This form is called a **complex fraction**; that is, it is a fraction containing a fraction in either the numerator or the denominator or both.

We can now rewrite a product or quotient of fractions as a single fraction in lowest terms.

Example

$$\frac{x^2 - 2x + 1}{x^2 + 2x - 3} \cdot \frac{x^2 + 3x}{x^2 + 2x} = \frac{(x-1)(x-1)}{(x+3)(x-1)} \cdot \frac{x(x+3)}{x(x+2)}$$

$$= \frac{(x-1)[(x-1)(x+3)x]}{(x+2)[(x-1)(x+3)x]}$$

$$= \frac{x-1}{x+2} \qquad (x \neq -3, -2, 0, 1)$$

Since the factors of the numerator and the denominator of the product of two fractions are just the factors of the numerators and the denominators, respectively, of the fractions, we can divide common factors out of the numerators and the denominators before writing the product as a single fraction. Thus, in the example above, we could write

$$\frac{x^2 - 2x + 1}{x^2 + 2x - 3} \cdot \frac{x^2 + 3x}{x^2 + 2x} = \frac{(\cancel{x-1})(x-1)}{(\cancel{x+3})(\cancel{x-1})} \cdot \frac{\cancel{x}(\cancel{x+3})}{\cancel{x}(x+2)}$$

$$= \frac{x-1}{x+2} \qquad (x \neq -3, -2, 0, 1).$$

Example

$$\frac{x^2 - x - 2}{x^2 + x - 2} \div \frac{x+1}{x-1} = \frac{(x-2)(\cancel{x+1})}{(x+2)(\cancel{x-1})} \cdot \frac{(\cancel{x-1})}{(\cancel{x+1})}$$

$$= \frac{x-2}{x+2} \qquad (x \neq -2, -1, 1)$$

REDUCTION OF COMPLEX FRACTIONS

When the quotient of two fractions is given in the form of a complex fraction, we have a choice of procedures for writing the quotient in the form of a simple (not complex) fraction.

Example Write $\dfrac{x + \dfrac{3}{4}}{x - \dfrac{1}{2}}$ as a simple fraction in lowest terms.

Solution 1 We multiply numerator and denominator by the least common denominator of the simple fractions involved. Thus,

$$\frac{\left(x + \dfrac{3}{4}\right)4}{\left(x - \dfrac{1}{2}\right)4} = \frac{4x + 3}{4x - 2} \qquad \left(x \neq \frac{1}{2}\right).$$

Solution 2 Alternatively, we can rewrite the complex fraction as follows:

$$\frac{x + \dfrac{3}{4}}{x - \dfrac{1}{2}} = \frac{\dfrac{4x+3}{4}}{\dfrac{2x-1}{2}} = \frac{4x+3}{4} \cdot \frac{2}{2x-1} = \frac{4x+3}{4x-2} \qquad \left(x \neq \frac{1}{2}\right).$$

If we have a more complicated expression involving a complex fraction, we can rewrite it by simplifying small parts at a time.

2.7 Products and Quotients of Rational Expressions

Example Write $\dfrac{1}{x + \dfrac{1}{x + \dfrac{1}{x}}}$ as a simple fraction in lowest terms.

Solution We can begin by concentrating on the lower right-hand expression, $\dfrac{1}{x + \dfrac{1}{x}}$.

We have

$$\frac{1}{x + \dfrac{1}{x}} = \frac{(1)x}{\left(x + \dfrac{1}{x}\right)x} = \frac{x}{x^2 + 1}.$$

Thus,

$$\frac{1}{x + \dfrac{1}{x + \dfrac{1}{x}}} = \frac{1}{x + \dfrac{x}{x^2 + 1}}.$$

Now we can apply either of the methods shown in the previous example to the right-hand expression above. Using the first method, we have

$$\frac{1}{x + \dfrac{1}{x + \dfrac{1}{x}}} = \frac{1(x^2 + 1)}{\left(x + \dfrac{x}{x^2 + 1}\right)(x^2 + 1)} = \frac{x^2 + 1}{x^3 + x + x} = \frac{x^2 + 1}{x^3 + 2x} \qquad (x \neq 0).$$

EXERCISE 2.7

Write each product or quotient as a single fraction in lowest terms. Assume that no denominator vanishes.

1. $\dfrac{-12a^2 b}{5c} \cdot \dfrac{10b^2 c}{24a^3 b}$

2. $\dfrac{a^2}{xy} \cdot \dfrac{3x^3 y}{4a}$

3. $\dfrac{xy}{a^2 b} \div \dfrac{x^3 y^2}{ab}$

4. $\dfrac{24a^3 b}{-6xy^2} \div \dfrac{3a^2 b}{12x}$

5. $\dfrac{x^2 - x - 20}{x^2 + 7x + 12} \cdot \dfrac{2x^2 + 6x}{x^2 - 25}$

6. $\dfrac{4x^2 + 8x + 3}{2x^2 - 5x + 3} \cdot \dfrac{6x^2 - 9x}{1 - 4x^2}$

7. $\dfrac{25a^2b^2 - 16}{4ab + 1} \div \dfrac{5ab + 4}{16a^2b^2 + 16ab + 3}$

8. $\dfrac{a^2 - 25}{a^2 - 16} \div \dfrac{a^2 + 2a - 15}{a^2 + a - 12}$

9. $\dfrac{x^2 - y^2}{x^2} \cdot \dfrac{x^2 - xy + y^2}{x^2} \div \dfrac{x^3 + y^3}{x^4}$

10. $\dfrac{a^3 + 8b^3}{a^2 - 64b^2} \cdot \dfrac{a^3 + 512b^3}{a^2 - ab - 6b^2} \div \dfrac{a^4 + 4a^2b^2 + 16b^4}{a^2 + 11ab + 24b^2}$

11. $\left(1 + \dfrac{1}{x}\right) \cdot \left(1 - \dfrac{1}{x}\right)$

12. $\left(x - \dfrac{1}{x}\right) \div \left(x + \dfrac{1}{x}\right)$

13. $\left(\dfrac{3}{x - 1} - \dfrac{2}{x + 1}\right) \cdot \dfrac{x - 1}{x}$

14. $\left(\dfrac{x}{x^2 - 9} + \dfrac{2}{x - 3}\right) \cdot \dfrac{x - 1}{x}$

15. $\left(\dfrac{2y}{2y - 1} - \dfrac{3}{y}\right) \div \dfrac{3}{2y^2 - y}$

16. $\left(\dfrac{y}{y^2 - 1} - \dfrac{y}{y^2 - 2y + 1}\right) \div \dfrac{y}{y - 1}$

17. $\dfrac{\dfrac{2}{a} + \dfrac{3}{2a}}{5 + \dfrac{1}{a}}$

18. $\dfrac{1 + \dfrac{1}{x}}{1 - \dfrac{1}{x}}$

19. $a - \dfrac{a}{a + \dfrac{1}{4}}$

20. $x - \dfrac{x}{1 - \dfrac{x}{1 - x}}$

21. $1 - \dfrac{1}{1 - \dfrac{1}{y - 2}}$

22. $2y + \dfrac{3}{3 - \dfrac{2y}{y - 1}}$

23. $\dfrac{1 + \dfrac{1}{1 - \dfrac{a}{b}}}{1 - \dfrac{3}{1 - \dfrac{a}{b}}}$

24. $\dfrac{1 - \dfrac{1}{\dfrac{a}{b} + 2}}{1 + \dfrac{3}{\dfrac{a}{2b} + 1}}$

25. $\dfrac{a + 2 - \dfrac{12}{a + 3}}{a - 5 + \dfrac{16}{a + 3}}$

26. $\dfrac{a + 4 - \dfrac{7}{a - 2}}{a - 1 + \dfrac{2}{a - 2}}$

27. $\dfrac{\dfrac{x^2 - 2xy + y^2 - z^2}{x^2 + 2xy + y^2 - z^2}}{\dfrac{x - y + z}{x + y - z}}$

28. $\dfrac{\dfrac{9x^2 - 6x}{6x^2 - 7x + 2} \cdot \dfrac{2x^2 + 13x - 7}{2x^2 + 6x}}{\dfrac{4x^2 - 8x + 3}{2x^2 + 3x - 9}}$

29. $\left(\dfrac{a^2b^2 + 2b^4}{27a^6} \div \dfrac{a^2 - 10ab + 25b^2}{3a^5 + 6a^3b^2}\right) \cdot \dfrac{15a^3b - 3a^4}{a^4 + 4a^2b^2 + 4b^4}$

30. $\left(\dfrac{x^2 - xy + y^2}{x^2 + xy + y^2} \cdot \dfrac{x^3 - y^3}{x^3 + y^3}\right) \div \dfrac{(x - y)^2}{(x + y)^2}$

2.8

POWERS WITH INTEGRAL EXPONENTS

So far we have considered powers of real numbers involving only positive-integer exponents. Powers with integral and rational exponents can be defined in a manner consistent with these properties. For example, we shall consider algebraic expressions such as

$$x^{1/2}, \qquad x^{-2}, \qquad x^{-1}(x^{-3} + y^{1/4}).$$

None of these expressions are polynomials.

REASON FOR DEFINING a^0 TO BE 1

We have seen that when m and n are positive integers, then for $m > n$,

(1) $$\frac{a^m}{a^n} = a^{m-n}.$$

If (1) is to hold also for $m = n$, then we must have

$$\frac{a^n}{a^n} = a^{n-n} = a^0.$$

Since $a^n/a^n = 1$ for $a \neq 0$, we can state the following.

> If $a \neq 0$, then
> $$a^0 = 1.$$

REASON FOR DEFINING a^{-n} TO BE $1/a^n$

In a similar way, if (1) is to hold for $m = 0$, then we must have

$$\frac{a^0}{a^n} = a^{0-n} = a^{-n}.$$

Since $a^0 = 1$, we can state the following.

> If $a \neq 0$, and n is a positive integer, then
> $$a^{-n} = \frac{1}{a^n}.$$

LAWS OF EXPONENTS FOR INTEGRAL EXPONENTS

The following summarizes, without proof, the laws of exponents for integral exponents.

> *If m and n are integers, then*
>
> $$\text{I} \quad a^m \cdot a^n = a^{m+n},$$
>
> $$\text{II} \quad \frac{a^m}{a^n} = a^{m-n},$$
>
> $$\text{III} \quad (a^m)^n = a^{mn},$$
>
> $$\text{IV} \quad (ab)^n = a^n b^n,$$
>
> $$\text{V} \quad \left(\frac{a}{b}\right)^n = \frac{a^n}{b^n},$$
>
> where $a, b \neq 0$.

Examples

a. $(x^2)^{-3} = x^{2 \cdot (-3)} = x^{-6}$

b. $(xy)^{-3} = x^{-3} y^{-3}$

c. $\left(\dfrac{x^2}{y}\right)^{-4} = \dfrac{(x^2)^{-4}}{(y)^{-4}}$

$\phantom{c.\left(\dfrac{x^2}{y}\right)^{-4}} = \dfrac{x^{-8}}{y^{-4}}$

$\phantom{c.\left(\dfrac{x^2}{y}\right)^{-4}} = \dfrac{1/x^8}{1/y^4}$

$\phantom{c.\left(\dfrac{x^2}{y}\right)^{-4}} = \dfrac{y^4}{x^8} \quad (x, y \neq 0)$

Although we shall not prove it here, parts III, IV, and V of the laws of exponents can be shown to apply to expressions involving more than two factors.

Examples

a. $\left(\dfrac{x^2 z^4}{5y}\right)^{-2} = \dfrac{x^{-4} z^{-8}}{5^{-2} y^{-2}}$

$\phantom{a.\left(\dfrac{x^2 z^4}{5y}\right)^{-2}} = \dfrac{5^2 y^2}{x^4 z^8}$

$\phantom{a.\left(\dfrac{x^2 z^4}{5y}\right)^{-2}} = \dfrac{25 y^2}{x^4 z^8}$

b. $\dfrac{xy^{-2}}{x^{-1} + y} = \dfrac{x \cdot \dfrac{1}{y^2}}{\dfrac{1}{x} + y} = \dfrac{\dfrac{x}{y^2}}{\dfrac{1 + xy}{x}}$

$\phantom{b.\dfrac{xy^{-2}}{x^{-1}+y}} = \dfrac{x}{y^2} \cdot \dfrac{x}{1 + xy} = \dfrac{x^2}{y^2(1 + xy)}$

2.8 Powers with Integral Exponents

To avoid the necessity of constantly noting exceptions, we shall assume that in the remaining exercises in this chapter the variables are restricted so that no denominator vanishes.

EXERCISE 2.8

For each expression write an equivalent basic numeral—that is, a numeral with exponent 1.

Examples a. $3 \cdot 5^{-2}$ b. $\dfrac{3}{2^{-3}}$ c. $4^2 + 4^{-2} - 4^0$

Solutions a. $3 \cdot \dfrac{1}{5^2} = 3 \cdot \dfrac{1}{25}$ b. $\dfrac{3}{\frac{1}{2^3}} = 3 \cdot 2^3$ c. $16 + \dfrac{1}{16} - 1 = 15 + \dfrac{1}{16}$

$\quad= \dfrac{3}{25}$ $= 24$ $= \dfrac{241}{16}$

1. 5^{-1}
2. 4^{-1}
3. 3^{-2}
4. 2^{-3}
5. $\dfrac{1}{3^{-3}}$
6. $\dfrac{1}{4^{-2}}$
7. $\dfrac{2^{-1} \cdot 3^0}{5}$
8. $\dfrac{2}{3^{-2} \cdot 4^0}$
9. $\left(\dfrac{3}{5}\right)^{-1}$
10. $\left(\dfrac{1}{3}\right)^{-2}$
11. $\dfrac{5^{-1}}{3^{-2}}$
12. $\dfrac{3^{-3}}{6^{-2}}$
13. $3^{-2} + 3^2$
14. $5^{-1} + 25^0$
15. $4^{-1} - 4^{-2}$
16. $8^{-2} - 2^0$

Write each given expression as a product or quotient in which each variable occurs at most once in the expression and involves positive exponents only.

Examples a. $x^{-3}x^5$ b. $(x^2 y^{-3})^{-1}$ c. $\left(\dfrac{x^{-1} y^2 z^0}{x^3 y^{-4} z^2}\right)^{-1}$

Solutions a. $x^{-3+5} = x^2$ b. $x^{-2} y^3 = \dfrac{1}{x^2} y^3$ c. $\dfrac{xy^{-2}z^0}{x^{-3}y^4 z^{-2}} = \dfrac{x \cdot x^3 \cdot 1 \cdot z^2}{y^4 \cdot y^2}$

$\quad= \dfrac{y^3}{x^2}$ $= \dfrac{x^4 z^2}{y^6}$

17. $\dfrac{x^2}{y^{-3}}$
18. $\dfrac{x^3}{y^{-2}}$
19. $(x^3 y^2)^2$
20. $(xy^2)^3$

21. $\left(\dfrac{3x}{y^3}\right)^2$ 22. $\left(\dfrac{x^2}{2y}\right)^3$ 23. $\left(\dfrac{x^2}{y}\right)^3\left(\dfrac{2y}{x}\right)^2$ 24. $\left(\dfrac{3x}{y^2}\right)^2\left(\dfrac{2y^3}{x}\right)^2$

25. $x^{-3}x^7$ 26. $\dfrac{x^3}{x^{-2}}$ 27. $(x^{-2}y^0)^3$ 28. $(x^{-2}y^3)^0$

29. $\dfrac{x^{-1}}{y^{-1}}$ 30. $\dfrac{x^{-3}}{y^{-2}}$ 31. $\dfrac{8^{-1}x^0y^{-3}}{(2xy)^{-5}}$ 32. $\left(\dfrac{x^{-1}y^3}{2x^0y^{-5}}\right)^{-2}$

Represent each expression as a single fraction involving positive exponents only.

Examples a. $x^{-1} + y^{-2}$ b. $(x^{-1} + x^{-2})^{-1}$ c. $\dfrac{x^{-1}}{x^{-1} + y^{-1}}$

Solutions a. $\dfrac{1}{x} + \dfrac{1}{y^2}$ b. $\left(\dfrac{1}{x} + \dfrac{1}{x^2}\right)^{-1}$ c. $\dfrac{\dfrac{1}{x}(xy)}{\left(\dfrac{1}{x} + \dfrac{1}{y}\right)(xy)}$

$= \dfrac{(y^2)1}{(y^2)x} + \dfrac{1(x)}{y^2(x)}$ $= \left(\dfrac{x+1}{x^2}\right)^{-1}$ $= \dfrac{y}{y+x}$

$= \dfrac{y^2 + x}{xy^2}$ $= \dfrac{x^2}{x+1}$

33. $x^{-1} - y^{-2}$ 34. $x^{-2} + y$ 35. $x^{-1}y + xy^{-1}$

36. $\dfrac{x}{y^{-1}} + \dfrac{x^{-1}}{y}$ 37. $x(x-y)^{-1}$ 38. $y(x+y)^{-2}$

39. $xy^{-1} + x^{-1}y$ 40. $x^{-1}y - xy^{-1}$ 41. $\dfrac{x^{-1} + y^{-1}}{(xy)^{-1}}$

42. $\dfrac{x}{y^{-1}} + \left(\dfrac{x}{y}\right)^{-1}$ 43. $(x^{-1} - y^{-1})^{-1}$ 44. $\dfrac{x^{-1} + y^{-1}}{x^{-1} - y^{-1}}$

For each expression, write an equivalent product in which each variable occurs only once.

Examples a. $\dfrac{x^n x^{n+1}}{x^{n-1}}$ b. $(y^{n-1})^{-3}$ c. $\dfrac{(y^{n-1})^2}{y^{n-2}}$

Solutions a. $x^{n+(n+1)-(n-1)} = x^{n+2}$ b. y^{-3n+3} c. $y^{(2n-2)-(n-2)} = y^n$

45. $x^n x^{n-1}$ 46. $x^{n-1} x^{2n}$ 47. $\dfrac{x^{n+1} x^{2n-1}}{x^{3n}}$

2.9 Powers with Rational Exponents

48. $\dfrac{y^{n+1} y^{n+2}}{y^n}$

49. $\dfrac{(y^{n+1} y)^2}{y^{2n}}$

50. $\left(\dfrac{y^{2n-1} y^n}{y^{2n}}\right)^3$

51. $\left(\dfrac{x^n}{x^{n-3}}\right)^2$

52. $\left(\dfrac{x^n}{x^{n-1}}\right)^3$

53. $\dfrac{x^n y^{n+1}}{x^{2n-1} y^n}$

54. $\dfrac{x^n y^{2n-1}}{x^{n+1} y^{2n}}$

55. $\left(\dfrac{x^{2n}}{x^{n+1}}\right)^{-2}$

56. $\left(\dfrac{x^{2n} y^{n-1}}{x^{n-1} y}\right)^{-2}$

*Write each number as the product of a number between 1 and 10 and a power of 10. (This exponential form is called **scientific notation**.)*

Examples   **a.** 680,000   **b.** 0.032   **c.** 0.0000431

Solutions   **a.** Factor.   **b.** Factor.   **c.** Factor.

$6.8 \times 100{,}000$ $3.2 \times \dfrac{1}{100}$ $4.31 \times \dfrac{1}{100{,}000}$

Write in exponential form.   Write in exponential form.   Write in exponential form.

6.8×10^5 3.2×10^{-2} 4.31×10^{-5}

57. 2,540
58. 38,421
59. 642,000
60. 2,541,000
61. 0.0014
62. 0.0000006
63. 0.0000230
64. 0.5020

2.9 POWERS WITH RATIONAL EXPONENTS

In Sections 2.2 and 2.8, we defined powers of real numbers with positive-integer, zero, and negative-integer exponents. Now we want to give meaning to powers of real numbers with rational numbers as exponents. We shall want any such definition to be consistent with all the laws of exponents for integers.

We know that when m and n are positive integers,

(1) $\qquad\qquad\qquad (a^m)^n = a^{mn}.$

If (1) is to hold for $m = 1/n$, and $a^{1/n}$ is a real number, then we must have

$$(a^{1/n})^n = a^{(1/n)(n)} = a^{n/n} = a^1 = a,$$

so the nth power of $a^{1/n}$ must be a. A number having a as its nth power is called an **nth root** of a. For $n = 2$ or $n = 3$, respectively, an nth root is called a **square root** or a **cube root**.

NUMBER OF ROOTS For n odd, each real number a has just one real nth root. Thus,

$$(-2)^3 = -8 \quad \text{and} \quad 2^3 = 8,$$

so that -2 is the cube root of -8, and 2 is the cube root of 8.

For n even and $a > 0$, a has two real nth roots. Thus,

$$(-2)^4 = 16 \quad \text{and} \quad 2^4 = 16,$$

so that -2 and 2 are both fourth roots of 16.

For n even and $a < 0$, a has no real nth root. Thus, -1 has no real square root, since the square of every real number is nonnegative.

If $a = 0$, then a has exactly one real nth root, namely 0.

If n is a positive integer, then $a^{1/n}$ is the real number if one exists, and the positive real number if two exist, such that,

$$(a^{1/n})^n = a.$$

Examples

a. $25^{1/2} = 5$

b. $-25^{1/2} = -5$

c. $(-25)^{1/2}$ is not a real number

d. $27^{1/3} = 3$

e. $-27^{1/3} = -3$

f. $(-27)^{1/3} = -3$

Notice in parts b and e that $-25^{1/2}$ and $-27^{1/3}$ denote $-(25^{1/2})$ and $-(27^{1/3})$, respectively.

To generalize from rational exponents of the form $1/n$, where n is a positive integer, to other rational exponents, we observe that any rational number can be written in the form m/n, for m an integer, and n a positive integer. By requiring n to be a positive integer we do not alter the fact that m/n can represent every rational number, since all that is done is to restrict the denominator of the fraction representing the rational number to a positive value. We need the following results before we define $a^{m/n}$.

If $a^{1/n}$ is a real number, m is an integer, and n is a positive integer, then

$$(a^{1/n})^m = (a^m)^{1/n}.$$

Observe that we require $a^{1/n}$ be a real number. This requirement is not satisfied if n is even and a is negative. If, for instance, $a = -3$, $m = 2$, and $n = 2$, then $a^{1/n} = (-3)^{1/2}$ is not a real number. Finally, we observe that if p is a positive integer and $a^{1/np}$ is a real number, then $(a^{1/n})^m$, $(a^m)^{1/n}$, $(a^{1/np})^{mp}$, and $(a^{mp})^{1/np}$ all represent the same number.

2.9 Powers with Rational Exponents

Examples
a. $(16^{1/2})^3 = 4^3 = 64$
b. $(16^3)^{1/2} = 4096^{1/2} = 64$
c. $(16^{1/4})^6 = 2^6 = 64$
d. $(16^6)^{1/4} = 16{,}777{,}216^{1/4} = 64$

We are now in a position to state the following.

> If $a^{1/n}$ is a real number, m is an integer, and n is a positive integer, then
> $$a^{m/n} = (a^{1/n})^m.$$

Examples
a. $8^{2/3} = (8^{1/3})^2 = 2^2 = 4$
b. $16^{-3/4} = (16^{1/4})^{-3} = 2^{-3} = \dfrac{1}{8}$

Because we define $a^{1/n}$ to be the positive nth root of a for a positive and n an even positive integer, and since a^m is positive for a negative and m an even positive integer, it follows that, for m and n even positive integers and a any real number,
$$(a^m)^{1/n} = |a|^{m/n}.$$

For the special case $m = n$ (m and n even),
$$(a^n)^{1/n} = |a|.$$

PROPERTIES OF RATIONAL POWERS

Powers with rational exponents have the same fundamental properties as powers with integral exponents, as long as the powers are real numbers. The laws of exponents for integers, appropriately reworded for rational-number exponents, can be used to rewrite exponential expressions involving rational exponents.

Examples
a. $\dfrac{x^{2/3}}{x^{1/3}} = x^{2/3 - 1/3}$
$= x^{1/3} \quad (x \neq 0)$

b. $\left(\dfrac{a^3 b^6}{c^{12}}\right)^{2/3} = \dfrac{(a^3)^{2/3}(b^6)^{2/3}}{(c^{12})^{2/3}}$
$= \dfrac{a^2 b^4}{c^8} \quad (c \neq 0)$

c. $(a^6)^{1/2} = |a|^{6/2}$
$= |a|^3$

Observe that in part c of the example it was necessary to use absolute-value notation because the expression has been defined to be positive for m and n even. For example, with $a = -2$, we have
$$[(-2)^6]^{1/2} = |-2|^{6/2} = |-2|^3 = 8,$$
whereas
$$(-2)^3 = -8.$$

EXERCISE 2.9

For each expression, write an equivalent basic numeral—that is, a numeral with exponent 1.

Examples   a. $64^{1/2}$ b. $\left(\dfrac{8}{27}\right)^{-2/3}$ c. $(-27)^{4/3}$

Solutions   a. $64^{1/2} = 8$ b. $\left[\left(\dfrac{8}{27}\right)^{1/3}\right]^{-2} = \left(\dfrac{2}{3}\right)^{-2} = \dfrac{9}{4}$ c. $[(-27)^{1/3}]^4 = (-3)^4 = 81$

1. $16^{1/2}$
2. $27^{1/3}$
3. $27^{2/3}$
4. $27^{4/3}$
5. $16^{-1/2}$
6. $8^{-1/3}$
7. $16^{-3/4}$
8. $27^{-2/3}$
9. $(-8)^{-2/3}$
10. $(-8)^{-4/3}$
11. $\left(\dfrac{4}{9}\right)^{3/2}$
12. $\left(\dfrac{4}{9}\right)^{-3/2}$

Write each expression as a product or quotient of powers in which each variable occurs only once, and all exponents are positive. Assume all variable bases are positive and all variable exponents are positive integers.

Examples   a. $\dfrac{x^{5/6}}{x^{2/3}}$ b. $\dfrac{(x^{1/2}y^2)^2}{(x^{2/3}y)^3}$ c. $(y^{2n} \cdot y^{n/2})^4$

Solutions   a. $x^{5/6 \cdot 2/3} = x^{5/6 - 4/6} = x^{1/6}$ b. $\dfrac{xy^4}{x^2y^3} = \dfrac{y}{x}$ c. $y^{8n} \cdot y^{2n} = y^{10n}$

13. $x^{2/3}x^{4/3}$
14. $x^{1/4}x^{5/4}$
15. $y^{1/2}y^{3/4}$
16. $y^{5/6}y^{1/3}$
17. $\dfrac{x^{5/6}}{x^{1/2}}$
18. $\dfrac{x^{3/4}}{x^{1/2}}$
19. $\dfrac{y^{2/3}}{y^{1/2}}$
20. $\dfrac{y^{3/5}}{y^{1/2}}$
21. $\left(\dfrac{x^{1/2}}{y^2}\right)^2 \left(\dfrac{y^4}{x^2}\right)^{1/2}$
22. $\left(\dfrac{y^{2/3}}{x}\right)^3 \left(\dfrac{x^2}{y^{1/2}}\right)^2$
23. $\left(\dfrac{x^5y^8}{y^{13}}\right)^{1/4}$
24. $\left(\dfrac{125x^3y^4}{27x^{-6}y}\right)^{1/3}$
25. $(x^2)^{n/2}(y^{2n})^{2/n}$
26. $(x^{n/2})^2(y^n)^{5/n}$
27. $\dfrac{x^{2n}}{x^{n/2}}$

2.9 Powers with Rational Exponents

28. $\left(\dfrac{a^n}{b}\right)^{1/2} \left(\dfrac{b}{a^{2n}}\right)^{3/2}$ **29.** $\dfrac{x^{3n}y^{2m-1}}{(x^n y^m)^{1/2}}$ **30.** $\left(\dfrac{x^{2n^2}}{x^{4n}}\right)^{1/n}$

31. $\left(\dfrac{xy^{-m}}{x^n y}\right)^{-1}$ **32.** $\left(\dfrac{x^{2n}y^n}{x^n y^{3n}}\right)^{-1/n}$

Apply the distributive law to write each product as a sum.

Examples   **a.** $x^{1/4}(x^{3/4} - x)$ **b.** $(x^{1/2} - x)(x^{1/2} + x)$

Solutions   **a.** $x^{1/4}x^{3/4} - x^{1/4}x = x - x^{5/4}$ **b.** $x^{1/2}x^{1/2} - x^2 = x - x^2$

33. $x^{1/2}(x^{1/2} - 1)$ **34.** $x^{1/3}(x^{1/3} + 2)$
35. $y^{2/3}(y - y^{1/3})$ **36.** $y^{1/2}(y^2 - y^{1/2})$
37. $(x^{1/2} - y^{1/2})(x^{1/2} + y^{1/2})$ **38.** $(x^{-1/2} + y^{1/2})(x^{-1/2} - y^{1/2})$
39. $(x + y)^{1/2}[(x + y)^{1/2} - (x + y)]$ **40.** $(x - y)^{2/3}[(x - y)^{-1/3} + (x - y)]$

Factor as indicated.

Examples   **a.** $y^{-1/2} + y^{1/2} = y^{-1/2}(?)$ **b.** $x^{3/2} - x^{-1/2} = x^{-1/2}(?)$

Solutions   **a.** $y^{-1/2}(1 + y)$ **b.** $x^{-1/2}(x^2 - 1)$

41. $x^{3/2} + x = x(?)$ **42.** $y - y^{2/3} = y^{1/3}(?)$
43. $x^{-3/2} + x^{-1/2} = x^{-1/2}(?)$ **44.** $y^{3/4} - y^{-1/4} = y^{-1/4}(?)$
45. $(x + 1)^{1/2} - (x + 1)^{-1/2} = (x + 1)^{-1/2}(?)$
46. $(y + 2)^{1/5} - (y + 2)^{-4/5} = (y + 2)^{-4/5}(?)$

In the previous problems, the variables were restricted to represent positive numbers. In Problems 47–52, consider each variable base to be any element of the set of real numbers and simplify.

Examples   **a.** $[(-3)^2]^{1/2}$ **b.** $[u^2(u + 5)]^{1/2}$

Solutions   **a.** $[(-3)^2]^{1/2} = |-3| = 3$ **b.** $[u^2(u + 5)]^{1/2} = |u|(u + 5)^{1/2}$

47. $[(-5)^2]^{1/2}$ **48.** $[(-3)^{12}]^{1/4}$ **49.** $(4x^2)^{1/2}$

50. $[x^2(x - 1)]^{1/2}$ **51.** $\dfrac{2}{[x^2(x + 1)]^{1/2}}$ **52.** $\left[\dfrac{9}{x^6(x^2 + 1)}\right]^{1/2}$

2.10

RADICAL EXPRESSIONS

Rational-number exponents are frequently denoted using the **radical sign**, $\sqrt{}$.

> If $a^{1/n}$ is a real number, n is a positive integer, and $n \neq 1$, then
> $$\sqrt[n]{a} = a^{1/n}.$$

Naturally, the radical expression on the left is not defined if the power on the right is not. In the symbolism $\sqrt[n]{a}$, a is called the **radicand** and n is called the **index** of the radical. The whole expression is called a **radical expression of order** n. If no index is shown with a radical expression, as, for example, in the case $\sqrt{a}$, then the index 2 is understood to apply. The symbol $\sqrt{a}$ denotes the nonnegative square root of a, where, of course, a cannot be negative. The symbol $\sqrt{x^2}$ provides us with an alternative means of writing $|x|$. That is, $\sqrt{x^2} = |x|$.

PROPERTIES OF RADICALS

The following properties of radical expressions are immediate consequences of the definition of $a^{1/n}$ and the laws of exponents.

> *For real values of a and b for which all the radical expressions in the equation denote real numbers, n a positive integer, and m an integer,*
>
> Ia $\sqrt[n]{a^n} = a$ (n an odd positive integer),
>
> Ib $\sqrt[n]{a^n} = |a|$ (n an even positive integer),
>
> II $\sqrt[n]{a^m} = (\sqrt[n]{a})^m$,
>
> III $\sqrt[n]{a} \cdot \sqrt[n]{b} = \sqrt[n]{ab}$,
>
> IV $\dfrac{\sqrt[n]{a}}{\sqrt[n]{b}} = \sqrt[n]{\dfrac{a}{b}}$ $(b \neq 0)$,
>
> V $\sqrt[cn]{a^{cm}} = \sqrt[n]{a^m}$ (c a positive integer).

It might be noted in parts III and IV that if $a, b < 0$ and n is even, then the radical on the left-hand side is not defined, even though the radical on the right-hand side is. Notice also that $n \neq 1$ is implied, since $\sqrt[1]{a}$ is not defined.

The several parts of this theorem can be used to rewrite radical expressions in various ways, and, in particular, to write them in what is called **standard form**. A radical expression is said to be in standard form if all the following hold:

2.10 Radical Expressions

1. The radicand contains no polynomial factor raised to a power equal to or greater than the index of the radical.
2. The radicand contains no fractions.
3. No radical expressions are contained in denominators of fractions.
4. The index of the radical is as small as possible.

Examples

a. $\sqrt[3]{24x^3y^2} = \sqrt[3]{8x^3}\sqrt[3]{3y^2} = 2x\sqrt[3]{3y^2}$

b. $\sqrt[6]{49} = \sqrt[3 \cdot 2]{7^{1 \cdot 2}} = \sqrt[3]{7}$

OPERATIONS ON RADICAL EXPRESSIONS

If we multiply the numerator and the denominator of the expression

$$\frac{1}{\sqrt{b}} \quad (b \neq 0)$$

by the fraction

$$\frac{\sqrt{b}}{\sqrt{b}} = 1,$$

then

$$\frac{1}{\sqrt{b}} = \frac{1 \cdot \sqrt{b}}{\sqrt{b} \cdot \sqrt{b}}$$

$$= \frac{\sqrt{b}}{b}.$$

This process of simplification is called **rationalizing the denominator**, because the result is a fraction with a denominator that is free of radicals. This does not exclude the possibility that the denominator is an irrational number. For example, if $b = \pi$, then the result above is equal to $\sqrt{\pi}/\pi$.

Example

$$\frac{2}{\sqrt{5}} = \frac{2 \cdot \sqrt{5}}{\sqrt{5} \cdot \sqrt{5}} = \frac{2\sqrt{5}}{5}$$

When the denominator contains a term other than a square root, we must multiply the numerator and the denominator by the term appropriate to clear the radical expression from the denominator. For example, to rationalize the denominator of

$$\frac{\sqrt[3]{4a^2}}{\sqrt[3]{b}}$$

we must multiply by

$$\frac{\sqrt[3]{b}\cdot\sqrt[3]{b}}{\sqrt[3]{b}\cdot\sqrt[3]{b}} = \frac{\sqrt[3]{b^2}}{\sqrt[3]{b^2}} = 1$$

so that

$$\frac{\sqrt[3]{4a^2}}{\sqrt[3]{b}} = \frac{\sqrt[3]{4a^2}\cdot\sqrt[3]{b^2}}{\sqrt[3]{b}\cdot\sqrt[3]{b^2}}$$

$$= \frac{\sqrt[3]{4a^2 b^2}}{\sqrt[3]{b^3}}$$

$$= \frac{\sqrt[3]{4a^2 b^2}}{b} \qquad (b \neq 0).$$

Example $\quad \dfrac{3}{\sqrt[4]{2}} = \dfrac{3\cdot\sqrt[4]{2^3}}{\sqrt[4]{2}\cdot\sqrt[4]{2^3}} = \dfrac{3\cdot\sqrt[4]{2^3}}{\sqrt[4]{2^4}} = \dfrac{3\sqrt[4]{8}}{2}$

It should be noted that it is not *always* preferable in working with fractions to rationalize denominators. Sometimes, in fact, it is desirable to rationalize the numerator. For example,

$$\frac{\sqrt[3]{2}}{\sqrt{3}} = \frac{\sqrt[3]{2}\cdot\sqrt[3]{2^2}}{\sqrt{3}\cdot\sqrt[3]{2^2}} = \frac{2}{\sqrt{3}\cdot\sqrt[3]{4}}.$$

BINOMIAL DENOMI-NATORS

Binomial denominators of fractions in which radicals occur in one or both of the two terms can also be rationalized. To accomplish this, we first observe that

$$(a + \sqrt{b})(a - \sqrt{b}) = a^2 - b,$$

and

$$(\sqrt{a} + \sqrt{b})(\sqrt{a} - \sqrt{b}) = a - b.$$

Each of the two factors in the products above is said to be the **conjugate** of the other, because the resulting right-hand member contains no radical term.

Now,

$$\frac{1}{a + \sqrt{b}} = \frac{a - \sqrt{b}}{(a + \sqrt{b})(a - \sqrt{b})}$$

$$= \frac{a - \sqrt{b}}{a^2 - b},$$

and

$$\frac{1}{\sqrt{a} + \sqrt{b}} = \frac{\sqrt{a} - \sqrt{b}}{(\sqrt{a} + \sqrt{b})(\sqrt{a} - \sqrt{b})}$$

$$= \frac{\sqrt{a} - \sqrt{b}}{a - b}.$$

2.10 Radical Expressions

Example $\quad \dfrac{1}{2-\sqrt{3}} = \dfrac{2+\sqrt{3}}{(2-\sqrt{3})(2+\sqrt{3})} = \dfrac{2+\sqrt{3}}{2^2-3} = \dfrac{2+\sqrt{3}}{4-3} = 2+\sqrt{3}$

Example $\quad \dfrac{\sqrt{5}}{\sqrt{5}-\sqrt{2}} = \dfrac{\sqrt{5}(\sqrt{5}+\sqrt{2})}{(\sqrt{5}-\sqrt{2})(\sqrt{5}+\sqrt{2})} = \dfrac{\sqrt{25}+\sqrt{10}}{5-2} = \dfrac{5+\sqrt{10}}{3}$

As noted above it is sometimes preferable, in working with fractions, to rationalize numerators.

Example $\quad \dfrac{\sqrt{7}+\sqrt{3}}{2} = \dfrac{(\sqrt{7}+\sqrt{3})(\sqrt{7}-\sqrt{3})}{2(\sqrt{7}-\sqrt{3})} = \dfrac{7-3}{2(\sqrt{7}-\sqrt{3})} = \dfrac{4}{2(\sqrt{7}-\sqrt{3})} = \dfrac{2}{\sqrt{7}-\sqrt{3}}$

The distributive law permits us to express certain sums as products.

Examples
a. $2\sqrt{5} + 4\sqrt{5} = (2+4)\sqrt{5} = 6\sqrt{5}$
b. $5\sqrt{2} - 9\sqrt{2} = (5-9)\sqrt{2} = -4\sqrt{2}$
c. $5\sqrt{2} + \sqrt{75} = 5\sqrt{2} + 5\sqrt{3} = 5(\sqrt{2}+\sqrt{3})$

The distributive law in the form $c(a+b) = ca + cb$ also permits us to write certain products as sums or differences.

Examples
a. $2(\sqrt{2} - 7) = 2\sqrt{2} - 14$
b. $(\sqrt{x} - 3)(\sqrt{x} + 3) = \sqrt{x}(\sqrt{x}+3) - 3(\sqrt{x}+3)$
$= x - 9$
c. $(\sqrt{x} + 2)(\sqrt{x} - 1) = \sqrt{x}(\sqrt{x}-1) + 2(\sqrt{x}-1)$
$= x - \sqrt{x} + 2\sqrt{x} - 2$
$= x + \sqrt{x} - 2$

EXERCISE 2.10

Assume that all variables represent positive real numbers and that all radicands are positive.

Write in radical form.

Examples a. $5^{1/2}$ b. $xy^{2/3}$ c. $(x - y^2)^{-1/2}$

Solutions a. $\sqrt{5}$ b. $x\sqrt[3]{y^2}$ c. $\dfrac{1}{\sqrt{x - y^2}}$

1. $2^{1/3}$
2. $3^{1/2}$
3. $3x^{1/3}$
4. $2y^{1/4}$
5. $(2y)^{1/2}$
6. $(3x)^{1/3}$
7. $x^{2/3}$
8. $y^{3/4}$
9. $xy^{2/3}$
10. $x^{1/2}y^{2/3}$
11. $(x^3y)^{1/4}$
12. $(xy)^{2/3}$
13. $(x - y)^{1/2}$
14. $(x^2 - y)^{1/3}$
15. $(x^2 + y)^{-2/3}$
16. $(2x + y)^{-3/4}$

Write an equivalent expression using positive fractional exponents in lowest terms.

Examples a. $\sqrt{x^3}$ b. $7\sqrt[3]{a^2}$ c. $\sqrt[4]{a - b}$

Solutions a. $x^{3/2}$ b. $7a^{2/3}$ c. $(a - b)^{1/4}$

17. $\sqrt[3]{y^2}$
18. $\sqrt[4]{y^3}$
19. $\sqrt[4]{2xy^2}$
20. $\sqrt[5]{x^2y^3}$
21. $3\sqrt[4]{x^3y}$
22. $\sqrt[7]{x^5y^2}$
23. $\sqrt[3]{a + b^2}$
24. $\sqrt{a^2 - b}$

Find the root indicated.

Examples a. $\sqrt[5]{-32}$ b. $\sqrt[3]{x^6y^3}$ c. $\sqrt{x^2y^6}$

Solutions a. -2 b. x^2y c. xy^3

25. $\sqrt{81}$
26. $\sqrt{64}$
27. $\sqrt[3]{-27}$
28. $\sqrt{x^4y^2}$
29. $\sqrt[3]{x^3y^6}$
30. $\sqrt[4]{\dfrac{4}{9}x^6y^{10}}$

Write in simplest form.

Examples a. $\sqrt[3]{2x^7y^3}$ b. $\sqrt{2xy}\sqrt{8x}$ c. $\dfrac{\sqrt{6a}\sqrt{5a}}{\sqrt{15}}$

2.10 Radical Expressions

Solutions a. $\sqrt[3]{x^6 y^3} \sqrt[3]{2x}$ b. $\sqrt{16x^2} \sqrt{y}$ c. $\sqrt{\dfrac{15}{15} \cdot 2a^2}$

$\qquad\qquad = x^2 y \sqrt[3]{2x}$ $\qquad = 4x\sqrt{y}$ $\qquad = a\sqrt{2}$

31. $\sqrt{4x^5}$ 32. $\sqrt{16y^3}$ 33. $\sqrt[4]{3x^5 y^5}$

34. $\sqrt[3]{-8x^6}$ 35. $\sqrt{3xy}\sqrt{6y}$ 36. $\sqrt[3]{y^4}\sqrt[3]{y^7}$

37. $\dfrac{\sqrt{8y}}{\sqrt{2y}}$ 38. $\dfrac{\sqrt{x^3 y^3}}{\sqrt{xy^2}}$ 39. $\dfrac{\sqrt[3]{2a^2 b^7}}{\sqrt[3]{ab^2}}$

Rationalize the denominator of each expression.

Examples a. $\dfrac{1}{\sqrt{2}}$ b. $\sqrt{\dfrac{3x}{7y}}$ c. $\sqrt[3]{\dfrac{2}{y}}$

Solutions a. $\dfrac{1 \cdot \sqrt{2}}{\sqrt{2} \cdot \sqrt{2}} = \dfrac{\sqrt{2}}{2}$ b. $\dfrac{\sqrt{3x} \cdot \sqrt{7y}}{\sqrt{7y} \cdot \sqrt{7y}} = \dfrac{\sqrt{21xy}}{7y}$ c. $\dfrac{\sqrt[3]{2} \cdot \sqrt[3]{y} \cdot \sqrt[3]{y}}{\sqrt[3]{y} \cdot \sqrt[3]{y} \cdot \sqrt[3]{y}} = \dfrac{\sqrt[3]{2y^2}}{y}$

40. $\dfrac{2}{\sqrt{3}}$ 41. $\dfrac{4}{\sqrt{5}}$ 42. $\sqrt{\dfrac{x}{2y}}$

43. $\sqrt[3]{\dfrac{2}{x^2}}$ 44. $\sqrt[3]{\dfrac{3}{y^2}}$ 45. $\sqrt[3]{\dfrac{3}{y}}$

Rationalize the numerator of each expression.

46. $\dfrac{\sqrt{3}}{3}$ 47. $\dfrac{\sqrt{2}}{4}$ 48. $\dfrac{\sqrt{x}}{\sqrt{y}}$

Rationalize denominators.

Examples a. $\dfrac{3}{\sqrt{2} - 1}$ b. $\dfrac{1}{\sqrt{x} - \sqrt{y}}$

Solutions a. $\dfrac{3(\sqrt{2} + 1)}{(\sqrt{2} - 1)(\sqrt{2} + 1)}$ b. $\dfrac{1(\sqrt{x} + \sqrt{y})}{(\sqrt{x} - \sqrt{y})(\sqrt{x} + \sqrt{y})}$

$\qquad\qquad = \dfrac{3\sqrt{2} + 3}{2 - 1}$ $\qquad = \dfrac{\sqrt{x} + \sqrt{y}}{x - y}$

$\qquad\qquad = 3\sqrt{2} + 3$

49. $\dfrac{-4}{1+\sqrt{3}}$
50. $\dfrac{1}{2-\sqrt{2}}$
51. $\dfrac{x}{\sqrt{x}-3}$

52. $\dfrac{\sqrt{x}}{\sqrt{x}-\sqrt{y}}$
53. $\dfrac{4\sqrt{2}-\sqrt{5}}{4\sqrt{2}+\sqrt{5}}$
54. $\dfrac{4\sqrt{a}-\sqrt{3a+1}}{\sqrt{a}+\sqrt{3a+1}}$

Rationalize numerators.

55. $\dfrac{1-\sqrt{2}}{2}$
56. $\dfrac{\sqrt{3}-2}{3}$
57. $\dfrac{1-\sqrt{x+a}}{\sqrt{x+a}}$
58. $\dfrac{\sqrt{x-a}+1}{\sqrt{x-a}}$

59. Find the value of $y^2 - 3y + 1$ for $y = 3 - \sqrt{2}$.

60. Find the value of $3x^2 - 4x - 2$ for $x = \dfrac{2-\sqrt{10}}{3}$.

Write each sum as a product.

Examples a. $4\sqrt{2} + 3\sqrt{2} - \sqrt{2}$ b. $2\sqrt{3} + 4\sqrt{12}$

Solutions a. $(4 + 3 - 1)\sqrt{2} = 6\sqrt{2}$ b. $2\sqrt{3} + 4 \cdot 2\sqrt{3} = 10\sqrt{3}$

61. $5\sqrt{2} + 3\sqrt{2}$
62. $8\sqrt{3} - 3\sqrt{3}$
63. $4\sqrt{5} + \sqrt{5} - 6\sqrt{5}$
64. $\sqrt{2} - 5\sqrt{2} + 2\sqrt{2}$
65. $4\sqrt{3} + 2\sqrt{12}$
66. $5\sqrt{5} - \sqrt{20}$

Write each product as a sum.

Examples a. $\sqrt{x}(\sqrt{2x} - \sqrt{x})$ b. $(\sqrt{x} - 2\sqrt{y})(2\sqrt{x} + \sqrt{y})$

Solutions a. $x\sqrt{2} - x$ b. $2x - 2y - 3\sqrt{xy}$

67. $\sqrt{2}(2 - \sqrt{2})$
68. $\sqrt{3}(4 + \sqrt{3})$
69. $\sqrt{3}(2 + \sqrt{2})$
70. $\sqrt{2}(3 - \sqrt{5})$
71. $(3 + \sqrt{5})(2 - \sqrt{5})$
72. $(\sqrt{3} - 5)(\sqrt{3} + 5)$

CHAPTER TEST

Simplify each expression.

1. $(2y^2 - 3y + 4) - (y^2 - 3y + 5)$
2. $2[3(x^2 + 2x - 1) + 4x] + (3 - 4x - 6x^2)$

Given $P(x) = 3x^2 - 2x + 1$, find the values.

3. $P(-2)$
4. $P(x)|_{-2}^{4}$

Simplify.

5. $2(3x - 1)(x + 2)$
6. $2\{z^2 - 2[z - 3(z^2 - 1) + 1] + 3z^2\}$

Factor completely.

7. $z^2 - 7z + 12$
8. $u^5 - 2u^3 + u$
9. $4y^{2n} - 1$
10. $8t^3 - 27$

Write each quotient as a polynomial.

11. $\dfrac{48x^5 y^2 z}{6x^4 yz}$
12. $\dfrac{24z^3 - 16z^2 + 72z}{8z}$
13. $\dfrac{2n^2 - 5n - 3}{n - 3}$
14. $\dfrac{2r^2 + 3r + 1}{2r + 1}$

Reduce to lowest terms where possible. State all restrictions on the variable or variables.

15. $\dfrac{8r^2 s^2 - 4rs^2 + 12rs}{2rs}$
16. $\dfrac{2x^2 - 2}{x + 1}$
17. $\dfrac{x^2 - 4x + 4}{x^2 - 4}$
18. $\dfrac{8n^2 + 40n + 32}{32 - 2n^2}$

Simplify each expression.

19. $\dfrac{x+1}{4} + \dfrac{x+2}{8}$

20. $\dfrac{3}{6x-3} - \dfrac{2}{2x-1}$

21. $\dfrac{2x-5}{2-x} + \dfrac{x}{2x-4}$

22. $\dfrac{4y-2}{y^2-3y+2} - \dfrac{3}{y-1}$

Simplify each expression.

23. $\dfrac{2x^2y^2}{9a^2b^2} \cdot \dfrac{45ab}{14xy}$

24. $\dfrac{x^2+3x+2}{x^2-1} \cdot \dfrac{x^2+2x-3}{x^2+5x+6}$

25. $\dfrac{13t^2}{20r^2} \div \dfrac{39t^3}{5r}$

26. $\dfrac{x^2-4}{ax-2a} \div \dfrac{x^2+4x+4}{x^2+2x}$

27. $\dfrac{\dfrac{a-b}{b}}{\dfrac{a+b}{3b}}$

28. $\dfrac{\dfrac{1}{xy} - \dfrac{1}{y}}{\dfrac{1}{y} - \dfrac{1}{xy}}$

Write each expression as a product or quotient in which each variable occurs at most once in the expression and all exponents are positive.

29. $(x^2y^4)^3$

30. $\dfrac{x^2y^{-3}}{x^{-1}y}$

31. $\dfrac{x^0 y^{-2}}{(xy)^{-2}}$

32. $\left(\dfrac{x^{-2}y^0}{y^{-3}}\right)^{-1}$

Represent each expression as a single fraction involving positive exponents only.

33. $x^{-1}y^{-1} + \dfrac{x}{y^{-1}}$

34. $\dfrac{x^{-2} - y^{-1}}{(xy)^{-1}}$

Write each expression as an equivalent expression in which each variable occurs only once.

35. $\dfrac{x^{2n}x^n}{x^{n+1}}$

36. $\left(\dfrac{x^n y^{n+1}}{xy^{n-1}}\right)^2$

Chapter Test

Write each number as the product of a number between 1 and 10 and a power of 10.

37. 47,300

38. 0.000045

Write each expression as a product or quotient of powers in which each variable occurs only once and all exponents are positive. Assume that all variable bases are positive and all variables in exponents are positive integers.

39. $x^{4/3} \cdot x^{1/2}$

40. $\left(\dfrac{x^2}{y^3}\right)^{-1/6}$

41. $(x^n y^{2n})^{-1/2}$

42. $\left(\dfrac{y^n}{x^n y^{2n}}\right)^{-1/n}$

Apply the distributive law to write each product as a sum.

43. $y^{1/2}(y^{1/2} - y)$

44. $(y - y^{1/2})(y + y^{1/2})$

Factor as indicated.

45. $x^{-1/4} + x^{1/2} = x^{-1/4}(?)$

46. $x^{2/3} - x^{-2/3} = x^{-2/3}(?)$

Write in simplest form.

47. $\sqrt{9x^3 y^5}$

48. $\sqrt{2xy}\sqrt{8x}$

49. $\dfrac{\sqrt{3y}\sqrt{2xy}}{\sqrt{3}}$

50. $\dfrac{\sqrt[3]{6x}\sqrt[3]{9x^2 y}}{\sqrt[3]{2y}}$

51. Rationalize the denominator: $\sqrt[3]{\dfrac{x}{y}}$.

52. Rationalize the numerator: $\dfrac{\sqrt{3y}}{6y}$.

Rationalize denominators.

53. $\dfrac{1}{2+\sqrt{3}}$

54. $\dfrac{2\sqrt{x}-1}{\sqrt{x}+1}$

Rationalize numerators.

55. $\dfrac{\sqrt{2}+3}{2}$

56. $\dfrac{\sqrt{y}-3}{\sqrt{y}+1}$

Write each sum as a product.

57. $3\sqrt{18} - 2\sqrt{18} + \sqrt{2}$

58. $4\sqrt[3]{24} + 2\sqrt[3]{3}$

Multiply factors and write all radicals in the result in simplest form.

59. $(\sqrt{5} - \sqrt{2})(\sqrt{5} + \sqrt{2})$

60. $(3\sqrt{x} - 4)(\sqrt{x} + 2)$

CUMULATIVE REVIEW OF PART I

Fill in the blanks with the appropriate answer without referring back to the text.

1. If $A = \{0, 6, -9, 7\}$ and $B = \{0, 1, 3, 8, 9\}$, then $A \cup B = $ _____ and $A \cap B = $ _____.
2. If $U = \{2, 4, 6, 8, 10, 12\}$ and $A' = \{2, 8, 12\}$, then $A = $ _____.
3. For any set A, $A \cup \emptyset = $ _____ and $A \cap \emptyset = $ _____.
4. $\emptyset = \{\emptyset\}$ (T or F) _____
5. 3/4, 1/2, -5 and 0.434343 ... are examples of _____ numbers.
6. For any real number x, $|-x| = x$. (T or F) _____
7. If $-3a < 27$, then $a > -9$. (T or F) _____
8. If $P(x) = 2x^2 - 3x + 4$, $P(0) = $ _____, $P(-3) = $ _____, $P(2) = $ _____.
9. If $P(x) = 5x - 4$ and $Q(x) = x^2 + 8$, then $P(Q(2)) = $ _____.
10. Multiply: $(4x - 3y)(x + 2y) = $ _____.

Factor the given expression.

11. $4x^2y^2 + 16xy^2 = $ _____
12. $2y^2 + y - 6 = $ _____
13. $r^3 - 9r = $ _____
14. $27z^3 + x^6y^3 = $ _____
15. Perform the long division: $\dfrac{6x^2 + 4x - 1}{x + 2} = $ _____.

Rationalize the denominators.

16. $\dfrac{1}{1 + \sqrt{5}} = $ _____
17. $\dfrac{1}{\sqrt[3]{2}} = $ _____
18. $\sqrt{a^2 + b^2} = a + b$ (T or F) _____
19. $\sqrt{36} = \pm 6$ (T or F) _____

20. Simplify and write with positive exponents: $\dfrac{(y^{3/4}x^{12})^{2/3}}{(x^2 y^{2/3})^3} = $ _____.

21. $\sqrt{75} - \sqrt{12} - \sqrt{27} = $ _____

22. $2^n \cdot 5^n = 10^n$ (T or F) _____

23. $(4^n)^{-1} = 4^{n-1}$ (T or F) _____

24. $5^{3+3n} = 125^{1+n}$ (T or F) _____

25. $[(2^{-1})^{-2}]^{-3} = (64)^{-1}$ (T or F) _____

Answers for Cumulative Review of Part I

1. $\{0, 1, 3, 6, 7, 8, 9, -9\}, \{0\}$
2. $\{4, 6, 10\}$
3. $A, \varnothing$
4. F
5. rational
6. F
7. T
8. 4, 31, 6
9. 56
10. $4x^2 + 5xy - 6y^2$
11. $4xy^2(x + 4)$
12. $(2y - 3)(y + 2)$
13. $r(r + 3)(r - 3)$
14. $(3z + x^2 y)(9z^2 - 3x^2 yz + x^4 y^2)$
15. $6x - 8 + \dfrac{15}{x+2}$
16. $-\dfrac{1 - \sqrt{5}}{4}$
17. $\dfrac{\sqrt[3]{4}}{2}$
18. F
19. F
20. $\dfrac{x^2}{y^{3/2}}$
21. 0
22. T
23. F
24. T
25. T

PART II

3 EQUATIONS AND INEQUALITIES IN ONE VARIABLE
4 RELATIONS AND FUNCTIONS
5 EXPONENTIAL AND LOGARITHMIC FUNCTIONS
6 SYSTEMS OF EQUATIONS
7 MATRICES AND DETERMINANTS

3
EQUATIONS AND INEQUALITIES IN ONE VARIABLE

Equations and inequalities that involve variables are called **open sentences**. Equations and inequalities involving only constants are referred to as **statements**. For example,

$$x + 2 = 7, \qquad x^2 - y \geq 4, \qquad |x - 3| < 2$$

are open sentences, whereas

$$3 + 2 = 5, \qquad 7 < 10 - 1, \qquad |3 + 2| > 0$$

are statements. Although a statement can be adjudged true or false, no such judgment is possible in the case of an open sentence. Thus, $3 - 2 = 1$ is a true statement, and $3 - 2 = 2$ is false, but we cannot assert that $x - 2 = 1$ is either true or false until we know something more about x. Open sentences can be looked upon as set selectors. Given any set of numbers, an equation such as $x - 2 = 1$ or an inequality such as $x - 2 > 1$ will serve to select certain numbers from the universal set U and reject others, depending on whether the numbers make the resulting statement true or false.

In this chapter, we shall be concerned with open sentences over the real numbers in one variable.

3.1

EQUIVALENT EQUATIONS; FIRST-DEGREE EQUATIONS

If we replace the variable x in $P(x) = Q(x)$ with an element from its replacement set U, and if the resulting statement is true, the element is a **solution**, or **root**, of the equation and is said to **satisfy** the equation. Thus, if x represents a real number, then 2 is a solution of $x + 3 = 5$, because $2 + 3 = 5$ is a true statement. On the other hand, 3 is not a solution of the equation, because $3 + 3 = 5$ is false. The subset of U consisting of all solutions of an equation is said to be the **solution set** of the equation. In the example we have been using here, $x + 3 = 5$, the solution set is $\{2\}$.

Equations that have the same solution set are called **equivalent equations**. For example, the equations

$$x + 3 = -3 \qquad \text{and} \qquad x = -6$$

are equivalent, because the solution set of each is $\{-6\}$.

ELEMENTARY TRANSFORMATIONS

To solve an equation, we usually determine the members of the solution set by inspection or generate a sequence of equivalent equations until we arrive at one with an obvious solution set. The following rules are direct consequences of the addition and multiplication laws for real numbers and are frequently used in generating equivalent equations. These rules are known as **elementary transformations**.

> *If $P(x)$, $Q(x)$, and $R(x)$ are expressions, then for all values of x for which $P(x)$, $Q(x)$, and $R(x)$ are real numbers, the sentence*
>
> $$P(x) = Q(x)$$
>
> *is equivalent to each of the following sentences:*
>
> I $P(x) + R(x) = Q(x) + R(x)$,
> II $P(x) - R(x) = Q(x) - R(x)$,
> III $P(x) \cdot R(x) = Q(x) \cdot R(x)$, $\qquad R(x) \neq 0$,
> IV $\dfrac{P(x)}{R(x)} = \dfrac{Q(x)}{R(x)}$ $\qquad R(x) \neq 0$.

Note that rules II and IV can be considered special cases of rules I and III, respectively.

An elementary transformation *always* produces an equivalent equation. However, care must be exercised in the application of rules III and IV for we have specifically excluded multiplication or division by zero. For example, to solve the equation

(1) $$\frac{x}{x-3} = \frac{3}{x-3} + 2,$$

we might first multiply each member by $x - 3$ to find an equation that is free of fractions. We then have

$$(x - 3)\frac{x}{x-3} = (x - 3)\frac{3}{x-3} + (x - 3)2,$$

or

(2) $$x = 3 + 2x - 6,$$

from which

$$x = 3.$$

Thus, 3 is a solution of (2). But, upon substituting 3 for x in (1), we have

$$\frac{3}{3-3} = \frac{3}{3-3} + 2 \quad \text{or} \quad \frac{3}{0} = \frac{3}{0} + 2,$$

and neither member is defined. In obtaining (2), each member of (1) was multiplied by $x - 3$; but if x is 3, then $x - 3$ is zero, and hence rule III is not applicable. Equation (2) is *not* equivalent to (1), and in fact (1) has no solution.

We can always ascertain whether what we think is a solution of an equation is such in reality by substituting the suggested solution in the original equation and determining whether the resulting statement is true. If each equation in a sequence is obtained by means of an elementary transformation, the sole purpose for such checking is to detect arithmetic errors. We shall dispense with checking solution sets in the examples that follow unless we apply what may be a nonelementary transformation—that is, unless we multiply or divide by an expression that vanishes for some value or values of the variable.

SOLUTION OF A LINEAR EQUATION

The equation

(3) $$ax + b = 0 \qquad (a \ne 0),$$

is called a **first-degree**, or **linear, equation**. Any equation that can be reduced to this form by elementary transformations is equivalent to a first-degree equation. We can show that such an equation always has one and only one solution. By rule I,

(4) $$ax = -b$$

is equivalent to (3); furthermore, by rule III, (4) is equivalent to

(5) $$x = -\frac{b}{a}.$$

This last equation, of course, has the unique solution $-b/a$. Since (3), (4), and (5) are equivalent, (3) has the unique solution $-b/a$.

An equation containing more than one variable, or containing symbols such as a, b, and c, representing constants, can often be solved for one of the symbols in terms of the remaining symbols by applying elementary transformations until the desired symbol is obtained by itself as one member of an equation.

Example Solve $ay = b + y$ for y.

Solution We generate the following sequence of equivalent equations:
$$ay - y = b,$$
$$y(a - 1) = b,$$
$$y = \frac{b}{a - 1} \qquad (a \neq 1).$$

Example The formula $S = P + Prt$ occurs in many problems involving simple interest. Solve this equation for the rate of interest r.

Solution
$$S = P(1 + rt)$$
$$\frac{S}{P} = 1 + rt$$
$$rt = \frac{S}{P} - 1$$
$$r = \frac{S}{Pt} - \frac{1}{t} = \frac{S - P}{Pt}$$

Example The value V of a computer, which initially cost \$200,000 and is undergoing linear depreciation over a period of 20 years, is given by $V = 200{,}000\left(1 - \frac{x}{20}\right)$. Assuming x represents years, determine when the value of the computer is one-half its initial cost.

Solution We must solve
$$100{,}000 = 200{,}000\left(1 - \frac{x}{20}\right)$$

for x. We have
$$1 - \frac{x}{20} = \frac{100{,}000}{200{,}000}$$
$$-\frac{x}{20} = \frac{1}{2} - 1$$
$$x = -20\left(-\frac{1}{2}\right)$$
$$x = 10.$$

3.1 Equivalent Equations; First-Degree Equations

EXERCISE 3.1

Solve

1. $6y - 2(2y + 5) = 6(5 + y)$
2. $3(7 - 2z) = 30 - 7(z + 1)$
3. $-3[x - (2x + 3) - 2x] = -9$
4. $-2[x - (x - 1)] = -3(x + 1)$
5. $4 - (x - 3)(x + 2) = 10 - x^2$
6. $6 + 3x - x^2 = 4 - (x - 2)(x + 3)$
7. $\dfrac{5x}{2} - 1 = x - \dfrac{1}{2}$
8. $\dfrac{x}{5} - \dfrac{x}{2} = 9$
9. $\dfrac{2x - 1}{5} = \dfrac{x + 1}{2}$
10. $\dfrac{2y}{3} - \dfrac{2y + 5}{6} = \dfrac{1}{2}$
11. $\dfrac{2}{x - 9} = \dfrac{9}{x + 12}$
12. $\dfrac{x}{x - 2} = \dfrac{2}{x - 2} + 7$
13. $\dfrac{5}{x - 3} = \dfrac{x + 2}{x - 3} + 3$
14. $\dfrac{2}{y + 1} + \dfrac{1}{3y + 3} = \dfrac{1}{6}$
15. $\dfrac{y}{y + 2} - \dfrac{3}{y - 2} = \dfrac{y^2 + 8}{y^2 - 4}$
16. $\dfrac{4}{2x - 3} + \dfrac{4x}{4x^2 - 9} = \dfrac{1}{2x + 3}$

Solve for the indicated variable.

Example Solve $l = a + (n - 1)d$ for d.

Solution Add $-a$ to each member.
$$l - a = (n - 1)d$$
Multiply each member by $\dfrac{1}{n - 1}$ $(n \neq 1)$.
$$\dfrac{l - a}{n - 1} = d \qquad (n \neq 1)$$
$$d = \dfrac{l - a}{n - 1} \qquad (n \neq 1)$$

17. $A = \dfrac{h}{2}(b + c)$ for c
18. $S = \dfrac{a}{1 - r}$ for r
19. $l = a + (n - 1)d$ for n
20. $\dfrac{1}{r} = \dfrac{1}{r_1} + \dfrac{1}{r_2}$ for r
21. $x^2 y' - 3x - 2y^3 y' = 1$ for y'
22. $2xy' - 3y' + x^2 = 0$ for y'
23. $x_1 x_2 - 2x_1 x_3 = x_4$ for x_1
24. $3x_1 x_3 + x_1 x_2 = x_4$ for x_1
25. $\dfrac{y - y_1}{x - x_1} = 6$ for y
26. $\dfrac{y - y_1}{x - x_1} = 2$ for x

27. For what value of k does the equation $2x - 3 = \dfrac{4+x}{k}$ have as its solution set $\{-1\}$?

28. Find a value of k so that the equation $3x - 1 = k$ is equivalent to the equation $2x + 5 = 1$.

Applications

29. *Biology* The power P produced by the heart in ergs per second is given by $P = QE$, where Q is the flow rate of blood and E is the energy per unit volume of blood. If the power produced by the left ventricle of the heart is 1.14×10^7 erg/sec and the flow rate is 83 cm^3/sec, what is the energy E?

30. *Business* The amount accrued S when a principal P is invested at simple interest at a rate r per period for t periods is given by $S = P + Prt$. Solve for P.

31. The amount accrued S when a principal P is invested at compund interest at an annual rate of interest r compounded annually for n years is given by $S = P(1 + r)^n$. Solve for P.

32. The number of units S of a commodity supplied by a manufacturer at a price x per unit is sometimes given by $S = ax + b$. Solve for x.

33. The number of units D demanded by consumers when a commodity sells for x dollars per unit is given by $D = -20x + 1000$. Determine the corresponding price when $D = 300$. When does $D = 0$?

34. When an item sells for p dollars the revenue R obtained by selling x items is $R = px$. Solve for p if $R = \$3000$ and $x = 15$.

35. The cost C of manufacturing x items of a commodity is given by $C = $ fixed costs $+$ variable costs. If the fixed costs are \$10,000 and the variable costs are $5x$, solve for x when $C = 12{,}125$.

36. The value of a computer is given by the formula $V = 200{,}000\left(1 - \dfrac{x}{20}\right)$. When is the computer worth nothing?

37. Determine when the computer described in Problem 36 is worth \$40,000.

38. The total cost, including 6% sales tax, of installing a new carpet in a home was \$2500. How much of this cost is attributable to the sales tax?

39. *Physics* The velocity v of a body that has accelerated for t seconds is given by $v = v_0 + at$, where a denotes the acceleration and v_0 is its initial velocity. A car, whose initial velocity is 30 m/sec, accelerates to 60 m/sec in 4 seconds. What is the acceleration of the car?

40. The relationship between temperature T_f, measured in degrees Fahrenheit, and temperature T_c, measured in degrees Celsius, is given by $T_f = \dfrac{9}{5} T_c + 32$. Solve for T_c when $T_f = 68$.

41. *Social Science* In one theory of learning, the probability p_n of a person responding in a certain manner on the nth attempt in a succession of trials is related to the probability p_{n-1} that the response given on the $(n-1)$st attempt is $p_n = p_{n-1} + a(1 - p_{n-1}) - bp_{n-1}$. Solve for p_{n-1}.

3.2

SECOND-DEGREE EQUATIONS

The equation
$$ax^2 + bx + c = 0 \qquad (a \neq 0),$$
is a **second-degree**, or **quadratic**, **equation**. Any equation that can be reduced to this form by elementary transformations is therefore equivalent to a quadratic equation. We shall designate the form shown above as the **standard form** for such equations.

SOLUTION BY FACTORING

The following property of real numbers is useful in solving some quadratic equations.

> For real numbers a and b, ab = 0 if and only if
> $$a = 0 \quad \text{or} \quad b = 0 \quad \text{or both.}$$

Example Find the solution set of $x^2 + 2x - 15 = 0$.

Solution The equation
$$x^2 + 2x - 15 = 0$$
is equivalent to
$$(x + 5)(x - 3) = 0.$$
Since $(x + 5)(x - 3) = 0$ is true if and only if
$$x + 5 = 0 \quad \text{or} \quad x - 3 = 0,$$
we can see by inspection that the only values of x that satisfy the original equation are -5 and 3. Hence, the solution set is $\{-5, 3\}$.

Example Find the solution set of $x^2 = 6x$.

Solution The equation is equivalent to
$$x^2 - 6x = 0$$
$$x(x - 6) = 0.$$
Hence the solution set is $\{0, 6\}$.

In solving an equation such as given in the last example the student is cautioned not to divide by x. The equations $x^2 = 6x$ and $x = 6$ are not equivalent.

NUMBER OF SOLUTIONS

In general, the solution set of a quadratic equation may contain two, one, or no real numbers as elements. The equation in the example above has two real solutions. Now, consider the equation

$$x^2 - 2x + 1 = 0.$$

Since $x^2 - 2x + 1 = 0$ is equivalent to

$$(x - 1)^2 = 0,$$

and since the only value of x for which $(x - 1)^2 = 0$ is 1, this is the only member of the solution set of $x^2 - 2x + 1 = 0$. We shall see that every quadratic equation must have two solutions, and accordingly we say that the solution of any quadratic equation having only one solution is of **multiplicity two**; that is, we shall count it twice as a solution.

SOLUTION BY SQUARE ROOTS

Now let us consider a special quadratic equation:

$$x^2 - q = 0 \qquad (q > 0).$$

Since $x^2 - q = 0$ is equivalent to $x^2 = q$, x must be a square root of q. Therefore, the solution set is $\{\sqrt{q}, -\sqrt{q}\}$.

Example Find the solution set of $x^2 - 5 = 0$.

Solution
$$x^2 = 5$$

The solution set is $\{\sqrt{5}, -\sqrt{5}\}$.

An example of a quadratic equation having no real solutions is $x^2 + 1 = 0$, since there exists no real number x such that its square is negative. The solution set in R of the given equation is $\emptyset$.

Quadratic equations of the form

(1) $$(x - p)^2 = q,$$

where $q \geq 0$, can be solved by observing that $x - p$ must be one of the square roots of q. That is, either

$$x - p = \sqrt{q} \qquad \text{or} \qquad x - p = -\sqrt{q},$$

and conversely. Thus, it is evident that for $q \geq 0$ the solution set of $(x - p)^2 = q$ is $\{p + \sqrt{q}, p - \sqrt{q}\}$.

SOLUTION BY COMPLETING THE SQUARE

Being able to find solution sets for quadratic equations of the form $(x - p)^2 = q$ enables us to find the solution set of any quadratic equation. Let us first consider the quadratic equation.

3.2 Second-Degree Equations

(2) $$x^2 + 6x - 1 = 0$$

or

(3) $$x^2 + 6x = 1.$$

Adding the square of one-half the coefficient of x (that is, $(6/2)^2 = 9$) to both members of (3) gives

$$x^2 + 6x + 9 = 10.$$

Since the left-hand member of the last equation is equal to $(x + 3)^2$ we obtain

(4) $$(x + 3)^2 = 10.$$

Therefore,

$$x + 3 = \sqrt{10} \quad \text{or} \quad x + 3 = -\sqrt{10}.$$

By solving each of these equations we find that the solution set of (2) is $\{-3 + \sqrt{10}, -3 - \sqrt{10}\}$.

The technique used to obtain (4) is called **completing the square.**

SOLUTION BY THE QUADRATIC FORMULA

The general quadratic equation

$$ax^2 + bx + c = 0 \quad (a \neq 0)$$

can be written in the form

$$x^2 + \frac{b}{a}x + \frac{c}{a} = 0.$$

or

$$x^2 + \frac{b}{a}x = -\frac{c}{a}.$$

We complete the square by adding the square of one-half the coefficient of x (that is, $(b/2a)^2 = b^2/4a^2$) to both members of the last equation. The left member is then a perfect square and may be written in the form $(x + b/2a)^2$:

$$x^2 + \frac{b}{a}x + \frac{b^2}{4a^2} = -\frac{c}{a} + \frac{b^2}{4a^2}$$

$$\left(x + \frac{b}{2a}\right)^2 = \frac{b^2 - 4ac}{4a^2}$$

$$x + \frac{b}{2a} = \pm \frac{\sqrt{b^2 - 4ac}}{2a}$$

Thus, we obtain the **quadratic formula,**

$$x = \frac{-b \pm \sqrt{b^2 - 4ac}}{2a},$$

where the solutions, or roots, of the general quadratic equation are expressed in terms of the coefficients. The symbol $\pm$ is used to condense the two equations

$$x = \frac{-b + \sqrt{b^2 - 4ac}}{2a} \quad \text{and} \quad x = \frac{-b - \sqrt{b^2 - 4ac}}{2a}$$

into a single equation. We need only substitute the coefficients a, b, and c of a given quadratic equation in the formula to find the solution set for the equation.

Example Find the solution set of $2x^2 - 3x + 1 = 0$.

Solution Substitute $a = 2$, $b = -3$, and $c = 1$ in the quadratic formula and simplify.

$$x = \frac{-(-3) \pm \sqrt{(-3)^2 - 4(2)(1)}}{2(2)} = \frac{3 \pm \sqrt{9 - 8}}{4} = \frac{3 \pm 1}{4}$$

The solution set is $\left\{1, \frac{1}{2}\right\}$.

DETERMINATION OF NUMBER OF REAL SOLUTIONS

An examination of the quadratic formula,

$$x = \frac{-b \pm \sqrt{b^2 - 4ac}}{2a},$$

shows that if $ax^2 + bx + c = 0$ is to have a nonempty solution set in the set of real numbers, then $\sqrt{b^2 - 4ac}$ must be real. This, in turn, implies that only those quadratic equations for which $b^2 - 4ac \geq 0$ have real solutions. The number represented by $b^2 - 4ac$ is called the **discriminant** of the quadratic equation $ax^2 + bx + c = 0$. It yields the following information about the nature of the solution set of the equation:

1. If $b^2 - 4ac = 0$, then there is precisely one real solution (multiplicity two).
2. If $b^2 - 4ac < 0$, then there are no real solutions.
3. If $b^2 - 4ac > 0$, then there are two real solutions.

In economics the price p, per unit, of a commodity can be expressed in terms of the number of units x of the commodity demanded by consumers. Very often this relationship is given by

$$p = ax + b$$

where a and b are constants. If revenue is

$$R = px$$

then

$$R = ax^2 + bx.$$

3.2 Second-Degree Equations

Example The revenue obtained through the sale of x units of a commodity is given by $R = -2x^2 + 1000x$. Determine the number of units that need to be sold in order to have $R = 125{,}000$.

Solution Solve

$$-2x^2 + 1000x = 125{,}000$$
$$0 = 2x^2 - 1000x + 125{,}000.$$

From the quadratic formula we find that $b^2 - 4ac = 0$. Hence there is only one solution

$$x = \frac{-(-1000)}{2(2)} = 250.$$

EXERCISE 3.2

Solve by factoring.

Example $x^2 + x = 30$

Solution Write an equivalent equation in standard form, and then factor the left-hand member.

$$x^2 + x - 30 = 0$$
$$(x + 6)(x - 5) = 0$$

Determine solutions by inspection, or set each factor equal to zero, and solve each linear equation

$$x + 6 = 0 \qquad x - 5 = 0$$
$$x = -6 \qquad x = 5$$

The solution set is $\{-6, 5\}$.

1. $x^2 + 2x = 0$
2. $x^2 - x = 5x$
3. $x^2 + 5x - 14 = 0$
4. $3x^2 - 6x = -3$
5. $2x^2 - 5x = 3$
6. $2y^2 = -(3y + 1)$
7. $6x^2 = 5x + 4$
8. $35 = t + 6t^2$
9. $x(2x - 3) = -1$
10. $(x - 2)(x + 1) = 4$
11. $3 = \dfrac{10}{x^2} - \dfrac{7}{x}$
12. $\dfrac{2}{x - 3} - \dfrac{6}{x - 8} = -1$

Solve by square roots.

Example $(x + 3)^2 = 7$

Solution Set $x + 3$ equal to each square root of 7.

$$x + 3 = \sqrt{7} \qquad x + 3 = -\sqrt{7}$$
$$x = -3 + \sqrt{7} \qquad x = -3 - \sqrt{7}$$

The solution set is $\{-3 + \sqrt{7}, -3 - \sqrt{7}\}$.

13. $x^2 = 4$
14. $9x^2 - 100 = 0$
15. $x^2 = 5$
16. $(x - 1)^2 = 4$
17. $(x - 6)^2 = 5$
18. $(x - a)^2 = 4$

Solve by completing the square.

Example $2x^2 + x - 1 = 0$

Solution Write an equivalent equation with the constant term as the right-hand member and the coefficient of x^2 equal to 1.

$$x^2 + \frac{1}{2}x = \frac{1}{2}$$

Add the square of one-half of the coefficient of x to each member of the equation.

$$x^2 + \frac{1}{2}x + \frac{1}{16} = \frac{1}{2} + \frac{1}{16}$$

Rewrite the left-hand member as a perfect square.

$$\left(x + \frac{1}{4}\right)^2 = \frac{9}{16}$$

Set $x + 1/4$ equal to each square root of 9/16.

$$x + \frac{1}{4} = \frac{3}{4} \qquad x + \frac{1}{4} = -\frac{3}{4}$$
$$x = \frac{1}{2} \qquad x = -1$$

The solution set is $\{1/2, -1\}$.

19. $x^2 + 4x - 12 = 0$
20. $x^2 - 2x + 1 = 0$
21. $x^2 + 9x + 20 = 0$
22. $x^2 - 2x - 1 = 0$
23. $2x^2 = 2 - 3x$
24. $2x^2 + 4x = -1$

Example $\dfrac{x^2}{4} + \dfrac{x}{4} = 3$

3.2 Second-Degree Equations

Solution Write an equivalent equation in standard form.
$$x^2 + x = 12$$
$$x^2 + x - 12 = 0$$

Substitute $a = 1$, $b = 1$, $c = -12$ in the quadratic formula and simplify.
$$x = \frac{-1 \pm \sqrt{1 + 48}}{2} = \frac{-1 \pm 7}{2}$$

The solution set is $\{3, -4\}$.

25. $x^2 - 3x + 2 = 0$
26. $x^2 + 4x + 4 = 0$
27. $2x^2 = 7x - 6$
28. $3x^2 = 5x - 1$
29. $\dfrac{x^2}{3} = \dfrac{1}{2}x + \dfrac{3}{2}$
30. $\dfrac{x^2 - 3}{2} + \dfrac{x}{4} = 1$
31. $x^2 - 2\sqrt{5}x + 5 = 0$
32. $x^2 + 2\sqrt{2}x + 2 = 0$
33. $2x^2 - \sqrt{3}x - 3 = 0$
34. $2x^2 + \sqrt{7}x - 7 = 0$
35. $x^2 - kx - 2k^2 = 0$
36. $2x^2 - kx + 3 = 0$
37. $ax^2 - x + c = 0$
38. $x^2 + 2x + c + 3 = 0$
39. Determine k so that the roots of $kx^2 + 4x + 1 = 0$ will be equal. [*Hint*: Use the discriminant.]
40. Determine k so that the roots of $x^2 - kx + 9 = 0$ will be equal.

Applications

41. *Biology* An estimation of the height h, measured in meters, that can be attained by a person pole-vaulting is given by
$$h = \frac{v^2}{2g} + 1.6$$
where v is the velocity of the person and $g = 9.8$ m/sec² is the acceleration of gravity. Determine v if the person has jumped 5 meters.

42. *Business* The amount of money S returned when a principal P is invested at an annual rate of interest r compounded quarterly for 2 years is $S = P\left(1 + \dfrac{r}{4}\right)^2$. Solve for r.

43. In the declining balance method of depreciation an item's value is obtained by reducing its book value by a constant percentage rate each year. If an item that costs A dollars initially is depreciated at a rate r each year, then at the end of 2 years its book value V is given by $V = A - rA - r(A - rA)$. Solve for r.

44. A company that sells calculators determines that the price per unit (in dollars) in the sale of x units is $p = -0.5x + 60$. Find the number of units that must be sold in order to generate a revenue of $1600.

45. The price per unit (in dollars) in the sale of x units of a commodity is $p = -0.1x + 10$. Find the number of units that must be sold in order to generate a revenue of $250.

46. The revenue that a company obtains in the sale of x units of a commodity is given by $R = -3x^2 + 180x$. Determine when $R = 0$ and $R = 2400$.

47. The profit P that a company obtains through the sale of x units of a commodity is given by $P = R - C$, where R and C are expressions representing revenue and cost, respectively. If $R = -2x^2 + 100x$ and $C = -2x + 100$, find x so that $P = 0$.

48. *Physics* The distance s above ground of a body projected upward is given by

$$s = -\frac{1}{2}gt^2 + v_0 t + s_0,$$

where $g = 32$ ft/sec^2 is the acceleration of gravity, v_0 is the initial velocity (measured in feet per second), and s_0 is the initial height of the body (measured in feet). An arrow shot upward from ground level at an initial velocity of 64 ft/sec is governed by the equation $s = -16t^2 + 64t$. Find the values of the time t for which $s = 0$.

49. A rock thrown upward from a height of 22 feet with an initial velocity of 96 ft/sec is governed by the equation $s = -16t^2 + 96t + 22$. Find the values of the time t for which $s = 102$.

3.3

EQUATIONS INVOLVING RADICALS

EQUALITY OF LIKE POWERS In order to find solution sets for equations containing radical expressions, we shall need the following result.

> If $U(x)$ and $V(x)$ are numerical expressions in x, then the solution set of $U(x) = V(x)$ is a subset of the solution set of $[U(x)]^n = [V(x)]^n$, for each positive integer n.

This result, which follows simply from the fact that products of equal numbers are equal numbers, permits us to raise both members of an equation to the same positive-integer power with the assurance that we do not lose any solutions of the original equation in the process. On the other hand, it does not assert that the resulting equation will be equivalent to the original equation, and indeed it will not always be. The equation

$$[U(x)]^n = [V(x)]^n$$

may have additional solutions, or **extraneous solutions**, that are not solutions of $U(x) = V(x)$. Thus, if $a = b$, then $a^4 = b^4$, but the converse does not necessarily hold. That is, a^4 and b^4 may be equal, but $a \neq b$. For example, $(3)^4 = (-3)^4$, but $3 \neq -3$. The solution set of the equation $x^4 = 81$, obtained from $x = 3$ by raising each member to the fourth power, contains -3 as an extraneous real solution, since -3 does not satisfy the original equation even though it does satisfy $x^4 = 81$.

3.3 Equations Involving Radicals

NECESSITY OF CHECKING SOLUTIONS Because the result of applying the above process is not always an equivalent equation, each solution obtained through its use must be substituted for the variable in the original equation to check its validity.

Example Find the solution set of $\sqrt[3]{x-1} = -1$.

Solution If we raise each member of $\sqrt[3]{x-1} = -1$ to the third power (cubing), we obtain

$$(\sqrt[3]{x-1})^3 = (-1)^3$$
$$x - 1 = -1,$$

which is equivalent to

$$x = 0.$$

Since $\sqrt[3]{0-1} = -1$, a solution of the original equation is 0. Moreover, 0 is the only real solution, since the solution set of the equation $\sqrt[3]{x-1} = -1$ is a subset of the solution set of $x = 0$.

Example Find the solution set of $\sqrt{x+2} + 4 = x$.

Solution We first write the equivalent equation

$$\sqrt{x+2} = x - 4$$

and then square each member

$$(\sqrt{x+2})^2 = (x-4)^2,$$

or

$$x + 2 = x^2 - 8x + 16.$$

This last equation is equivalent to

$$x^2 - 9x + 14 = 0,$$

or

$$(x-2)(x-7) = 0,$$

which clearly has solutions 2 and 7. Upon replacing x by 2 in the original equation, however, we obtain

$$\sqrt{2+2} + 4 = 2,$$

or

$$6 = 2,$$

which is false. Hence, 2 is not a solution of the original equation; it is an extraneous root. On the other hand, 7 does satisfy the original equation, so the solution set we seek is {7}.

After raising both members of an equation to a power, it may be necessary to raise the members of the resulting equation to a power a second time in order to obtain an equation clear of radical expressions.

Example Find the solution set of $\sqrt{x+4} + \sqrt{9-x} = 5$.

Solution It is helpful if this equation is first transformed so that each member of the equivalent equation contains only one of the radical expressions. Thus, by adding $-\sqrt{9-x}$ to each member, we obtain

$$\sqrt{x+4} = 5 - \sqrt{9-x}.$$

We then square each member

$$(\sqrt{x+4})^2 = (5 - \sqrt{9-x})^2$$
$$x + 4 = 25 - 10\sqrt{9-x} + 9 - x$$
$$2x - 30 = -10\sqrt{9-x}$$
$$x - 15 = -5\sqrt{9-x}.$$

Squaring again yields

$$(x - 15)^2 = (-5\sqrt{9-x})^2$$
$$x^2 - 30x + 225 = 25(9 - x)$$
$$x^2 - 5x = 0$$
$$x(x - 5) = 0.$$

It is clear that this last equation has 0 and 5 as solutions. Since both of these satisfy the original equation, the solution set we seek is {0, 5}.

EXERCISE 3.3

Solve and check. If there is no solution, so state.

1. $\sqrt{x} = 8$
2. $4\sqrt{x} = 9$
3. $\sqrt{y+8} = 1$
4. $\sqrt{y+3} = 2$
5. $x - 1 = \sqrt{2x+1}$
6. $2y + 3 = \sqrt{y+2}$
7. $2x - 1 = \sqrt{x+1}$
8. $\sqrt[5]{7-x} = 2$
9. $\sqrt[3]{2-y} = 3$

3.3 Equations Involving Radicals

10. $\sqrt{x+3}\sqrt{x-9} = 8$
11. $\sqrt{x}\sqrt{x+9} = 20$
12. $x - 2 = \sqrt{2x^2 - 3x + 2}$
13. $\sqrt{y^2 - y} = \sqrt{y^2 + 2y - 3}$
14. $\sqrt{x} + \sqrt{2} = \sqrt{x+2}$
15. $\sqrt{y+4} = \sqrt{y+20} - 2$
16. $\sqrt{13 + \sqrt{x}} + \sqrt{x+1}$
17. $\sqrt{5 + \sqrt{x}} = \sqrt{x} - 1$
18. $(y+7)^{1/2} + (y+4)^{1/2} = 3$
19. $(5+x)^{1/2} + x^{1/2} = 5$
20. $(z-3)^{1/2} + (z+5)^{1/2} = 4$
21. $(y^2 - 3y + 5)^{1/2} - (y+2)^{1/2} = 0$
22. $\sqrt{2x-4} - \sqrt{x-1} = 1$

Solve for the indicated variable.

23. $r = \sqrt{\dfrac{A}{\pi}}$ for A
24. $t = \sqrt{\dfrac{2s}{g}}$ for g
25. $x\sqrt{xy} = 1$ for y
26. $T = 2\pi\sqrt{\dfrac{2h}{3g}}$ for g
27. $x = \sqrt{a^2 - y^2}$ for y
28. $y = \dfrac{1}{\sqrt{1-x}}$ for x

Applications

29. *Business* A manufacturer determines that the revenue from and cost of making x units of a commodity are given by $R = 10\sqrt{2x}$ and $C = 4x + 8$, respectively. Find the values of x for which $R = C$.

30. In the design of an inventory system, the optimal size Q of a particular order is given by the formula

$$Q = \sqrt{\dfrac{2DS}{H}}\sqrt{\dfrac{H+P}{P}}.$$

Solve for H when $D = 10{,}000$, $S = 150$, $P = 5$, and $Q = 1000$.

31. *Physics* The period T of oscillation of a plane pendulum of length L is given by

$$T = 2\pi\sqrt{\dfrac{L}{g}}.$$

Determine the length of a pendulum whose period of oscillation is 2 seconds. Use $g = 980$ cm/sec^2.

3.4

SUBSTITUTION IN SOLVING EQUATIONS

Some equations that are not polynomial can nevertheless be solved by means of related polynomial equations. For example, though $y + 2\sqrt{y} - 8 = 0$ is not a polynomial equation, if the variable p is substituted for the radical expression $\sqrt{y}$, we then have $p^2 + 2p - 8 = 0$, which is a polynomial equation in p.

Example Find the solution set of $y + 2\sqrt{y} - 8 = 0$.

Solution If we set $p = \sqrt{y}$ and substitute in the given equation, we have

$$p^2 + 2p - 8 = 0$$
$$(p + 4)(p - 2) = 0,$$

which has -4 and 2 as solutions. Since $-4 < 0$, we must reject it as a source for solutions because $p = \sqrt{y}$, and $\sqrt{y}$ is always nonnegative. The other value, $p = 2$, leads to $\sqrt{y} = 2$ and hence $y = 4$. The solution set we seek is $\{4\}$.

The technique of substituting one variable for another—or, more generally, a variable for an expression—is not limited to cases involving radicals, but is useful in any situation in which an equation is polynomial in form. For example, in the equation

$$\left(x + \frac{1}{x}\right)^{-2} + 6\left(x + \frac{1}{x}\right)^{-1} + 8 = 0,$$

we would set

$$p = \left(x + \frac{1}{x}\right)^{-1}.$$

Similarly, in the equation

$$(y + 3)^{1/2} - 4(y + 3)^{1/4} + 4 = 0,$$

we would set

$$p = (y + 3)^{1/4}.$$

EXERCISE 3.4

Solve for x, y, or z.

Example $x^4 - 10x^2 + 9 = 0$

Solution Set $x^2 = p$; then $x^4 = p^2$. Solve for p.
$$p^2 - 10p + 9 = 0$$
$$(p - 9)(p - 1) = 0$$
$$p = 9 \qquad p = 1$$

Set each value of $p = x^2$ and solve for x.

$$x^2 = 9 \qquad\qquad x^2 = 1$$
$$x = 3 \quad x = -3 \qquad x = 1 \quad x = -1$$

The solution set is $\{3, -3, 1, -1\}$.

1. $x - 2\sqrt{x} - 15 = 0$
2. $x^4 - 5x^2 + 4 = 0$
3. $2x^4 + 17x^2 - 9 = 0$
4. $z^4 - 2z^2 - 24 = 0$
5. $(y^2 + 5y)^2 - 8y(y + 5) - 84 = 0$
6. $y^2 - 5 - 5\sqrt{y^2 - 5} + 6 = 0$
7. $y^{2/3} - 2y^{1/3} - 8 = 0$
8. $z^{2/3} - 2z^{1/3} - 35 = 0$
9. $y^{-2} - y^{-1} - 12 = 0$
10. $z^{-2} + 9z^{-1} - 10 = 0$
11. $(x - 1)^{1/2} - 2(x - 1)^{1/4} - 15 = 0$
12. $8x^{-6} + 7x^{-3} - 1 = 0$
13. $\left(x + \dfrac{4}{x}\right)^2 + \left(x + \dfrac{4}{x}\right) = 20$
14. $\left(\dfrac{y + 2}{y - 1}\right)^2 - 5\left(\dfrac{y + 2}{y - 1}\right) = -6$
15. $\dfrac{9y^2}{(y + 2)^2} - \dfrac{9y}{y + 2} + 2 = 0$
16. $\dfrac{y + 1}{2y^2} + \dfrac{36y^2}{y + 1} - 9 = 0$
17. $(y + 3)^{1/2} - 4(y + 3)^{1/4} = -4$
18. $\left(x + \dfrac{1}{x}\right)^{-2} + 6\left(x + \dfrac{1}{x}\right)^{-1} = -8$
19. $(y^4 - 4y^3 + 4y^2) - 23(y^2 - 2y) + 120 = 0$
20. $(4x^4 - 12x^3 + 9x^2) - 3(2x^2 - 3x) + 2 = 0$

3.5

SOLUTION OF LINEAR INEQUALITIES

Open sentences such as

(1) $$x + 3 \geq 10$$

and

(2) $$\frac{-2y - 3}{3} < 5$$

are called **inequalities**. For appropriate values of the variable, one member of an inequality represents a real number that is less than ($<$), less than or equal to ($\leq$), greater than or equal to ($\geq$), or greater than ($>$) the real number represented by the other member.

Any element of the replacement set of the variable for which an inequality is valid is called a **solution**, and the set of all solutions of an inequality is called the **solution set** of the inequality. Inequalities that are true for every element in the replacement set of the variable—such as $x^2 + 1 > 0$, for x real—are called **absolute** or **unconditional inequalities**. Inequalities that are not true for every element of the replacement set are called **conditional inequalities**—for example, (1) and (2) above.

ELEMENTARY TRANSFORMATIONS

As in the case with equations, we shall solve a given inequality by generating a series of **equivalent inequalities** (inequalities having the same solution set) until we arrive at one for which the solution set is obvious. To do this, we shall need the following rules, or **elementary transformations**.

> If $P(x)$, $Q(x)$, and $R(x)$ are expressions, then for all values of x for which $P(x)$, $Q(x)$, and $R(x)$ are real numbers, the sentence
>
> $$P(x) < Q(x)$$
>
> is equivalent to each of the following sentences:
>
> I $\quad P(x) + R(x) < Q(x) + R(x)$,
>
> II $\quad P(x) - R(x) < Q(x) - R(x)$,
>
> III $\quad P(x) \cdot R(x) < Q(x) \cdot R(x) \qquad R(x) > 0$,
>
> IV $\quad \dfrac{P(x)}{R(x)} < \dfrac{Q(x)}{R(x)} \qquad R(x) > 0$,
>
> V $\quad P(x) \cdot R(x) > Q(x) \cdot R(x) \qquad R(x) < 0$,
>
> VI $\quad \dfrac{P(x)}{R(x)} > \dfrac{Q(x)}{R(x)} \qquad R(x) < 0$.

Similarly, the sentence

$$P(x) \leq Q(x)$$

3.5 Solutions of Linear Inequalities

is equivalent to sentences of the form I–VI, with < (or >) replaced by ≤ (or ≥) under the same conditions as above. Note also that rules II, IV, and VI can be considered special cases of rules I, III, and V, respectively.

The foregoing can be interpreted as follows:

I and II. The addition or subtraction of the same expression to or from each member of an inequality produces an equivalent inequality in the same sense.

III and IV. If each member of an inequality is multiplied or divided by the same expression representing a positive number, the result is an equivalent inequality in the same sense.

V and VI. If each member of an inequality is multiplied or divided by the same expression representing a negative number, the result is an equivalent inequality in the opposite sense.

Observe that the variables in multipliers and divisors are restricted from values for which the expression vanishes; that is, multiplying or dividing by zero is not permitted.

SOLUTION OF A LINEAR INEQUALITY

The above rules can be used to solve inequalities in the same way that the properties of equality are applied to solve equations.

Example Find the solution set of $\dfrac{-x+3}{4} > -\dfrac{2}{3}$.

Solution Multiplying each member by 12 yields

$$3(-x+3) > -8,$$
$$-3x + 9 > -8.$$

Adding -9 to each member, we have

$$-3x > -17.$$

Finally, dividing each member by -3, we obtain

$$x < \frac{17}{3},$$

and the solution set is written $\{x \mid x < 17/3\}$.

GRAPHICAL REPRESENTATION

The solution set in the above example can be pictured on a line graph as shown in Figure 3.1. The heavy line indicates points with coordinates in the solution set.

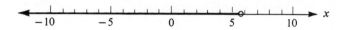

Figure 3.1

Inequalities sometimes appear in a form such as

(3) $\qquad -6 < 3x \leq 15,$

where an expression is bracketed between two inequality symbols. This means $-6 < 3x$ and $3x \leq 15$. The solution set of such an inequality is obtained in the same manner as the solution set of any other inequality. In (3) above, each expression may be divided by 3 to obtain

$$-2 < x \leq 5.$$

The solution set, $\{x \mid -2 < x \leq 5\}$, is shown on a line graph in Figure 3.2. Note here that the open dot at the left-hand endpoint of the interval indicates that -2 *is not* a member of the solution set, whereas the solid dot at the other end shows that 5 *is* a member of the solution set.

Figure 3.2

EXERCISE 3.5

Solve each inequality and represent the solution set on a line graph.

1. $x + 7 > 8$
2. $x - 5 \leq 7$
3. $3x - 2 > 1 + 2x$
4. $2x + 3 \leq x - 1$
5. $\dfrac{2x - 3}{2} \leq 5$
6. $\dfrac{3x + 4}{3} > 12$
7. $\dfrac{x}{3} + 2 < \dfrac{x}{4}$
8. $\dfrac{1}{2}(4 - x) > \dfrac{1}{3}x$
9. $-6 \leq 3x < 12$
10. $3 < x + 2 \leq 8$
11. $1 < 3x - 1 < 5$
12. $0 < 4 - 3x < 16$

Graph each set and designate each in simpler set notation.

3.5 Solutions of Linear Inequalities

Example $\{x \mid x + 2 \geq 0\} \cap \{x \mid x - 3 < 1\}$

Solution Solve each inequality and graph. Indicate the region where the graphs overlap.

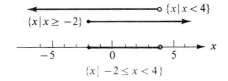

13. $\{x \mid x - 2 < 3\} \cap \{x \mid x + 4 > 2\}$
14. $\left\{ x \mid \dfrac{1 + x}{2} \leq 3 \right\} \cap \{x \mid x \leq 6\}$
15. $\{x \mid 2x - 1 > 5\} \cap \left\{ x \mid \dfrac{x - 1}{3} \geq 4 \right\}$
16. $\{x \mid 4 - x < 2\} \cap \left\{ x \mid \dfrac{2x + 5}{2} < 0 \right\}$
17. $\left\{ x \mid \dfrac{3 - x}{4} < -2 \right\} \cup \left\{ x \mid \dfrac{2 + 3x}{3} \leq 1 \right\}$
18. $\left\{ x \mid \dfrac{3x + 2x}{3} < 0 \right\} \cup \left\{ x \mid \dfrac{x + 3}{4} > 0 \right\}$
19. $\{x \mid 2 < x < 5\} \cup \{x \mid x \geq 3\} \cup \{x \mid x < 0\}$
20. $\{x \mid -3 \leq x < 4\} \cup \{x \mid x \geq 5\} \cup \{x \mid x > 8\}$

Example Determine k so that the solutions of $x^2 - 5x + k + 3 = 0$ are not real numbers.

Solution The solutions are not real if $b^2 - 4ac < 0$. Substituting -5 for b, 1 for a, and $k + 3$ for c in $b^2 - 4ac$, we have

$$(-5)^2 - 4(1)(k + 3) < 0$$
$$25 - 4k - 12 < 0$$
$$-4k < -13$$
$$k > \dfrac{13}{4}.$$

21. Determine k so that the roots of $x^2 + 2x + k + 3 = 0$ will be real numbers.
22. Determine k so that the roots of $x^2 + 9x + k = 2$ will be real numbers.
23. Determine k so that $x^2 - 2x + 1 = k$ will have no real roots.
24. Determine k so that the expression $\sqrt{-2k + 10}$ represents a real number.

Applications

25. *Biology* The arterial blood pressure P, measured in torr,* of a person standing erect is given by

$$P = P_h - \rho g h / 1.33 \times 10^3$$

where P_h is the average pressure in an artery at heart level, ρ is the density of blood, g is the acceleration of gravity, and h is the distance above or below the level of the heart. If $P_h = 100$ torr, $\rho = 1.05$ g/cm^3, and $g = 980$ cm/sec^2, determine the range of P when $0 \leq h \leq 50$. [*Hint*: Solve for h.]

---
*1 torr is the pressure necessary to support a column of mercury 1 mm high.

26. *Business* The revenue and cost for producing x units of a commodity are $R = 1.5x$ and $C = 0.7x + 1200$, respectively. Determine the values of x for which $R > C$.

27. The profit a company obtains through the sales of x units of an item is given by $P = 8x - 3600$. Determine the values of x for which $P \geq 1000$.

28. *Physics* The relationship between degrees Fahrenheit T_f and degrees Celsius T_c is $T_f = 9/5\ T_c + 32$. Find the range on the Celsius scale corresponding to $32 \leq T_f \leq 212$.

29. Use the relation between T_c and T_f in Problem 28 to find the range on the Fahrenheit scale corresponding to $-10 \leq T_c \leq 20$.

3.6

SOLUTION OF QUADRATIC INEQUALITIES

Solving quadratic inequalities offers somewhat different problems than solving linear inequalities. For example, consider the inequality

$$x^2 + 4x < 5.$$

To determine values of x for which this condition holds, we might first rewrite the sentence equivalently as

$$x^2 + 4x - 5 < 0,$$

and then as

$$(x + 5)(x - 1) < 0.$$

It is clear here that those values and only those values of x for which the factors $x + 5$ and $x - 1$ are opposite in sign will be in the solution set. Thus, the solution set can be determined analytically by finding the values of x such that

$$x + 5 < 0 \quad \text{and} \quad x - 1 > 0,$$

or else

$$x + 5 > 0 \quad \text{and} \quad x - 1 < 0.$$

Each of these two cases can be considered separately.

First, $x + 5 < 0$ and $x - 1 > 0$ imply $x < -5$ and $x > 1$, conditions which are not satisfied by any values of x. But $x + 5 > 0$ and $x - 1 < 0$ imply $x > -5$ and $x < 1$, which lead to the solution set $\{x | -5 < x < 1\}$. An alternative set notation is $\{x | x > -5\} \cap \{x | x < 1\}$.

USE OF SIGN GRAPHS One relatively easy way to visualize the solution set of a quadratic inequality is to indicate on a number line the signs associated with each factor for number replacements for the variable. Figure 3.3 shows such an arrangement, a **sign graph**, for the example above. This picture is constructed by first showing on the number line the

3.6 Solution of Quadratic Inequalities

places where $x + 5$ is positive ($x > -5$) and the places where it is negative ($x < -5$), and then showing on a second line the places where $x - 1$ is positive ($x > 1$) and those where it is negative ($x < 1$). The third line can then be marked by observing the parts of the first two lines where the signs are alike and the parts where the signs are opposite. It is desired that the product $(x + 5)(x - 1)$ be negative. The third line shows clearly that this occurs where $-5 < x < 1$, so the solution set of the inequality is $\{x \mid -5 < x < 1\}$. This set can be graphed as in Figure 3.4.

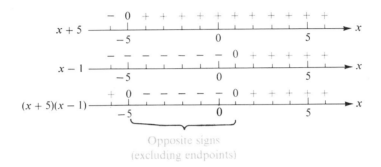

Figure 3.3

Figure 3.4

MULTIPLYING MEMBERS BY A VARIABLE   Inequalities involving fractions have to be approached with care if any fraction contains a variable in the denominator. If each member of an inequality is multiplied by an expression containing the variable, we have to be careful to distinguish between those values of the variable for which the expression denotes a positive number and those for which it denotes a negative number. Alternatively, we can make sure that the expression by which we multiply is always positive.

Example   Find the solution set of $\dfrac{3x}{x - 4} \geq 5$.

Solution 1   First we note that $x = 4$ is not a solution. Next, we multiply each member of

$$\frac{3x}{x - 4} \geq 5$$

by $x - 4$, observing, as we do, that two cases arise for $x \neq 4$. First, we obtain

and, second,
$$3x \geq 5(x - 4) \quad \text{for} \quad x - 4 > 0,$$

$$3x \leq 5(x - 4) \quad \text{for} \quad x - 4 < 0.$$

We can now find the solution for each case. Let us look at the first case, in which $x - 4 > 0$, or, equivalently, $x > 4$. We have

$$3x \geq 5(x - 4)$$
$$3x \geq 5x - 20$$
$$20 \geq 2x$$
$$10 \geq x,$$

which shows that $x \leq 10$ will satisfy the inequality as long as $x > 4$. In terms of sets, this means that $\{x | x \leq 10\} \cap \{x | x > 4\}$ is in the solution set of the inequality. Another expression describing this intersection is $\{x | 4 < x \leq 10\}$. Now, examining the case $x - 4 < 0$, or equivalently $x < 4$, we have

$$3x \leq 5(x - 4)$$
$$3x \leq 5x - 20$$
$$20 \leq 2x$$
$$10 \leq x.$$

But x cannot be greater than or equal to 10 and at the same time be less than 4. That is, $\{x | x \geq 10\} \cap \{x | x < 4\} = \emptyset$. This means that the entire solution set of the inequality is the set we obtained for the first case, $\{x | 4 < x \leq 10\}$. What we have accomplished here is to determine that the solution set is given by

$$[\{x | x \leq 10\} \cap \{x | x > 4\}] \cup [\{x | x \geq 10\} \cap \{x | x < 4\}].$$

which we express more simply as

$$\{x | 4 < x \leq 10\} \cup \emptyset \quad \text{or} \quad \{x | 4 < x \leq 10\}.$$

Solution 2 Let us approach this directly by means of a sign graph. We can rewrite the given inequality equivalently as

$$\frac{3x}{x - 4} - 5 \geq 0,$$

from which we obtain

$$\frac{-2x + 20}{x - 4} \geq 0.$$

3.6 Solution of Quadratic Inequalities

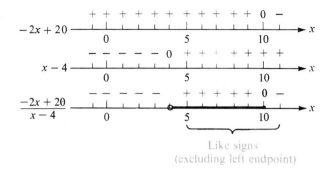

For this to be valid, $x - 4$ must not be 0, and the numerator and the denominator must be of like sign. The figure shows that the quotient $(-2x + 20)/(x - 4)$ is positive or zero for x between 4 and 10, including 10 but excluding 4, a value of x for which the denominator is 0. The desired solution set is therefore $\{x \mid 4 < x \leq 10\}$, which has the graph shown on the bottom line of the figure.

EXERCISE 3.6

Solve and represent each solution set on a line graph.

1. $(x + 1)(x - 2) > 0$
2. $(x + 2)(x + 5) < 0$
3. $x(x - 2) \leq 0$
4. $x(x + 3) \geq 0$
5. $x^2 - 3x - 4 > 0$
6. $x^2 - 5x - 6 \geq 0$
7. $x^2 < 5$
8. $4x^2 + 1 < 0$
9. $x^2 > -5$
10. $x^2 + 1 > 0$
11. $\dfrac{2}{x} \leq 4$
12. $\dfrac{3}{x - 6} > 8$
13. $x^2 - 16 \geq 0$
14. $25 - x^2 < 0$

Example $\dfrac{x}{x - 2} \geq 5$

Solution We can approach this directly by means of a sign graph. We first rewrite the given inequality equivalently as

$$\frac{x}{x - 2} - 5 \geq 0,$$

from which we obtain

$$\frac{-4x + 10}{x - 2} \geq 0.$$

For this to be valid, $x - 2$ must not be 0, and the numerator and the denominator must be of like sign. The sign graph shows that the quotient $(-4x + 10)/(x - 2)$ is positive or

$$-4x + 10 \quad \underset{-3 \quad 0 \quad 3}{\overset{+\ +\ +\ +\ +\ +\ +\ +\ +\ +\ 0\ -}{\longrightarrow}} x$$

$$x - 2 \quad \underset{-3 \quad 0 \quad 3}{\overset{-\ -\ -\ -\ -\ -\ -\ -\ -\ 0\ +\ +}{\longrightarrow}} x$$

$$\frac{-4x + 10}{x - 2} \quad \underset{-3 \quad 0 \quad 3}{\overset{-\ -\ -\ -\ -\ -\ -\ -\ -\ +0\ -}{\longrightarrow}} x$$

Like signs
(excluding left endpoint)

zero for x between 2 and 5/2, including 5/2 but excluding 2, a value of x for which the denominator is 0. The desired solution set is therefore $\{x \mid 2 < x \leq 5/2\}$, which has the graph shown on the last line of the figure.

15. $\dfrac{x}{x + 2} > 4$

16. $\dfrac{x + 2}{x - 2} \geq 6$

17. $\dfrac{2}{x - 2} \geq \dfrac{4}{x}$

18. $\dfrac{3}{4x + 1} > \dfrac{2}{x - 5}$

19. $x(x - 2)(x + 3) > 0$

20. $x^3 - 4x \leq 0$

21. Show that if a and b are real and $a, b > 0$, then $(a + b)^2 > a^2 + b^2$.

22. Show that if x and y are real and $x + y = 6$, then $xy \leq 9$. [*Hint*: Assume $xy > 9$, and consider $x(6 - x) > 9$.]

23. Show that for any real x, $x^2 + 1 \geq 2x$. [*Hint*: $(x - 1)^2 \geq 0$ for all real numbers.]

24. Show that the sum of any positive number and its reciprocal is greater than or equal to 2. [*Hint*: Use Problem 23.]

Applications

25. *Business* The revenue from and cost of producing x units of a commodity are $R = -2x^2 + 24x$ and $C = 4x + 32$, respectively. Determine the values of x for which $R > C$.

26. The profit a company obtains through the sale of x units of an item is given by $P = -x^2 + 8x - 15$. Determine the values of x for which $P > 0$.

27. *Physics* The distance, in feet, above ground level when a projectile is shot upward at an initial velocity of 128 ft/sec is given by $s = -16t^2 + 128t$, where t represents time in seconds. Determine the time interval for which the projectile is above a height of 192 feet. Determine the time interval during which the projectile is in the air.

28. *Social Science* In one theory of learning, the proportion of successful acts is related to the number of times x that an act is attempted by the formula $y = (ax + b)/(x + c)$, where a, b, and c are constants. If x is a positive integer and $a = 2$, $b = 4$, and $c = 20$, determine the times that will yield $y > 1/2$.

3.7
EQUATIONS AND INEQUALITIES INVOLVING ABSOLUTE VALUES

EQUATIONS In Section 1.2, we defined the absolute value of a real number x by

$$|x| = \begin{cases} x & \text{if } x \geq 0, \\ -x & \text{if } x < 0, \end{cases}$$

and interpreted it in terms of distance on a number line. For example, $|-5| = 5$ by definition, but 5 also denotes the distance the graph of -5 is located from the origin (Figure 3.5).

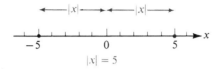

Figure 3.5

More generally, the expression $|x - a|$ satisfies

$$|x - a| = \begin{cases} x - a & \text{if } x - a \geq 0 \text{ or, equivalently, if } x \geq a, \\ -(x - a) & \text{if } x - a < 0 \text{ or, equivalently, if } x < a, \end{cases}$$

and can be interpreted on the number line as denoting the distance the graph of x is located from the graph of a (Figure 3.6).
Since

$$\sqrt{x^2} = \begin{cases} x & \text{if } x \geq 0, \\ -x & \text{if } x < 0, \end{cases}$$

or $|x| = \sqrt{x^2}$ we can assert that

$$|x - a| = \sqrt{(x - a)^2}.$$

Since $(x - a)^2 = (a - x)^2$, it follows that $|x - a| = |a - x|$.

All the foregoing facts can be used effectively in finding solution sets for equations involving absolute values. Several approaches can be used.

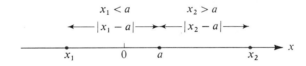

Figure 3.6

Example Find the solution set of $|x - 3| = 5$.

Solution 1 This can be solved by inspection. Since $x - 3$ represents the distance the graph of x is located from the graph of 3, and since by the equation this distance is 5, the two solutions of the equation are $3 + 5$, or 8, and $3 - 5$, or -2. Thus, the solution set is $\{8, -2\}$.

Solution 2 By definition, $|x - 3| = 5$ implies that

$$x - 3 = 5 \quad \text{for } x - 3 > 0,$$
$$-(x - 3) = 5 \quad \text{for } x - 3 < 0.$$

Equivalently we obtain

$$x = 8 \quad \text{and} \quad x = -2.$$

The solution set is $\{8, -2\}$.

Solution 3 Since $|x - 3| = \sqrt{(x - 3)^2}$, we can write the equation $|x - 3| = 5$ as

$$\sqrt{(x - 3)^2} = 5.$$

Then

$$(x - 3)^2 = 25$$
$$x^2 - 6x + 9 = 25$$
$$x^2 - 6x - 16 = 0$$
$$(x - 8)(x + 2) = 0,$$

and again we are led to the solution set $\{8, -2\}$.

The choice of approach depends on the situation and is largely a matter of which is most convenient.

INEQUALITIES Inequalities involving absolute-value notation require some additional discussion.

Example Find the solution set of $|x + 1| < 3$.

Solution 1 From the definition of absolute value, this inequality is equivalent to

$$x + 1 < 3 \quad \text{for} \quad x + 1 \geq 0,$$

and

$$-(x + 1) < 3 \quad \text{for} \quad x + 1 < 0,$$

so the solution set is given by

$$\{x | x < 2 \quad \text{for} \quad x \geq -1\} \cup \{x | x > -4 \quad \text{for} \quad x < -1\}.$$

or simply by $\{x| -4 < x < 2\}$. The graph is shown here.

Solution 2 The inequality can be written equivalently as

$$\sqrt{(x+1)^2} < 3,$$

which, for *positive* expressions, is equivalent to

$$(x+1)^2 < 9,$$

or

$$x^2 + 2x + 1 < 9$$
$$x^2 + 2x - 8 < 0$$
$$(x+4)(x-2) < 0.$$

Solving the last quadratic inequality by means of sign graphs or otherwise yields $\{x| -4 < x < 2\}$.

Example Find the solution set of $|x + 1| > 3$.

Solution By the definition of absolute value, this inequality is equivalent to

$$x + 1 > 3 \quad \text{for} \quad x + 1 \geq 0,$$

and

$$-(x + 1) > 3 \quad \text{for} \quad x + 1 < 0,$$

so the solution set is given by

(1) $$\{x|x > 2 \quad \text{or} \quad x < -4\}.$$

Alternatively, we could write (1) as

$$\{x|x > 2\} \cup \{x|x < -4\},$$

where the union gives a precise meaning to the word "or." The graph of the solution set is shown here.

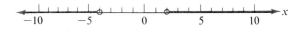

EXERCISE 3.7

Solve.

Example $|x + 5| = 8$

Solution Determine the solution set by inspection or write as two first-degree equations and solve each equation. Alternatively, use the method of Solution 3 shown on page 96.

$$x + 5 = 8 \qquad -(x + 5) = 8$$
$$x = 3 \qquad -x - 5 = 8$$
$$x = -13$$

The solution set is $\{3, -13\}$.

1. $|x| = 6$
2. $|x| = 3$
3. $|x - 1| = 4$
4. $|x - 6| = 3$
5. $\left|x - \dfrac{2}{3}\right| = \dfrac{1}{3}$
6. $\left|x - \dfrac{3}{4}\right| = \dfrac{1}{2}$
7. $|2x + 5| = 2$
8. $|3x + 7| = 1$
9. $\left|1 - \dfrac{1}{2}x\right| = \dfrac{3}{4}$
10. $\left|1 + \dfrac{3}{2}x\right| = \dfrac{1}{2}$
11. $\left|\dfrac{3x + 2}{4}\right| = 6$
12. $\left|\dfrac{2x - 1}{3}\right| = 3$

Solve and represent each solution set on a line graph.

Example $|2x - 1| \leq 7$

Solution Rewrite without absolute-value symbol.

$2x - 1 \leq 7$ for $2x - 1 \geq 0$

or

$-(2x - 1) \leq 7$ for $2x - 1 < 0$

Then, $x \leq 4$ for $x \geq 1/2$, and $x \geq -3$ for $x < 1/2$.

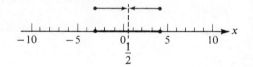

The solution set is $\{x \mid -3 \leq x \leq 4\}$.

13. $|x| < 2$
14. $|x - 1| > 2$
15. $|x + 3| \leq 4$
16. $2|x + 1| \leq 8$
17. $|2x - 5| < -3$
18. $|2x + 4| < -1$
19. $|x - 3| > 4$
20. $|2x + 4| > 2$
21. $|2x + 1| \geq 5$
22. $|3x - 5| \geq 4$
23. $\left|\dfrac{2x + 1}{3}\right| < 4$
24. $\left|\dfrac{3x - 2}{4}\right| \leq 5$

3.8 Word Problems

Replace each sentence with a single inequality by using absolute-value notation.

Example $-3 < x < 7$

Solution $7 + (-3) = 4$
Therefore, values of x are centered about 4/2, or 2. Adding -2 to each member, we have
$$-5 < x - 2 < 5,$$
$$|x - 2| < 5.$$

25. $1 < x < 3$ 26. $-5 \le x \le 9$ 27. $-9 \le x \le -7$
28. $5 < x < 13$ 29. $-7 \le 2x \le 12$ 30. $-5 < 3x < 10$
31. Show that $|-a| = |a|$. [*Hint*: Consider two possible cases, a nonnegative and a negative.]
32. Show that $|a - b| = |b - a|$. [*Hint*: Consider two possible cases, $a - b \ge 0$ and $a - b < 0$.]
33. Show that $|a^2| = |a|^2 = a^2$.
34. Show that $|ab| = |a| \cdot |b|$. [*Hint*: Consider four possible cases.]
35. Show that $|x - a| < b$ is equivalent to $-b < x - a < b$.
36. Show that $|x - a| > b$ is equivalent to $x - a > b$ or $x - a < -b$.

3.8

WORD PROBLEMS

Equations and inequalities can be used to express symbolically quantitative relations in word problems. The problem may be explicitly concerned with numbers, or it may be concerned with numerical measures of physical quantities. In either event, we seek the set of numbers (the solution set) for which the stated relationship holds. The following suggestions are frequently helpful in expressing the conditions of the problem symbolically:

1. Determine the quantities asked for and represent them by symbols. Since at this point we are using only one variable, all relevant quantities should be represented in terms of this variable.

2. Where applicable, draw a sketch and label all known quantities; label the unknown quantities in terms of symbols.

3. Find a quantity that can be represented in two different ways and write this representation as an equation. The equation may derive from:
   (a) the problem itself, which may state a relationship explicitly, for example, "What number added to 4 gives 7?" produces the equation $x + 4 = 7$;

(b) formulas or relationships that are part of your general mathematical background, for example, the area of a circle is $A = \pi r^2$ and the distance covered in time t when traveling at a constant rate r is $d = rt$.

4. Solve the resulting equation.

5. Check the results against the original problem. It is not sufficient to check the result in the equation, because the equation itself may be in error.

In some cases, the mathematical model we obtain for a physical situation is a quadratic equation that has two real solutions. It may be that one but not both of the solutions fits the physical situation. For example, if we were asked to find two consecutive positive integers of which the product is 72, we would write the equation

$$x(x + 1) = 72$$

as our model. Solving this equation, we have

$$x^2 + x - 72 = 0$$
$$(x + 9)(x - 8) = 0,$$

with solution set $\{8, -9\}$. Since -9 is not a positive integer, we must reject it as a possible answer to our original question; the solution 8, however, leads to the consecutive integers 8 and 9. As additional examples, observe that we would not accept -6 feet as the height of a man, or $27/4$ for the number of persons in a room.

A quadratic equation used as a model for a physical situation may have two, one, or no *meaningful* solutions—meaningful, that is, in a physical sense. Answers to word problems should always be checked against the universal set of meaningful numbers for the original problem.

Example A man has an annual income of $6000 from two investments. He has $10,000 more invested at 4% than he has invested at 3%. How much does he have invested at each rate?

Solution a. Represent the amount invested at each rate symbolically.

Let A represent the amount in dollars invested at 3%.

Then $A + 10,000$ represents the amount invested at 4%.

Write an equation relating the interest from each investment and the total interest.

$$\left[\begin{array}{c}\text{Interest from}\\ 3\% \text{ investment}\end{array}\right] + \left[\begin{array}{c}\text{Interest from}\\ 4\% \text{ investment}\end{array}\right] = [\text{Total interest}]$$

$$0.03A + 0.04(A + 10,000) = 6000$$

3.8 Word Problems

b. Solve for A.

$$3A + 4A + 40{,}000 = 600{,}000$$
$$7A = 560{,}000$$
$$A = 80{,}000 \quad \text{therefore} \quad A + 10{,}000 = 90{,}000$$

The man had $80,000 invested at 3% and $90,000 invested at 4%.

Check Does 3% of 80,000 (2400) added to 4% of 90,000 (3600) equal 6000? Yes.

Example A woman has invested $800 at compound interest. After a period of 2 years she receives $882. What is the annual rate of interest, r?

Solution a. The amount obtained after 1 year is the amount deposited plus the interest, that is,

$$800 + 800r.$$

Then, after 2 years,

$$882 = 800 + 800r + r(800 + 800r)$$
$$= 800(1 + r)^2.$$

b. Divide both members by 800.

$$(1 + r)^2 = 1.1025$$

Solve for r by taking the square roots.

$$1 + r = 1.05 \quad \text{or} \quad 1 + r = -1.05$$

Since the rate of interest must be positive, it follows that $r = 0.05$, or 5%.

Check The return on the investment after 1 year would be $800 (1.05) = $840. Is the return after 2 years, $840 (1.05), equal to $882? Yes.

Example A man owns $40,000 worth of stocks and bonds. Each bond has a value of $100 and each share of stock is valued at $200. If the total number of shares of stock and bonds is 250, then how many of each kind does the man own?

Solution Let x represent the number of shares of stock the man owns. The value of these shares is then $200x$. Similarly, if $(250 - x)$ represents the number of bonds the man owns, then their total value is $100(250 - x)$.

The combined value is

$$200x + 100(250 - x) = 40{,}000.$$

Now,

$$200x + 25{,}000 - 100x = 40{,}000$$
$$100x = 40{,}000 - 25{,}000$$
$$100x = 15{,}000$$
$$x = 150,$$

and $(250 - x) = 250 - 150 = 100$. That is, the man owns 150 shares of stock and 100 bonds.

Check Is the value of 150 shares of $200 stock plus 100 bonds worth $100 each equal to $40,000? Yes.

EXERCISE 3.8

Solve the following word problems.

In Problems 1–20, use a mathematical model in the form of a first-degree equation in one variable.

Example A collection of coins consisting of dimes and quarters has a value of $11.60. How many dimes and quarters are in the collection if there are 32 more dimes than quarters?

Solution a. Represent the unknown quantities symbolically.

Let x represent the number of quarters.

Then $x + 32$ represents the number of dimes.

Write an equation relating the value of the quarters and the value of the dimes to the value of the entire collection.

$$\begin{bmatrix} \text{Value of} \\ \text{quarters} \\ \text{in cents} \end{bmatrix} + \begin{bmatrix} \text{Value of} \\ \text{dimes} \\ \text{in cents} \end{bmatrix} = \begin{bmatrix} \text{Value of} \\ \text{collection} \\ \text{in cents} \end{bmatrix}$$
$$25x \quad + \quad 10(x + 32) \quad = \quad 1160$$

b. Solve for x.

$$25x + 10x + 320 = 1160$$
$$35x = 840$$
$$x = 24 \quad \text{therefore} \quad x + 32 = 56$$

There are 24 quarters and 56 dimes in the collection.

Check Do 24 quarters and 56 dimes have a value of $11.60? Yes.

1. A woman has $1.15 in change consisting of two more nickels than dimes. How many dimes and how many nickels does she have?

3.8 Word Problems

2. A collection of 300 dimes and nickels has a value of $17.50. How many nickels are there?
3. An amusement park sells two types of ride coupon books. Book A sells for $5.00 and book B sells for $3.40. At the end of 1 week a total of 140,000 books is sold, yielding a revenue of $676,000. How many A coupon books were sold during the week?
4. The admission at a baseball game was $2.00 for adults and $1.25 for children. The receipts were $103.75 for 68 paid admissions. How many adults, and how many children, attended the game?
5. How many pounds of an alloy containing 32% silver must be melted with 25 pounds of an alloy containing 48% silver to obtain an alloy containing 42% silver?
6. Together, two investors own 300 shares of stock in a company. If one investor owns 4 times as many shares as the other, how many shares does each own?
7. A sum of $2000 is invested, part at 3% and the remainder at 4%. Find the amount invested at each rate if the yearly income from the two investments is $66.
8. A sum of $2700 is invested, part at 4% and the remainder at 5%. Find the total yearly interest if the interest on each investment is the same.
9. A man has 3 times as much money invested in 3% bonds as he has in stocks paying 5%. How much does he have invested in each if his yearly income from the investments is $1680?
10. A woman has $1000 more invested at 5% than she has invested at 4%. If her annual income from the two investments is $698, how much does she have invested at each rate?
11. Find three consecutive integers whose sum is 15.
12. The sum of the digits of a two-digit number is 9. When the digits are reversed the number is less than the original. Find the first number.
13. In a course there are three tests worth 50 points each, 10 quizzes worth 10 points each, and a final examination worth 100 points. A student has 192 points going into the final examination. What score must the student obtain on the final examination in order to have an overall average of 80%.
14. In a course there are 16 quizzes worth 10 points each. For the first 10 quizzes a student averages 70%. How many points must the student obtain in the remaining six quizzes in order to raise that average to 75%?
15. The relationship between temperature T_f, measured in degrees Fahrenheit, and temperature T_c, measured in degrees Celsius, is given by $T_f = 9/5\, T_c + 32$. Determine the temperature that is the same on both scales.
16. What temperature on the Fahrenheit scale reads twice that on the Celsius scale?

Example An express train travels 150 miles in the same time that a freight train travels 100 miles. If the express goes 20 miles per hour faster than the freight train, find each rate.

Solution a. Represent the unknown quantities symbolically.

　　　　　Let r represent a rate for the freight train.

　　　　　Then $r + 20$ represents the rate of the express train.

The fact that the times, t, are equal is the significant equality in the problem.

$$[t \text{ of freight}] = [t \text{ of express}]$$

Express the time of each train in terms of r (time = distance/rate).

$$\frac{100}{r} = \frac{150}{r+20}$$

b. Solve for r.

$$(r+20)100 = (r)150$$

$$100r + 2000 = 150r$$

$$-50r = -2000$$

$$r = 40 \quad \text{therefore} \quad r+20 = 60$$

The freight train's rate is 40 miles per hour and the express train's rate is 60 miles per hour.

Check Does the time of the freight train (100/40) equal the time of the express train (150/60)? Yes.

17. An airplane travels 1260 miles in the same time that an automobile travels 420 miles. If the rate of the airplane is 120 miles per hour greater than the rate of the automobile, find the rate of each.

18. Two cars start together and travel in the same direction; one goes twice as fast as the other. At the end of 3 hours they are 96 miles apart. How fast is each traveling?

19. A freight train leaves town A for town B, traveling at an average rate of 40 miles per hour. Three hours later a passenger train leaves town A for town B on a parallel track, traveling at an average rate of 80 miles per hour. How far from town A does the passenger train pass the freight train?

20. A child walked to a friend's house at the rate of 4 miles per hour and ran back home at the rate of 6 miles per hour. How far apart are the two houses if the round trip took 20 minutes?

In Problems 21–34 use a mathematical model in the form of a second-degree equation in one variable.

21. Find two numbers whose sum is 15 and whose product is 56.
22. Find two numbers whose sum is -12 and whose product is 32.
23. Find two consecutive integers whose product is 42.
24. Find two consecutive positive integers such that the sum of their squares is 85.
25. Two airplanes with lines of flight at right angles to each other pass each other (at slightly different altitudes) at noon. One is flying at 140 miles per hour and the other at 180 miles per hour. How far apart are they at 12:30 PM?
26. What is the annual rate of interest if $22,898 is returned on a deposit of $20,000 for 2 years at compound interest?
27. A ball thrown vertically upward reaches a height s in feet given by the equation $s = -16t^2 + 56t$, where t is the time in seconds after the throw. How long will it take the ball to reach

3.8 Word Problems

a height of 24 feet on its way up? How long after the throw will the ball return to the height from which it was thrown?

28. If the radius of a circle is doubled, the area is increased by 48π square centimeters. What is the original radius?

29. What change in the length of a side of a square, originally 3 centimeters on a side, will decrease the area by 5 square centimeters?

30. A box without a top is to be made from a square piece of tin by cutting a 2 inch square from each corner and folding up the sides. If the box is to hold 128 cubic inches, what should be the length of each side of the original square?

31. Two tanks, each cylindrical in shape and 10 feet in length, are to be replaced by a single tank of the same length. If the original tanks have radii that measure 6 feet and 8 feet, respectively, what must be the radius of the single tank replacing them if it is to hold the same volume of liquid?

32. A theater that is rectangular in shape seats 720 people. The number of rows needed to seat the people would be four fewer if each row held six more seats. How many seats would then be in each row?

33. A man and his son working together can paint their house in 4 days. The man can do the job alone in 6 days less than the son can do it alone. How long would it take each of them to paint the house alone? [*Hint*: What part of the job could each of them do in 1 day?]

34. A rancher wishes to build a rectangular corral with 1000 feet of fencing. What should the dimensions of the corral be if the area enclosed is to be 40,000 square feet?

In Problems 35 and 36 use a mathematical model in the form of an inequality in one variable.

Example   A student must have an average of 80–90% on five tests to receive a grade of B in a course. The student's grades on the first four tests were 98%, 76%, 86%, and 92%. What grade on the fifth test would qualify the student for a B in the course?

Solution   a. Represent the unknown quantity symbolically.

Let x represent a grade (in percent) on the last test.

Write an inequality expressing the word sentence.

$$80 \leq \frac{98 + 76 + 86 + 92 + x}{5} \leq 90$$

b. Solve for x.

$$400 \leq 352 + x \leq 450$$

$$48 \leq x \leq 98$$

Any grade equal to or greater than 48% and less than or equal to 98%.

35. In the above example, what grade on the fifth test would qualify the student for a B if the grades on the first four tests were 78%, 64%, 88%, and 76%?

36. In the above example, what grade on the fifth test would qualify the student for a B if the grades on the first four tests are as given but the fifth test is weighted twice as much as any one test?

CHAPTER TEST

Solve each equation.

1. $2 + \dfrac{x}{3} = \dfrac{5}{6}$
2. $\dfrac{2x}{3} - \dfrac{2x+5}{6} = \dfrac{1}{2}$
3. $\dfrac{x}{x+1} + \dfrac{4}{5} = 6$
4. $1 - \dfrac{y-2}{y-3} = \dfrac{3}{y-1}$
5. Solve $\dfrac{x+y}{5} = \dfrac{x-y}{3}$ for y in terms of x.
6. Solve $\dfrac{x+y}{5} = \dfrac{x-y}{3}$ for x in terms of y.

Solve by factoring.

7. $x(2x - 3) = -1$
8. $2x(x - 2) = x + 3$

Solve by square roots.

9. $3x^2 = 21$
10. $(x + 4)^2 = 3$

Solve by completing the square.

11. $x^2 + 3x - 1 = 0$
12. $2x^2 + x - 1 = 0$

Solve for x by using the quadratic formula.

13. $2x^2 - x - 2 = 0$
14. $x^2 + kx - 4 = 0$
15. Determine k so that $x^2 - 6x + k = 0$ will have exactly one real solution.
16. Determine k so that $x^2 + kx + 4 = 0$ will have exactly one real solution.

Solve each equation.

17. $x - 3\sqrt{x} + 2 = 0$
18. $\sqrt{x+1} + \sqrt{x+8} = 7$

Chapter Test

Solve each equation.

19. $y^4 - 5y^2 + 4 = 0$

20. $y^{-2} - y^{-1} - 20 = 0$

Solve each inequality.

21. $\dfrac{x-3}{4} \leq 6$

22. $2(x-1) > \dfrac{2}{3}x$

Graph each set intersection.

23. $\{x \mid x + 3 < 11\} \cap \{x \mid 2x - 4 > 0\}$

24. $\left\{x \mid \left|\dfrac{x+4}{3}\right| \leq 6 + x\right\} \cap \left\{x \mid \left|\dfrac{2x-1}{3}\right| < 1\right\}$

Solve each inequality.

25. $x^2 + 4x - 5 < 0$

26. $\dfrac{2}{x-3} < 4$

Solve each equation.

27. $|2x + 1| = 7$

28. $\left|x - \dfrac{1}{3}\right| = \dfrac{2}{3}$

29. Solve $|x - 2| > 5$, and graph the solution set on a number line.

30. Write $-3 \leq x \leq 11$ using a single inequality involving an absolute-value symbol.

31. Purchases of VTRs (video tape recorders) are expected to increase next year by 25% to an annual sale of 600,000 units. Approximately how many units were sold this year?

32. The monthly price of a particular magazine when received by subscription is 60 cents. This price is a 40% discount from the monthly newstand price. What is the newstand price?

33. In a recent election, the winning candidate received 150 votes more than the losing candidate. How many votes did each receive if there were 4376 votes cast?

34. When the length of each square is increased by 5 inches, the area is increased by 85 square inches. Find the length of a side of the original square.

35. Clerk A can process 50 applications in 4 hours, and clerk B can process 50 applications in 8 hours. How long will it take both clerks working together to process 100 applications?

36. A boat sailed across a lake and back in $2\frac{1}{2}$ hours. If the rate returning was 2 miles per hour less than the rate going, and if the distance each way was 6 miles, find the rate each way.

37. When $1000 is invested at 5% simple interest, the amount returned after t periods is given by $S = 1000 + 50t$. Over how many periods should the $1000 remain in the investment in order to guarantee a return of between $1600 and $2200?

4
RELATIONS AND FUNCTIONS

4.1
ORDERED PAIRS OF REAL NUMBERS

When the order in which the numbers of a number pair are to be considered is specified, the pair is called an **ordered pair**, and the pair is denoted by a symbol such as (3, 2), (2, 3), (−1, 5), or (0, 3). Each of the two numbers in an ordered pair is called a **component** of the ordered pair; the first is called the **first component** and the second is called the **second component**. Having established what is meant by an ordered pair, we are ready to consider a set operation involving such pairs.

> The **Cartesian product** of two sets A and B, denoted by $A \times B$, is the set of all ordered pairs (x, y) such that $x \in A$ and $y \in B$.

$R \times R$, OR R^2, AND THE GEOMETRIC PLANE

The most important Cartesian product with which we shall be concerned is that formed from the set R of real numbers. The product $R \times R$, which is often denoted by R^2, is the set of all possible ordered pairs of real numbers. The fact that each member of R^2 corresponds to a point in the geometric plane, and the coordinates of each point in the geometric plane are the components of a member of R^2, is the basis for all plane graphing. As you probably recall from your earlier study of algebra, the correspondence between points in the plane and ordered pairs of real numbers is usually established through a **Cartesian** (or **rectangular**) **coordinate system**, as shown in Figure 4.1.

SOLUTIONS OF AN EQUATION IN TWO VARIABLES

Equations in two variables, such as

$$3x + 2y = 12 \qquad x^2 y + 3x = y^5 \qquad \sqrt{xy} = y^2 - 5,$$

have ordered pairs of real numbers as solutions. For example, if the components of (2, 3) are substituted for the variables x and y, in that order, in the equation

109

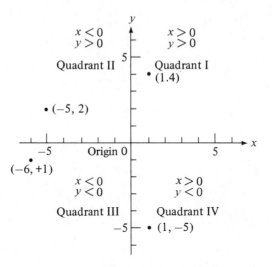

Figure 4.1

(1) $$3x + 2y = 12,$$

the result is

$$3(2) + 2(3) = 12,$$

which is true. Hence, (2, 3) is a solution of (1). In this book, the first component of an ordered pair is a value for the **abscissa**, x, and the second component is a value for the **ordinate**, y. Since many equations (and inequalities) in two variables have an infinite number of solutions, we shall use the set-builder notation

$$\{(x, y) | \text{condition on } x \text{ and } y\}$$

to represent the set of all solutions.

SUBSETS OF $R \times R$

Any condition on x and y—that is, any sentence in two variables, x and y—expresses a relationship between elements in the replacement sets of the two variables, and this relationship is precisely represented by the solution set of the sentence in two variables. When x and y are real numbers, the solution set is always a subset of $R \times R$. This leads us to the following.

> Any subset of $R \times R$ is called a **relation** in R.

The relation is said to be in R because the components of the ordered pairs in the relation are elements of R. Alternatively, we refer to the relationship as being in $R \times R$. To avoid needless repetition, we shall assume throughout our discussion that all relations are in R.

4.1 Ordered Pairs of Real Numbers

The set of all first components in the ordered pairs in a relation is called the **domain** of the relation, and the set of all second components is called the **range** of the relation. Thus,

$$\{(\overset{\downarrow}{2}, 5), (\overset{\downarrow}{3}, 10), (\overset{\downarrow}{4}, 15)\} \quad \text{Elements in the domain}$$
$$\underset{\uparrow \qquad \uparrow \qquad \uparrow}{} \quad \text{Elements in the range}$$

is a relation with domain $\{2, 3, 4\}$ and range $\{5, 10, 15\}$.

If a relation is defined by an equation and the domain is not specified, we shall understand that the domain is *the set of all real numbers for which a real number exists in the range*. For example, the domain of

$$S = \left\{ (x, y) \,\middle|\, y = \frac{1}{x - 2} \right\}$$

is $\{x \,|\, x \in R, \ x \neq 2\}$, because for every real number x except 2, the expression $1/(x - 2)$ represents a real number.

In a relation, each element in the domain is said to be *paired with* or *mapped onto* an element in the range. A relation, as in the above example, may be a one-to-one **mapping** (each element in the domain is mapped onto one element in the range), as illustrated in Figure 4.2. A relation may be a mapping in which one or more elements in the domain are mapped onto more than one element in the range, as shown in Figure 4.3, part a, or in which more than one element in the domain is mapped onto the same element in the range, as shown in Figure 4.3, part b.

A special kind of relation that is very important in mathematics is a **function**.

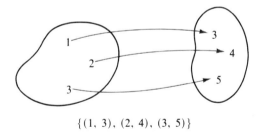

$\{(1, 3), (2, 4), (3, 5)\}$

Figure 4.2

> A **function** is a relation in which no two ordered pairs have the same first component and different second components.

For example, the relation $\{(1, 3), (2, 3), (3, 4)\}$ given in Figure 4.3, part b, is a function, whereas the relation given in Figure 4.3, part a, is not a function.

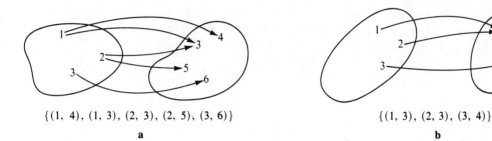

$\{(1, 4), (1, 3), (2, 3), (2, 5), (3, 6)\}$

a

$\{(1, 3), (2, 3), (3, 4)\}$

b

Figure 4.3

GRAPHICAL CHARACTERIZATION OF A FUNCTION

A function, therefore, associates each element in its domain with one and only one element in its range. In a graphical sense, this implies that no two of the ordered pairs in a function graph into points on the same vertical line.

As you should recall from your earlier study of algebra, graphs are often continuous lines and curves. Figure 4.4 shows three such graphs. Imagine a vertical line moving across each of these from left to right. If the line at any position cuts the graph of the relation in more than one point, then the relation is *not* a function. Thus, although parts a and c of Figure 4.4 show the graphs of relations that are functions, part b shows the graph of a relation that is not a function, because the vertical line meets the graph in two places. This means that for the particular value of x involved, the relation associates two distinct values for y.

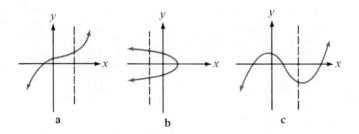

Figure 4.4

ALGEBRAIC CHARACTERIZATION OF A FUNCTION

When a relation is defined by an equation, one way to find out whether the relation is a function is to solve the equation explicitly for the variable y representing an element in the range. This will show if more than one value of y is associated with any single value of x.

4.1 Ordered Pairs of Real Numbers

Example Is the relation $\{(x, y) \mid y^2 = 1 + x^2\}$ a function?

Solution Since $y^2 = 1 + x^2$ implies either $y = \sqrt{1 + x^2}$ or $y = -\sqrt{1 + x^2}$, the assignment of a real value to x will result in two different values for y, and hence the relation is not a function.

FUNCTION NOTATION

In Section 2.1 you became acquainted with a notation that is widely used in discussing functions. In general, functions are denoted by single symbols; for example, f, g, h, and C, D, F and so on, might designate functions. The symbol for a function can be used in conjunction with the variable representing an element in the domain to represent the associated element in the range. Thus, $f(x)$, read "f of x" or "the value of f at x," is the element in the range of f associated with the element x in the domain.

Suppose
$$f = \{(x, y) \mid y = x + 3\}.$$

The alternative notation
$$f = \{(x, f(x)) \mid f(x) = x + 3\}$$

can be used, where $f(x)$ plays the same role as y. In practice, however, a function is simply denoted by $y = f(x)$; it is understood that this notation is equivalent to some set of ordered pairs.

The total revenue that a company obtains through the sale of x units of a commodity very often is given by a function such as $R(x) = -2x^2 + 1000x$.

Example If $R(x) = -2x^2 + 1000x$, find $R(4)$.

Solution Substitute 4 for x.
$$R(4) = -2(4)^2 + 1000(4)$$
$$= -32 + 4000$$
$$= 3680.$$

In the preceding example, when the company sells four units of an item the corresponding revenue is 3680. This latter number is in units of dollars; it could represent simply $3680.

A function such as $C(x) = 0.5x + 300$ could represent the total cost to a company when x units of a commodity are produced.

Example For some corporations recent tax laws require that they pay a rate of 22% on income up to $25,000 and 50% on all income over $25,000. If x denotes income then the tax T can be written as the function

(1) $$T(x) = \begin{cases} 0.22x, & 0 \le x \le 25{,}000 \\ 0.50x, & x > 25{,}000. \end{cases}$$

For example, the tax on an income of $10,000 would be

$$T(10{,}000) = 0.22(10{,}000) = \$2200$$

whereas the tax on an income of $100,000 would be

$$T(100{,}000) = 0.5(100{,}000) = \$50{,}000.$$

Example Express the area A of a square as a function of its perimeter P.

Solution As shown in the figure, let x represent the length of a side in the square. The perimeter is then

$$P = 4x.$$

The area of the square is

$$A = x^2.$$

Since $x = P/4$ we can express A as a function of P,

$$A(P) = \left(\frac{P}{4}\right)^2$$

$$= \frac{P^2}{16}.$$

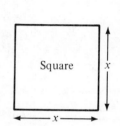

Square

EXERCISE 4.1

Supply the missing components so that the ordered pairs a–e below are solutions of the given equations.

a. (0, ?)   **b.** (1, ?)   **c.** (2, ?)   **d.** (−3, ?)   **e.** $\left(\frac{2}{3}, ?\right)$

1. $2x + y = 6$
2. $y = 9 - x^2$
3. $y = \dfrac{3x}{x^2 - 2}$
4. $y = 0$
5. $y = \sqrt{3x + 11}$
6. $y = |x - 1|$

a. Specify the domain and the range of each relation.
b. State whether each relation is a function.

4.1 Ordered Pairs of Real Numbers

Example $\{(3, 5), (4, 8), (4, 9), (5, 10)\}$

Solution
a. Domain (the set of first components): $\{3, 4, 5\}$
Range (the set of second components): $\{5, 8, 9, 10\}$
b. The relation is not a function, because two ordered pairs, $(4, 8)$ and $(4, 9)$, have the same first components.

7. $\{(2, 3), (5, 7), (7, 8)\}$
8. $\{(-1, 6), (0, 2), (3, 3)\}$
9. $\{(2, -1), (3, 4), (3, 6)\}$
10. $\{(-4, 7), (-4, 8), (3, 2)\}$
11. $\{(5, 5), (6, 6), (7, 7)\}$
12. $\{(0, 0), (2, 4), (4, 2)\}$

Specify the maximum domain that would yield real numbers y for elements in the range of the relation defined by each equation.

Examples a. $y = \sqrt{4-x}$ b. $y = \dfrac{1}{x(x+2)}$

Solutions
a. For what values of x is $4 - x \geq 0$?
The domain is $\{x \mid x \leq 4\}$.
b. For what values of x is $x(x+2) \neq 0$?
The domain is $\{x \mid x \neq 0, -2\}$.

13. $y = x + 7$
14. $y = 2x - 3$
15. $y = x^2$
16. $y = \dfrac{1}{x}$
17. $y = \dfrac{1}{x-2}$
18. $y = \dfrac{1}{x^2+1}$
19. $y = \sqrt{x-3}$
20. $y = \sqrt{2x+5}$
21. $y = \sqrt{4-x^2}$
22. $y = \sqrt{x^2-9}$
23. $y = \dfrac{4}{x(x-1)}$
24. $y = \dfrac{x}{(x-1)(x+2)}$

State whether the given equation defines a function.

Examples a. $x^2 y = 3$ b. $x^2 + y^2 = 36$

Solutions
a. Solve explicitly for y.
$$y = \frac{3}{x^2}$$
Yes. There is only one value of y associated with each value of x ($x \neq 0$).
b. Solve explicitly for y.
$$y = \pm\sqrt{36 - x^2}$$
No. There are two values of y associated with each value of x satisfying $|x| < 6$.

25. $x + y = 3$
26. $y = -x^2$
27. $y = \sqrt{x^2 - 5}$
28. $y = \sqrt{16 - x^2}$
29. $x^2 + y^2 = 16$
30. $y = \pm\sqrt{x^2}$
31. $y^2 = x^3$
32. $y = ax^n$

If $f(x) = x + 2$, find the given element in the range.

Example $f(3)$

Solution Substitute 3 for x.
$$f(3) = 3 + 2 = 5$$
The element is 5.

33. $f(0)$ **34.** $f(1)$ **35.** $f(-3)$ **36.** $f(a)$

If $g(x) = x^2 - 2x + 1$, find the given element in the range.

37. $g(-2)$ **38.** $g(0)$ **39.** $g(a + 1)$ **40.** $g(a - 1)$

If $f(x) = x + 2$ defines a function, find the element in the domain of f associated with the given element in the range.

Example 5

Solution Replacing $f(x)$ with 5, we have
$$5 = x + 2 \quad \text{or} \quad x = 3.$$
The element is 3.

41. 3 **42.** -2 **43.** a **44.** $a + 2$

If $g(x) = x^2 - 1$, find all elements in the domain of g associated with the given element in the range.

45. $g(x) = 0$ **46.** $g(x) = 3$ **47.** $g(x) = 8$ **48.** $g(x) = 5$

If $f(x) = x + 2$ and $g(x) = x - 4$, find each of the following.

49. $f(0)$ **50.** $g(2)$ **51.** $f(g(2))$ **52.** $f(g(x))$

53. If $f(x) = x^2 - x + 1$, find $\dfrac{f(x+h) - f(x)}{h}$.

54. If $f(x) = 3x^2 + 4x$, find $\dfrac{f(x+h) - f(x)}{h}$.

4.2 Linear Functions

Any function satisfying the condition that $f(-x) = f(x)$ for all x in the domain is called an **even function**. Any function satisfying the condition that $f(-x) = -f(x)$ for all x in the domain is called an **odd function**. Which of the following functions are even and which are odd?

55. $f(x) = x^2$
56. $f(x) = x^3$
57. $f(x) = x^4 - x^2$
58. $f(x) = x^3 - x$

Applications

59. Express the perimeter P of a square as a function of its area A.
60. Express the area A of a circle as a function of its circumference C.
61. Express the volume V of a cube as a function of the area A of its base.
62. *Biology* It is often assumed that the volume of a body, and hence its weight W, is a function of its length s. This assumption is given by $W = ks^3$, where k is a constant. For a certain kind of lizard $k = 500$ g/m^3. Find the weight of a lizard 0.4 meter long.
63. *Business* A revenue function $R(x)$ can be defined by
$$R(x) = px$$
where x is the number of units of a commodity, and p, which may itself be a function of x, represents price.
   a. If $p = 200 - 5x$, what is $R(10)$?
   b. If $p = 400 - \dfrac{x^2}{1,000,000}$, what is $R(100)$?
64. A revenue function is given by
$$R(x) = \begin{cases} 300x, & \text{if } 0 \le x < 100 \\ -x^2 + 400x, & \text{if } x \ge 100. \end{cases}$$
Find $R(20)$, $R(100)$, and $R(200)$.
65. Use (1) on page 114 to determine the income tax a corporation would pay on an income of \$20,000. What is the tax on \$150,000 of income?

4.2

LINEAR FUNCTIONS

A **first-degree**, or **linear**, **equation** in x and y is an equation that can be written equivalently in the form

(1) $\qquad Ax + By + C = 0 \qquad$ (A and B not both 0).

GRAPHS OF FIRST-DEGREE EQUATIONS

We shall call (1) the **standard form** for a linear equation. The graph of any such equation (technically, of its solution set) is a straight line, although we do not prove this here.

Since two distinct points determine a straight line, it is evident that we need find only two solutions of such an equation to determine its graph—that is, the graph of the solution set of the equation. In practice, the two solutions easiest to find are usually those with first and second components, respectively, equal to zero—the solutions $(0, y_1)$ and $(x_1, 0)$.

The numbers x_1 and y_1 are called the **x-** and **y-intercepts** of the graph. As an example, consider the equation

(2) $$3x + 4y = 12.$$

If $y = 0$, we have $x = 4$, and the x-intercept is 4. If $x = 0$, then $y = 3$, and the y-intercept is 3. Thus, the graph of (2) appears as shown in Figure 4.5.

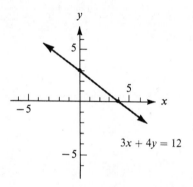

Figure 4.5

If the graph intersects both axes at or near the origin, then the intercepts either do not represent two separate points, or the points are too close together to be of much use in drawing the graph. It is then necessary to plot at least one other point at a distance far enough removed from the origin to establish the line with pictorial accuracy.

LINEAR FUNCTION When $B \neq 0$, we can solve (1) for y in terms of x. Thus, we obtain the general form of a linear function

$$f(x) = ax + b$$

where a and b are constants. The student should verify that (2) could also be expressed as

$$f(x) = -\frac{3}{4}x + 3.$$

EQUATIONS OF HORIZONTAL LINES Two special cases of linear equations are worth noting. First, an equation such as

$$y = 4$$

may be considered an equation in two variables,

4.2 Linear Functions

$$0x + y = 4.$$

For each x, this equation assigns $y = 4$. That is, any ordered pair of the form $(x, 4)$ is a solution of the equation. For instance,

$$(-3, 4), \quad (-1, 4), \quad (3, 4), \quad (4, 4),$$

are all solutions of the equation. If we graph these points and connect them with a straight line, we have the graph shown in Figure 4.6, part a.

Since the equation

$$y = 4$$

assigns to each x the same value for y, the function defined by this equation is called a **constant function**. Every constant function is also a linear function.

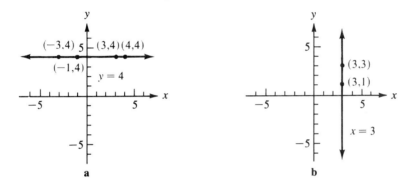

Figure 4.6

EQUATIONS OF VERTICAL LINES

The other special case of the linear equation is of the type

$$x = 3,$$

which may be considered as

$$x + 0y = 3.$$

Here, only one value is permissible for x, namely 3, whereas any value may be assigned to y. That is, any ordered pair of the form $(3, y)$ is a solution of this equation. If we choose two solutions, say $(3, 1)$ and $(3, 3)$, we can draw the graph shown in Figure 4.6, part b. It is clear that this equation does *not* define a function.

Any two distinct points in a plane can be looked upon as the endpoints of a line segment. Two fundamental properties of a line segment are its **length** and its **inclination** with respect to the x-axis.

Let P_1, with coordinates (x_1, y_1), and P_2, with coordinates (x_2, y_2), be endpoints of a line segment. If we construct through P_2 a line parallel to the y-axis, and through P_1 a line parallel to the x-axis, the lines will meet at a point P_3, as shown in Figure 4.7. The x-coordinate of P_3 is evidently the same as the x-coordinate of P_2, and the y-coordinate of P_3 is the same as that of P_1; hence the coordinates of

P_3 are (x_2, y_1). By inspection, we observe that the distance from P_2 to P_3 is simply the absolute value of the difference in the y-coordinates of the two points, $|y_2 - y_1|$, and the distance between P_1 and P_3 is the absolute value of the difference of the x-coordinates of these points, $|x_2 - x_1|$.

In general, since $y_2 - y_1$ is positive or negative as $y_2 > y_1$ or $y_2 < y_1$, respectively, and $x_2 - x_1$ is positive or negative as $x_2 > x_1$ or $x_2 < x_1$, respectively, it is also convenient to designate the distances represented by $x_2 - x_1$ and $y_2 - y_1$ as positive or negative. Such distances are called **directed distances**.

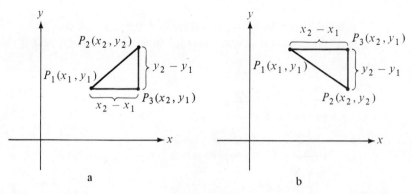

Figure 4.7

DISTANCE FORMULA The Pythagorean theorem can be used to find the length, d, of the line segment from P_1 to P_2. This theorem asserts that the square of the length of the hypotenuse of any right triangle is equal to the sum of the squares of the lengths of the sides. Thus, we have

$$d^2 = (x_2 - x_1)^2 + (y_2 - y_1)^2,$$

and by considering only the positive (or nonnegative) square root of the right-hand member, we obtain

(3) $$d = \sqrt{(x_2 - x_1)^2 + (y_2 - y_1)^2}.$$

Since the distances $x_2 - x_1$ and $y_2 - y_1$ are squared, it makes no difference here whether they are positive or negative—the result is the same. Equation (3) is a formula for the **distance** between any two points in the plane in terms of the coordinates of the points. The distance is always taken as positive—or 0 if the points coincide. If the points P_1 and P_2 lie on the same horizontal line, we have observed that the directed distance between them is

$$x_2 - x_1,$$

4.2 Linear Functions

and if they lie on the same vertical line, then the directed distance is

$$y_2 - y_1.$$

If we are concerned only with distance and not direction, then these become

$$d = |x_2 - x_1| \quad \text{and} \quad d = |y_2 - y_1|.$$

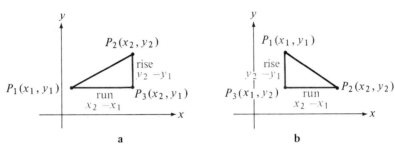

Figure 4.8

SLOPE OF A LINE SEGMENT
The inclination of a line segment joining two points can be measured by comparing the *rise* of the segment with a given *run*, as shown in Figure 4.8. The ratio of rise to run is called the **slope**, m, of the line segment. Since the rise is simply $y_2 - y_1$ and the run is $x_2 - x_1$, the slope of the line segment joining P_1 and P_2 is given by

$$m = \frac{y_2 - y_1}{x_2 - x_1}.$$

If P_2 is to the right of P_1, $x_2 - x_1$ will necessarily be positive, and the slope will be positive or negative as $y_2 - y_1$ is positive or negative. Thus, positive slope indicates that a line rises to the right; negative slope indicates that it falls to the right. Since

$$\frac{y_2 - y_1}{x_2 - x_1} = \frac{-(y_1 - y_2)}{-(x_1 - x_2)} = \frac{y_1 - y_2}{x_1 - x_2},$$

the restriction that P_2 be to the right of P_1 is not necessary, and the order in which the points are considered is immaterial in determining slope.

If a line segment is parallel to the x-axis, then $y_2 - y_1 = 0$, and the line has slope 0; but if it is parallel to the y-axis, then $x_2 - x_1 = 0$, and its slope is not defined. These two special cases are shown in Figure 4.9.

In the next section, we shall see how the slope concept is applied in discussing linear functions and their graphs.

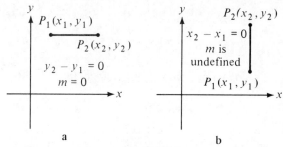

Figure 4.9

Linear functions occur in many applications.

REVENUE If an item sells at a constant price p then the total revenue obtained in selling x items is given by the linear function

$$R(x) = px.$$

COST The total cost to a company for producing x units of an item is given by

$$\text{cost} = \text{variable cost} + \text{fixed cost}.$$

A fixed cost is that cost to a company that remains constant regardless of how many units are produced (for example, the insurance that is paid on a building). A variable cost often depends directly on the number of items produced. If a is the cost per unit, then the variable cost for x units is ax. Thus, if b represents the fixed cost, the total cost C is given by the linear function

$$C(x) = ax + b.$$

SIMPLE INTEREST When a principal P is invested at a fixed rate r per period then the interest I obtained after t periods is

(4) $$I = Prt.$$

Interest calculated by (4) is said to be **simple interest**. The amount S received after t periods is given by the linear function of t

$$S = P + I$$
$$= P + Prt$$
$$= P(1 + rt).$$

MARINE BIOLOGY At the turn of the century, the weight W of a blue whale, measured in long tons, was given by the simple formula

$$W = L$$

where L was its length. For example, if the blue whale were 80 feet long then its weight was said to be 80 long tons. At the current time, the International Whaling

4.2 Linear Functions

Commission has specified that the weight of a mature blue whale be defined by the linear function

(5) $$W = 3.51L - 192$$

where L is its length, *provided* $L \geq 70$ feet.

Example From (5) we see that the weight of an 80-foot blue whale is given by

$$W = 3.51(80) - 192$$
$$= 88.8 \text{ long tons.}$$

For lengths $0 < L < 70$, blue whales are considered immature and (5) is not used. (It is of interest to note that at birth a blue whale is approximately 24 feet long.)

FAHRENHEIT/ CELSIUS The relationship between temperature T_f, measured in degrees Fahrenheit, and temperature T_c, measured in degrees Celsius, is linear:

(6) $$T_f = \frac{9}{5} T_c + 32$$

Example From (6) we see that 20°C is equivalent to

$$T_f = \frac{9}{5}(20) + 32$$
$$= 36 + 32$$
$$= 68°F.$$

PHYSIOLOGY In Problem 25 of Section 3.5 we saw that the arterial blood pressure P of an erect person is a linear function of the distance h (in centimeters) measured from the level of the heart:

(7) $$P = P_h - \rho g h / 1.33 \times 10^3.$$

P_h is the average blood pressure in an artery at heart level ($h = 0$) and is usually taken to be 100 torr; g is the acceleration of gravity (980 cm/sec²) and ρ is the density of blood (1.5 g/cm³). Above the level of the heart the variable h is taken to be positive; below heart level h is negative.

LIFE INSURANCE The growth of purchases of life insurance between 1973 and 1977 was closely approximated by a linear function. The amount A of life insurance in force, measured in trillions of dollars, was given by*

(8) $$A = 0.21(x - 1973) + 1.78.$$

In (8) the variable x represents a year starting with 1973.

*Adapted from the American Council of Life Insurance's "Life Insurance Fact Book."

EXERCISE 4.2

Graph.

Example $3x + 4y = 24$

Solution   Determine the intercepts.

If $x = 0$, then $y = 6$.

If $y = 0$, then $x = 8$.

Sketch the line through $(0, 6)$ and $(8, 0)$, as shown.

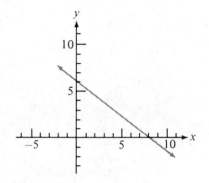

1. $y = 3x + 1$
2. $y = x - 5$
3. $y = -2x$
4. $2x + y = 3$
5. $3x - y = -2$
6. $3x = 2y$
7. $2x + 3y = 6$
8. $3x - 2y = 8$
9. $2x + 5y = 10$
10. $y = 5$
11. $x = -2$
12. $x = -3$

Example   Graph $f(x) = x - 1$. Represent $f(5)$ and $f(3)$ by drawing line segments from $(5, 0)$ to $(5, f(5))$ and from $(3, 0)$ to $(3, f(3))$.

Solution $f(5) = 5 - 1 = 4$

The ordinate at $x = 5$ is 4.

$f(3) = 3 - 1 = 2$

The ordinate at $x = 3$ is 2.

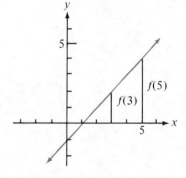

13. Plot the graph of $f(x) = 2x + 4$. Represent $f(0)$ and $f(4)$ by drawing line segments from $(0, 0)$ to $(0, f(0))$ and from $(4, 0)$ to $(4, f(4))$.

14. Plot the graph of $f(x) = 2x + 1$. Represent $f(3)$ and $f(-2)$ by drawing line segments from $(3, 0)$ to $(3, f(3))$ and from $(-2, 0)$ to $(-2, f(-2))$.

15. Suppose the function defined by $y = 2x + 1$ has as domain the set of real numbers between and including -1 and 1. Plot the graph of the function on a rectangular coordinate system. What is the range of the function?

16. Graph $x + y = 6$ and $5x - y = 0$ on the same set of axes. Estimate the coordinates of the point of intersection. What can you say about the coordinates of this point in relation to the two linear equations?

4.2 Linear Functions

Find the distance between each of the given pairs of points, and find the slope of the line segment joining them.

Example $(3, -5), (2, 4)$

Solution Consider $(3, -5)$ as P_1 and $(2, 4)$ as P_2.

$$d = \sqrt{(x_2 - x_1)^2 + (y_2 - y_1)^2}$$
$$= \sqrt{[2 - 3]^2 + [4 - (-5)]^2}$$
$$= \sqrt{1 + 81} = \sqrt{82}$$

$$m = \frac{y_2 - y_1}{x_2 - x_1}$$
$$= \frac{4 - (-5)}{2 - 3} = \frac{9}{-1} = -9$$

Distance, $\sqrt{82}$; slope, -9.

17. $(1, 1), (4, 5)$ **18.** $(-1, 1), (5, 9)$ **19.** $(-3, 2), (2, 14)$
20. $(-4, -3), (1, 9)$ **21.** $(2, 1), (1, 0)$ **22.** $(-3, 2), (0, 0)$
23. $(5, 4), (-1, 1)$ **24.** $(2, -3), (-2, -1)$ **25.** $(3, 5), (-2, 5)$
26. $(2, 0), (-2, 0)$ **27.** $(0, 5), (0, -5)$ **28.** $(-2, -5), (-2, 3)$

Find the lengths of the sides of the triangle having vertices as given.

29. $(10, 1), (3, 1), (5, 9)$ **30.** $(0, 6), (9, -6), (-3, 0)$
31. $(5, 6), (11, -2), (-10, -2)$ **32.** $(-1, 5), (8, -7), (4, 1)$

33. Show that the triangle described in Problem 30 is a right triangle. [*Hint:* Use the converse of the Pythagorean theorem.]

34. The two line segments with endpoints at $(0, -7)$, $(8, -5)$ and $(5, 7)$, $(8, -5)$ are perpendicular. Find the slope of each line segment. Compare the slopes. Do the same for the perpendicular line segments with endpoints at $(8, 0)$, $(6, 6)$ and $(-3, 3)$, $(6, 6)$. Can you make a conjecture about the slopes of perpendicular line segments?

35. The graph of a linear function contains the points $(2, -3)$ and $(6, -1)$. Find an equation that defines the function.

36. Determine algebraically whether the points lie on the same line.

  a. $(2, 7), (-2, -5), (0, 1)$ **b.** $(9, 5), (-3, -1), (0, 1)$

37. Show by similar triangles that the coordinates of the midpoint of the line segment joining the points $P_1(x_1, y_1)$ and $P_2(x_2, y_2)$ are given by

$$x = \frac{x_1 + x_2}{2} \quad \text{and} \quad y = \frac{y_1 + y_2}{2}.$$

38. Using the results of Problem 37, find the coordinates of the midpoint of the line segment joining the following points:

   a. (2, 4) and (6, 8) b. $(-4, 6)$ and $(6, -10)$

Applications

39. *Biology* Use the function given in (5) to find the weight of a 100-foot blue whale.
40. The weight of an adult blue whale determined by (5) is 115 long tons. What is its length?
41. Use the function given in (7) to find the arterial blood pressure in the head 50 centimeters above the heart level.
42. Use the function given in (7) to find the arterial blood pressure in a leg 140 centimeters below heart level.
43. The length L of a man's foot, measured in inches, is related to the size S of his shoe by the linear function $L = S/3 + 8$. What is the length of a man's foot if he wears a size 12 shoe?
44. Use the function in Problem 43 to find the size of a man's shoe if the length of his foot is 11 inches.
45. *Business* An item sells for $1500 per unit. What is a function giving the revenue from the sale of x units?
46. The total revenue obtained from the sale of 100 items is $42,000. What is a function giving the revenue from the sale of x units?
47. A small company's yearly fixed costs consist of $50,000 rent for a building, $25,000 for insurance, and $400,000 in salaries. If it costs $50 per unit to manufacture a tool, what is a function giving the total cost for making x units a year?
48. It is determined that the total cost to a company for manufacturing x units of a product is a linear function. Find the cost if it is known that the company's fixed cost is $100,000 and that the total cost for producing five units is $160,000.
49. A cost function, $C(x)$, in dollars, is given by $C(x) = 5x + 20$, where x is the number of units of a product that are made. How many units can be made at a total cost of $470?
50. A company's revenue and cost function for making and selling x units of a commodity are $R(x) = 300x$ and $C(x) = 20x + 5400$, respectively. What is the company's profit function? [*Hint*: Profit is revenue minus cost.]
51. A man lends $1000 to a friend at a simple interest rate of 5% per year. In other words, if the friend pays back the loan in 2 years, he then pays $1100. Express the amount that the friend has to pay back as a linear function of the number of years the loan is outstanding.
52. The number of units demanded of a commodity selling for x dollars per unit is given by

$$D(x) = -6x + 144.$$

   Find the following:

   a. $D(1)$, $D(5)$, $D(20)$ b. $D(x+1) - D(x)$

   c. The value of x for which $D(x) = 0$.

53. Use the linear function given in (8) to find the amount of life insurance in force in 1973. In 1977.
54. *Physics* Express the following temperatures in the Celsius scale.
   a. $-20°F$ b. $32°F$ c. $212°F$

4.3 Forms of Linear Equations

55. A news magazine recently reported that in a test of backup cooling systems in a nuclear reactor the temperature of the core jumped to 516°C after two of the cooling systems were purposely caused to fail. When a third cooling system was automatically engaged, the temperature of the core was reduced to 149°C. Express both of these temperatures in degrees Fahrenheit.

56. Temperature in degrees Kelvin (°K) is related to the Celsius scale by the simple linear function $T_k = T_c + 273.15$.

    a. Convert 40°C to the absolute scale.

    b. Convert 250°K to the Celsius scale.

    d. Absolute zero is defined as 0°K. What is this temperature in degrees Celsius?

57. According to the law of Guy-Lussac, the relationship between volume and temperature of a gas under constant pressure is given by the linear function $V = V_0(1 + \gamma t)$, where V_0 is the volume of the gas at 0°C and γ is a constant. The temperature t is measured in degrees Celsius. Find the volume occupied by helium at 100°C when $V_0 = 20$ cubic meters and $\gamma = 0.00366$.

4.3

FORMS OF LINEAR EQUATIONS

POINT-SLOPE FORM Assuming that the slope of the line segment joining any two points on a line does not depend on the points (as can be shown by considering similar triangles), consider a line in the plane with given slope m that passes through a given point (x_1, y_1), as shown in Figure 4.10. If we choose any other point on the line and assign to it the coordinates (x, y), it is evident that the slope of the line is given by

$$\frac{y - y_1}{x - x_1} = m,$$

from which

(1) $$y - y_1 = m(x - x_1).$$

Note that (1) is satisfied also by $(x, y) = (x_1, y_1)$. Since now x and y are the coordinates of *any* point on the line, (1) is an equation of the line passing through (x_1, y_1) with slope m. This is called the **point-slope form** for a linear equation.

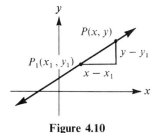

Figure 4.10

SLOPE-INTERCEPT FORM

Now consider the equation of the line with slope m passing through a given point on the y-axis having coordinates $(0, b)$, as shown in Figure 4.11. Substituting the components of $(0, b)$ in the point-slope form of a linear equation,

$$y - y_1 = m(x - x_1),$$

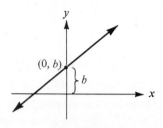

Figure 4.11

we obtain

$$y - b = m(x - 0),$$

from which

(2) $$y = mx + b.$$

Equation (2) is called the **slope-intercept form** for a linear equation. Any linear equation in standard form can be written equivalently in the slope-intercept form by solving for y in terms of x if $B \neq 0$. For example,

$$2x + 3y - 6 = 0$$

can be written equivalently as

$$y = -\frac{2}{3}x + 2.$$

The slope of the line, $-2/3$, and the y-intercept, 2, can now be read directly from the slope-intercept form of the equation.

INTERCEPT FORM

If the x- and y-intercepts of the graph of

(3) $$y = mx + b$$

are a and b $(a, b \neq 0)$, respectively, as shown in Figure 4.12, then the slope m is clearly equal to $-b/a$. Replacing m in (3) with $-b/a$, we have

$$y = -\frac{b}{a}x + b$$

$$ay = -bx + ab$$

$$bx + ay = ab,$$

4.3 Forms of Linear Equations

and multiplying each member by $1/ab$ produces

$$\frac{x}{a} + \frac{y}{b} = 1.$$

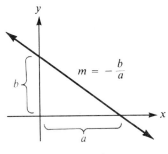

Figure 4.12

This is called the **intercept form** for a linear equation.

Any of the three forms discussed in this section may be used in working with linear functions and their graphs.

Example Find the equation, in standard form, of the line passing through $(3, -5)$ and having slope -2.

Solution Substitute given values in the point-slope form of the linear equation.

$$y - y_1 = m(x - x_1)$$
$$y - (-5) = -2(x - 3)$$
$$y + 5 = -2x + 6$$
$$2x + y - 1 = 0$$

Example When a certain commodity sells at a price of x dollars per unit, the number of units demanded, $D(x)$, is given by the linear equation

$$D(x) = -4x + 100.$$

Does the demand increase or decrease as the price increases?

Solution If we compute the value of $D(x)$ at particular values of x, such as

$$D(1) = 96$$
$$D(5) = 80$$
$$D(20) = 20,$$

we see that $D(x)$ decreases as x increases. Actually, it is easily seen that the function decreases by four units for each increase in x; this is because the slope is a negative number, -4.

Practically, we are not concerned with values of x for which $D(x) < 0$, so the demand decreases until $D(x) = 0$. The solution to this latter equation is, of course, the x-intercept, 25.

DEPRECIATION When a car is purchased, everyone is aware that the value of the car decreases, or **depreciates**, as time goes on. Also, when a company purchases equipment, such as office furniture, a machine, or a computer, the income tax laws allow the company to depreciate the purchase over a number of years. One form of depreciation is **straight-line**, or **linear depreciation**. An item loses *all* its initial worth of A dollars over a period of n years by an amount of A/n each year.* For example, if an item costs \$20,000 when it is new and it is depreciated linearly over 25 years, then

$$\frac{\$20{,}000}{25} = \$800$$

is the annual rate of depreciation. After x years the amount of depreciation is
$$800x,$$
so that if V is the value of the item at any time, we must have

$$V = 20{,}000 - 800x$$
$$= 20{,}000 - \frac{20{,}000}{25} x$$
$$= 20{,}000 \left(1 - \frac{x}{25}\right).$$

Here it only makes sense to consider $0 \le X \le 25$ since the item is worthless after 25 years. (Note, if x is chosen larger than 25, then y would be negative—which is impossible.) The graph of the above equation is the line segment shown in Figure 4.13.

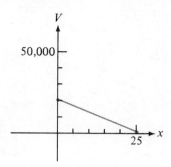

Figure 4.13

*At the end of n years an item may have some *salvage value*. See Problem 46.

4.3 Forms of Linear Equations

In general, if an item costs A dollars initially, then its subsequent value, when depreciated linearly over n years, is given by

$$V = A - \frac{A}{n}x$$

$$= A\left(1 - \frac{x}{n}\right) \qquad (0 \leq x \leq n).$$

The negative slope, $-A/n$, means that the value V decreases as x increases.

EXERCISE 4.3

Find the equation in standard form of the line that passes through each of the given points and has the given slope.

1. $(2, 1);\quad m = 4$
2. $(-2, 3);\quad m = 5$
3. $(5, 5);\quad m = -1$
4. $(-3, -2);\quad m = \frac{1}{2}$
5. $(0, 0);\quad m = 3$
6. $(-1, 0);\quad m = 1$
7. $(0, -1);\quad m = -\frac{1}{2}$
8. $(2, -1);\quad m = \frac{3}{4}$
9. $(-2, -2);\quad m = -\frac{3}{4}$
10. $(2, -3);\quad m = 0$
11. $(-4, 2);\quad m = 0$
12. $(-1, -2);\quad$ parallel to y-axis

Write each equation in slope-intercept form; specify the slope of the line and the y-intercept.

Example $2x - 3y = 5$

Solution Solve explicitly for y.

$$-3y = 5 - 2x$$
$$3y = 2x - 5$$
$$y = \frac{2}{3}x - \frac{5}{3}$$

Compare with the general slope-intercept form $y = mx + b$.

Slope, $2/3$; y-intercept, $-5/3$.

13. $x + y = 3$
14. $2x + y = -1$
15. $3x + 2y = 1$
16. $3x - y = 7$
17. $x - 3y = 2$
18. $2x - 3y = 0$
19. $4x + 6y + 12 = 0$
20. $-5x + 2y - 6 = 0$

Find the equation in standard form of the line with the given intercepts.

Example $x = 3$; $y = -\dfrac{1}{2}$

Solution Substitute 3 and $-1/2$ for a and b, respectively, in the intercept form $x/a + y/b = 1$.

$$\dfrac{x}{3} + \dfrac{y}{-1/2} = 1$$

$$x - 6y - 3 = 0$$

21. $x = 2$; $y = 3$
22. $x = 4$; $y = -1$
23. $x = -2$; $y = -5$
24. $x = -1$; $y = 7$
25. $x = -\dfrac{1}{2}$; $y = \dfrac{3}{2}$
26. $x = \dfrac{2}{3}$; $y = -\dfrac{3}{4}$

Write the equation in standard form of the line passing through the given point and parallel to the graph of the given equation. [*Hint: Parallel lines have the same slope.*]

27. $(2, 1)$; $y = 3x + 4$
28. $(-3, 2)$; $y = -4x + 9$
29. $(-2, 5)$; $2x - 3y = 6$
30. $(-5, -1)$; $5x - 4y = 7$
31. $(2, 3)$; $x = 6$
32. $(5, -4)$; $y = 2$
33. Show that for $x_2 \neq x_1$,

$$y - y_1 = \left(\dfrac{y_2 - y_1}{x_2 - x_1}\right)(x - x_1)$$

is an equation of the line joining the points (x_1, y_1) and (x_2, y_2). This is the **two-point form** of the linear equation.

34. Use the two-point form given in Problem 33 to find the equation of the line through the given points.

    a. $(2, 1)$ and $(-1, 3)$
    b. $(3, 0)$ and $(5, 0)$
    c. $(-2, 1)$ and $(3, -2)$
    d. $(-1, -1)$ and $(1, 1)$

35. Consider the linear function $y = f(x)$. Assuming that $(2, 3)$ and $(-4, 4)$ are known to be on the graph of f, find $f(x)$.

36. Find a number k such that the graph of $3kx + 4y = 18$ has x-intercept -1.

Applications

37. *Business* When a company makes 100 units of a product its costs are $25,000 and when it makes 600 units its costs are $50,000. If the total cost C is given by a linear function, what is the cost for manufacturing x units?

38. The demand D for a commodity is a linear function of its price x. Find D if the demand decreases by 40 for every $2 increase in price and $D = 0$ when $x = 1500$.

39. A can of beans cost 10 cents in 1966 and 25 cents in 1976. If the increase in cost is assumed to be a constant, determine the price of the beans in any year after 1966. What is the yearly increase in cost? What is the cost of the can of beans in 1986?

40. At LM University the tuition was $1600 a year in 1968 and $3250 in 1978. If the tuition is assumed to grow linearly, what would the tuition be in 1988?

4.3 Forms of Linear Equations

Example The purchase of a house is considered an investment since its value usually grows subsequently; that is to say, the value **appreciates**. If a house is purchased outright for $80,000 and if the rate of **appreciation** is 7% of its original cost per year, determine the equation that gives the value of the house for any year after the purchase.

Solution Let x be the number of years after the purchase of the house. The increase in value each year is

$$80{,}000(0.07) = 5600.$$

After x years the increase will be

$$5600x.$$

Thus, the subsequent value V of the house is given by

$$V = 5600x + 80{,}000.$$

41. A diamond that cost $100,000 when purchased is expected to appreciate at a rate of 12% per year. What is its expected value 20 years after its purchase?

42. A salesperson receives a salary of $200 per week and a 5.5% commission on all items that she sells during the week. What is her salary S if she sells x items in 1 week?

43. Suppose a sports car that initially costs $6000 depreciates linearly over its useful life span of 10 years. If x ($0 \leq x \leq 10$) is the number of years after its purchase, and V is the value of the car after any year, then express V as a function of x. When is the car worth $3600?

44. A printing company owns a press that cost $100,000 when new. If it is projected that the press has a useful life span of 25 years, what is the annual rate of linear depreciation? How much has the press depreciated in 15 years? What is the value of the press in 15 years?

45. A car that costs $5000 when new depreciates at a rate of 20% of its original value each year for the first 2 years, and then at a rate of 10% of its original value after the second year. Express the value of the car in terms of its age. What is the value of the car when it is 2 years old? 5 years old? At what age is the car worth nothing? [*Hint*: Express the function in two parts, one valid for $0 \leq x \leq 2$, and the other for $x > 2$.]

46. In straight-line depreciation the annual amount of depreciation can be defined as

$$\frac{\text{initial cost} - \text{salvage value}}{\text{estimated useful life}}.$$

A machine that cost $20,000 when new has a salvage value of $2000 in 10 years. Determine the value V of the machine at the end of x years. What is the value of the machine in 4 years?

47. *Physics* The relationship between temperature T_f on the Fahrenheit scale and temperature T_c on the Celsius scale is linear. Derive (6) of Section 4.2 if (0°C, 32°F) and (100°C, 212°F) are points on the graph of T_f.

4.4

QUADRATIC FUNCTIONS

GRAPH OF A QUADRATIC FUNCTION

Consider the quadratic equation in two variables,

(1) $$y = x^2 - 4.$$

As with linear equations in two variables, solutions of this equation must be ordered pairs (x, y). We need replacements for both x and y in order to obtain a statement we may judge to be true or false. As before, such ordered pairs can be found by arbitrarily assigning values to x and computing related values for y. For instance, assigning the value -3 to x in (1), we obtain

$$y = (-3)^2 - 4$$
$$y = 5,$$

and $(-3, 5)$ is a solution. Similarly, we find that

$$(-2, 0), \quad (-1, -3), \quad (0, -4), \quad (1, -3), \quad (2, 0), \quad (3, 5)$$

are also solutions of (1). If we locate the corresponding points on the plane, we have the graph shown in Figure 4.14, part a. Clearly, these points do not lie on a straight line, and we might reasonably inquire whether the graph of the solution set of (1), $\{(x, y) | y = x^2 - 4\}$, forms any kind of a meaningful pattern on the plane. By graphing additional solutions of (1)—solutions with x-components between those already found—we may be able to obtain a clearer picture. Accordingly, we find the solutions

$$\left(\frac{-5}{2}, \frac{9}{4}\right), \quad \left(\frac{-3}{2}, \frac{-7}{4}\right), \quad \left(\frac{-1}{2}, \frac{-15}{4}\right), \quad \left(\frac{1}{2}, \frac{-15}{4}\right), \quad \left(\frac{3}{2}, \frac{-7}{4}\right), \quad \left(\frac{5}{2}, \frac{9}{4}\right),$$

and by graphing these points in addition to those found earlier, we have the graph shown in Figure 4.14, part b. It now appears reasonable to connect these points in sequence, say from left to right, by a smooth curve as in Figure 4.15, and to assume that the resulting curve is a good approximation to the graph of (1).

QUADRATIC FUNCTION

Any quadratic equation of the form $y = ax^2 + bx + c$, where a, b, and c are real and $a \neq 0$, will determine only one y for each x and thus defines a function having as domain the entire set of real numbers and as range some subset of the real numbers. For example, we observe from the graph in Figure 4.15 that the range of the function defined by (1) is the set of real numbers $\{y | y \geq -4\}$. The general form of a **quadratic function** is then

(2) $$f(x) = ax^2 + bx + c, \quad a \neq 0.$$

The graph of a quadratic function is called a **parabola**.

A parabola that is the graph of a quadratic function (2) will have a lowest (minimum) point or a highest (maximum) point, depending on whether $a > 0$ or $a < 0$,

4.4 Quadratic Functions

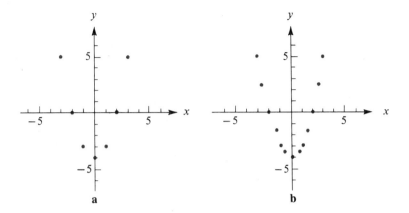

Figure 4.14

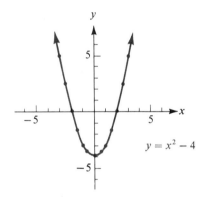

Figure 4.15

respectively. Such a point is called the **vertex** of the parabola. The line through the vertex and parallel to the y-axis is called the **axis of symmetry**, or simply the **axis**, of the parabola; it separates the parabola into two parts, each the mirror image of the other in the axis. If we observe that the graphs of

(3) $$y = ax^2 + bx + c$$

and

(4) $$y = ax^2 + bx$$

have the same axis (Figure 4.16), we can find an equation for the axis of (3) by inspecting (4). Factoring the right-hand member of $y = ax^2 + bx$ yields

(5) $$y = x(ax + b),$$

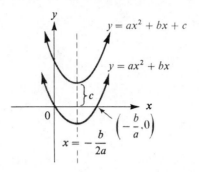

Figure 4.16

and thus we can see that 0 and $-b/a$ are the x-intercepts of the graph. Since the axis of symmetry bisects the segment with these endpoints, an equation for the axis of symmetry is

$$x = -\frac{b}{2a}.$$

Example Find an equation for the axis of symmetry of the graph of $y = 2x^2 - 5x + 7$.

Solution By comparing the given equation to $y = ax^2 + bx + c$, we see that $a = 2$ and $b = -5$. Hence, we have

$$x = -\frac{b}{2a} = -\frac{-5}{2(2)} = \frac{5}{4}.$$

and $x = 5/4$ is the desired equation for the axis.

Since the vertex of a parabola lies on its axis of symmetry, obtaining an equation for this axis will give us the x-coordinate of the vertex. The y-coordinate can easily be obtained by substitution in the equation for the parabola. The coordinates of the vertex are then

(6) $$\left(-\frac{b}{2a}, f\left(-\frac{b}{2a}\right)\right).$$

When graphing a quadratic function, it is desirable first to select components for the ordered pairs that ensure that the more significant parts of the graph are displayed. For a parabola, these parts include the intercepts, if they exist, and the maximum or minimum point on the curve.

Example Graph $y = x^2 - 3x - 4$.

4.4 Quadratic Functions

Solution When $x = 0$ we see that the y-intercept is -4. Setting $y = 0$, we have

$$0 = x^2 - 3x - 4 = (x - 4)(x + 1),$$

and the x-intercepts are 4 and -1. Since $a = 1 > 0$ the vertex will be a minimum point. The x-coordinate of the vertex is

$$x = -\frac{b}{2a} = -\frac{-3}{2(1)} = \frac{3}{2}.$$

By substituting $3/2$ for x in $y = x^2 - 3x - 4$, we obtain the y-coordinate of the vertex,

$$y = \left(\frac{3}{2}\right)^2 - 3\left(\frac{3}{2}\right) - 4 = \frac{9}{4} - \frac{9}{2} - 4 = -\frac{25}{4}.$$

Graphing the intercepts and the coordinates of the vertex and then sketching the curve produce the graph shown.

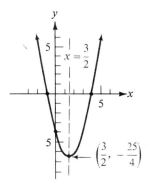

SOLUTIONS, ZEROS, AND x-INTERCEPTS

Consider the graph of the function

(7) $$f(x) = ax^2 + bx + c,$$

for $a \neq 0$, and the solution set of the equation

(8) $$ax^2 + bx + c = 0.$$

Any value of x for which $f(x) = 0$ in (7) will be a solution of (8). Since any point on the x-axis has y-coordinate zero [that is, $f(x) = 0$], the x-intercepts of the graph of (7) are the real solutions of (8). Values of x for which $f(x) = 0$ are called the **zeros of the function**. Thus, we have three different names for a single idea:

1. The *real numbers in the solution set* of the equation $ax^2 + bx + c = 0$. These are called the *solutions* or *roots* of the equation.
2. The *real zeros of the function* defined by $f(x) = ax^2 + bx + c$.
3. The *x-intercepts* of the graph of the equation $f(x) = ax^2 + bx + c$.

We recall from Chapter 3 that a quadratic equation may have no real solution, one real solution, or two real solutions. If the equation has no real solution, we find that the graph of the related quadratic equation in two variables does not touch the x-axis: if there is one real solution, the graph is tangent to the x-axis; if there are two real solutions, the graph crosses the x-axis in two distinct points. These cases are illustrated in Figure 4.17.

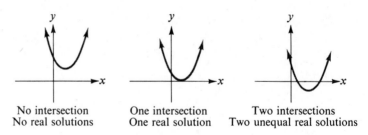

Figure 4.17

REVENUE In Section 4.2 we saw that when a product sells for p dollars per unit the **revenue** obtained through the sale of x units is

(9) $$R(x) = px.$$

In economics the price p itself often depends on the number of units sold in a linear fashion

(10) $$p = ax + b.$$

Substituting (10) in (9) then gives a quadratic revenue function

$$R(x) = (ax + b)x$$
$$= ax^2 + bx$$

identical with (5).

PROFIT If R and C are revenue and cost functions, respectively, then

$$P(x) = R(x) - C(x)$$

is called a **profit** function.

Example If $R(x) = -x^2 + 400x$ and $C(x) = 300x + 1000$, find the profit function and determine the maximum profit.

Solution The profit is given by the quadratic function

$$P(x) = (-x^2 + 400x) - (300x + 1000)$$
$$= -x^2 + 100x - 1000.$$

Since the coefficient of x^2 is negative, it follows that the graph of P will have a maximum point. From (6) the maximum will occur at $x = -100/(-2) = 50$. The maximum profit is then

$$P(50) = -(50)^2 + 100(50) - 1000$$
$$= 1500.$$

4.4 Quadratic Functions

FLOW OF BLOOD In 1842 the French physician Jean Louis Poiseuille discovered that the velocity v (in centimeters per second) of blood flowing in an artery whose cross section is a circle of fixed radius R is given by the quadratic function

(11) $$v(r) = k(R^2 - r^2), \qquad 0 \leq r \leq R,$$

where k is a positive constant. As shown in Figure 4.18 the variable r is the distance from the center of the artery out to a cylindrical layer or *lamina* of blood. When there is no turbulence or vorticing, the motion of the blood is said to be **laminar flow**. Observe that (11) implies $v(0) = kR^2$ and that $v(R) = 0$. In other words, the maximum rate of flow occurs at the center of the artery whereas the minimum velocity occurs at the outer wall of the artery.

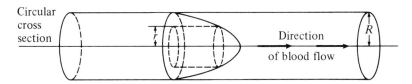

Figure 4.18

FALLING BODIES When an object such as a stone or ball is thrown with an initial velocity of v_0 ft/sec from a building s_0 feet high (Figure 4.19), the distance s above ground level at any time t is given by the quadratic function

$$s(t) = -\frac{1}{2}gt^2 + v_0 t + s_0$$

where $g = 32$ ft/sec^2.

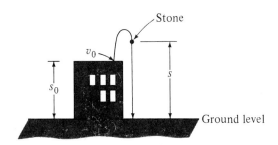

Figure 4.19

EXERCISE 4.4

Graph. (Obtain the intercepts and vertex, and then sketch the curve.)

Example $y = x^2 - 7x + 6$

Solution The y-intercept is clearly 6. Since the solutions of

$$x^2 - 7x + 6 = (x - 1)(x - 6) = 0$$

are 1 and 6, these are the x-intercepts. The axis of symmetry has the equation

$$x = -\frac{b}{2a} = \frac{7}{2}.$$

Writing 7/2 for x in $y = x^2 - 7x + 6$, we obtain

$$y = -\frac{25}{4}.$$

Hence, the vertex, in this case a minimum point, is $(7/2, -25/4)$. Using these points, the graph is sketched as shown.

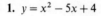

1. $y = x^2 - 5x + 4$
2. $y = x^2 - 3x + 2$
3. $y = x^2 - 6x - 7$
4. $y = x^2 - 3x + 2$
5. $f(x) = -x^2 + 5x - 4$
6. $f(x) = -x^2 - 8x + 9$
7. $g(x) = \frac{1}{2}x^2 + 2$
8. $g(x) = -\frac{1}{2}x^2 - 2$
9. $y = x^2 - 3x$
10. $y = 2x^2 + 4x$
11. Graph $f(x) = x^2 + 1$. Represent $f(0)$ and $f(4)$ by drawing line segments from $(0, 0)$ to $(0, f(0))$ and from $(4, 0)$ to $(4, f(4))$.
12. Graph $g(x) = x^2 + 1$. Represent $g(-3)$ and $g(2)$ by drawing line segments from $(-3, 0)$ to $(-3, g(-3))$ and from $(2, 0)$ to $(2, g(2))$.
13. On a single set of axes, sketch the family of four curves that are the graphs of

$$y = x^2 + k \qquad (k = -2, 0, 2, 4).$$

What effect does varying k have on the graph?

14. On a single set of axes, sketch the family of six curves that are the graphs of

$$y = kx^2 \qquad \left(k = \frac{1}{2}, 1, 2, -\frac{1}{2}, -1, -2\right).$$

What effect does varying k have on the graph?

15. Graph the relation $\{(x, y) | x = y^2\}$.

4.4 Quadratic Functions

a. What kind of curve is the graph?
b. Is the given relation a function? Why or why not?
c. Does the graph of this relation have a maximum or minimum point?

16. Graph the relation $\{(x, y) | x = y^2 - 2y\}$.
    a. What kind of curve is the graph?
    b. Is the given relation a function? Why or why not?
    c. Does the graph of this relation have a maximum or minimum point?

Graph each relation. (The graphs are parabolas.)

17. $\{(x, y) | x = y^2 - 4\}$
18. $\{(x, y) | x = y^2 - 2y - 3\}$
19. $\{(x, y) | x = y^2 - 4y + 4\}$
20. $\{(x, y) | x = 2y^2 + 3y - 2\}$

Applications

21. Find two numbers whose sum is 8 and whose product is as great as possible.
22. Find the maximum possible area of a rectangle with perimeter of 100 inches.
23. Find the minimum value of the squared distance between the point $(2, -1)$ and an arbitrary point (x, y) on the graph of $y = x$.
24. The area A of a circle is a quadratic function of the radius r: $A = \pi r^2$. Determine the increase in area if the radius of a circle increases from 1 centimeter to 3 centimeters.
25. *Biology* If the radius of an artery is 0.3 centimeter and $k = 1000$, what is the velocity of blood flowing at $r = 0.1$? At $r = 0.2$? What are the maximum and minimum rates of flow? Graph the velocity function $v(r)$.
26. *Business* Find the maximum revenue when $R(x) = x(-2x + 100)$.
27. Find the maximum revenue when $R(x) = -3x^2 + 600x$.
28. Find the maximum profit $R(x) = x(-2x + 100)$ and $C(x) = 20x + 80$.
29. Find the maximum profit when $R(x) = -x^2 + 500x$ and $C(x) = 100x + 20,000$.

Example A $80,000 house appreciates in value at the same rate for 2 years. Express the value of the house after 2 years as a function of the rate of appreciation.

Solution Let x be the rate of appreciation and y be the value of the house after 2 years. After 1 year the value of the house is

$$80,000x + 80,000.$$

Thus, after 2 years we have

$$y = (80,000x + 80,000)x + 80,000x + 80,000$$
$$= 80,000x^2 + 80,000x + 80,000x + 80,000$$
$$= 80,000(x^2 + 2x + 1)$$
$$= 80,000(x + 1)^2.$$

It is readily apparent that the graph of this parabola has its vertex at $(-1, 0)$.

30. A company makes open cardboard boxes from a square 10 inches on a side by cutting out a square from each corner and then turning up the sides. Express the area A of the base of the resulting box as a function of the length of the side of the square cut out of each corner.

31. A rancher wants to build a three-sided rectangular corral onto an existing side of a barn. The rancher has only $150 to spend on lumber for fencing, and wants to fence in the maximum area. If the fence costs $2 per foot, determine the function that expresses the area A of the corral as a function of the length of one side.

32. A deposit of $2500 draws compound interest at the same annual rate for 2 years. Express the value of the account after 2 years as a function of the annual rate of interest.

33. An airline company has a fleet of planes, with a total of 1500 seats, which fly daily between Los Angeles and New York. When the fares are set so that the company makes a profit of $100 on each fare, the planes fly completely full. Over the years the company observes that for each $5 increase in profit by increasing the fare rate, it loses 50 passengers each day. Express the company's daily profit P as a function of the number of fare increases.

34. A $5000 car depreciates in value at the same rate $-r, r > 0$, for 2 years. Express the value V of the car after 2 years as a function of the rate of depreciation.

35. An open box, in the form of a cube, is to be constructed from two different materials. The material for the sides costs 1.5 cents per square inch and the material for the bottom of the box costs 2 cents per square inch. Find the total cost C of construction as a function of the length of side.

36. A park, 300 by 400 feet, is to be surrounded by a walkway of uniform width. Express the total area A of the walkway as a function of its width.

37. *Physics* The distance above ground level, in feet, of a ball thrown up from a rooftop is given by $s(t) = -16t^2 + 80t + 224$. Evaluate $s(0)$, $s(5)$, and $s(7)$, and interpret the results.

38. The distance above ground level, in feet, of a ball thrown up from a rooftop is given by $s(t) = -16t^2 + 64t + 100$. Find the maximum height that the ball will attain.

4.5

POLYNOMIAL FUNCTIONS

GRAPHS OF POLYNOMIAL FUNCTIONS

In Section 4.2, we graphed linear functions

$$f(x) = a_0 x + a_1;$$

and in Section 4.4, we graphed quadratic functions

$$f(x) = a_0 x^2 + a_1 x + a_2.$$

We can graph any real polynomial function

$$P(x) = a_0 x^n + a_1 x^{n-1} + \cdots + a_n$$

by similar methods—that is, by combining the plotting of points with a consideration of certain general properties of the defining equations. In the case of the general polynomial equation, we shall lean more heavily on the use of plotted points. There is one fact about polynomial equations, however, that can be useful. This involves

4.5 Polynomial Functions

turning points, or **local maximum** and **minimum** values of y. For example, in Figure 4.20, part a, there is one local maximum as well as one local minimum, or a total of two turning points. In Figure 4.20, part c, there are one local maximum and two local minima, for a total of three turning points. In general, we have the following result.

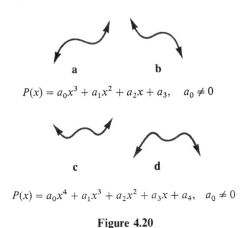

$$P(x) = a_0 x^3 + a_1 x^2 + a_2 x + a_3, \quad a_0 \neq 0$$

$$P(x) = a_0 x^4 + a_1 x^3 + a_2 x^2 + a_3 x + a_4, \quad a_0 \neq 0$$

Figure 4.20

The graph of a polynomial function of degree n.

$$P(x) = a_0 x^n + a_1 x^{n-1} + \cdots + a_n, \quad a_0 \neq 0,$$

is a smooth curve that has at most $n - 1$ turning points.

As a consequence of this result the graphs of third- and fourth-degree polynomial functions might appear as in Figure 4.20.

GENERAL FORM OF A POLYNOMIAL GRAPH To ascertain whether a graph ultimately goes up to the right, taking the general form shown in Figure 4.20, parts a and c, rather than the form shown in parts b and d, we can examine the leading coefficient, a_0, of the right-hand member of the defining equation. If $a_0 > 0$, then we can expect the graph to go up at the right in a form similar to parts a or c: if $a_0 < 0$, we can expect the graph to go down at the right similar to parts b or d. If $a_0 > 0$, then the graph ultimately goes down to the left, as in parts a and d, or up to the left, as in parts b and c, according as n is odd or even. In each case, then, the leading term $a_0 x^n$ governs the behavior of the graph of the polynomial function for large values of $|x|$.

For the actual graphing process, we can obtain ordered pairs $(x, f(x))$ for any polynomial function by direct substitution.

Example Graph $P(x) = 2x^3 + 13x^2 + 6x$.

Solution Since P is a cubic polynomial function with positive leading coefficient, we expect to find a graph of form similar to that in Figure 4.20, part a. Since we do not know where we should look for turning points, let us start with $x = 0$. By inspection, we have $P(0) = 0$, so the graph includes the origin. For $x = 1$, $x = 2$, and $x = -1$ we have, respectively, $P(1) = 21$, $P(2) = 80$, and $P(-1) = 5$. Thus, $(1, 21)$, $(2, 80)$, and $(-1, 5)$ are points on the graph. In similar fashion, we find that the following points are also on the graph:

$$(-2, 24), \quad (-3, 45), \quad (-4, 56), \quad (-5, 45), \quad (-6, 0).$$

The graphs of the nine ordered pairs are shown in part a of the accompanying figure. These points make the general appearance of the graph clear, and there remains only the question of whether the function has a zero between -1 and 0. If we let x have values $-3/4$, $-1/4$, and $-1/2$, we obtain the additional pairs $(-3/4, 63/32)$, $(-1/4, -23/32)$, $(-1/2, 0)$. Then for $-1/2 < x < 0$, we have $P(x) < 0$, and the graph can be sketched as shown in part b.

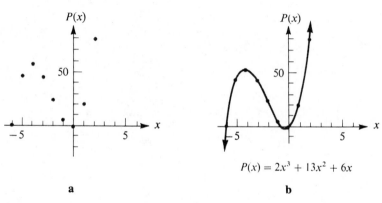

a b

Although the graph of a cubic polynomial function can have at most two turning points, the next example shows that it may not have any.

Example Graph $P(x) = x^3$.

Solution For $x = -2$, $x = -1$, $x = 0$, $x = 1$, $x = 2$, we obtain respectively, the following points on the graph

$$(-2, -8), (-1, -1), (0, 0), (1, 1), (2, 8).$$

Connect these points with a smooth graph as shown in the figure.

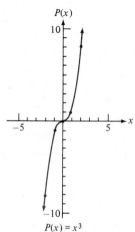

4.6 Special Functions and Variation

EXERCISE 4.5

Graph. Use enough of the domain to include all turning points.

1. $P(x) = x^3 + 1$
2. $P(x) = 1 - x^3$
3. $P(x) = x^3 - 4x^2 + 3x$
4. $P(x) = x^3 - 2x^2 + 1$
5. $P(x) = 2x^3 + 9x^2 + 7x - 6$
6. $P(x) = 3x^3 + 2x^2 - x + 1$
7. $P(x) = x^4$
8. $P(x) = -x^4 + x$
9. $P(x) = x^4 - x^3 - 2x^2 + 3x - 3$
10. $P(x) = x^4 - 4x^3 - 4x + 12$

Applications

11. The volume V of a sphere is a cubic function of its radius r: $V = \frac{4}{3}\pi r^3$. Determine the increase in volume if the radius of a sphere increases from 1 centimeter to 3 centimeters.

12. *Biology* Poiseuille's law states that the volume of blood F flowing through a small artery of radius R per unit time is given by the fourth-degree function $F = (\pi k/8)R^4$ where k is a positive constant. Determine the effect on the volume flow of blood per unit time if the radius of the artery is constricted from 1 to 0.5 centimeter.

13. *Business* A company makes open cardboard boxes from a square 10 inches on a side by cutting a square from each corner and then turning up the sides. Express the volume V of the resulting box as a function of the length of the side of the square cut out of each corner.

14. A container is constructed in the form of a right circular cylinder whose height is twice its radius. Express the volume of the container as a function of its height.

4.6
SPECIAL FUNCTIONS AND VARIATION

ABSOLUTE-VALUE FUNCTIONS

Functions involving the absolute value of one or both of the variables are not only useful in more advanced courses in mathematics, they also offer interesting properties in their own right. We recall that

$$|x| = \begin{cases} x & \text{if } x \geq 0, \\ -x & \text{if } x < 0. \end{cases}$$

Now consider the function defined by

(1) $\qquad y = |x|.$

From the definition of $|x|$, for nonnegative x we have

(2) $\qquad y = x,$

and for negative x,

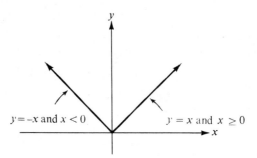

Figure 4.21

(3) $$y = -x.$$

If we graph (2) and (3) on the same set of axes, we have the graph of $y = |x|$ shown in Figure 4.21. In the case of any equation involving $|x|$ or $|f(x)|$, we can always plot individual points to deduce the graph. For instance, if

(4) $$y = |x| + 1,$$

we can find solutions by assigning values to x and computing values for y. Some solutions of (4) are

$$(-2, 3), \quad (-1, 2), \quad (0, 1), \quad (1, 2), \quad (2, 3),$$

which can be graphed as shown in Figure 4.22, part a. The graph of $y = |x| + 1$ appears in Figure 4.22, part b. As an alternative approach, the definition of $|x|$ implies that

$$y = |x| + 1$$

is equivalent to

$$y = x + 1 \quad \text{for } x \geq 0,$$
$$y = -x + 1 \quad \text{for } x < 0,$$

and these can be graphed separately over the specified domains.

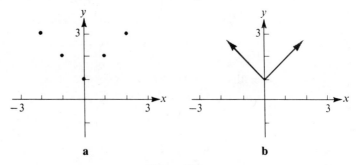

Figure 4.22

4.6 Special Functions and Variation

RELATED GRAPHS

It is usually advisable, where possible, to avoid graphing many points. For instance, comparing equations (1) and (4),

(1) $$y = |x|$$

and

(4) $$y = |x| + 1,$$

we observe that for each x the ordinate in (4) is one unit greater than that in (1); consequently, the graph of (4) is simply the graph of (1) with each ordinate increased by 1. Whenever we can, we should use such considerations to help us graph equations.

BRACKET FUNCTION

Another interesting function (sometimes called the **bracket function**) is defined by the equation

(5) $$f(x) = [x],$$

where the brackets denote "the greatest integer not greater than." Thus,

$$[2] = 2 \qquad \left[\frac{7}{4}\right] = 1 \qquad [-2] = -2 \qquad \left[\frac{-3}{2}\right] = -2 \qquad \left[-\frac{5}{2}\right] = -3.$$

To graph (5), we consider unit intervals along the x-axis. If $0 \leq x < 1$, $[x]$ is 0, since the greatest integer contained in any number between 0 and 1 is 0. Similarly, if $1 \leq x < 2$, $[x]$ is 1; for $-2 \leq x < -1$, $[x]$ is -2; etc. The graph of (5), therefore, is as shown in Figure 4.23. The heavy dot on the left-hand endpoint of each line segment indicates that the endpoint is a part of the graph. The function defined by (5) is sometimes called a **step function**, for obvious reasons.

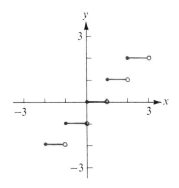

Figure 4.23

POWER FUNCTION

It can be shown that for any real number a and any real number n

(6) $$f(x) = ax^n$$

is a function. Equation (6), called the **power function**, reduces to a polynomial function when n is a nonnegative integer. For example, when $n = 2$, (6) is a quadratic function whose graph is a parabola having vertex at the origin. The graph of (6), when $a = 1$ and $n = 3$, is given on page 144. For $a = 1$ and $n = -1$ and $n = 1/2$, the power functions are $y = x^{-1} = 1/x$ and $y = x^{1/2} = \sqrt{x}$, respectively. The graphs of these two functions are given in Figure 4.24.

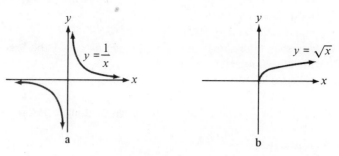

Figure 4.24

PHYSIOLOGY

In the analysis of walking, the leg is considered a physical pendulum. Assuming that the center of mass is located at the center of the leg, it can be shown that the period of oscillation is given by

(7) $$T = 2\pi \sqrt{\frac{2}{3}\frac{L}{g}}$$

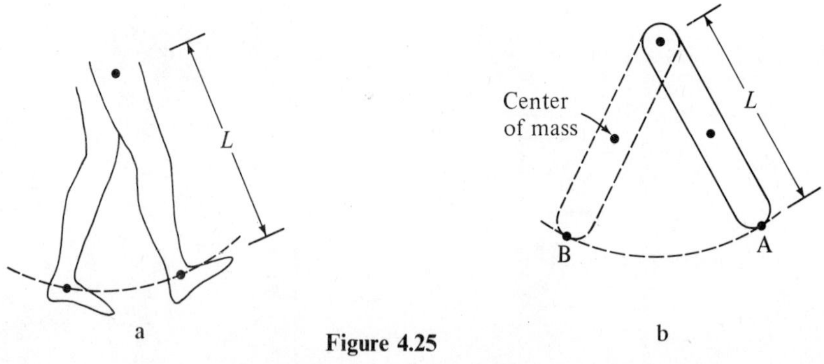

Figure 4.25

where g is the constant acceleration of gravity. We note that since T can be written as

$$T = 2\pi \sqrt{\frac{2}{3g}} \sqrt{L},$$

(7) is a power function having the form $y = a\sqrt{x}$. As shown in Figure 4.25, part b, T is

4.6 Special Functions and Variation

the total time it takes the swinging pendulum to move from B to A and then from A back to B.

VARIATION When $k > 0$ and $n > 0$ in (6) we say that y varies **directly** as the nth power of x. In the form $y = k/x^n$, $k > 0$, $n > 0$ (or $y = kx^m$, $m = -n < 0$) we say that y varies **inversely** as a power of x. Another way of describing direct and inverse variation is to say that "y is directly proportional to x^n" and "y is inversely proportional to x^n."

Examples of power function occur in many diverse sciences. In biology it is assumed that the volume of a body, and thus its weight W, is proportional to the third power of length s; that is, $W = ks^3$. Under this latter assumption it can be shown that the surface area S of an animal is related to its weight by $S = cW^{2/3}$, where c is a constant. In physics the intensity I of sound is inversely proportional to the square of the distance d from the source: $I = k/d^2$. In operations management the number of man-hours M needed to produce x units of a commodity is often predicted by the power function $M = kx^n$.

Example A building contractor knows that the total cost, C, for building a house varies directly with the cost, x, of the material used. If the total cost for construction of a house is $14,000 when the cost of material is $8000, what would be the cost of the house if the material is $9000?

Solution We use $C = kx$ and $C = 14,000$ when $x = 8000$ to determine k. We have

$$14{,}000 = k(8000)$$

$$k = \frac{14}{8}$$

$$= \frac{7}{8}$$

$$= 1.75.$$

Thus, with $C = 1.75x$ and $x = 9000$,

$$C = 1.75(9000)$$

$$= \$15{,}750.$$

EXERCISE 4.6

Graph the function defined by the given equation over the domain $\{x \mid -5 \leq x \leq 5\}$.

1. $y = |x| + 2$
2. $y = -|x| + 3$
3. $f(x) = |x + 1|$
4. $F(x) = |x - 2|$
5. $y = -|2x - 1|$
6. $y = |3x + 2|$

7. $g(x) = |2x| - 3$
8. $y = |3x| + 2$
9. $y = |3x| - |x|$
10. $f(x) = |2x| + |x|$
11. $y = 3|x| - x$
12. $y = -2|x| + x$
13. $H(x) = |x^2|$
14. $y = |x|^2$
15. $y = |x + 1| - |x|$
16. $g(x) = |x + 1| - x$
17. $y = [x]$
18. $f(x) = [x] - 1$
19. $y = [x + 1]$
20. $y = [2x]$
21. $F(x) = [x] + x$
22. $y = \left[\dfrac{1}{2}x\right] + x$
23. $y = [x] - x$
24. $y = |[x]|$

Graph the given power function.

25. $y = -x^{1/2}$
26. $y = x^{1/3}$
27. $y = x^{2/3}$
28. $y = -\dfrac{1}{x}$
29. $y = \dfrac{1}{x^2}$
30. $y = \dfrac{1}{x^3}$

31. If y varies directly as x^2, and $y = 9$ when $x = 3$, find y when $x = 4$.
32. If y varies inversely as x^3, and $y = 4$ when $x = 1/2$, find y when $x = -1$.

Applications

33. *Biology* Use (7) to find the period of a swinging leg 84 centimeters long. Use a calculator and the fact that $g = 980$ cm/sec².

34. The kinetic energy K of a body moving at velocity v is given by the power function $K = \dfrac{1}{2}mv^2$, where m represents mass in kilograms. Find the kinetic energy of 0.098 kilogram of blood ejected from the heart (during the systole or contraction stage) at a rate of 0.15 m/sec into an aorta.

35. The weight W of a person is directly proportional to the cube of the person's length s. Find the constant of proportionality for a person 60 inches tall who weighs 120 pounds. Find the weight when the same person attains the height of 72 inches.

36. The surface area S (in square meters) of an animal can be approximated from its weight W (in kilograms) by $S = kW^{2/3}$, where k is a constant of proportionality depending on the animal. For humans, k is usually taken to be 0.11. Deduce the surface area of a person whose weight is 64 kilograms.

37. *Business* The postage on a letter sent by first-class mail is c cents per ounce or fraction thereof. Write an equation relating the cost, C, of mailing a letter and the weight of the letter in ounces, x.

38. Profit, P, varies directly with the total number, N, of sales. Find P when $N = 7$ if $P = 10$ when $N = 2$.

39. The demand, D, for a commodity varies inversely as $\sqrt{x}$, where x is the price of the item. Find D when $x = 0.0001$ if $D = 5$ when $x = 0.01$.

40. A manufacturer determines that the demand D for a commodity varies inversely as $\sqrt{x+3}$, where x represents the price of the item. If the demand is 500 units when its price is one unit, what is the demand when the price is six units?

41. The cost, C, to a manufacturer for making x units varies directly as $100 + 4x$. Find C when $x = 50$ if $C = 600$ when $x = 25$.

4.7 Inverse Relations and Functions

42. *Physics* The distance a body falls under the influence of gravity varies directly as the square of time. If a body falls 16 feet in 1 second, how far will it fall in 4 seconds?

43. The period T of an ordinary plane pendulum of length L (such as the pendulum in a grandfather clock) is given by $T = 2\pi \sqrt{L/g}$. How much must the length increase in order to double the period of a pendulum?

44. *Social Science* The number of man-hours M needed to assemble x units of a product is given by $M = 200x^{0.341}$. Find M when $x = 16$. Use a calculator.

45. According to a theory advanced by G. K. Zipf, an approximation to the number N of cities in the United States that have populations in excess of x million is given by the formula $N = 24x^{-1.43}$. Estimate the number of cities with a population of over 1 million. Of over 2 million. Use a calculator.

4.7

INVERSE RELATIONS AND FUNCTIONS

INVERSE RELATIONS

If the two components of each ordered pair in a relation r are interchanged, then the resulting relation is called the **inverse relation of r** and is denoted by r^{-1}. For example, the relations

$$r = \{(1, 2), (3, 4), (5, 5)\}$$

and

$$r^{-1} = \{(2, 1), (4, 3), (5, 5)\}$$

are inverses of each other.

INVERSE FUNCTIONS

Recall that, a function is a set of ordered pairs (x, y) such that no two have the same first components and different second components. When the components of every ordered pair in a function f are interchanged, the resulting relation may or may not be a function. For example, if

(1, 5),   (2, 5),   (3, 6)

are ordered pairs in f, then

(5, 1),   (5, 2),   (6, 3)

are members of the relation formed by interchanging the components of these ordered pairs. Clearly, these latter pairs cannot be members of a function since two of them, (5, 1) and (5, 2), have the same first component and different second components. If, however, a function f is **one-to-one**, that is, if no two different ordered pairs in f have the same second component (same value for y), then the relation obtained by interchanging the first and second components of every pair in the function will also be a function. This function is called the **inverse function of** f, which is usually denoted by f^{-1} (read "f inverse").

> If the function f is such that no two of its ordered pairs with different first components have the same second component, then the **inverse function** f^{-1} is the set of ordered pairs obtained from f by interchanging the first and second components of each ordered pair in f.

It is evident from this definition that the domain and range of f^{-1} are just the range and domain, respectively, of f. If $y = f(x)$ defines a function f, and if f is one-to-one, then $x = f(y)$ defines the inverse of f. For example, the inverse of the one-to-one function defined by

(1) $$y = 3x + 2$$

is defined by

(2) $$x = 3y + 2,$$

or, when y is expressed in terms of x, by

(3) $$y = \frac{1}{3}x - \frac{2}{3}.$$

Equations (2) and (3) are equivalent. In general, the equation $x = f(y)$ is equivalent to $y = f^{-1}(x)$.

GRAPHS The graphs of inverse relations are related in an interesting way. To see this, we first observe in Figure 4.26 that the graphs of the ordered pairs (a, b) and (b, a) are always located symmetrically with respect to the graph of $y = x$. Therefore, because for every ordered pair (a, b) in a one-to-one function f the ordered pair (b, a) is in f^{-1}, the graphs of $y = f^{-1}(x)$ and $y = f(x)$ are reflections of each other about the graph of $y = x$.

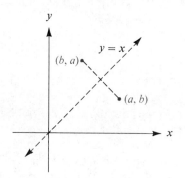

Figure 4.26

4.7 Inverse Relations and Functions

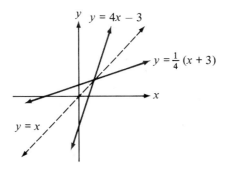

Figure 4.27

Figure 4.27 shows the graphs of the one-to-one function $y = f(x)$ defined by

(4) $$y = 4x - 3.$$

and the inverse function $y = f^{-1}(x)$ defined by

(5) $$y = \frac{1}{4}(x + 3),$$

together with the graph of $y = x$.

Since an element in the domain of the inverse f^{-1} of a one-to-one function f is the range element in the corresponding ordered pair of f, and vice versa, it follows that for every x in the domain of f,

$$f^{-1}[f(x)] = x$$

(read "f inverse of f of x is equal to x"), and, for every x in the domain of f^{-1},

$$f[f^{-1}(x)] = x$$

(read "f of f inverse of x is equal to x"). Using (4) and (5) as an example, we have

$$f(x) = 4x - 3,$$

and

$$f^{-1}(x) = \frac{1}{4}(x + 3).$$

It follows that

$$f^{-1}[f(x)] = \frac{1}{4}[(4x - 3) + 3] = x$$

and

$$f[f^{-1}(x)] = 4\left[\frac{1}{4}(x + 3)\right] - 3 = x.$$

We observe that every linear function of the form $f(x) = ax + b$, $a \neq 0$, is one-to-one and hence possesses an inverse function. Also, it should be noted that because every function is a relation, every function has an inverse relation, but this inverse is not always a function. For example, the function $y = f(x)$ defined by

$$y = x^2$$

possesses the inverse relation

$$\{(x, y) | x = y^2\} = \{(x, y) | y = \pm \sqrt{x}\},$$

Since the inverse relation associates two different y's with each x for all but one value in its domain, this inverse is not a function.

EXERCISE 4.7

In Problems 1–12 graph each given function and its inverse function using the same set of axes.

Example $y = -\dfrac{1}{3}x + 2$

Solution Interchange the variables in the defining equation.

$$x = -\frac{1}{3}y + 2.$$

Solve for y.

$$y = -3x + 6.$$

Graph both equations.

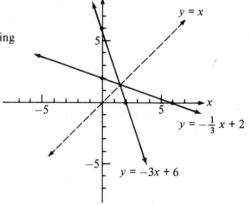

1. $f = \{(-2, 3), (4, 7), (5, 9)\}$
2. $f = \{(-3, 1), (2, -1), (3, 4)\}$
3. $f = \{(-1, -1), (2, 2), (3, 3)\}$
4. $f = \{(-4, 4), (0, 0), (4, -4)\}$
5. $y = 2x + 6$
6. $y = 3x - 6$
7. $y = 4 - 2x$
8. $y = 6 + 3x$
9. $3x - 4y = 12$
10. $x - 6y = 6$
11. $4x + y = 4$
12. $2x - 3y = 12$

4.7 Inverse Relations and Functions

In Problems 13–18 each function is one-to-one. Find $f^{-1}(x)$.

13. $f(x) = x^3$
14. $f(x) = x^3 + 1$
15. $f(x) = 2x^3 + 4$
16. $f(x) = -4x^3 + 9$
17. $f(x) = \sqrt[3]{x} + 2$
18. $f(x) = 5\sqrt[3]{x} - 7$

In Problems 19–26 each equation defines a one-to-one function. Find an equation defining f^{-1} and show that $f[f^{-1}(x)] = f^{-1}[f(x)] = x$.

19. $y = x$
20. $y = -x$
21. $2x + y = 4$
22. $x - 2y = 4$
23. $3x - 4y = 12$
24. $3x + 4y = 12$
25. $y = \sqrt{x + 3}$
26. $y = \sqrt{5 - 2x}$

CHAPTER TEST

Specify the domain and the range of each relation.

1. $\{(4, -1), (2, -4), (3, -5)\}$
2. $\{(1, 2), (1, 3), (1, 6)\}$
3. $\left\{(x, y) \mid y = \dfrac{1}{x+4}\right\}$
4. $\{(x, y) \mid y = \sqrt{x-6}\}$

Let $f(x) = x - 3$ and $g(x) = x^2 + 4$. Find each of the following.

5. $g(3)$
6. $f(-2)$
7. $f(g(0))$
8. $g(x + h)$

Find the distance between each of the given pairs of points, and find the slope of the line segment joining them.

9. $(2, 0)$ and $(-3, 4)$
10. $(-6, 1)$ and $(-8, 2)$

Find the equation, in standard form, of the line passing through each of the given points and having the given slope.

11. $(2, -7); m = 4$
12. $(-6, 3); m = \dfrac{1}{2}$

Write each equation in slope-intercept form; specify the slope and the y-intercept of the line.

13. $4x + y = 6$
14. $3x - 2y = 16$

Find the equation in standard form of the line with the given intercepts.

15. $x = -3; y = 2$
16. $x = \dfrac{1}{3}; y = -4$

17. Before a particular computer could be manufactured, it cost the IBC company $5 million in research and development. The cost to manufacture one of these computers is estimated to be $15,000. What is the total cost $C(x)$ to the company for producing x computers?

Chapter Test

18. The value of a machine that is depreciated linearly over 15 years is given by $V = 50{,}000(1 - x/20)$. What is the initial cost of the machine? What is its value after 8 years? What is its salvage value (value after 15 years)?

Find the x-intercepts, the axis of symmetry, vertex, and graph of each function.

19. $f(x) = x^2 - x - 6$ 20. $f(x) = -x^2 + 7x - 10$

21. A company finds that the revenue obtained through the sale of x items is given by $R(x) = x(1000 - x/10)$. Determine the sales level that will yield the maximum revenue. What is the maximum revenue?

22. A projectile is shot upward from ground level. Its distance above the ground, measured in feet, is given by $s(t) = -16t^2 + 160t$. Determine the maximum height attained by the projectile. At what time will the projectile hit the ground?

23. Find ordered pairs in the function P defined by $P(x) = x^3 + 4x^2 + x - 6$ for the following replacements for x: $-4, -3, -2, -1, 0, 1, 2, 3, 4$.

24. Graph the function in Problem 23.

Graph the function defined by the given equation over the domain $\{x \mid -5 \leq x \leq 5\}$.

25. $f(x) = |x| + 4$ 26. $g(x) = |x - 4|$

27. $g(x) = |x| + 1$ 28. $h(x) = |x - 1|$

29. If y varies directly as x^2, and $y = 20$ when $x = 2$, find y when $x = 5$.

30. The number of posts needed to string a telephone line over a given distance varies inversely as the distance between posts. If it takes 80 posts separated by 120 feet to string a wire between two points, how many posts would be required if the posts were 150 feet apart?

31. The supply S of a commodity available on the market at a price x varies directly as $2x + 1$. Find S when $x = 4$ if $S = 25$ when $x = 2$.

32. The demand D of a commodity varies inversely as $x + 4$, where x represents price. If $D = 10$ when $x = 1$, what is the price corresponding to a demand of $D = 2$?

Graph each function and its inverse function, using the same set of axes.

33. $f = \{(-3, 1), (-1, 3), (2, 4)\}$ 34. $f(x) = 2x - 6$

5
EXPONENTIAL AND LOGARITHMIC FUNCTIONS

5.1
THE EXPONENTIAL FUNCTION

Powers of the form b^x, where $b > 0$, and $b \neq 1$, can be used to define functions. Notice that the **base** b must be restricted to positive values to ensure that b^x be real for all rational numbers x; for example, $(-1)^{1/2}$ is not a real number. Now, since it can be shown that for each real x there is one and only one number b^x, the equation

(1) $$f(x) = b^x \qquad (b > 0)$$

defines a function. Because $1^x = 1$ for all real x, (1) defines a constant function if $b = 1$. If $b \neq 1$, we say that (1) defines an **exponential function**.

Exponential functions can be visualized most clearly by considering their graphs. We illustrate two typical examples, in which $0 < b < 1$ and $b > 1$, respectively.

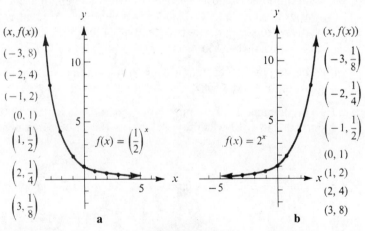

Figure 5.1

158

5.1 The Exponential Function

Assigning values to x in the equations

$$f(x) = \left(\frac{1}{2}\right)^x \quad \text{and} \quad f(x) = 2^x,$$

we find some ordered pairs in each solution set and sketch the graphs as shown in parts a and b of Figure 5.1.

INCREASING AND DECREASING FUNCTIONS

Notice that the graph of the function determined by $f(x) = (1/2)^x$ goes *down* to the right, and the graph of the function determined by $f(x) = 2^x$ goes *up* to the right. For this reason, we say that the former function is a **decreasing function** and that the latter is an **increasing function**. In general, (1) is a decreasing function for $0 < b < 1$, and an increasing function for $b > 1$. In either case, the domain of an exponential function is the set of real numbers and the range is the set of positive real numbers.

Observe that by the laws of exponents the function $f(x) = \left(\frac{1}{2}\right)^x$ can be written in an alternative form:

$$f(x) = \left(\frac{1}{2}\right)^x$$
$$= (2^{-1})^x$$
$$= 2^{-x}.$$

THE NUMBER e

Many applications involve an exponential function with the irrational number.

$$2.7182818\ldots$$

as its base. This number occurs so often that it is denoted by the letter e. Since $e > 1$ and $1/e < 1$, the graphs of $y = e^x$ and $y = e^{-x}$ are similar to those given in Figure 5.1, parts a and b, respectively. The values of e^x and e^{-x} can be obtained from a scientific calculator or from Table II on page 552.

INTEREST

In Section 11.4 we shall see that when an amount of money P is invested at a yearly rate of interest r compounded continuously, the return S after m years is given by

(2) $$S = Pe^{rm}.$$

S is called the **future value** of P.

BEHAVIOR

The graph of the function

(3) $$f(x) = a - ae^{-kx}, \quad x \geq 0$$

where a and k are positive constants, is sometimes called a **learning curve**. For example, $f(x)$ could represent the strength of a habit acquired after x repetitions of an act. Observe that for $k > 0$, e^{-kx} decreases as x increases, hence we conclude that (3) is an increasing function; in other words, the strength of a habit increases with repetition.

Example The number of motors installed each day by a new automobile assembly worker after x days on the job is estimated by the formula

(4) $$N(x) = 60 - 60e^{-0.3x}, \qquad x \geq 0.$$

Utilizing Table II, we find that on the fifth day on the job ($x = 5$) the worker is expected to install

$$N = 60 - 60e^{-1.5} = 60 - 60(0.2231) \approx 47$$

motors. As shown in the accompanying figure, the model of performance given by (4) indicates that the worker will be expected to install 60 motors each day after learning the job.

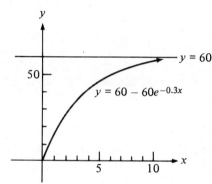

EXERCISE 5.1

Find the second components of the ordered pairs to make the pairs solutions of the given equation.

1. $y = 3^x$; $(0, \underline{\ ?\ }), (1, \underline{\ ?\ }), (2, \underline{\ ?\ })$
2. $y = -2^x$; $(-2, \underline{\ ?\ }), (0, \underline{\ ?\ }), (2, \underline{\ ?\ })$
3. $y = -5^x$; $(-2, \underline{\ ?\ }), (0, \underline{\ ?\ }), (2, \underline{\ ?\ })$
4. $y = 4^x$; $(0, \underline{\ ?\ }), (1, \underline{\ ?\ }), (2, \underline{\ ?\ })$
5. $f(x) = \left(\dfrac{1}{2}\right)^x$; $(-3, \underline{\ ?\ }), (0, \underline{\ ?\ }), (3, \underline{\ ?\ })$
6. $f(x) = \left(\dfrac{1}{3}\right)^x$; $(-3, \underline{\ ?\ }), (0, \underline{\ ?\ }), (3, \underline{\ ?\ })$
7. $g(x) = 10^x$; $(-2, \underline{\ ?\ }), (-1, \underline{\ ?\ }), (0, \underline{\ ?\ })$
8. $g(x) = 10^{-x}$; $(0, \underline{\ ?\ }), (1, \underline{\ ?\ }), (2, \underline{\ ?\ })$

Graph each function.

9. $y = 4^x$
10. $y = 5^x$
11. $y = 10^x$
12. $y = 10^{-x}$
13. $y = 2^{-x}$
14. $y = 3^{-x}$
15. $y = \left(\dfrac{1}{3}\right)^x$
16. $y = \left(\dfrac{1}{4}\right)^x$
17. $y = \left(\dfrac{1}{2}\right)^{-x}$
18. $y = \left(\dfrac{1}{3}\right)^{-x}$

5.1 The Exponential Function

19. $y = 2^x + 2$
20. $y = 2^x - 1$
21. $y = -2^x$
22. $y = 1 - 2^x$
23. $y = 2^{1-x}$
24. $y = 4^{x-1}$
25. Graph $f(x) = 1^x$. Is this an exponential function?
26. Graph $y = 10^x$, $x > 0$ and $x = 10^y$, $x > 0$ on the same set of axes.

Solve each equation by inspection.

27. $10^x = \dfrac{1}{100}$
28. $\left(\dfrac{1}{2}\right)^x = 16$
29. $16^x = 8$

Determine an integer n such that $n < x < n + 1$.

30. $3^x = 16.2$
31. $4^x = 87.1$
32. $10^x = 0.016$

Applications

33. *Biology* The number of bacteria present in a culture after t hours is given by $N(t) = 1000(3/2)^t$. Determine the number of bacteria present at $t = 0$, $t = 1$, $t = 2$, and $t = 3$.

34. Use a calculator to determine the number of bacteria present in the culture described in Problem 33 at $t = 2.71$.

35. The number of bacteria present in a culture is related to time by $N = N_0 e^{0.04t}$, where N_0 represents the amount of bacteria present at time $t = 0$, and t is time in hours. If 10,000 bacteria are present 10 hours after the beginning of the experiment, how many were present when $t = 0$?

36. The population P in a community after t years is predicted by $P(t) = P_0 2^{kt}$, where P_0 is the population of the community measured at $t = 0$. If the initial population is known to have doubled in 15 years, what is the value of the constant k.

37. The concentration C of a certain drug in the bloodstream at any time t (in minutes) is given by $C(t) = 100 - 100e^{-0.5t}$. Use Table II or a calculator to determine the concentration at $t = 0$, $t = 1$, $t = 2$, and $t = 3$. Graph $C(t)$.

38. *Business* Use (2) and Table II to find the value of S when $P = 10{,}000$, $r = 0.08$, and $m = 10$.

39. Use (2) and Table II to find the yearly rate of interest r when $S = 1349.90$, $P = 1000$, and $m = 5$.

40. A company determines that its revenue and cost functions in selling a product can be expressed as functions of time t measured in days. If $R(t) = 1000(1 - e^{-0.03t})$ and $C(t) = 200 - e^{-0.02t}$, what profit does the company realize when $t = 20$ days? Use Table II or a calculator.

41. *Physics* According to the **Bouguer-Lambert law**, the intensity I of a vertical beam of light of intensity I_0 passing through a transparent substance decreases according to the exponential function $I(x) = I_0 e^{-kx}$, $k > 0$, where x is depth measured in meters. If the intensity of light 1 meter below the surface of water is 30% of I_0, what is the intensity 3 meters below the surface? Use Table II or a calculator.

42. The barometric (or atmospheric) pressure P, measured in bar, at a height h kilometers above the surface of the earth is given by $P(h) = 1.013e^{-0.102h}$. Use Table II or a calculator to find the barometric pressure at a height of 10 kilometers.

43. The amount of a certain radioactive element remaining at any time t is given by $A = A_0 e^{-0.4t}$, where t is measured in seconds and A_0 is the amount present initially. How much of the element would remain after 3 seconds if 40 grams were present initially?

44. *Social Science* Graph the learning curve defined by $f(x) = 50(1 - e^{-0.2x})$.

45. The function $f(x) = e^{-x^2}$ occurs frequently in the study of statistics. Use Table II or a calculator to complete the following table.

x	-2	-1	-0.5	0	0.5	1	2
$f(x)$							

Graph the function.

46. The function

$$f(x) = \frac{1}{\sigma\sqrt{2\pi}} e^{-(x-\mu)^2/2\sigma^2}$$

is called the **normal**, or **Gaussian, distribution** function; μ is called the mean and σ is a positive constant called the standard deviation. Graph $f(x)$ with $\sigma = \mu = 1$.

5.2
THE LOGARITHMIC FUNCTION

INVERSE OF AN EXPONENTIAL FUNCTION

In the exponential function

(1) $\qquad y = b^x, \quad b > 0, \quad b \neq 1,$

There is only one x associated with each y (see the graphs of this function in Figure 5.1 for $b = 1/2$ and $b = 2$). Since (1) is one-to-one, its inverse function is defined by

(2) $\qquad x = b^y \qquad (b > 0, b \neq 1).$

Observe that the relation $x > 0$ is implied by (2), because there is no real number y for which b^y is not positive.

The graphs of functions of this form can be illustrated by the example

$$x = 10^y.$$

5.2 The Logarithmic Function

We consider x the variable denoting an element in the domain and, in the defining equation, assign arbitrary values for x, say,

$$(0.01, \ ), \quad (0.1, \ ), \quad (1, \ ), \quad (10, \ ), \quad (100, \ ),$$

to obtain the ordered pairs

$$(0.01, -2), \quad (0.1, -1), \quad (1, 0), \quad (10, 1), \quad (100, 2).$$

These can be plotted and connected with a smooth curve as in Figure 5.2.

It is always useful to be able to express the variable denoting an element in the range explicitly in terms of the variable denoting an element in the domain. To do this in an equation such as that defining (2), we use the notation

(3) $$y = \log_b x \qquad (x > 0, \ b > 0, \ b \neq 1).$$

Functions defined by such equations are called **logarithmic functions**.

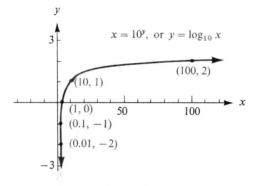

Figure 5.2

PROPERTIES OF LOGARITHMIC FUNCTIONS

From the graph in Figure 5.2, we generalize from $\log_{10} x$ to $\log_b x$ and observe that, for $b > 1$, a logarithmic function has the following properties:

1. The domain is the set of positive real numbers, and the range is the set of all real numbers.
2. For $0 < x < 1$, $\log_b x < 0$. For $x = 1$, $\log_b x = 0$. For $x > 1$, $\log_b x > 0$.

LOGARITHMIC AND EXPONENTIAL STATEMENTS

It should be recognized that the equations appearing in (2) and (3) determine the same function in the same way that $x = y + 4$ and $y = x - 4$ determine the same function, and we may use whichever equation suits our purpose. Thus, exponential statements may be written in logarithmic form, and logarithmic statements may be written in exponential form. We summarize this result for $b > 0$, $b \neq 1$ as follows:

> $y = \log_b x$ and $x = b^y$ are equivalent.

Examples Write each of the following statements in logarithmic form.

a. $5^2 = 25$ b. $8^{1/3} = 2$ c. $3^{-2} = \dfrac{1}{9}$

Solutions a. $\log_5 25 = 2$ b. $\log_8 2 = \dfrac{1}{3}$ c. $\log_3 \dfrac{1}{9} = -2$

Examples Write each of the following statements in exponential form.

a. $\log_{10} 100 = 2$ b. $\log_3 81 = 4$ c. $\log_2 \dfrac{1}{2} = -1$

Solutions a. $10^2 = 100$ b. $3^4 = 81$ c. $2^{-1} = \dfrac{1}{2}$

The logarithmic function associates with each number x the exponent y such that the power $b^y = x$. In other words, we can think of $\log_b x$ as an exponent on b. Thus, by substituting (3) in (2) we obtain

(4) $$b^{\log_b x} = x.$$

LAWS OF LOGARITHMS

Since a logarithm is an exponent, the following theorem follows directly from the properties of powers with real-number exponents.

> If x_1, x_2, and b are positive real numbers, $b \neq 1$, and m is any real number, then
>
> I $\log_b(x_1 x_2) = \log_b x_1 + \log_b x_2,$
>
> II $\log_b \dfrac{x_1}{x_2} = \log_b x_1 - \log_b x_2,$
>
> III $\log_b (x_1)^m = m \log_b x_1.$

The validity of I, II, and III can be established from the fact that (4) enables us to write positive numbers x_1 and x_2 as $x_1 = b^{\log_b x_1}$ and $x_2 = b^{\log_b x_2}$. For example, to obtain *I* we note

$$x_1 x_2 = b^{\log_b x_1} \cdot b^{\log_b x_2}$$
$$= b^{\log_b x_1 + \log_b x_2}.$$

5.2 The Logarithmic Function

Converting the latter exponential statement to an equivalent logarithm statement yields

$$\log_b(x_1 x_2) = \log_b x_1 + \log_b x_2.$$

EXERCISE 5.2

Express in logarithmic notation.

1. $4^2 = 16$
2. $5^3 = 125$
3. $3^3 = 27$
4. $8^2 = 64$
5. $\left(\dfrac{1}{2}\right)^2 = \dfrac{1}{4}$
6. $\left(\dfrac{1}{3}\right)^2 = \dfrac{1}{9}$
7. $8^{-1/3} = \dfrac{1}{2}$
8. $64^{-1/6} = \dfrac{1}{2}$
9. $10^2 = 100$
10. $10^0 = 1$
11. $10^{-1} = 0.1$
12. $10^{-2} = 0.01$

Express in exponential notation.

13. $\log_2 64 = 6$
14. $\log_5 25 = 2$
15. $\log_3 9 = 2$
16. $\log_{16} 256 = 2$
17. $\log_{1/3} 9 = -2$
18. $\log_{1/2} 8 = -3$
19. $\log_{10} 1000 = 3$
20. $\log_{10} 1 = 0$
21. $\log_{10} 0.01 = -2$

Find the value of each logarithm.

22. $\log_5 5$
23. $\log_7 49$
24. $\log_2 32$
25. $\log_4 64$
26. $\log_5 \sqrt{5}$
27. $\log_3 \sqrt{3}$
28. $\log_3 \dfrac{1}{3}$
29. $\log_5 \dfrac{1}{5}$
30. $\log_3 3$
31. $\log_2 2$
32. $\log_{10} 10$
33. $\log_{10} 100$
34. $\log_{10} 1$
35. $\log_{10} 0.1$
36. $\log_{10} 0.01$

Solve each equation for x, y, or b.

Examples a. $\log_2 x = 3$ b. $\log_b 2 = \dfrac{1}{2}$

Solutions Determine the solution by inspection, or write in exponential form and then determine the solution.

 a. $2^3 = x$
 $x = 8$

 b. $b^{1/2} = 2$
 $(b^{1/2})^2 = (2)^2$
 $b = 4$

37. $\log_3 9 = y$ **38.** $\log_5 125 = y$ **39.** $\log_b 8 = 3$ **40.** $\log_b 625 = 4$

41. $\log_4 x = 3$ **42.** $\log_{1/2} x = -5$ **43.** $\log_2 \dfrac{1}{8} = y$ **44.** $\log_5 5 = y$

45. $\log_b 10 = \dfrac{1}{2}$ **46.** $\log_b 0.1 = -1$ **47.** $\log_2 x = 2$ **48.** $\log_{10} x = -3$

49. Show that $\log_b 1 = 0$ for $b > 0$. [*Hint*: Express the statement in exponential form.]

50. Show that $\log_b b = 1$ for $b > 0$.

51. Show that $\log_b b^x = x$ for $b > 0$.

52. Graph $y = \log_2 x$. By examining the graph, what can you assert about $\log_2 a$ and $\log_2 b$ if $a < b$?

Express as the sum or difference of simpler logarithmic quantities.

Example $\log_b \left(\dfrac{xy}{z}\right)^{1/2}$

Solution By the laws of logarithms, part III,

$$\log_b \left(\dfrac{xy}{z}\right)^{1/2} = \dfrac{1}{2} \log_b \left(\dfrac{xy}{z}\right).$$

By parts I and II,

$$\dfrac{1}{2} \log_b \left(\dfrac{xy}{z}\right) = \dfrac{1}{2} (\log_b x + \log_b y - \log_b z).$$

53. $\log_b(xy)$ **54.** $\log_b(xyz)$ **55.** $\log_b \left(\dfrac{x}{y}\right)$

56. $\log_b \left(\dfrac{xy}{z}\right)$ **57.** $\log_b x^5$ **58.** $\log_b x^{1/2}$

59. $\log_b \sqrt[3]{x}$ **60.** $\log_b \sqrt[3]{x^2}$ **61.** $\log_b \sqrt{\dfrac{x}{z}}$

62. $\log_b \sqrt{xy}$ **63.** $\log_{10} \sqrt[3]{\dfrac{xy^2}{z}}$ **64.** $\log_{10} \sqrt[5]{\dfrac{x^2 y}{z^3}}$

65. $\log_{10}\left(2\pi\sqrt{\dfrac{l}{g}}\right)$ **66.** $\log_{10}\sqrt{s(s-a)(s-b)(s-c)}$

Express as a single logarithm with coefficient 1.

Example $\dfrac{1}{2}(\log_b x - \log_b y)$

Solution By the laws of logarithms, parts II and III,

$$\dfrac{1}{2}(\log_b x - \log_b y) = \dfrac{1}{2}\log_b\left(\dfrac{x}{y}\right) = \log_b\left(\dfrac{x}{y}\right)^{1/2}.$$

67. $\log_b x + \log_b y$ **68.** $\log_b x - \log_b y$

69. $2\log_b x + 3\log_b y$ **70.** $\dfrac{1}{4}\log_b x - \dfrac{3}{4}\log_b y$

71. $3\log_b x + \log_b y - 2\log_b z$ **72.** $\dfrac{1}{3}(\log_b x + \log_b y - 2\log_b z)$

73. $\log_{10}(x-2) + \log_{10} x - 2\log_{10} z$ **74.** $\dfrac{1}{2}(\log_{10} x - 3\log_{10} y - 5\log_{10} z)$

75. Show that $\dfrac{1}{4}\log_{10} 8 + \dfrac{1}{4}\log_{10} 2 = \log_{10} 2$.

76. Show that $4\log_{10} 3 - 2\log_{10} 3 + 1 = \log_{10} 90$.

77. Show that $\log_{10}[\log_3(\log_5 125)] = 0$.

78. Use (4) and the laws of exponents, to establish part II of the laws of logarithms.

5.3

LOGARITHMS TO THE BASE 10

There are two logarithmic functions of special interest in mathematics; one is defined by

(1) $$y = \log_{10} x,$$

and the other by

(2) $$y = \log_e x,$$

where we recall, e is an irrational number with the decimal approximation 2.7182818.

DETERMINATION OF $\log_{10} x$

Because these functions possess similar properties, and because we are more familiar with the number 10, we shall, for the present, confine our attention to (1).

Values for $\log_{10} x$ are called **logarithms to the base 10**, or **common logarithms**. The problem with which we are concerned in this section is that of finding $\log_{10} x$ for each positive x. Recall

(3) $$y = \log_{10} x \quad \text{and} \quad x = 10^y$$

are equivalent and consequently $\log_{10} x$ can easily be determined for all values of x that are integral powers of 10:

$$\log_{10} 10 = 1, \quad \text{since } 10^1 = 10$$
$$\log_{10} 100 = 2, \quad \text{since } 10^2 = 100;$$

and similarly,

$$\log_{10} 1 = 0, \quad \text{since } 10^0 = 1$$
$$\log_{10} 0.1 = -1, \quad \text{since } 10^{-1} = 0.1$$
$$\log_{10} 0.01 = -2, \quad \text{since } 10^{-2} = 0.01$$
$$\log_{10} 0.001 = -3, \quad \text{since } 10^{-3} = 0.001.$$

A table of logarithms such as Table I in Appendix II is used to find $\log_{10} x$ for $1 \leq x \leq 10$. Consider the excerpt from this table shown in Figure 5.3. Each number in the column headed by x represents the first two significant digits of x, while each number in the row opposite x contains the third significant digit of x. The digits located at the intersection of a row and a column specify the logarithm of x. For example, to find $\log_{10} 4.25$, we find 4.2 in the x column, and look across the row

Column ↓

x	0	1	2	3	4	5	6	7	8	9
3.8	.5798	.5809	.5821	.5832	.5843	.5855	.5866	.5877	.5888	.5899
3.9	.5911	.5922	.5933	.5944	.5955	.5966	.5977	.5988	.5999	.6010
4.0	.6021	.6031	.6042	.6053	.6064	.6075	.6085	.6096	.6107	.6117
4.1	.6128	.6138	.6149	.6160	.6170	.6180	.6191	.6201	.6212	.6222
4.2	.6232	.6243	.6253	.6263	.6274	(.6284)	.6294	.6304	.6314	.6325
4.3	.6335	.6345	.6355	.6365	.6375	.6385	.6395	.6405	.6415	.6425
4.4	.6435	.6444	.6454	.6464	.6474	.6484	.6493	.6503	.6513	.6522
4.5	.6532	.6542	.6551	.6561	.6571	.6580	.6590	.6599	.6609	.6618
4.6	.6628	.6637	.6646	.6656	.6665	.6675	.6684	.6693	.6702	.6712

Row → (points to row 4.2)

Figure 5.3

5.3 Logarithms to the Base 10

to where it intersects the column headed by 5. Thus, we see that

$$\log_{10} 4.25 = 0.6284.$$

The equality sign is used here in an inexact sense. More properly, we should write $\log_{10} 4.25 \approx 0.6284$; $\log_{10} 4.25$ is *irrational* and does not equal the *rational number* 0.6284. We shall follow customary usage, however, writing = instead of $\approx$, and leave the intent to the context.

CHARAC-TERISTIC AND MANTISSA

Now suppose we wish to find $\log_{10} x$ for values of x outside the range of the table—that is, for $0 < x < 1$ or $x > 10$. This can be done quite readily by first representing the number in scientific notation—that is, as the product of a number between 1 and 10 and a power of 10—and applying the laws of logarithms. For example,

$$\log_{10} 42.5 = \log_{10}(4.25 \times 10^1) = \log_{10} 4.25 + \log_{10} 10^1$$
$$= 0.6284 + 1 = 1.6284,$$
$$\log_{10} 4250 = \log_{10}(4.25 \times 10^3) = \log_{10} 4.25 + \log_{10} 10^3$$
$$= 0.6284 + 3 = 3.6284.$$

Observe that the decimal portion of the logarithm is always 0.6284, and the integral portion is just the exponent on 10 when the number is written in scientific notation.

This process can be reduced to a mechanical one by considering $\log_{10} x$ to consist of two parts, an integral part (called the **characteristic**) and a nonnegative decimal fraction part (called the **mantissa**). Thus, the table of values for $\log_{10} x$ for $1 < x < 10$ can be looked upon as a table of mantissas for $\log_{10} x$ for all $x > 0$.

To find $\log_{10} 43,700$, we first write

$$\log_{10} 43,700 = \log_{10}(4.37 \times 10^4) = \log_{10} 4.37 + 4$$

Upon examining the table of logarithms, we find that $\log_{10} 4.37 = 0.6405$, so that

$$\log_{10} 43,700 = 4.6405,$$

where we have prefixed the characteristic 4.

Now consider an example of $\log_{10} x$ for $0 < x < 1$. To find $\log_{10} 0.00402$, we write

$$\log_{10} 0.00402 = \log_{10}(4.02 \times 10^{-3}) = \log_{10} 4.02 - 3$$

Examining the table, we find $\log_{10} 4.02$ is 0.6042. Upon adding 0.6042 to the characteristic -3, we obtain

$$\log_{10} 0.00402 = -2.3958,$$

where the decimal portion of the logarithm is no longer 0.6042 as it is in the case of all numbers $x > 1$ for which the first three significant digits of x are 402. To circumvent this situation, it is customary to write the logarithm in a form in which the fractional part is positive. In the above example, we write

$$\log_{10} 0.00402 = 0.6042 - 3$$
$$= 0.6042 + (7 - 10)$$
$$= 7.6042 - 10,$$

and the fractional part is positive. The logarithms

$$6.6042 - 9,$$
$$12.6042 - 15,$$

etc., are equally valid representations, but $7.6042 - 10$ is customary.

DETERMINATION OF antilog$_{10}$ x

It is possible to reverse the process described in this section and, being given $\log_{10} x$, to find x. In this event, x is referred to as the **antilogarithm** (antilog$_{10}$) of $\log_{10} x$. For example, antilog$_{10}$ 1.6395 can be obtained by locating the mantissa, 0.6395, in the body of the $\log_{10}$ tables and observing that the associated antilog$_{10}$ is 4.36. Thus,

$$\text{antilog}_{10} 1.6395 = 4.36 \times 10^1 = 43.6.$$

LINEAR INTERPOLATION

If we seek the common logarithm of a number that is not an entry in the table (for example, $\log_{10} 3712$), or if we seek x when $\log_{10} x$ is not an entry in the table, it is customary to use a procedure called **linear interpolation**.

A table of common logarithms is a set of ordered pairs. For each number x there is an associated number $\log_{10} x$, and we have $\{(x, \log_{10} x)\}$ displayed in convenient tabular form. Because of space limitations, only three digits for the number x and four digits for the number $\log_{10} x$ appear in our table. By means of linear interpolation, however, the table can be used to find approximations to logarithms for four-digit numbers.

Let us examine geometrically the concepts involved. A portion of the graph of

$$y = \log_{10} x$$

is shown in Figure 5.4, part a. The curvature is exaggerated to illustrate the principle involved. We propose to use the line segment joining the points P_1 and P_2 as an approximation to the curve passing through the two points. If a large graph of $y = \log_{10} x$ were available, the value of, say, $\log_{10} 4.257$ could be found by using the ordinate of the point T on the curve for $x = 4.257$. Since there is no way to accomplish this with a table of values only, we shall instead use the ordinate of the point S on the line segment as an approximation to the ordinate of the point T on the curve.

This can be accomplished directly from the set of numbers available in the table of logarithms. Consider Figure 5.4, part b, where line segments $P_2 P_3$ and $P_4 P_5$ are perpendicular to line segment $P_1 P_3$.

5.3 Logarithms to the Base 10

From geometry, we have triangle $P_1P_4P_5$ similar to triangle $P_1P_2P_3$, where the lengths of the corresponding sides are proportional, and hence

(4) $$\frac{x}{X} = \frac{y}{Y}.$$

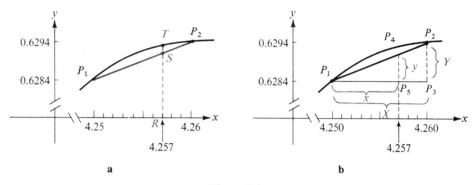

Figure 5.4

If we know any three of these numbers, we can determine the fourth. For the purpose of interpolation, we assume all of our members have four-digit numerals; that is, we consider 4.250 instead of 4.25, and 4.260 instead of 4.26. We note that the point corresponding to 4.257 is located just 7/10 of the distance between the points corresponding to 4.250 and 4.260, respectively, and the value Y, namely 0.0010, is just the difference between the logarithms 0.6284 and 0.6294. It follows from (4) that

$$\frac{7}{10} = \frac{y}{0.0010},$$

from which

$$y = \frac{7}{10}(0.0010) = 0.0007.$$

We now add 0.0007 to 0.6284 and thus obtain a good approximation to the required logarithm. That is, we have

$$\log_{10} 4.257 = 0.6291.$$

An example in Exercise 5.3 shows a convenient arrangement for the calculations involved in this illustration. The antilogarithm of a number can be found by a similar procedure. With practice, however, it is possible to interpolate mentally in both procedures.

EXERCISE 5.3

Write the characteristic of each logarithm.

Examples   **a.** $\log_{10} 248$ **b.** $\log_{10} 0.0057$

Solutions   **a.** Represent the number in scientific notation.

$$\log_{10}(2.48 \times 10^2)$$

The exponent on the base 10 is the characteristic.

$$2$$

b. Represent the number in scientific notation.

$$\log_{10}(5.7 \times 10^{-3})$$

The exponent on the base 10 is the characteristic.

$$-3 \quad \text{or} \quad 7 - 10$$

1. $\log_{10} 312$
2. $\log_{10} 0.02$
3. $\log_{10} 0.00851$
4. $\log_{10} 8.012$
5. $\log_{10} 0.00031$
6. $\log_{10} 0.0004$
7. $\log_{10}(15 \times 10^3)$
8. $\log_{10}(820 \times 10^4)$

Find each logarithm.

Examples   **a.** $\log_{10} 16.8$ **b.** $\log_{10} 0.043$

Solutions   **a.** Represent the number in scientific notation.

$$\log_{10}(1.68 \times 10^1)$$

Determine the mantissa from Table I, Appendix II.

$$0.2253$$

Add the characteristic as determined by the exponent on the base 10.

$$1.2253$$

b. Represent the number in scientific notation.

$$\log_{10}(4.3 \times 10^{-2})$$

Determine the mantissa from Table I, Appendix II.

$$0.6335$$

Add the characteristic as determined by the exponent on the base 10.

$$0.6335 - 2$$
or
$$8.6335 - 10$$

9. $\log_{10} 6.73$
10. $\log_{10} 891$
11. $\log_{10} 0.813$
12. $\log_{10} 0.00214$
13. $\log_{10} 0.08$
14. $\log_{10} 0.000413$
15. $\log_{10}(2.48 \times 10^2)$
16. $\log_{10}(5.39 \times 10^{-3})$

5.3 Logarithms to the Base 10

Find each antilogarithm.

Example antilog$_{10}$ 2.7364

Solution Locate the mantissa in the body of the table of mantissas and determine the associated antilog$_{10}$ (a number between 1 and 10); write the characteristic as an exponent on the base 10.

$$5.45 \times 10^2 = 545$$

17. antilog$_{10}$ 0.6128
18. antilog$_{10}$ 0.2504
19. antilog$_{10}$ 0.5647
20. antilog$_{10}$ 3.9258
21. antilog$_{10}$ (8.8075 − 10)
22. antilog$_{10}$ (3.9722 − 5)
23. antilog$_{10}$ 3.7388
24. antilog$_{10}$ 2.0086
25. antilog$_{10}$ (6.8561 − 10)
26. antilog$_{10}$ (1.8156 − 4)

Find each logarithm.

Example log$_{10}$ 4.257

Solution Interpolate mentally or use the following procedure.

$$0.010 \left\{ 0.007 \left\{ \begin{array}{c|c} x & \log_{10} x \\ 4.250 & 0.6284 \\ 4.257 & ? \\ 4.260 & 0.6294 \end{array} \right\} y \right\} 0.0010$$

Set up a proportion and solve for y.

$$\frac{0.007}{0.010} = \frac{7}{10} = \frac{y}{0.0010}$$

$$y = 0.0007$$

Add 0.0007, the value of y, to 0.6284.

$$\log_{10} 4.257 = 0.6284 + 0.0007 = 0.6291$$

27. log$_{10}$ 4.213
28. log$_{10}$ 8.184
29. log$_{10}$ 1522
30. log$_{10}$ 203.4
31. log$_{10}$ 37,110
32. log$_{10}$ 72.36
33. log$_{10}$ 0.5123
34. log$_{10}$ 0.008351

Find each antilogarithm.

Example antilog$_{10}$ 0.6446

Solution Interpolate mentally or use the following procedure.

$$0.0010 \left\{ 0.0002 \begin{pmatrix} x & \text{antilog}_{10} x \\ 0.6444 & 4.410 \\ 0.6446 & ? \\ 0.6454 & 4.420 \end{pmatrix} y \right\} 0.010$$

Set up a proportion and solve for y.

$$\frac{0.0002}{0.0010} = \frac{y}{0.010}$$

$$y = 0.002$$

Add 0.002, the value of y, to 4.410.

$$\text{antilog}_{10} 0.6446 = 4.410 + 0.002 = 4.412$$

35. antilog$_{10}$ 0.5085   **36.** antilog$_{10}$ 0.8087   **37.** antilog$_{10}$ 1.0220
38. antilog$_{10}$ 3.0759   **39.** antilog$_{10}$(8.7055 − 10)   **40.** antilog$_{10}$(9.8742 − 10)
41. antilog$_{10}$(2.8748 − 3)   **42.** antilog$_{10}$(7.7397 − 10)

43. If we use linear interpolation to find log$_{10}$ 3.751 and log$_{10}$ 3.755, which of the resulting approximations should we expect to be more nearly correct? Why?

44. If we use linear interpolation to find log$_{10}$ 1.025 and log$_{10}$ 9.025, which of the resulting approximations should we expect to be more nearly correct? Why?

Find each of the following by means of Table I, Appendix II.

Example $10^{0.6263}$

Solution The logarithmic function is the inverse of the exponential function. Hence, the element $10^{0.6263}$ in the range of the exponential function is antilog$_{10}$ 0.6263 in the domain of the logarithmic function. From Table I, we find

$$10^{0.6263} = \text{antilog}_{10} 0.6263 = 4.23.$$

45. $10^{0.9590}$ **46.** $10^{0.8241}$ **47.** $10^{3.6990}$
48. $10^{2.3874}$ **49.** $10^{2.0531}$ **50.** $10^{1.7396}$

51. Write the following logarithms to the base 10 in a form in which both the integral and fractional parts are negative.

   a. 8.7321 − 10   **b.** 6.4187 − 10

52. Write the following logarithms to the base 10 in a form in which the fractional part is positive.

   a. 2 − 0.7113   **b.** −4.6621

5.4

APPLICATIONS OF COMMON LOGARITHMS

pH OF A SOLUTION

The **pH**, or hydrogen potential, of a solution is defined by

(1) $$pH = -\log_{10}[H^+]$$

where $[H^+]$ is the concentration of hydrogen ions in an aqueous solution in moles per liter. When $0 < pH < 7$ the solution is said to be *acid*; for $pH > 7$ the solution is *base* or *alkaline*; for $pH = 7$ the solution is *neutral* (for example, water). A strongly acid solution such as lemon juice has a pH in the range $pH \leq 3$. Human urine averages around $pH = 6$. Note that (1) can also be written $pH = \log_{10} 1/[H^+]$.

Example In a healthy person it is found that the concentration of hydrogen ions in blood is $[H^+] = 3.98 \times 10^{-8}$ moles/liter. Determine the pH of blood.

Solution From (1) we find that the pH of blood is given by

$$pH = -\log_{10} 3.98 \times 10^{-8}$$
$$= -[\log_{10} 3.98 + \log_{10} 10^{-8}]$$
$$= -[\log_{10} 3.98 - 8]$$
$$= -[0.5999 - 8]$$
$$\approx 7.4.$$

Severe illness, or even death, could result when a person's blood pH falls outside the narrow limits $7.2 \leq pH \leq 7.6$. We note that values of pH are usually given to the nearest tenth of a unit.

THE ENVIRONMENT

The **intensity level** of sound measured in decibels (dB)* is given by

(2) $$b = 10 \log_{10} \frac{I}{I_0}.$$

Here I_0 is a reference intensity of 10^{-16} watt/cm² corresponding to approximately the faintest sound that can be heard. When $I = I_0$ then (2) gives $b = 0$ dB. The intensity levels of frequently occurring sounds are given in the table on page 176.

---

*The decibel is (1/10) bell. This latter unit, named for Alexander Graham Bell (1847–1922), proved to be too large in practice.

Source	Intensity level (dB)
Threshold of hearing	0
Whisper	20
Normal talking	40–60
Some TV commercials	65
Smoke detector alarm	70
Jet airplane taking off	80–100
Threshold of pain	120

Example Determine the intensity level of a sound having intensity 10^{-4} watt/cm^2.

Solution From (2) we see that the intensity level is

$$b = 10 \log_{10} \frac{10^{-4}}{10^{-16}} = 10 \log_{10} 10^{12}$$

$$= 120 \text{ dB}.$$

As indicated in the table, sound at an intensity level of around 120 dB can cause pain. Prolonged exposure to levels around 90 dB (easily produced by rock and roll groups) can cause temporary deafness.

Equation (2) can be used to obtain the intensity level b_2 of a sound at a distance d_2 from the source if one level b_1 is measured at a distance d_1. Now it is known that the intensity of sound I is inversely proportional to the square of the distance d from the source:

(3) $$I = \frac{k}{d^2}$$

where k is the intensity at a unit distance from the source. Substituting (3) into (2) and using properties II and III of logarithms gives

$$b = 10 \log_{10} \frac{k/d^2}{I_0} = \log_{10} \frac{k/I_0}{d^2}$$

$$= 10 \left[\log_{10} k/I_0 - \log_{10} d^2 \right]$$

$$= 10 \left[\log_{10} k/I_0 - 2 \log_{10} d \right].$$

Hence at d_2 and d_1,

(4) $$b_2 = 10 \left[\log_{10} k/I_0 - 2 \log_{10} d_2 \right]$$

(5) $$b_1 = 10 \left[\log_{10} k/I_0 - 2 \log_{10} d_1 \right].$$

Subtracting (5) from (4) then yields

5.4 Applications of Common Logarithms

$$b_2 - b_1 = 10 \, [2 \log_{10} d_1 - 2 \log_{10} d_2]$$
$$= 20 \log_{10} d_1/d_2$$

or

(6) $$b_2 = b_1 + 20 \log_{10} \frac{d_1}{d_2}.$$

AIRPORT NOISE REDUCTION In an experimental two-segment approach to an airport a plane starts a 6-degree glide path at 5.5 miles out from the runway and switches to a 3-degree glide path 1.5 miles out. A normal approach consists of a 3-degree glide path starting 5.5 miles out from the runway. The obvious purpose of the two-segment approach is that one plane causes less noise simply because it is higher. It is readily shown that at the 5.5-mile point the higher plane P_2, shown in Figure 5.5, part a, is at an altitude of 2635 feet whereas P_1, shown in Figure 5.5, part b, at the same point, has an altitude of 1522 feet. Equation (3) then implies that at this starting point of both glide paths, the sound intensity I of P_2 is one-third that of P_1. However, this does *not* mean that the *intensity level* of P_2 is one-third that of the intensity level of P_1.

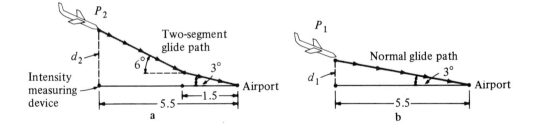

Figure 5.5

Example The intensity level b_1 of plane P_1 5.5 miles out from the runway is measured on ground level at 80 dB. Determine the intensity level b_2 of plane P_2 at the same point.

Solution Identifying $d_1 = 1522$ and $d_2 = 2635$, it follows from (6) that

$$b_2 = 80 + 20 \log_{10} \frac{1522}{2635}$$
$$= 80 + 20 \log_{10} 0.58$$
$$= 80 - 4.7314$$
$$\approx 75 \text{ dB}.$$

EXERCISE 5.4

Calculate the pH of a solution with given hydrogen-ion concentration.

1. $[H^+] = 10^{-7}$
2. $[H^+] = 4.0 \times 10^{-5}$
3. $[H^+] = 2.0 \times 10^{-8}$
4. $[H^+] = 8.5 \times 10^{-3}$
5. $[H^+] = 6.3 \times 10^{-7}$
6. $[H^+] = 5.7 \times 10^{-7}$

Calculate the hydrogen-ion concentration $[H^+]$ of a solution with given pH.

Example pH = 4.8

Solution Substitute 4.8 for pH in the relationship $pH = \log_{10} \dfrac{1}{[H^+]}$.

$$\log_{10} \dfrac{1}{[H^+]} = 4.8$$

Equate antilog$_{10}$ of each member and solve for $[H^+]$.

$$\dfrac{1}{[H^+]} = \text{antilog}_{10}\ 4.8 = 6.3 \times 10^4$$

$$[H^+] = \dfrac{1}{6.3 \times 10^4} = 0.16 \times 10^{-4} = 1.6 \times 10^{-5}$$

7. pH = 3.0
8. pH = 4.2
9. pH = 5.6
10. pH = 8.3
11. pH = 7.2
12. pH = 6.9

Calculate the intensity level of sound with given intensity.

13. 10^{-7}
14. 10^{-16}
15. 10^{-10}
16. 5×10^{-6}
17. 3×10^{-9}
18. Show that the intensity level (2) can be expressed as $b = 160 + 10 \log_{10} I$.

Applications

19. *Biology–Chemistry* The concentration of hydrogen ions in the stomach fluid is given by $[H^+] = 9.3 \times 10^{-3}$. Calculate the pH of the fluid.
20. A particular brand of beer has a pH of 4.4. Calculate the concentration of hydrogen ions $[H^+]$.
21. *Physics* At 10 meters away, the intensity level of a jet airplane engine is measured at 130 dB. What is the intensity level 500 meters from the engine?
22. The intensity level of normal talking is 40 dB measured 3 feet from its source. What is the intensity level 12 feet from the source?

23. The magnitude R of an earthquake of intensity I measured on the Richter scale is given by $R = \log_{10} I/I_0$, where I_0 is a reference level corresponding to $R = 0$. The intensity of an earthquake in the San Francisco area in 1979 was measured as $10^{5.9} I_0$. What was its magnitude on the Richter scale?

24. The San Francisco earthquake of 1906 had a magnitude of 8.25 on the Richter scale. How much greater was the intensity of the 1906 earthquake compared with the earthquake that occurred in 1979? [*Hint*: Determine I_2/I_1, where I_2 and I_1 are the intensities of the 1906 and 1979 earthquakes, respectively.]

25. The absolute magnitudes M_A and M_B of two stars A and B are related to their absolute luminosities L_A and L_B by $M_B - M_A = 2.5 \log_{10} L_A/L_B$. Calculate the absolute magnitude of the star Betelgeuse if it is known that the absolute magnitude of the sun is 4.7 and that the absolute luminosity of Betelgeuse is 10^5 times the absolute luminosity of the sun.

5.5

COMPUTATIONS WITH LOGARITHMS

The use of hand calculators has almost eliminated the need to perform routine numerical computations with pencil and paper and logarithms. For example, to compute $\sqrt{264}$ on a typical calculator we need only enter the number 264 and depress the key marked $\sqrt{}$. We obtain

There is no question but that the use of logarithms in computations cannot compare in either speed or accuracy to the use of a calculator. Nevertheless, we feel that there is something to be be *learned* by doing some routine computations by means of logarithms. The equations involved in these computations shed light on the basic properties of the logarithmic function. We reproduce here these properties, or laws, of logarithms using the base 10.

If x_1 and x_2 are positive real numbers, then

$$\text{I} \quad \log_{10}(x_1 x_2) = \log_{10} x_1 + \log_{10} x_2,$$

$$\text{II} \quad \log_{10} \frac{x_1}{x_2} = \log_{10} x_1 - \log_{10} x_2,$$

$$\text{III} \quad \log_{10}(x_1)^m = m \log_{10} x_1.$$

Before illustrating the use of these laws, let us make two further observations:

> **L-1** If $M = N$ ($M, N > 0$), then $\log_b M = \log_b N$.
>
> **L-2** If $\log_b M = \log_b N$, then $M = N$.

Both L-1 and L-2 follow from the fact that the variables in a logarithmic function are in one-to-one correspondence.

Example Find the product of 3.825 and 0.00729, using logarithms.

Solution Let $N = (3.825)(0.00729)$. By property L-1,
$$\log_{10} N = \log_{10}[(3.825)(0.00729)].$$
Now, by I,
$$\log_{10} N = \log_{10} 3.825 + \log_{10} 0.00729,$$
and from Table I we obtain
$$\log_{10} 3.825 = 0.5826 \qquad \log_{10} 0.00729 = 7.8627 - 10,$$
so that
$$\log_{10} N = (0.5826) + (7.8627 - 10)$$
$$= 8.4453 - 10.$$
The computation is completed by referring to the table for
$$N = \operatorname{antilog}_{10}(8.4453 - 10) = 2.788 \times 10^{-2}$$
$$= 0.02788.$$
Thus, we have
$$N = (3.825)(0.00729) = 0.02788.$$

Actual computation on a calculator shows the product to be 0.02788425, so the result obtained by use of logarithms is correct to four significant digits. Some error should be expected, because we are using approximations to irrational numbers when we employ a table of logarithms.

Example Compute $\dfrac{(8.21)^{1/2}(2.17)^{2/3}}{(3.14)^3}$.

Solution Setting
$$N = \frac{(8.21)^{1/2}(2.17)^{2/3}}{(3.14)^3},$$

5.5 Computations with Logarithms

we obtain

$$\log_{10} N = \log_{10} \frac{(8.21)^{1/2}(2.17)^{2/3}}{(3.14)^3}$$

$$= \log_{10}(8.21)^{1/2} + \log_{10}(2.17)^{2/3} - \log_{10}(3.14)^3$$

$$= \frac{1}{2}\log_{10} 8.21 + \frac{2}{3}\log_{10} 2.17 - 3\log_{10} 3.14.$$

Table I provides values for the logarithms involved here, and the remainder of the computation is routine. In order to avoid confusion in computations of this sort, a systematic approach of some kind is desirable (see the example in Exercise 5.4).

As observed on page 169, in computations involving numbers less than 1, it is sometimes convenient to use expressions other than such differences as $7._____ - 10$ and $8._____ - 10$ for negative characteristics.

Example Compute $\sqrt[3]{0.0235}$.

Solution Setting

$$N = \sqrt[3]{0.0235} = (0.0235)^{1/3},$$

we have

$$\log_{10} N = \log_{10}(0.0235)^{1/3} = \frac{1}{3}\log_{10}(0.0235)$$

From Table I, we see that the mantissa of this logarithm is 0.3711 and its characteristic is -2. Because we wish to multiply by $1/3$, we select $7.3711 - 9$ instead of $8.3711 - 10$ for the logarithm. Thus,

$$\log_{10} N = \frac{1}{3}(7.3711 - 9) = 2.4570 - 3,$$

where we have avoided obtaining a nonintegral negative part of the logarithm. From the table,

$$N = \text{antilog}_{10}(2.4570 - 3) = 0.2864.$$

Thus,

$$\sqrt[3]{0.0235} = 0.2864.$$

Example The future value of $1000 invested at 6% interest compounded annually for 25 years is given by

$$S = 1000(1.06)^{25}.$$

Use logarithms to compute S.

Solution

$$\log_{10} S = \log_{10} 1000 + \log_{10}(1.06)^{25}$$
$$= \log_{10} 10^3 + \log_{10}(1.06)^{25}$$
$$= 3 + 25 \log_{10} 1.06$$
$$= 3 + 25(0.0253)$$
$$= 0.6325 + 3.$$

Therefore,

$$S = \text{antilog}_{10}(0.6325 + 3) = 4.29 \times 10^3$$
$$= \$4290.$$

EXERCISE 5.5

Compute each of the following, using logarithms.

Example $\sqrt{\dfrac{(23.4)(0.681)}{4.13}}$

Solution Let $N = \sqrt{\dfrac{(23.4)(0.681)}{4.13}}$. Then

$$\log_{10} N = \frac{1}{2}(\log_{10} 23.4 + \log_{10} 0.681 - \log_{10} 4.13)$$

$$\left. \begin{array}{l} \log_{10} 23.4 = 1.3692 \\ \log_{10} 0.681 = 9.8331 - 10 \end{array} \right\} \text{add}$$

$$\left. \begin{array}{l} \log_{10}(23.4)(0.681) = 11.2023 - 10 \\ \log_{10} 4.13 = \underline{0.6160} \\ \phantom{\log_{10} 4.13 =\ }10.5863 - 10 \end{array} \right\} \text{subtract}$$

$$\log_{10} \frac{(23.4)(0.681)}{4.13} = 10.5863 - 10 = 0.5863$$

$$\frac{1}{2} \log_{10} \frac{(23.4)(0.681)}{4.13} = \frac{1}{2}(0.5863) = 0.2931.$$

5.5 Computations with Logarithms

Hence, $\log_{10} N = 0.2931$, from which

$$N = \text{antilog}_{10}\, 0.2931 = 1.964.$$

1. $(2.32)(1.73)$
2. $(83.2)(6.12)$
3. $\dfrac{3.15}{1.37}$
4. $\dfrac{1.38}{2.52}$
5. $\dfrac{0.0149}{32.3}$
6. $\dfrac{0.00214}{3.17}$
7. $(2.3)^5$.
8. $(4.62)^3$
9. $\sqrt[3]{8.12}$
10. $\sqrt[5]{75}$
11. $(0.0128)^5$
12. $(0.0021)^6$
13. $\sqrt{0.0021}$
14. $\sqrt[5]{0.0471}$
15. $\sqrt[3]{0.0214}$
16. $\sqrt[4]{0.0018}$
17. $\dfrac{(8.12)(8.74)}{7.19}$
18. $\dfrac{(0.412)^2(84.3)}{\sqrt{21.7}}$
19. $\dfrac{(6.49)^2 \sqrt[3]{8.21}}{17.9}$
20. $\dfrac{(2.61)^2(4.32)}{\sqrt{7.83}}$
21. $\dfrac{(0.3498)(27.16)}{6.814}$
22. $\dfrac{(4.813)^2(20.14)}{3.612}$
23. $\sqrt{\dfrac{(4.71)(0.00481)}{(0.0432)^2}}$
24. $\sqrt{\dfrac{(2.85)^3(0.97)}{(0.035)}}$
25. $500(1.07)^{30}$
26. $30(1.09)^{35}$
27. $1000(1.08)^{-25}$
28. $800(1.05)^{-20}$
29. Find an approximate value for $2^{\sqrt{2}}$.
30. Find an approximate value for 2^{π}.

Applications

31. The area A of a circle is given by $A = \pi r^2$. Use logarithms to find the area of a circle with a radius of 5.63 centimeters.

32. The area, A, of a triangle in terms of the sides is given by the formula

$$A = \sqrt{s(s-a)(s-b)(s-c)},$$

where a, b, and c are the lengths of the sides of the triangle and s equals one-half of the perimeter. Use logarithms to find the area of a triangle in which the lengths of the three sides are 2.314 inches, 4.217 inches, and 5.618 inches.

33. *Biology* The surface area S of an animal is related to its weight W by $S = 0.3\, W^{2/3}$. Use logarithms to find the surface area of an animal weighing 50 kilograms.

34. The quantity Q of trees in a forest in n years is predicted by $Q = N(1 + r/100)^n$, where N is the initial number of trees and r represents, in percent, the yearly increase. Use logarithms to find Q when $N = 10{,}000$, $r = 5\%$, and $n = 20$ years.

35. *Business* If \$1 had been invested at the start of the year 0 A.D. at only 2% annual interest compounded annually, its value at the end of 1985 would be $(1.02)^{1985}$. Use logarithms to find this value.

36. *Physics* The period, T, of a simple pendulum is given by the formula $T = 2\pi \sqrt{L/g}$, where T is in seconds, L is the length of the pendulum in feet, $g \approx 32$ feet per second per second, and $\pi \approx 3.14$. Use logarithms to find the period of a pendulum 1 foot long.

5.6
EXPONENTIAL AND LOGARITHMIC EQUATIONS

An equation in one variable in which the variable occurs in an exponent is called an **exponential equation**. Solution sets of some such equations can be found by means of logarithms.

Example Find the solution set of $5^x = 7$.

Solution Since $5^x > 0$ for all x, we can apply property L-1 from Section 5.4 and write

$$\log_{10} 5^x = \log_{10} 7,$$

and from property III of the same section,

$$x \log_{10} 5 = \log_{10} 7.$$

Dividing each member by $\log_{10} 5$, we obtain

$$x = \frac{\log_{10} 7}{\log_{10} 5} = \frac{0.8451}{0.6990} \approx 1.209.$$

The solution set is

$$\left\{ \frac{\log_{10} 7}{\log_{10} 5} \right\},$$

and a decimal approximation for the single solution is 1.209. Note that for this approximation, the logarithms are *divided*, not subtracted.

COMPOUND INTEREST In Section 11.1, we shall see that when an annual interest rate r is **compounded**, say, t times a year, then the amount S returned on an initial deposit P after m years is given by

$$S = P\left(1 + \frac{r}{t}\right)^{tm}$$

For a given value of t we can use logarithms to determine either r or m.

Example Find the rate of interest r such that

$$4800 = 2500\left(1 + \frac{r}{4}\right)^{48}$$

Solution We first divide by 2500.

5.6 Exponential and Logarithmic Equations

$$1.92 = \left(1 + \frac{r}{4}\right)^{48}$$

$$1.92^{1/48} = \left(1 + \frac{r}{4}\right)$$

Solving for r by means of logarithms, we obtain

$$\log_{10}\left(1 + \frac{r}{4}\right) = \log_{10} 1.92^{1/48}$$

$$= \frac{1}{48} \log_{10} 1.92$$

$$= \frac{1}{48} (0.2833),$$

$$\log_{10}\left(1 + \frac{r}{4}\right) = 0.0059$$

$$\text{antilog}_{10}\, 0.0059 = 1 + \frac{r}{4}$$

$$= 1.014.$$

Thus, we have

$$r = 4(1.014 - 1)$$

$$= 0.056.$$

The required annual rate of interest is about 5.6%.

Example Find m (the number of years since the time of deposit) such that

$$40(1 + 0.02)^m = 51.74$$

Solution Divide each member by 40; equate $\log_{10}$ of each member; and apply properties II and III of Section 5.4.

$$(1.02)^m = \frac{51.74}{40}$$

$$\log_{10}(1.02)^m = \log_{10} \frac{51.74}{40}$$

$$m \log_{10}(1.02) = \log_{10} 51.74 - \log_{10} 40$$

$$m(0.0086) = 1.7138 - 1.6021$$

$$m = \frac{0.1117}{0.0086}$$

$$= 13 \text{ years}$$

EXERCISE 5.6

Solve. Leave each solution in logarithmic form using the base 10.

Example $3^{x-2} = 16$

Solution By property L-1,

$$\log_{10} 3^{x-2} = \log_{10} 16.$$

By property III,

$$(x - 2)\log_{10} 3 = \log_{10} 16$$

$$x - 2 = \frac{\log_{10} 16}{\log_{10} 3}$$

$$x = \frac{\log_{10} 16}{\log_{10} 3} + 2.$$

The solution set is $\left\{ \dfrac{\log_{10} 16}{\log_{10} 3} + 2 \right\}$.

1. $2^x = 7$
2. $3^x = 4$
3. $3^{x+1} = 8$
4. $2^{x-1} = 9$
5. $7^{2x-1} = 3$
6. $3^{x+2} = 10$
7. $4^{x^2} = 15$
8. $8^{x^2} = 21$
9. $3^{-x} = 10$
10. $2.13^{-x} = 8.1$
11. $3^{1-x} = 15$
12. $4^{2-x} = 10$

Solve for the indicated variable. Leave each result in the form of an equation equivalent to the given equation.

13. $y = x^n$ for n
14. $y = Cx^{-n}$ for n
15. $y = e^{kt}$ for t
16. $y = Ce^{-kt}$ for t

Solve.

Example $\log_{10}(x + 9) + \log_{10} x = 1$

Solution By property I, the given equation is equivalent to $\log_{10}(x+9)x = 1$ for $x+9$ and x greater than 0. Write this equation in exponential form, and solve for x.

$$x^2 + 9x = 10^1$$
$$x^2 + 9x - 10 = 0$$
$$(x + 10)(x - 1) = 0$$
$$x = -10 \qquad x = 1$$

Since neither $\log_{10}(-10 + 9)$ nor $\log_{10}(-10)$ is defined, -10 does not satisfy the **original** equation; the solution set, therefore, is simply $\{1\}$.

5.6 Exponential and Logarithmic Equations

17. $\log_{10} x + \log_{10} 2 = 3$
18. $\log_{10}(x - 1) - \log_{10} 4 = 2$
19. $\log_{10} x + \log_{10}(x + 21) = 2$
20. $\log_{10}(x + 3) + \log_{10} x = 1$
21. $\log_{10}(x + 2) + \log_{10}(x - 1) = 1$
22. $\log_{10}(x - 3) - \log_{10}(x + 1) = 1$

Solve for the variable m (nearest year), r (nearest 1/2%), or S (accuracy obtainable on a four-place table of mantissas).

Example $(1 + r)^{12} = 1.127$

Solution Equate $\log_{10}$ of each member and apply property III.
$$12 \log_{10}(1 + r) = \log_{10} 1.127 = 0.0519$$
Multiply each member by 1/12.
$$\log_{10}(1 + r) = \frac{1}{12}(0.0519) = 0.0043$$
Determine $\text{antilog}_{10}\, 0.0043$ and solve for r.
$$\text{antilog}_{10}\, 0.0043 = 1.01$$
$$1 + r = 1.01$$
$$r = 0.01 \quad \text{or } 1\%$$

23. $(1 + 0.03)^{10} = S$
24. $(1 + 0.04)^8 = S$
25. $(1 + r)^6 = 1.34$
26. $(1 + r)^{10} = 1.48$
27. $100\left(1 + \dfrac{r}{2}\right)^{10} = 113$
28. $40\left(1 + \dfrac{r}{4}\right)^{12} = 50.9$
29. $(1 + 0.04)^m = 2.19$
30. $(1 + 0.04)^m = 1.60$
31. $150(1 + 0.01)^{4m} = 240$
32. $60(1 + 0.02)^{2m} = 116$

Applications

33. *Biology* The number of pheasants in a game refuge after t years is predicted by the formula $P(t) = 100(5/4)^t$. Determine when the pheasant population will triple in number.

34. *Biology–Chemistry* Recall from Section 5.4 that the pH of a solution is given by $\text{pH} = -\log_{10}[H^+]$. Solve this equation for $[H^+]$.

35. *Business* The amount S accumulated when an amount R is invested in an annuity at an interest rate i per period for n periods is given by $S = R[(1 + i)^n - 1]/i$. The amount S accumulated when an amount P is invested for the same number of periods at compound interest is $S = P(1 + i)^n$. Determine the value of n for which the amounts accumulated are the same. (This is sometimes called the **capital-recovery period**.)

36. *Physics* In Section 5.4 we saw that the intensity level of sound is given by $b = 10 \log_{10} I/I_0$. Solve this formula for I.

37. The absolute magnitudes M_A and M_B of two stars A and B are related to their absolute luminosities L_A and L_B by $M_B - M_A = 2.5 \log_{10} L_A/L_B$. Solve this formula for L_B.

5.7

LOGARITHMS TO THE BASE e

The number e ($e \approx 2.7182818$) mentioned in Section 5.3 is of great mathematical interest and importance, and logarithms to the base e are frequently encountered in practical situations. Logarithms to the base e are called **natural logarithms**. It is possible to determine $\log_e x$ provided we have a table of $\log_{10} x$. Indeed, given a table of $\log_b x$, we can always find $\log_a x$ for any $a > 0$, $a \neq 1$. For example, if we wanted to express $\log_8 9$ in terms of $\log_{10}$, we could proceed as follows. First we let

$$\log_8 9 = N.$$

Then, from our definition of a logarithm, we have

$$8^N = 9.$$

Using property III of logarithms to the base 10, we obtain

$$N \log_{10} 8 = \log_{10} 9,$$

from which

$$N = \frac{\log_{10} 9}{\log_{10} 8}.$$

Hence,

$$\log_8 9 = \frac{\log_{10} 9}{\log_{10} 8}.$$

CONVERSION TO ANOTHER BASE We can do this for the general case. If $N = \log_a x$, we have by definition

$$a^N = x.$$

We can equate the logarithms to the base b, $b > 0$ and $b \neq 1$, of each member of (1) to obtain

$$\log_b a^N = \log_b x.$$

By property III of logarithms this yields

$$\log_b x = N \log_b a \quad \text{or} \quad \log_b x = \log_a x \cdot \log_b a$$

from which we obtain

(2)
$$\log_a x = \frac{\log_b x}{\log_b a}.$$

CONVERSION TO BASE e Equation (2) gives us a means of finding $\log_a x$ when we have a table of logarithms to the base b. In particular, if $a = e$ and $b = 10$, we have

5.7 Logarithms to the Base e

(3) $$\log_e x = \frac{\log_{10} x}{\log_{10} e}.$$

Since $\log_{10} e \approx 0.4343$, equation (3) can be written

(4) $$\log_e x = \frac{\log_{10} x}{0.4343},$$

or

(5) $$\log_e x = 2.3026 \log_{10} x,$$

which gives us a direct means of approximating $\log_e x$ when we have a table of logarithms to the base 10.

Table III in Appendix II is a table of $\log_e x$. We cannot, however, use this table in the same way we used Table I—that is, simply list mantissas for logarithms over a certain interval and then manipulate characteristics to take care of all other intervals. Our numeration system is based on 10 and not on e. With the help of logarithms, though, we can use equation (5) and Table I of $\log_{10}$ to find $\log_e x$.

Example Find $\log_e 278$.

Solution 1 Since

$$\log_e x = 2.3026 \log_{10} x,$$

using Table I, we have

$$\log_e 278 = 2.3026 \log_{10} 278$$
$$= 2.3026(2.4440)$$
$$= 5.63$$

Solution 2 Alternatively, we observe that $278 = 2.78 \times 10^2$. Hence, using the $\log_e$ table, we have

$$\log_e 278 = \log_e(2.78 \times 10^2) = \log_e 2.78 + 2 \log_e 10$$
$$= 1.0225 + 2(2.3026)$$
$$= 1.0225 + 4.6052$$
$$= 5.6277 \approx 5.63.$$

Example Find $\log_e 0.278$.

Solution This time, let us simply observe that $0.278 = 2.78 \times 10^{-1}$. Hence,

$$\log_e 0.278 = \log_e(2.78 \times 10^{-1}) = \log_e 2.78 + \log_e 10^{-1}$$
$$= \log_e 2.78 - \log_e 10$$
$$= 1.0225 - 2.3026$$
$$= -1.2801.$$

As one would expect, the logarithm is negative.

INTEREST We saw in Section 5.1 that when an annual rate of interest, r, is compounded continuously over a period of m years, then the future value S of an investment P is given by the formula

$$S = Pe^{rm}.$$

Example Use natural logarithms to compute the future value of \$100 invested at 5% interest compounded continuously for 10 years.

Solution From the above formula we have

$$S = 100e^{(0.05)10}$$
$$= 100e^{0.5}$$
$$\log_e S = \log_e 100 + \log_e e^{0.5}$$
$$= 4.6052 + 0.5$$
$$= 5.1052.$$

Therefore, with interpolation in Table III, we have

$$S = \text{antilog}_e\, 5.1052$$
$$= 165.$$

Table II can also be used to find $\text{antilog}_e\, x$ because $\text{antilog}_e\, x = e^x$.

Examples **a.** Find $\text{antilog}_e\, 3.2$ **b.** Find $\text{antilog}_e(-3.2)$.

Solutions The values can be read directly from Table II.

a. $\text{antilog}_e\, 3.2 = e^{3.2}$ **b.** $\text{antilog}_e(-3.2) = e^{-3.2}$
$\quad = 24.533$ $\quad = 0.0408.$

CONCENTRATION OF A DRUG IN THE BLOOD Under some conditions the concentration C of a drug in the bloodstream at any time t can be shown to be given by

(6) $$C(t) = \frac{a}{b} + \left(C_0 - \frac{a}{b}\right)e^{-bt}.$$

5.7 Logarithms to the Base e

C_0 is the concentration of the drug measured at $t=0$ and a and b are positive constants. As t increases the term e^{-bt} decreases. Thus, for large values of time the concentration C is close to the value a/b; this latter value is known as the **steady-state** concentration.

Example Find the time at which $C(t)$ defined by (6) is one-half the steady-state concentration.

Solution We must find t such that $C(t) = \dfrac{a}{2b}$:

$$\frac{a}{2b} = \frac{a}{b} + \left(C_0 - \frac{a}{b}\right)e^{-bt}$$

$$-\frac{a}{2b} = \left(C_0 - \frac{a}{b}\right)e^{-bt}$$

$$e^{-bt} = \frac{-\dfrac{a}{2b}}{C_0 - \dfrac{a}{b}}$$

$$= \frac{a}{2(a - bC_0)}$$

$$-bt = \log_e \frac{a}{2(a - bC_0)}$$

$$t = -\frac{1}{b}\log_e \frac{a}{2(a - bC_0)}$$

EXERCISE 5.7

Find each logarithm to the nearest hundredth.

Example $\log_3 7$

Solution Use logarithms to the base 10 to evaluate.

$$\log_3 7 = \frac{\log_{10} 7}{\log_{10} 3} = \frac{0.8451}{0.4771} = 1.77$$

1. $\log_2 10$
2. $\log_2 5$
3. $\log_5 240$
4. $\log_3 18$
5. $\log_7 8.1$
6. $\log_5 60$
7. $\log_{100} 38$
8. $\log_{100} 240$

Find each logarithm directly from Table III.

9. $\log_e 3$
10. $\log_e 8$
11. $\log_e 17$
12. $\log_e 98$
13. $\log_e 327$
14. $\log_e 107$
15. $\log_e 450$
16. $\log_e 605$

Find each antilogarithm directly from Table II.

17. $\operatorname{antilog}_e 0.50$
18. $\operatorname{antilog}_e 1.5$
19. $\operatorname{antilog}_e 3.4$
20. $\operatorname{antilog}_e 4.5$
21. $\operatorname{antilog}_e 0.231$
22. $\operatorname{antilog}_e 1.43$
23. $\operatorname{antilog}_e (-0.15)$
24. $\operatorname{antilog}_e (-0.95)$
25. $\operatorname{antilog}_e (-2.5)$
26. $\operatorname{antilog}_e (-4.2)$
27. $\operatorname{antilog}_e (-0.255)$
28. $\operatorname{antilog}_e (-3.65)$

Solve without using the tables of logarithms.

29. If $\log_2 8 = 3$, find $\log_8 2$.
30. If $\log_4 16 = 2$, find $\log_{16} 4$.
31. If $\log_{10} 3 = 0.4771$, find $\log_3 10$.
32. If $\log_{10} e = 0.4343$, find $\log_e 10$.
33. If $\log_{10} 5 = 0.6990$, find $\log_5 100$.
34. If $\log_{10} 3 = 0.4771$, find $\log_3 100$.
35. a. Show that $\log_9 7 = (1/2)\log_3 7$.
    b. By a similar argument, show that for $a, b < 0$, $a \neq 1$, $\log_{a^2} b = (1/2)\log_a b$.
    [*Hint:* In the first case, $\log_9 7 = 1/\log_7 9$ and $9 = 3^2$.]
36. Explain why $y = \log_x x$, $x > 1$, defines a function. What is its domain? What is its range? Graph the function.
37. Show that $(\log_{10} 4 - \log_{10} 2)\log_2 10 = 1$.
38. Show that $(2 \log_2 3)(\log_9 2 + \log_9 4) = 3$.

Applications

39. *Biology* The population P of a town in t years is predicted by the formula $P = 500\, e^{0.06t}$. In how many years will the population be 3000?
40. Use (6) to find the time at which $C(t)$ is two-thirds the steady-state concentration.
41. In many instances the population P of certain kinds of animals and insects is given by the **logistic function**

$$P(t) = \frac{a}{b + e^{-at}}$$

where a and b are positive constants. Solve for t.

42. *Business* Use natural logarithms to find the future value of $500 invested at 8% annual interest compounded continuously for 30 years.
43. Use natural logarithms to find the future value of $1000 invested at 6% annual interest compounded continuously for 25 years.
44. *Physics* The amount A of radioactive material remaining after t years is given by $A(t) = A_0 e^{-kt}$, where A_0 is the amount of material present at $t = 0$ and k is a positive constant.

5.7 Logarithms to the Base e

If $A_1 = A(t_1)$ and $A_2 = A(t_2)$, $t_1 < t_2$, show that the half-life of the material is given by

$$t = \frac{(t_1 - t_2)\log_e 2}{\log_e A_1/A_2}.$$

The half-life of a radioactive substance is the time required for one-half of the substance to decay.

45. The intensity I (in lumens) of a light beam, after passing through a thickness x (in centimeters) of a medium having an absorption coefficient of 0.1, is given by $I = 1000e^{-0.1x}$. How many centimeters of the material would reduce the illumination to 800 lumens?

46. Using the information given in Problem 45, find the thickness of a medium that would reduce the illumination to 600 lumens.

47. The current i in a certain kind of series circuit is given by

$$i(t) = \frac{E}{R}(1 - e^{-Rt/L}).$$

where E, R, and L are positive constants. Solve for t.

48. *Social Science* The number of motors N installed each day by a new automobile assembly worker after x days on the job is estimated by the formula $N = 60 - 60e^{-0.3x}$. In how many days will $N = 40$?

CHAPTER TEST

1. Graph the function $y = 3^x$.
2. Solve by inspection: $10^x = 1/1000$.

Express in logarithmic notation.

3. $3^4 = 81$
4. $3^{-2} = \dfrac{1}{9}$

Express in exponential notation.

5. $\log_2 8 = 3$
6. $\log_{10} 0.0001 = -4$

Solve.

7. $\log_2 16 = y$
8. $\log_{10} x = 3$

Express as the sum or difference of simpler logarithmic quantities.

9. $\log_{10} \sqrt[3]{xy^2}$
10. $\log_{10} \dfrac{2R^3}{\sqrt{PQ}}$

Express as a single logarithm with coefficient 1.

11. $2 \log_b x - \dfrac{1}{3} \log_b y$
12. $\dfrac{1}{3}(2 \log_{10} x + \log_{10} y) - 3 \log_{10} z$

Find each logarithm. Interpolate as required.

13. $\log_{10} 42$
14. $\log_{10} 0.00314$
15. $\log_{10} 682.4$
16. $\log_{10} 0.04142$

Chapter Test

Find each antilogarithm. Interpolate as required.

17. $\text{antilog}_{10}\, 1.8287$
18. $\text{antilog}_{10}(8.6684 - 10)$
19. $\text{antilog}_{10}\, 0.4240$
20. $\text{antilog}_{10}(9.8224 - 10)$

Calculate the pH of a solution with given hydrogen-ion concentration.

21. $[H^+] = 3.5 \times 10^{-4}$
22. $[H^+] = 8.8 \times 10^{-7}$

Calculate the hydrogen-ion concentration $[H^+]$ of a solution with given pH.

23. $\text{pH} = 2.6$
24. $\text{pH} = 6.4$

Calculate the intensity level of sound with given intensity.

25. 10^{-8}
26. 2×10^{-5}

27. At a distance of 4 feet, the sound emitting from a smoke detector alarm has an intensity level of 70 dB. What is the intensity level 40 feet from the alarm?

28. At an altitude of 5000 feet the noise of a jet aircraft has an intensity level of 70 dB. What is the intensity level at 100 feet?

Compute by means of logarithms.

29. $(8.73)(41.6)$
30. $\dfrac{85.9}{1.73}$
31. $\dfrac{(4.78)(62.3)^2}{12.8}$
32. $\sqrt[3]{43}\, \sqrt[5]{0.024}$

Solve. Leave solutions in logarithmic form using the base 10.

33. $5^x = 7$
34. $3^{x+1} = 80$
35. Solve $\log_{10}(x - 3) + \log_{10} x = 1$.
36. Solve $(1 + .05)^m = 1.215$ for m to the nearest integer.

Find each value using Table I.

37. $\log_3 11$
38. $\log_2 120$
39. Find $\log_e 130$ using Table III.
40. Find antilog$_e$ 4.8 using Table II.
41. Find antilog$_e(-0.24)$ using Table II.
42. Given $A = A_0 e^{-0.4t}$, find A if $A_0 = 30$ grams and $t = 5$ seconds.
43. Given $S = Pe^{rm}$, find S if $P = \$500$, $r = 8\%$, and $m = 25$ years.
44. The population of a community in t years is predicted by the formula $P = P_0 e^{kt}$, where k is a positive constant and P_0 is the initial population. Show that the time required to double the population is $t = (\log_e 2)/k$. This is called the **law of Malthus**.
45. The amount of radioactive substance remaining at time t is $A = A_0 e^{-0.007t}$, where A_0 is the amount present initially. Determine the time at which $A = 0.03 A_0$.

6
SYSTEMS OF EQUATIONS

6.1
SYSTEMS OF LINEAR EQUATIONS IN TWO VARIABLES

SOLUTION SET OF A SYSTEM

We shall begin by considering the system

(1) $$a_1 x + b_1 y + c_1 = 0 \qquad (a_1, b_1 \text{ not both } 0)$$

(2) $$a_2 x + b_2 y + c_2 = 0 \qquad (a_2, b_2 \text{ not both } 0)$$

In a geometric sense, because the graphs of both equations are straight lines, we are confronted with three possibilities, as illustrated in Figure 6.1:

a. The graphs are the same line.

b. The graphs are parallel but distinct lines.

c. The graphs intersect in one and only one point.

These possibilities lead, correspondingly, to the conclusion that one and only one of the following is true for any given system of two such linear equations in x and y:

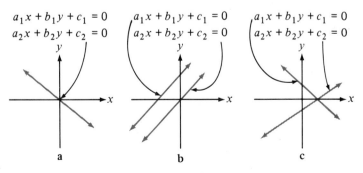

Figure 6.1

a. The intersection of the solution sets of (1) and (2) contain all those ordered pairs found in either one of the given solution sets; that is, the solution sets are equal.

b. The intersection of the solution sets is the null set.

c. The intersection of the solution sets contains exactly one ordered pair.

The intersection of the solution sets of (1) and (2) is called the **solution set of the system**.

DEPENDENCY AND CONSISTENCY

In case a, the left-hand members of the two equations in standard form are said to be **linearly dependent**, and the equations are **consistent**; in case b, the left-hand members are **linearly independent**, and the equations are **inconsistent**; and in case c, the left-hand members are linearly independent and the equations are consistent. If the two left-hand members are linearly dependent, then one of them can be obtained from the other through multiplication by a constant. For example,

$$2x + 4y - 8 \quad \text{and} \quad 6x + 12y - 24$$

are linearly dependent, but

$$2x + 4y - 8 \quad \text{and} \quad 6x + 12y - 23$$

are not.

You will recall that equivalent open sentences are defined to be open sentences that have the same solution set. Systems of open sentences may be said to be equivalent in a similar sense.

> *If the solution set of one system of open sentences is equal to (the same as) the solution set of another system of open sentences, then the systems are* **equivalent**.

GENERATION OF EQUIVALENT SYSTEMS

In seeking the solution set of a system of equations, our procedure will be to generate equivalent systems until we arrive at a system of which the solution set is obvious. The following result establishes one way to do this.

> *Any ordered pair that satisfies both the equations*
> $$f(x, y) = 0$$
> $$g(x, y) = 0$$
> *also satisfies the equation*
> $$a \cdot f(x, y) + b \cdot g(x, y) = 0$$
> *for all real numbers a and b,*

6.1 Systems of Linear Equations in Two Variables

To verify this result, let us assume (x_1, y_1) is in the solution set of the system
$$f(x, y) = 0$$
$$g(x, y) = 0.$$
It follows that $f(x_1, y_1) = 0$ and $g(x_1, y_1) = 0$ and, for any real numbers a and b, that
$$a \cdot f(x_1, y_1) + b \cdot g(x_1, y_1) = 0.$$
The polynomial $a \cdot f(x, y) + b \cdot g(x, y)$, where a and b are not both 0, is said to be a **linear combination** of the polynomials $f(x, y)$ and $g(x, y)$. We sometimes refer to the equation $a \cdot f(x, y) + b \cdot g(x, y) = 0$ as a linear combination of the equations $f(x, y) = 0$ and $g(x, y) = 0$.

In solving linear systems of the form (1) and (2) it is always possible to find appropriate choices of multipliers a and b, such that the linear combination

(3) $\qquad a(a_1 x + b_1 y + c_1) + b(a_2 x + b_2 y + c_2) = 0$

will be free of one variable; that is, the coefficient of one variable will be 0. We can then form an equivalent system by substituting equation (3) for either (1) or (2).

Example Solve

(4) $\qquad \begin{aligned} x - 3y + 5 &= 0 \\ 2x + y - 4 &= 0. \end{aligned}$

Solution We first form the linear combination
$$1(x - 3y + 5) + 3(2x + y - 4) = 0,$$
or
$$7x - 7 = 0,$$
from which
$$x - 1 = 0.$$
Replacing $x - 3y + 5 = 0$ with $x - 1 = 0$, we then have the equivalent system

(5) $\qquad \begin{aligned} x \phantom{{} + y} - 1 &= 0 \\ 2x + y - 4 &= 0. \end{aligned}$

We can next replace the second equation in (5) with the linear combination
$$-2(x - 1) + 1(2x + y - 4) = 0,$$
or
$$y - 2 = 0,$$
to obtain the equivalent system

(6) $\qquad \begin{aligned} x - 1 &= 0 \\ y - 2 &= 0, \end{aligned}$

from which, by inspection, we can obtain the unique solution (1, 2). Hence, the solution set of (4) is $\{(1, 2)\}$. Graphs of the systems (4), (5), and (6) appear in Figure 6.2. The point of intersection in each case has coordinates (1, 2).

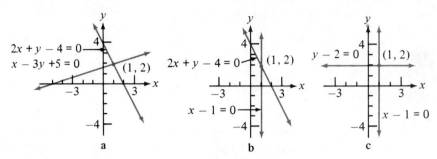

Figure 6.2

SOLUTION OF A SYSTEM BY SUBSTITUTION

Note that we could have proceeded somewhat differently from (4), as follows: The second equation implies $y = -2x + 4$. By substituting this equation in the first we obtain $x - 3(-2x + 4) + 5 = 0$ or $7x - 7 = 0$ or $x = 1$. Correspondingly, we then have $y = -2(1) + 4 = 2$. Hence, again the solution set of (4) is $\{(1, 2)\}$. This latter way of proceeding to the solution is called the **method of substitution**.

CHOICE OF MULTIPLIERS FOR LINEAR COMBINATIONS

Observe that the multipliers used in the above example were first 3 and 1, and later -2 and 1. These were chosen because they produced coefficients that were additive inverses, first for the terms in y and later for the terms in x. In general, the linear combination

$$b_2(a_1 x + b_1 y + c_1) - b_1(a_2 x + b_2 y + c_2) = 0$$

will always be free of y, and

$$a_2(a_1 x + b_1 y + c_1) - a_1(a_2 x + b_2 y + c_2) = 0$$

will be free of x. For example, to form a useful linear combination of the equations in the system

$$2x + 6y + 7 = 0$$
$$5x + 4y + 3 = 0,$$

we might use 5 and -2, or -5 and 2, if we wish to obtain an equation that is free of x; and we might use 2 and -3, or -2 and 3, if we wish to obtain an equation that is free of y.

CRITERIA FOR DEPENDENT OR INCONSISTENT EQUATIONS

If the coefficients of the variables in one equation in a system are proportional to the corresponding coefficients in the other equation, then the equations might be either dependent or inconsistent. It can be shown that the equations in the system

$$a_1 x + b_1 y + c_1 = 0$$
$$a_2 x + b_2 y + c_2 = 0 \qquad (a_2, b_2, c_2 \neq 0)$$

are dependent if

6.1 Systems of Linear Equations in Two Variables

$$\frac{a_1}{a_2} = \frac{b_1}{b_2} = \frac{c_1}{c_2},$$

and inconsistent if

$$\frac{a_1}{a_2} = \frac{b_1}{b_2} \neq \frac{c_1}{c_2}.$$

Systems of linear equations are quite useful in expressing relationships in practical applications. By assigning separate variables to represent separate physical quantities, we can usually decrease the difficulty in symbolically representing these relationships.

EXERCISE 6.1

Solve each system.

Example (1) $\quad \frac{2}{3}x - y = 2$

(2) $\quad x + \frac{1}{2}y = 7$

Solution Multiply each member of equation (1) by 3 and each member of equation (2) by 2.

(1') $\quad\quad\quad\quad 2x - 3y = 6$
(2') $\quad\quad\quad\quad 2x + y = 14$

Form a linear combination by adding -1 times (1') to 1 times (2') and solve for y.

(3) $\quad\quad\quad\quad \begin{aligned} 4y &= 8 \\ y &= 2 \end{aligned}$

Equations (2) and (3) constitute an equivalent system for (1) and (2).

Substitute 2 for y in any of (1), (2), (1'), or (2'), and solve for x. In this example, (2) is used.

$$x + \frac{1}{2}(2) = 7$$

$$x = 6$$

The solution set is $\{(6, 2)\}$.

1. $x - y = 1$
   $x + y = 5$

2. $2x - 3y = 6$
   $x + 3y = 3$

3. $3x + y = 7$
   $2x - 5y = -1$

4. $2x - y = 7$
   $3x + 2y = 14$

5. $5x - y = -29$
   $2x + 3y = 2$

6. $6x + 4y = 12$
   $3x + 2y = 12$

7. $5x + 2y = 3$
   $x = 0$

8. $2x - y = 0$
   $x = -3$

9. $3x - 2y = 4$
   $y = -1$

10. $x + 2y = 6$
    $x = 2$

11. $\frac{1}{4}x - \frac{1}{3}y = -\frac{5}{12}$
    $\frac{1}{10}x + \frac{1}{5}y = \frac{1}{2}$

12. $\frac{2}{3}x - y = 4$
    $x - \frac{3}{4}y = 6$

13. $\frac{1}{7}x - \frac{3}{7}y = 1$
    $2x - y = -4$

14. $\frac{1}{3}x - \frac{2}{3}y = 2$
    $x - 2y = 6$

15. $\frac{1}{x} + \frac{1}{y} = 6$
    $\frac{1}{x} - \frac{1}{y} = -2$

    [*Hint*: Let $X = \frac{1}{x}$ and $Y = \frac{1}{y}$.]

16. $\frac{4}{x} - \frac{1}{y} = -2$
    $\frac{8}{x} + \frac{3}{y} = 1$

17. $\frac{2}{x+1} - \frac{6}{y-5} = 8$
    $\frac{1}{x+1} - \frac{3}{y-5} = 4$

18. $\frac{3}{x-1} + \frac{9}{y-1} = 12$
    $\frac{7}{x-1} - \frac{5}{y-1} = -24$

19. Find *a* and *b* so that the graph of $ax + by + 3 = 0$ passes through the points $(-1, 2)$ and $(-3, 0)$.

20. Find *a* and *b* so that the solution set of the system

    $ax + by = 4$
    $bx - ay = -3$

    is $\{(1, 2)\}$.

21. Recall that the slope-intercept form of the equation of a straight line is given by $y = mx + b$. Find an equation of the line that passes through the points $(0, 2)$ and $(3, -8)$.

22. Find an equation of the line that passes through the points $(-6, 2)$ and $(4, 1)$.

Applications

23. *Chemistry* How many pounds of an alloy containing 45% silver must be melted with an alloy containing 60% silver to obtain 40 pounds of a 48% silver alloy?

24. A chemist wishes to obtain 100 cubic centimeters of a solution containing 50% acid by combining a 35% acid solution with a 75% solution. How much of each type of solution should be used?

25. *Business* A woman has $1000 more invested at 5% than she has invested at 4%. If her annual income from the two investments together is $698, how much does she have invested at each rate?

26. A man has three times as much money invested in 3% bonds as he has in stocks paying 5%. How much does he have invested in each if his yearly income from the investments is $1680?

27. Tickets to a rock concert cost $10 and $20. A total of 10,000 tickets were sold, which generated a revenue of $158,000. How much of each type of ticket were sold?

28. A woman has $1.80 in nickels and dimes, with three more dimes than nickels. How many dimes and nickels does she have?

29. The following table indicates the number of units of forms X and Y that printing machines A and B can produce each hour.

	A	B
X	2	4
Y	5	3

If both machines start at the same time, how many hours must each run in order to print 32 units of form X and 59 units of form Y?

30. *Physics* An airplane travels 1260 miles in the same time that an automobile travels 420 miles. If the rate of the airplane is 120 miles per hour greater than the rate of the automobile, find the rate of each.

31. Two cars start together and travel in the same direction, one going twice as fast as the other. At the end of 3 hours, they are 96 miles apart. How fast is each traveling?

6.2

SYSTEMS OF LINEAR EQUATIONS IN THREE VARIABLES

SOLUTIONS OF AN EQUATION IN THREE VARIABLES

A solution of an equation in three variables, such as

(1) $$x + 2y - 3z + 4 = 0,$$

is an ordered triple of numbers (x, y, z), because all three of the variables must be replaced before we can decide whether the sentence is true. Thus, $(0, -2, 0)$ and $(-1, 0, 1)$ are solutions of (1), whereas $(1, 1, 1)$ is not. There are, of course, infinitely many members in the solution set of such an equation.

SOLUTION OF A SYSTEM

The solution set of a system of linear (first-degree) equations in three variables is the intersection of the solution sets of the separate equations in the system. We are primarily interested in systems involving three equations, such as

(1) $$x + 2y - 3z + 4 = 0$$
(2) $$2x - y + z - 3 = 0$$
(3) $$3x + 2y + z - 10 = 0.$$

The solution set of this system can be found by methods analogous to those used in Section 6.1.

> *If any equation in the system*
> $$f(x, y, z) = 0$$
> $$g(x, y, z) = 0$$
> $$h(x, y, z) = 0$$
> *is replaced by a linear combination, with nonzero coefficients, of itself and any one of the other equations in the system, then the result is an equivalent system.*

To see how this result applies, let us examine the system (1), (2), and (3). If we begin by replacing (2) with the linear combination formed by multiplying (1) by -2 and (2) by 1, we have

$$-2(x + 2y - 3z + 4) + 1(2x - y + z - 3) = 0,$$

or

$$-5y + 7z - 11 = 0,$$

and we obtain the equivalent system

(1) $\qquad\qquad x + 2y - 3z + 4 = 0$
(2') $\qquad\qquad\qquad -5y + 7z - 11 = 0$
(3) $\qquad\qquad 3x + 2y + z - 10 = 0,$

where (2') is free of x. Next, if we replace (3) by the sum of -3 times (1) and 1 times (3), we have

(1) $\qquad\qquad x + 2y - 3z + 4 = 0$
(2') $\qquad\qquad\qquad -5y + 7z - 11 = 0$
(3') $\qquad\qquad\qquad -4y + 10z - 22 = 0,$

where both (2') and (3') are free of x. If now (3') is replaced by the sum of 4 times (2') and -5 times (3'), we have

(1) $\qquad\qquad x + 2y - 3z + 4 = 0$
(2') $\qquad\qquad\qquad -5y + 7z - 11 = 0$
(3'') $\qquad\qquad\qquad\qquad -22z + 66 = 0,$

which is equivalent to the original (1), (2), and (3). But, from (3''), we see that for any solution of (1), (2'), and (3''), $z = 3$; that is, the solution will be of the form $(x, y, 3)$. If 3 is substituted for z in (2'), we have

$$-5y + 7(3) - 11 = 0 \qquad \text{or} \qquad y = 2,$$

6.2 Systems of Linear Equations in Three Variables

and any solution of the system must be of the form $(x, 2, 3)$. Substituting 2 for y and 3 for z in (1) gives

$$x + 2(2) - 3(3) + 4 = 0 \quad \text{or} \quad x = 1,$$

so the single member of the solution set of (1), (2′), and (3″) is (1, 2, 3). Therefore, the solution set of the original system is $\{(1, 2, 3)\}$.

This process of solving a system of linear equations can be reduced to a series of mechanical procedures as follows.

Example   Solve.

(1) $\quad x + 2y - z + 1 = 0$
(2) $\quad x - 3y + z - 2 = 0$
(3) $\quad 2x + y + 2z - 6 = 0$

Solution   Multiply (1) by -1 and add the result to 1 times (2); multiply (1) by -2 and add the result to 1 times (3).

(4) $\quad -5y + 2z - 3 = 0$
(5) $\quad -3y + 4z - 8 = 0$

Multiply (4) by -3 and add the result to 5 times (5).

(6) $\quad 14z - 31 = 0$

Now (1), (4), and (6) constitute a set of equations equivalent to (1), (2), and (3).

(1) $\quad x + 2y - z + 1 = 0$
(4) $\quad -5y + 2z - 3 = 0$
(6) $\quad 14z - 31 = 0$

Solve for z in (6).

$$z = \frac{31}{14}$$

Substitute 31/14 for z in (4).

$$-5y + 2\left(\frac{31}{14}\right) - 3 = 0$$

$$y = \frac{2}{7}$$

Substitute 2/7 for y and 31/14 for z in either (1), (2), or (3); we use (1).

$$x + 2\left(\frac{2}{7}\right) - \frac{31}{14} + 1 = 0$$

$$x = \frac{9}{14}$$

The solution set is $\left\{\left(\dfrac{9}{14}, \dfrac{2}{7}, \dfrac{31}{14}\right)\right\}$.

LINEAR DEPENDENCE AND INCONSISTENT EQUATIONS

If at any step in the procedure used in the examples above, the resulting linear combination vanishes or yields a contradiction, the system contains linearly dependent left-hand members or else inconsistent equations, or both, and it either has an infinite number of members or else has no member in its solution set.

GRAPHS IN THREE DIMENSIONS

By establishing a three-dimensional Cartesian coordinate system as shown in Figure 6.3, a one-to-one correspondence can be established between the points in a three-dimensional space and ordered triples of real numbers. If this is done, it can be shown that the graph of a linear equation in three variables is a plane. For example, the equation

$$x + y + z = 1$$

represents the plane through the points (1, 0, 0), (0, 1, 0), and (0, 0, 1), as illustrated in Figure 6.3. Hence, the solution set of a system of three linear equations in three variables consists of the coordinates of the common intersection of three planes. Figure 6.4 shows the possibilities for the relative positions of the plane graphs of three linear equations in three variables. In case a, the common intersection consists of a single point, and hence the solution set of the corresponding system of three equations contains a single member. In cases b, c, and d, the intersection is a line or a plane, and the solution of the corresponding system has infinitely many members. In cases e, f, g, and h, the three planes have no common intersection, and the solution set of the corresponding system is the null set. Linear equations corresponding to cases a, b, c, and d are consistent, and the others inconsistent.

The use of linear combinations to solve systems of linear equations can be extended to cover cases of n equations in n variables. Clearly, as n grows larger, the time and effort necessary to find the solution set of a system increase correspondingly. Fortunately, modern high-speed computers can handle such computations in stride

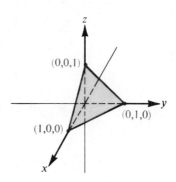

Figure 6.3

6.2 Systems of Linear Equations in Three Variables

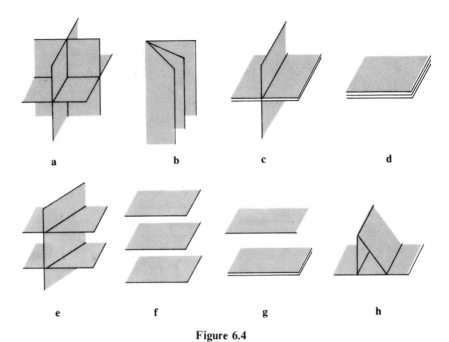

Figure 6.4

for fairly large values of n. There are analytic methods other than those exhibited here which are preferable for very large systems.

EXERCISE 6.2

Solve.

1. $x + y + z = 2$
   $2x - y + z = -1$
   $x - y - z = 0$

2. $x + y + z = 1$
   $2x - y + 3z = 2$
   $2x - y - z = 2$

3. $x + y + 2z = 0$
   $2x - 2y + z = 8$
   $3x + 2y + z = 2$

4. $2x - 3y + z = 3$
   $x - y - 3z = -1$
   $-x + 2y - 3z = -4$

5. $x - 2y + z = -1$
   $2x + y - 3z = 3$
   $3x + 3y - 2z = 10$

6. $x - 2y + 4z = -3$
   $3x + y - 2z = 12$
   $2x + y - 3z = 11$

7. $4x - 2y + 3z = 4$
   $2x - y + z = 1$
   $3x - 3y + 4z = 5$

8. $x + 5y - z = 2$
   $3x - 9y + 3z = 6$
   $x - 3y + z = 4$

9. $x + z = 5$
   $y - z = -4$
   $x + y = 1$

10. $5y - 8z = -19$
    $5x - 8z = 6$
    $3x - 2y = 12$

11. $x - \frac{1}{2}y - \frac{1}{2}z = 4$
    $x - \frac{3}{2}y - 2z = 3$
    $\frac{1}{4}x + \frac{1}{4}y - \frac{1}{4}z = 0$

12. $x + 2y + \frac{1}{2}z = 0$
    $x + \frac{3}{5}y - \frac{2}{5}z = \frac{1}{5}$
    $4x - 7y - 7z = 6$

13. The equation for a circle can be written $x^2 + y^2 + ax + by + c = 0$. Find the equation of the circle with graph containing the points $(2, 3)$, $(3, 2)$, and $(-4, -5)$.

14. Find values for a, b, and c so that the graph of $x^2 + y^2 + ax + by + c = 0$ will contain the points $(0, 0)$, $(6, 0)$, and $(0, 8)$.

16. Find values for a, b, and c so that the graph of $y = ax^2 + bx + c$ contains the points $(-1, 2)$, $(1, 6)$, and $(2, 11)$.

15. Three solutions of the equation $ax + by + cz = 1$ are $(0, 2, 1)$, $(6, -1, 2)$, and $(0, 2, 0)$. Find the coefficients a, b, and c.

17. Three solutions of the equation $ax + by + cz = 1$ are $(2, 1, 0)$, $(-1, 3, 2)$, and $(3, 0, 0)$. Find the coefficients a, b, and c.

18. Show that the system

$$x + y + 2z = 2$$
$$2x - y - z = 3$$

has an infinite number of members in its solution set. List two ordered triples that are solutions. [*Hint*: Express x in terms of z alone, and express y in terms of z alone.]

19. Give a geometric argument to show that any system of two consistent equations in three variables has an infinite number of members in its solution set.

20. Show that the system

   (1) $\quad x + y + z = 3$
   (2) $\quad x - 2y - z = -2$
   (3) $\quad x + y + 2z = 4$
   (4) $\quad 2x - y + 2z = 4$

has $\emptyset$ as its solution set. [*Hint*: Does the solution set of the system (1), (2), and (3) have any member that satisfies (4)?]

Applications

21. The sum of three numbers is 15. The second equals 2 times the first and the third equals the second. Find the numbers.

22. The sum of three numbers is 2. The first number is equal to the sum of the other two, and the third number is the result of subtracting the first from the second. Find the numbers.

23. A box contains $6.25 in nickels, dimes, and quarters. There are 85 coins in all, with 3 times as many nickels as dimes. How many coins of each kind are there?

24. The perimeter of a triangle is 15 inches. The side x is 20 inches shorter than the side y, and the side y is 5 inches longer than the side z. Find the lengths of the sides of the triangle.

25. A woman had $446 in ten-dollar, five-dollar, and one-dollar bills. There were 94 bills in all and 10 more five-dollar bills than ten-dollar bills. How many bills of each kind did she have?

26. *Biology* The following table shows the percentage of protein, carbohydrate, and fat in each ounce of food groups A, B, and C.

	A	B	C
Protein	60%	20%	10%
Carbohydrate	10%	50%	4%
Fat	10%	2%	0

If a diet is to consist of 8.6 ounces of protein, 4.28 ounces of carbohydrate, and 1.32 ounces of fat each day, how many ounces of each food group must be consumed?

27. *Social Science* The Government distributes $680,000 among 100 scientists who belong to three different research organizations, A, B, and C. Each scientist in organization A receives $10,000, each in B receives $4000, and each in C receives $5000. Organization A receives 5 times as much as that received by organization B. Determine how many scientists belong to each research organization. How much is received by each organization?

6.3

SYSTEMS OF NONLINEAR EQUATIONS

SUBSTI-TUTION The substitution method is sometimes a convenient means of finding solution sets of systems in which nonlinear equations are present.

Example Solve.

(1) $$x^2 + y^2 = 25$$
(2) $$x + y = 7$$

Solution Equation (2) can be written equivalently in the form

(3) $$y = 7 - x,$$

and we can replace y in (1) by $(7 - x)$ from (3). This produces

(4) $$x^2 + (7 - x)^2 = 25.$$

The solution set of (4) contains all values of x for which the ordered pair (x, y) is a common solution of (1) and (2). We can now find the solution set of (4):

$$x^2 + 49 - 14x + x^2 = 25$$
$$x^2 - 7x + 12 = 0$$
$$(x - 3)(x - 4) = 0,$$

which is satisfied if x is either 3 or 4. Now, replacing x in the *first-degree equation* (3) by each of these numbers, we obtain

$$y = 7 - 3 = 4 \quad \text{and} \quad y = 7 - 4 = 3,$$

so the solution set of the system (1) and (2) is $\{(3, 4), (4, 3)\}$.

If, in the above example, we had substituted the values 3 and 4 that we obtained for x in the *second-degree equation* (1), we would have obtained some extraneous solutions that do not satisfy (2).

LINEAR COMBINATION If both the equations in a system of two equations in two variables are second-degree in both variables, the use of linear combinations of the equations often provides a simpler means of solution than does substitution.

Example Solve.

(5) $\qquad\qquad\qquad 3x^2 - 7y^2 + 15 = 0$
(6) $\qquad\qquad\qquad 3x^2 - 4y^2 - 12 = 0$

Solution By forming the linear combination of -1 times (5) and 1 times (6), we obtain

$$3y^2 - 27 = 0$$
$$y^2 = 9,$$

from which

$$y = 3 \quad \text{or} \quad y = -3,$$

and we have the y-components of the members of the solution set of the system (5) and (6). Substituting 3 for y in either (5) or (6), say (5), we have

$$3x^2 - 7(3)^2 + 15 = 0$$
$$x^2 = 16,$$

from which

$$x = 4 \quad \text{or} \quad x = -4.$$

Thus, the ordered pairs $(4, 3)$ and $(-4, 3)$ are solutions of the system. Substituting -3 for y in (5) or (6) [this time we shall use (6)] gives us

$$3x^2 - 4(-3)^2 - 12 = 0$$
$$x^2 = 16,$$

so that

6.3 Systems of Nonlinear Equations

$$x = 4 \quad \text{or} \quad x = -4.$$

Thus, the ordered pairs $(4, -3)$ and $(-4, -3)$ are solutions of the system. Accordingly, the complete solution set is $\{(4, 3), (4, -3), (-4, 3), (-4, -3)\}$.

LINEAR COMBINATION AND SUBSTITUTION

The solution of some systems requires the application of both linear combinations and substitution.

Example Solve.

(7) $$x^2 + y^2 = 5$$
(8) $$x^2 - 2xy + y^2 = 1$$

Solution By forming the linear combination of 1 times (7) and -1 times (8), we obtain

(9) $$2xy = 4$$
$$xy = 2.$$

The system (7) and (8) is equivalent to the system (7) and (9), which can be solved by substitution. From (9), we have

$$y = \frac{2}{x}.$$

Replacing y in (7) by $2/x$, we obtain

$$x^2 + \left(\frac{2}{x}\right)^2 = 5$$

(10) $$x^2 + \frac{4}{x^2} = 5$$

$$x^4 + 4 = 5x^2$$
(11) $$x^4 - 5x^2 + 4 = 0,$$

which is a quadratic in x^2. The left-hand member of (11), when factored, yields

$$(x^2 - 1)(x^2 - 4) = 0,$$

from which we obtain

$$x^2 - 1 = 0 \quad \text{or} \quad x^2 - 4 = 0,$$

so that

$$x = 1, \quad x = -1, \quad x = 2, \quad x = -2.$$

Since the step from (10) to (11) was not an elementary transformation, we are careful to note that these values of x all satisfy (10). Substituting $1, -1, 2,$ and -2 in turn for x in (9), we obtain the corresponding values for y, and thus the solution set of the system (7) and (9) or, equivalently, the solution set of the system (7) and (8), is $\{(1, 2), (-1, -2), (2, 1), (-2, -1)\}$.

PERCEPTION In the study of perception, psychologists have maintained that a rectangle that is most pleasing to the human eye is one whose dimensions satisfy the following: the ratio of the length of the short side is to the length of the long side as the length of the long side is to the sum of the lengths of the short and long sides. This proportion is called the **golden ratio**, or **golden section**, and often occurs in architecture and paintings of antiquity. If the lengths of sides are as indicated in Figure 6.5, then the golden ratio is given by

(12) $$\frac{x}{y} = \frac{y}{x+y}.$$

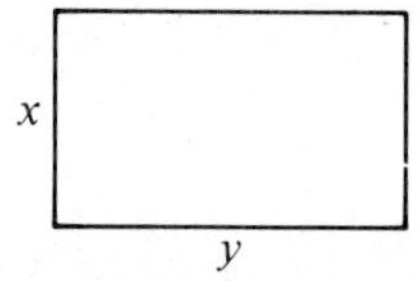

Figure 6.5

Example A sheet of paper is to contain 100 square inches. Determine the dimensions of the page that satisfy the specifications of the golden ratio.

Solution Multiplying both members of (12) by $y(x+y)$ yields $x^2 + xy = y^2$. Now since the area of the page is 100 square inches, we must also have $xy = 100$. Thus, we obtain the system of equations

(13) $$x^2 + xy = y^2$$

(14) $$xy = 100.$$

Using (14) to write $y = 100/x$ in (13) then gives

$$x^2 + 100 = \left(\frac{100}{x}\right)^2$$

$$x^4 + 100x^2 = 10{,}000$$

(15) $$x^4 + 100x^2 - 10{,}000 = 0.$$

From the quadratic formula we find that the only real solutions of (15) are obtained from

$$x^2 = -50 + 50\sqrt{5}.$$

6.3 Systems of Nonlinear Equations

Since $x > 0$ we obtain

$$x = \sqrt{-50 + 50\sqrt{5}} \approx 7.9$$

$$y = \frac{100}{\sqrt{-50 + 50\sqrt{5}}} \approx 12.7$$

We leave it as an exercise (Problem 32) to show that the proportion of the length of the short side to the length of the long side in a so-called golden rectangle is 1:1.618 (approximately).

APPROXI-MATIONS TO SOLUTIONS

When the degree of any equation in a set of equations is greater than two, or when the left-hand member of one of the equations of the form $f(x, y) = 0$ is not a polynomial, it may be very difficult or impossible to find common solutions analytically. In this case, we can at least obtain approximations to any real solutions by graphical methods. Consider the system of equations

(16)
$$y = 2^{-x}$$
$$y = \sqrt{0.01x},$$

and the graphs of these equations in Figure 6.6.

Examining Figure 6.6, part b, which is an enlargement of a section of part a, we observe that the curves intersect at approximately (2.6, 0.16). From geometric considerations we conclude that this ordered pair approximates the only member in the solution set of (16).

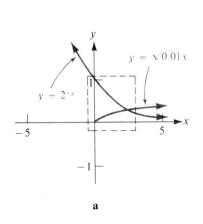
a

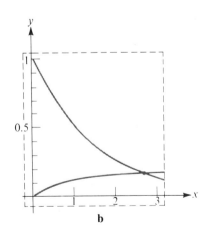
b

Figure 6.6

This result shows, in particular, that the equation

$$2^{-x} - \sqrt{0.01x} = 0$$

has just one root, approximately 2.6, and suggests a method of approximating solutions of similar equations.

EXERCISE 6.3

Solve by the method of substitution. Check the solutions by sketching the graphs of the equations and estimating the coordinates of any points of intersection.

Example (1) $\qquad y = x^2 + 2x + 1$
(2) $\qquad y - x = 3$

Solution Solve (2) explicitly for y.

(2') $\qquad y = x + 3$

Substitute $x + 3$ for y in (1).

(3) $\qquad x + 3 = x^2 + 2x + 1$

Solve for x.

$$x^2 + x - 2 = 0$$
$$(x + 2)(x - 1) = 0$$
$$x = -2 \qquad x = 1$$

Substitute each of these values in turn in (2') to determine values for y.

If $x = -2$, then $y = 1$.
If $x = 1$, then $y = 4$.

The solution set is $\{(-2, 1), (1, 4)\}$.

1. $y = x^2 - 5$
   $y = 4x$
2. $y = x^2 - 2x + 1$
   $y + x = 3$
3. $x^2 + y^2 = 13$
   $x + y = 5$
4. $x^2 + 2y^2 = 12$
   $2x - y = 2$
5. $x + y = 1$
   $xy = -12$
6. $2x - y = 9$
   $xy = -4$

Solve using linear combinations.

6.3 Systems of Nonlinear Equations

7. $x^2 + y^2 = 10$
   $9x^2 + y^2 = 18$

8. $x^2 + 4y^2 = 52$
   $x^2 + y^2 = 25$

9. $x^2 - y^2 = 7$
   $2x^2 + 3y^2 = 24$

10. $x^2 + 4y^2 = 25$
    $4x^2 + y^2 = 25$

11. $4x^2 - 9y^2 + 132 = 0$
    $x^2 + 4y^2 - 67 = 0$

12. $16y^2 + 5x^2 - 26 = 0$
    $25y^2 - 4x^2 - 17 = 0$

13. $x^2 - xy + y^2 = 7$
    $x^2 + y^2 = 5$

14. $3x^2 - 2xy + 3y^2 = 34$
    $x^2 + y^2 = 17$

15. $3x^2 + 3xy - y^2 = 35$
    $x^2 - xy - 6y^2 = 0$

16. $x^2 - xy + y^2 = 21$
    $x^2 + 2xy - 8y^2 = 0$

Solve by graphing. Approximate components of solutions to the nearest half unit.

17. $y = 10^x + 1$
    $x + y = 2$

18. $y = 2^x$
    $y - x = 2$

19. $y = \log_{10} x$
    $y = x^2 - 2x + 1$

20. $y = 10^x$
    $y = x^2$

21. $y = 10^{-x}$
    $y = \log_{10} x$

22. $y = 2^{x-3}$
    $y = \log_2 x$

23. How many real solutions are possible for simultaneous systems of linearly independent equations that consist of:
    a. two linear equations in two variables?
    b. one linear equation and one quadratic equation in two variables?
    c. two quadratic equations in two variables?
    Support each of your answers with sketches.

24. What relationships must exist between the numbers a and b so that the solution set of the system
    $$x^2 + y^2 = 25$$
    $$y = ax + b$$
    has two ordered pairs of real numbers? One ordered pair of real numbers? No ordered pairs of real numbers? [*Hint*: Use substitution and consider the nature of the roots of the resulting quadratic equation.]

Applications

25. The sum of two numbers is 6 and their product is 35/4. Find the numbers.

26. The sum of the squares of two positive numbers is 13. If twice the first number is added to the second, the sum is 7. Find the numbers.

27. The hypotenuse of a right triangle is 400 centimeters. Find the length of sides if the shorter side is one-half the longer.

28. *Business* The annual income from an investment is $32. If the amount invested were $200 more and the rate 1/2% less, the annual income would be $35. What are the amount and rate of the investment?

29. The area of a rectangular field is 4500 square feet. The cost of fencing on the longer side is $2 per foot and on the shorter side is $1 per foot. What are the dimensions of the field if the total cost for fencing is $660?

30. *Physics* At a constant temperature, the pressure P and volume V of a gas are related by the equation $PV = k$. The product of the pressure (in pounds per square inch) and the volume (in cubic inches) of a certain gas is 30 inch-pounds. If the temperature remains constant as the pressure is increased 4 pounds per square inch, the volume is decreased by 2 cubic inches. Find the original pressure and volume of the gas.

31. *Social Science* Find the dimensions of a golden rectangle whose area is 500 square centimeters.

32. Show that the proportion of the length of the short side to the length of the long side in a golden rectangle is 1:1.618 (approximately). [*Hint*: Write (12) as $y/x = (x+y)/y$ and solve for x/y.]

6.4

BREAK-EVEN POINT AND EQUILIBRIUM POINT

BREAK-EVEN POINT If $C(x)$ is the cost to a company for producing x units of an item, and if $R(x)$ is the total revenue obtained by the sale of these items, then the point where $C(x) = R(x)$, for $x \geq 0$, is called the **break-even point**. That is, the break-even point is the point, or number of units manufactured, after which the company will show either a profit or a loss. Geometrically it is simply the point of intersection, if any, of the revenue curve with the cost curve. We note that $x \geq 0$, since it is implicit that the company should produce some or none of the units.

Example A small manufacturer of leather goods sells coats for $200. If the cost to the company is $C(x) = 125x + 300$, find the number of coats that must be sold before the company shows a profit.

Solution If x denotes the number of coats sold, then $R(x) = 200x$ is the revenue taken in by the company. Equate $C(x)$ and $R(x)$.

$$125x + 300 = 200x.$$

This gives

$$x = 4.$$

The figure shows that the break-even point is (4,800), and for $x > 4$ the company shows a profit.

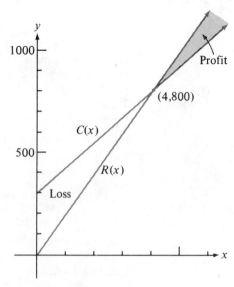

6.4 Break-Even Point and Equilibrium Point

Alternatively, we could define a profit function $P(x) = R(x) - C(x)$. The break-even point is then the x-intercept for $x \geq 0$, as shown in Figure 6.7. For the above example, we have $P(x) = 75x - 300$.

We observe that:

1. the solution of $C(x) = R(x)$ or $P(x) = 0$ does not have to be an integer,
2. neither $C(x)$ nor $R(x)$ have to be linear functions.

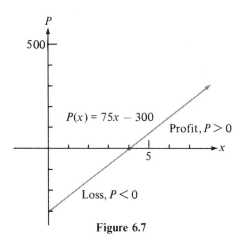

Figure 6.7

Example Find the break-even point(s) when $R(x) = 20x - \dfrac{x^2}{20}$ and $C(x) = 4x + 280$.

Solution The equation $R(x) = C(x)$ gives

$$20x - \frac{x^2}{10} = 4x + 280$$

$$0 = x^2 - 160x + 2800$$

$$0 = (x - 20)(x - 140).$$

Thus, there are two break-even points: $(20, R(20)) = (20, 360)$ and $(140, R(140)) = (140, 840)$. The accompanying figure shows the graphs of $R(x)$ and $C(x)$ and the regions of profit and loss.

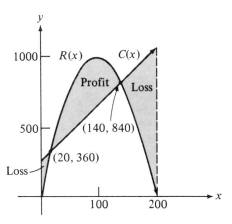

EQUILIBRIUM POINT In economic theory it is said that an **equilibrium point** is attained, or the market is in equilibrium, when the demand for a certain commodity equals the supply on hand. If $D(x)$ and $S(x)$ are the respective demand and supply functions corresponding to a price x in dollars, we must solve $D(x) = S(x)$ for $x \geq 0$. At this price, there is neither a shortage nor a surplus of the commodity.

Example Find the equilibrium point when $D(x) = 20 - x^2$ and $S(x) = x^2 + 3x$.

Solution The equation $D(x) = S(x)$ yields

$$20 - x^2 = x^2 + 3x$$
$$0 = 2x^2 + 3x - 20$$
$$0 = (2x - 5)(x + 4)$$

has the solutions $x = 2.5$ and $x = -4$. Since x must be nonnegative, the equilibrium point is (2.5, 13.75).

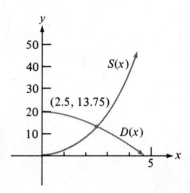

We know intuitively that $D(x)$ should be a decreasing function, since the demand for a commodity goes down as the price goes up.

EXERCISE 6.4

Find the break-even point(s).

1. $C(x) = 800x$, $R(x) = 400x + 36{,}000$
2. $C(x) = 500x$, $R(x) = 200x + 9000$
3. $C(x) = 30x + 100$, $R(x) = 5x + 650$
4. $C(x) = 60x + 850$, $R(x) = 10x + 1700$
5. $C(x) = 5x + 250$, $R(x) = 60x - x^2$
6. $C(x) = 27x + 44$, $R(x) = 40x - 0.5x^2$

Find the equilibrium point(s).

7. $D(x) = 18 - x$, $S(x) = x + 4$
8. $D(x) = 10 - 2x$, $S(x) = 0.4x + 4$
9. $D(x) = 900 - x^2$, $S(x) = x^2 + 70x$
10. $D(x) = 850 - 0.5x^2$, $S(x) = x^2 + 250$

6.4 Break-Even Point and Equilibrium Point

11. $D(x) = \dfrac{1500}{x}$, $S(x) = 2x + 10$

12. $D(x) = \dfrac{3}{x^2}$, $S(x) = x + 2$

Applications

13. *Business* Given that the supply function $S(x)$ is a linear function passing through the origin and the equilibrium point $(40, 50)$, and that the graph of the demand function $D(x)$ is a straight line perpendicular to the graph of $S(x)$, determine the equations of $S(x)$ and $D(x)$.

14. Let x be the price of a commodity, and suppose the demand for the commodity is given by $D(x) = 1000x^2 - 6000x + 9000$. Calculate $D(2)$, $D(3)$, and $D(4)$. Explain why such a demand does not make economic sense for $x > 3$.

15. A company finds that its fixed cost is \$50,000 and that it costs \$200 per unit to manufacture an item. If the item sells for \$700 per unit, find the
    a. cost and revenue functions,
    b. profit function,
    c. break-even point.

16. A company produces x units of a commodity, where x is a positive integer. If the cost and revenue functions are $C(x) = 50x + 5700$ and $R(x) = 450x$, what is the smallest number of units that must be made before the company realizes a profit?

Example A vendor drives a truck through residential neighborhoods to sell ice cream bars. Each day his initial cost for the ice cream is \$40. The cost for gas, oil, and general upkeep of the truck is figured to be 20 cents per mile. If the vendor averages 15 sales at 20 cents apiece each mile, how many miles and how many ice cream bars must he sell in order to show a daily profit?

Solution If x denotes miles, then the daily cost function in terms of cents is

$$C(x) = 20x + 4000.$$

The revenue function is

$$R(x) = 300x.$$

Equate $C(x)$ and $R(x)$,

$$20x + 4000 = 300x$$
$$280x = 4000$$

This gives

$$x = 14\tfrac{2}{7} \text{ miles}.$$

The number of bars that must be sold before a profit is shown is

$$15\left(14\tfrac{2}{7}\right) = 214\tfrac{2}{7}$$

rounded to the next largest integer, namely, 215.

17. A small service station sells gas at $1.35 per gallon. At the start of every week the initial cost for a supply of gas is $600. If the owner figures the incidental cost per gallon at 20 cents then how many gallons of gas must be pumped per week before a profit is shown?

18. A drive-in movie theater presently has 200 parking spaces. The owner decides to increase the number of parking spaces to a number not to exceed 500. The revenue per space is now $4, but it will decrease by 1 cent for each additional space over 200. If the owner determines that the total cost will be $2 per space, because of the cost of the added land and more employees, find the break-even point. Should the owner increase the size of the theater? [*Hint*: Graph the profit function.]

19. A car rental agency has a fleet of 200 cars and charges $14 for each car per day. For each $1 increase in the daily charge, the agency determines that it will rent four fewer cars. It costs the agency $6 per day for each rented car for general maintenance. If x denotes the amount of increase in daily charge, determine
    a. the revenue and cost functions $R(x)$ and $C(x)$,
    b. the break-even point,
    c. the profit function $P(x)$,
    d. the interval for which $P > 0$.

CHAPTER TEST

Solve each system.

1. $x + 5y = 18$
   $x - y = -3$

2. $x + 5y = 11$
   $2x + 3y = 8$

3. $2x - 3y = 8$
   $3x + 2y = 7$

4. Find values for a and b so that the graph of $ax + by = 19$ passes through the points $(2, 3)$ and $(-3, 5)$.

Solve each system.

5. $x + 3y - z = 3$
   $2x - y + 3z = 1$
   $3x + 2y + z = 5$

6. $x + y + z = 2$
   $3x - y + z = 4$
   $2x + y + 2z = 3$

7. $2x + 3y - z = -2$
   $x - y + z = 6$
   $3x - y + z = 10$

8. Find values for a, b, and c so that the graph of $y = ax^2 + bx + c$ contains the points $(-1, 9)$, $(0, 4)$, and $(1, 3)$.

Solve each system.

9. $x^2 + y = 3$
   $5x + y = 7$

10. $x^2 + 3xy + x = -12$
    $2x - y = 7$

11. $2x^2 + 5y^2 - 53 = 0$
    $4x^2 + 3y^2 - 43 = 0$

12. Approximate the solution of the system
    $$y = 3^x$$
    $$y = 3 - x$$
    by graphical methods.

13. A biologist wishes to make a nutrient solution containing 56 milligrams of food A and 39 milligrams of food B by combining two solutions X and Y. Solution X is known to consist of 40% food A and 15% food B; solution Y consists of 20% food A and 30% food B. How much of solutions X and Y must be used in order to obtain the desired nutrient solution?

14. Determine the break-even point if $R(x) = 50x$ and $C(x) = 10x + 70$, where x is the whole number of units produced.

15. Find the equilibrium point if the supply and demand curves are determined by the functions $S(x) = 2x + 1$ and $D(x) = \dfrac{14}{x + 2}$.

7
MATRICES AND DETERMINANTS

7.1
DEFINITIONS; MATRIX ADDITION

Matrices are a common tool today in mathematics and business, physical, social, and life sciences.

If a company sells x units of a commodity, then the information concerning revenue, cost, and profit can be represented by the fourtuple $(x, r(x), c(x), p(x))$. Similarly, if A, B, and C are stores selling a single commodity, then the information about the revenue, cost, and profit for 1 year could be given in tabular form as

Stores	A	B	C	
Units sold	4	7	12	
Yearly revenue	16	19	37	in units of dollars
Yearly costs	9	8	10	
Yearly profit	7	11	27	

or simply by the array of numbers

$$\begin{bmatrix} 4 & 7 & 12 \\ 16 & 19 & 37 \\ 9 & 8 & 10 \\ 7 & 11 & 27 \end{bmatrix}.$$

If the revenue, cost, and profit information for the same stores for a second year is given by the array

$$\begin{bmatrix} 5 & 8 & 11 \\ 17 & 20 & 35 \\ 10 & 10 & 12 \\ 7 & 10 & 23 \end{bmatrix}$$

7.1 Definitions; Matrix Addition

then the array that gives the *combined* number of units sold for two successive years, along with the total revenue, cost, and profit information is

$$\begin{bmatrix} 9 & 15 & 23 \\ 33 & 39 & 72 \\ 19 & 18 & 22 \\ 14 & 21 & 50 \end{bmatrix}$$

We note that the numbers in this last array are obtained by adding the corresponding numbers in the preceding two arrays. In the following discussion we shall generalize the concept of arrays of numbers as well as the addition of two (or more) arrays.

RMINOLOGY Any rectangular array of numbers (or other suitable entries) is called a **matrix**. A matrix is customarily displayed in a pair of brackets or parentheses, as

$$\begin{bmatrix} 1 & 2 & 3 \\ 4 & 5 & 6 \end{bmatrix} \quad \text{or} \quad (x \quad r(x) \quad c(x) \quad p(x)).$$

In this book, we shall use the bracket notation.

The entries in a matrix are called **elements**. The **order**, or **dimension**, of a matrix is the ordered pair having as first component the number of (horizontal) **rows** and as second component the number of (vertical) **columns** in the matrix. Thus,

$$\begin{bmatrix} 1 & 2 & 3 \\ 4 & 5 & 6 \end{bmatrix}, \quad \begin{bmatrix} 1 \\ 2 \\ 3 \end{bmatrix}, \quad \begin{bmatrix} a_1 & a_2 & a_3 & a_4 \\ b_1 & b_2 & b_3 & b_4 \\ c_1 & c_2 & c_3 & c_4 \\ d_1 & d_2 & d_3 & d_4 \end{bmatrix}$$

are 2×3 (read "two by three"), 3×1 (read "three by one"), and 4×4 (read "four by four") matrices, respectively. Note that the number of *rows* is given first, and then the number of *columns*. A matrix consisting of a single row is called a **row matrix** or a **row vector**, whereas a matrix consisting of a single column is called a **column matrix** or a **column vector**.

Matrices are frequently represented by capital letters. Thus, we might want to talk about the matrices A and B, where

$$A = \begin{bmatrix} a_1 & a_2 \\ b_1 & b_2 \end{bmatrix} \quad \text{and} \quad B = [b_1 \quad b_2].$$

To show that A is a 2×2 matrix, we can write $A_{2 \times 2}$. Similarly, $B_{1 \times 2}$ is a matrix with one row and two columns.

To represent the entries of a matrix, either single or double subscript notation is employed. Consider any 3×3 matrix, A. We can represent A by

$$A = \begin{bmatrix} a_1 & a_2 & a_3 \\ b_1 & b_2 & b_3 \\ c_1 & c_2 & c_3 \end{bmatrix},$$

where a different letter is used for each row and a single subscript denotes the column in which each entry is located. Alternatively, we can use a different letter for each column, and let the subscript denote the row. Thus, we might write

$$A = \begin{bmatrix} a_1 & b_1 & c_1 \\ a_2 & b_2 & c_2 \\ a_3 & b_3 & c_3 \end{bmatrix}.$$

In either event, problems would clearly arise if we wanted to talk about a matrix containing a large number of rows or columns, because we would run out of letters.

A much more useful convention involves double subscripts, where a single letter, say a, is used to denote an entry in a matrix, and then *two* subscripts are appended, the first subscript telling in which *row* the entry occurs, and the second telling in which *column*. Thus, we write

$$A = \begin{bmatrix} a_{11} & a_{12} & a_{13} \\ a_{21} & a_{22} & a_{23} \\ a_{31} & a_{32} & a_{33} \end{bmatrix},$$

where a_{21} is the element in the second *row* and first *column*, a_{33} is the element in the third *row* and third *column*, and, if we wish to generalize, a_{ij} is the element in the ith *row* and jth *column*.

Two matrices, A and B, are **equal** if and only if both matrices are of the same order and $a_{ij} = b_{ij}$ for each i, j.

Thus,

$$\begin{bmatrix} 2 & 1 \\ 3 & 0 \end{bmatrix} = \begin{bmatrix} \frac{4}{2} & 2-1 \\ \sqrt{9} & 0 \end{bmatrix} \quad \text{but} \quad \begin{bmatrix} 2 & 1 \\ 3 & 0 \end{bmatrix} \neq \begin{bmatrix} 2 & 3 \\ 1 & 0 \end{bmatrix}.$$

The **transpose** of a matrix A, denoted by A^t, is the matrix in which the rows are the columns of A and the columns are the rows of A.

Thus,

$$\begin{bmatrix} 2 & 1 \\ 3 & 0 \end{bmatrix}^t = \begin{bmatrix} 2 & 3 \\ 1 & 0 \end{bmatrix} \quad \text{and} \quad \begin{bmatrix} 1 & 2 & 3 \\ 4 & 5 & 6 \end{bmatrix}^t = \begin{bmatrix} 1 & 4 \\ 2 & 5 \\ 3 & 6 \end{bmatrix}.$$

The **sum** of two matrices of the same order, $A_{m \times n}$ and $B_{m \times n}$, is the matrix $(A + B)_{m \times n}$, in which the entry in the ith row and jth column is $a_{ij} + b_{ij}$, for $i = 1, 2, 3, \ldots, m$ and $j = 1, 2, 3, \ldots, n$.

7.1 Definitions; Matrix Addition

For example, from our initial discussion on page 222 we have seen

$$\begin{bmatrix} 4 & 7 & 12 \\ 16 & 19 & 37 \\ 9 & 8 & 10 \\ 7 & 11 & 27 \end{bmatrix} + \begin{bmatrix} 5 & 8 & 11 \\ 17 & 20 & 35 \\ 10 & 10 & 12 \\ 7 & 10 & 23 \end{bmatrix} = \begin{bmatrix} 4+5 & 7+8 & 12+11 \\ 16+17 & 19+20 & 37+35 \\ 9+10 & 8+10 & 10+12 \\ 7+7 & 11+10 & 27+23 \end{bmatrix}$$

$$= \begin{bmatrix} 9 & 15 & 23 \\ 33 & 39 & 72 \\ 19 & 18 & 22 \\ 14 & 21 & 50 \end{bmatrix}$$

The sum of two matrices of different order is not defined. For example, the sum of

$$A = \begin{bmatrix} 1 & 3 & 4 \end{bmatrix}$$

and

$$B = \begin{bmatrix} 1 \\ 5 \end{bmatrix}$$

is not defined.

*A matrix with each entry equal to 0 is a **zero matrix**.*

Zero matrices are generally denoted by the symbol **0**. This distinguishes the zero matrix from the real number 0. For example,

$$\mathbf{0}_{2 \times 4} = \begin{bmatrix} 0 & 0 & 0 & 0 \\ 0 & 0 & 0 & 0 \end{bmatrix} \quad \text{and} \quad \mathbf{0}_{3 \times 3} = \begin{bmatrix} 0 & 0 & 0 \\ 0 & 0 & 0 \\ 0 & 0 & 0 \end{bmatrix}$$

are 2 × 4 and 3 × 3 zero matrices.

*The **negative of a matrix** $A_{m \times n}$, denoted by $-A_{m \times n}$, is formed by replacing each entry in the matrix $A_{m \times n}$ with its negative.*

For example, if

$$A_{3 \times 2} = \begin{bmatrix} 3 & -1 \\ 2 & -2 \\ -4 & 5 \end{bmatrix}, \quad \text{then} \quad -A_{3 \times 2} = \begin{bmatrix} -3 & 1 \\ -2 & 2 \\ 4 & -5 \end{bmatrix}.$$

The sum $B_{m \times n} + (-A_{m \times n})$ is called the **difference** of $B_{m \times n}$ and $A_{m \times n}$ and is denoted $B_{m \times n} - A_{m \times n}$.

PROPERTIES OF SUMS

At this point, we are able to establish the following facts concerning sums of matrices with real-number entries.

> If A, B, and C are $m \times n$ matrices with real-number entries, then:
>
> **I** $(A + B)_{m \times n}$ is a matrix with real-number entries,
>
> **II** $(A + B) + C = A + (B + C)$,
>
> **III** $A + B = B + A$,
>
> **IV** the matrix $\mathbf{0}_{m \times n}$ has the property that for every matrix $A_{m \times n}$,
> $$A + \mathbf{0} = A \quad \text{and} \quad \mathbf{0} + A = A,$$
>
> **V** for every matrix $A_{m \times n}$, the matrix $-A_{m \times n}$ has the property that
> $$A + (-A) = \mathbf{0} \quad \text{and} \quad (-A) + A = \mathbf{0}.$$

To establish property IV, we note that since each entry of the zero matrix is 0, it follows that the entries of $A_{m \times n} + \mathbf{0}_{m \times n}$ are $a_{ij} + 0 = a_{ij}$. Similarly, the entries of $\mathbf{0}_{m \times n} + A_{m \times n}$ are $0 + a_{ij} = a_{ij}$.

For example,

$$\begin{bmatrix} a_{11} & a_{12} \\ a_{21} & a_{22} \end{bmatrix} + \begin{bmatrix} 0 & 0 \\ 0 & 0 \end{bmatrix} = \begin{bmatrix} a_{11} & a_{12} \\ a_{21} & a_{22} \end{bmatrix}.$$

To establish property V, we let the entries of $A_{m \times n}$ and $-A_{m \times n}$ be a_{ij} and $-a_{ij}$, respectively. Since each entry of $A + (-A)$ is $a_{ij} - a_{ij}$, or 0, we have $A + (-A) = \mathbf{0}$. Similarly, $(-A) + A = \mathbf{0}$.

For example, if

$$A = \begin{bmatrix} 1 & -1 & 2 \\ 3 & -1 & 1 \end{bmatrix},$$

then

$$A + (-A) = \begin{bmatrix} 1 & -1 & 2 \\ 3 & -1 & 1 \end{bmatrix} + \begin{bmatrix} -1 & 1 & -2 \\ -3 & 1 & -1 \end{bmatrix} = \begin{bmatrix} 0 & 0 & 0 \\ 0 & 0 & 0 \end{bmatrix} = \mathbf{0}.$$

The verification of properties I, II, and III is left as an exercise.

7.1 Definitions; Matrix Addition

EXERCISE 7.1

State the order and find the transpose of each matrix.

Example $\begin{bmatrix} 2 & 4 \\ 1 & -3 \\ 6 & 0 \end{bmatrix}$

Solution 3×2 matrix; $\begin{bmatrix} 2 & 4 \\ 1 & -3 \\ 6 & 0 \end{bmatrix}^t = \begin{bmatrix} 2 & 1 & 6 \\ 4 & -3 & 0 \end{bmatrix}$

1. $\begin{bmatrix} 6 & -1 \\ 2 & 3 \end{bmatrix}$
2. $\begin{bmatrix} 4 & 1 \\ 0 & -2 \end{bmatrix}$
3. $\begin{bmatrix} 2 & -7 & 3 \\ 1 & 4 & 0 \end{bmatrix}$

4. $\begin{bmatrix} -3 & 1 \\ 6 & 0 \\ 0 & 2 \end{bmatrix}$
5. $\begin{bmatrix} 2 & 3 & -1 \\ 4 & 0 & 1 \\ -2 & 3 & 1 \end{bmatrix}$
6. $\begin{bmatrix} 4 & -1 & -2 \\ 3 & 0 & 0 \\ 2 & 1 & 1 \end{bmatrix}$

7. $\begin{bmatrix} 4 & -3 & -1 & 0 \\ 2 & 1 & 1 & 6 \end{bmatrix}$
8. $\begin{bmatrix} -2 & 1 & 3 & 2 \\ 4 & 0 & 0 & -2 \\ -1 & 3 & 2 & 4 \end{bmatrix}$

Write each sum or difference as a single matrix.

Example $\begin{bmatrix} 2 & 1 & 4 \\ 3 & -1 & 0 \end{bmatrix} + \begin{bmatrix} 6 & 3 & 0 \\ -2 & 1 & 0 \end{bmatrix}$

Solution $\begin{bmatrix} 2 & 1 & 4 \\ 3 & -1 & 0 \end{bmatrix} + \begin{bmatrix} 6 & 3 & 0 \\ -2 & 1 & 0 \end{bmatrix} = \begin{bmatrix} 2+6 & 1+3 & 4+0 \\ 3-2 & -1+1 & 0+0 \end{bmatrix} = \begin{bmatrix} 8 & 4 & 4 \\ 1 & 0 & 0 \end{bmatrix}$

9. $\begin{bmatrix} 2 & 3 \\ 1 & 6 \end{bmatrix} + \begin{bmatrix} 1 & -2 \\ 2 & 3 \end{bmatrix}$
10. $\begin{bmatrix} 4 & -1 & 3 \\ 2 & 1 & 0 \end{bmatrix} + \begin{bmatrix} 3 & -1 & 0 \\ 4 & 0 & -2 \end{bmatrix}$

11. $\begin{bmatrix} 3 & 0 & -1 \\ 2 & 1 & 2 \end{bmatrix} + \begin{bmatrix} 6 & -1 & 0 \\ 0 & 2 & 4 \end{bmatrix}$
12. $[1 \ 3 \ 5 \ 7] + [0 \ -2 \ 1 \ 3]$

13. $\begin{bmatrix} 4 & -3 \\ 2 & 1 \end{bmatrix} - \begin{bmatrix} 6 & 0 \\ -2 & 1 \end{bmatrix}$
14. $\begin{bmatrix} 4 & -1 & 2 \\ 3 & 1 & -4 \end{bmatrix} - \begin{bmatrix} -1 & -1 & 2 \\ 3 & 1 & 4 \end{bmatrix}$

15. $\begin{bmatrix} 10 & 3 & 2 \\ 5 & 1 & 7 \\ 6 & 1 & 9 \end{bmatrix} - \begin{bmatrix} 8 & 12 & 15 \\ -2 & 5 & 6 \\ -3 & 1 & 9 \end{bmatrix}$
16. $\begin{bmatrix} 3 & -1 & 2 \\ 4 & -2 & 1 \\ 6 & 3 & 2 \end{bmatrix} - \begin{bmatrix} 2 & -1 & 2 \\ 4 & -1 & 1 \\ 6 & 3 & 1 \end{bmatrix}$

17. $\begin{bmatrix} 4 \\ 3 \\ -1 \end{bmatrix} + \begin{bmatrix} 6 \\ 0 \\ -2 \end{bmatrix}$
18. $\begin{bmatrix} 2 & 3 \\ 1 & 0 \\ -1 & 2 \end{bmatrix} + \begin{bmatrix} -2 & 0 \\ -3 & 0 \\ 4 & -1 \end{bmatrix}$

19. $\begin{bmatrix} 2 & 3 & 4 \\ -1 & 6 & 2 \\ 1 & 0 & 3 \end{bmatrix} + \begin{bmatrix} 0 & 0 & 0 \\ 0 & 0 & 0 \\ 0 & 0 & 0 \end{bmatrix}$

20. $\begin{bmatrix} 2 & -3 \\ 4 & -1 \\ -2 & 1 \end{bmatrix} + \begin{bmatrix} -2 & 3 \\ -4 & 1 \\ 2 & -1 \end{bmatrix}$

21. Use properties I–IV to argue that $X + A = B$ and $X = B - A$ are equivalent matrix equations in the system of 2×2 matrices.

Solve each of the following matrix equations.

Example $X + \begin{bmatrix} 2 & 3 \\ 1 & 7 \end{bmatrix} = \begin{bmatrix} 9 & -4 \\ 2 & 0 \end{bmatrix}$

Solution $X = \begin{bmatrix} 9 & -4 \\ 2 & 0 \end{bmatrix} - \begin{bmatrix} 2 & 3 \\ 1 & 7 \end{bmatrix} = \begin{bmatrix} 7 & -7 \\ 1 & -7 \end{bmatrix}$

22. $X + \begin{bmatrix} 3 & -1 \\ 2 & 1 \end{bmatrix} = \begin{bmatrix} 5 & 1 \\ -3 & 5 \end{bmatrix}$

23. $X - \begin{bmatrix} -1 & 0 \\ 0 & 0 \end{bmatrix} = \begin{bmatrix} 3 & -1 \\ 2 & 1 \end{bmatrix}^t$

24. $X + \begin{bmatrix} 3 & 2 \\ -1 & 4 \end{bmatrix} = \begin{bmatrix} 2 & 6 \\ 1 & 5 \end{bmatrix} + \begin{bmatrix} -4 & -8 \\ -2 & 0 \end{bmatrix}$

25. $\begin{bmatrix} 1 & 3 \\ -1 & 0 \end{bmatrix}^t - \begin{bmatrix} 0 & 1 \\ 1 & 0 \end{bmatrix} = \begin{bmatrix} 2 & -2 \\ -1 & 3 \end{bmatrix}^t - X$

26. $\begin{bmatrix} 2 & 4 \\ 1 & -1 \end{bmatrix} - \begin{bmatrix} 6 & -2 \\ -3 & 1 \end{bmatrix}^t = \begin{bmatrix} 7 & 3 \\ 0 & -2 \end{bmatrix} - X$

27. Show that $[A_{2 \times 2} + B_{2 \times 2}]^t = A^t_{2 \times 2} + B^t_{2 \times 2}$. Does an analogous result seem valid for $n \times n$ matrices?

28. Verify property I of matrices.

29. Verify property II of matrices.

30. Verify property III of matrices.

Applications

31. *Biology* The label on a can of beans contains the following nutritional information: 260 calories, 44 grams carbohydrate, 12 grams protein, 4 grams fat. For a can of corn of the same size there are: 150 calories, 30 grams carbohydrate, 4 grams protein, and 1 gram fat. Write all this information as a 2×4 matrix.

32. *Business* The revenues for 1 week in thousands of dollars, for each of five stores in a grocery chain are represented by the entries in the 1×5 matrix

$$R = [100 \quad 500 \quad 350 \quad 425 \quad 380].$$

The costs for the week of each of the five stores are the entries in the matrix

$$C = [85 \quad 390 \quad 225 \quad 340 \quad 274].$$

Compute the matrix $R - C$ and interpret the entries.

33. Three stores A, B, and C, each selling a single commodity, represent the number of units sold, revenue, cost, and profit over 3 successive years by the rows in three 4×3 matrices:

$$\begin{bmatrix} 20 & 14 & 6 \\ 11 & 12 & 9 \\ 6 & 8 & 7 \\ 5 & 4 & 2 \end{bmatrix}, \begin{bmatrix} 15 & 17 & 8 \\ 9 & 10 & 6 \\ 5 & 7 & 8 \\ 4 & 3 & -2 \end{bmatrix}, \begin{bmatrix} 24 & 20 & 12 \\ 13 & 14 & 10 \\ 16 & 7 & 18 \\ -3 & 7 & -8 \end{bmatrix}.$$

   a. What is the revenue for store A in year 3?
   b. What is the cost for store B in year 1?
   c. What is the profit for store C in year 2?
   d. How many units of the commodity does store C sell in year 3?
   e. Represent the total number of units sold, the total revenue, the total cost, and the total profit for stores A, B, and C over the 3 years as a single 3×4 matrix.

34. *Social Science* For three psychology students, A, B, and C, respectively, the scores on two examinations are given by the 3×2 matrix:

$$\begin{bmatrix} 75 & 60 \\ 80 & 79 \\ 91 & 88 \end{bmatrix}.$$

What did student A receive on the second examination? What did student C receive on the first examination?

7.2

MATRIX MULTIPLICATION

We shall be interested in two kinds of products involving matrices: (1) the product of a matrix and a real number, and (2) the product of two matrices.

> The **product** of a real number c and an $m \times n$ matrix A with entries a_{ij} is the matrix cA with corresponding entries ca_{ij}, where $i = 1, 2, 3, ..., m$ and $j = 1, 2, 3, ..., n$.

For example,

$$3 \begin{bmatrix} 2 & 1 \\ 0 & 5 \end{bmatrix} = \begin{bmatrix} 3 \times 2 & 3 \times 1 \\ 3 \times 0 & 3 \times 5 \end{bmatrix} = \begin{bmatrix} 6 & 3 \\ 0 & 15 \end{bmatrix}.$$

PRODUCTS WITH REAL NUMBERS   The following properties state some algebraic laws for the multiplication of matrices by real numbers.

> If A and B are $m \times n$ matrices, and c and d are real numbers, then
>
> I cA is an $m \times n$ matrix,
> II $c(dA) = (cd)A$,
> III $(c + d)A = cA + dA$,
> IV $c(A + B) = cA + cB$,
> V $1A = A$,
> VI $(-1)A = -A$,
> VII $0A = \mathbf{0}$,
> VIII $c\mathbf{0} = \mathbf{0}$.

Since the results are fairly obvious, we shall verify only property IV. The elements of $A + B$ are of the form $a_{ij} + b_{ij}$, so it follows, by definition, that the elements of $c(A + B)$ are of the form $c(a_{ij} + b_{ij})$. But, since a_{ij}, b_{ij}, and c denote real numbers, $c(a_{ij} + b_{ij}) = ca_{ij} + cb_{ij}$. Now, the elements of cA are of the form ca_{ij}, and those of cB are of the form cb_{ij}, so that the elements of $cA + cB$ are of the form $ca_{ij} + cb_{ij}$.

PRODUCTS OF MATRICES

Turning now to the *product of two matrices*, we have the following definition.

> The **product** of the matrices $A_{m \times p}$ and $B_{p \times n}$ is the matrix $(AB)_{m \times n}$ with entries determined as follows: The entry c_{ij} in the ith row and jth column of $(AB)_{m \times n}$ is found by multiplying the first element in the ith row of A by the first element of the jth column of B, to this product adding the product of the second element in the ith row of A with the second element in the jth column of B, to this sum adding the product of the third element in the ith row of A with the third element in the jth column of B, and so on.

For example, the product of the matrices

$$A_{2 \times 2} = \begin{bmatrix} a_{11} & a_{12} \\ a_{21} & a_{22} \end{bmatrix} \quad \text{and} \quad B_{2 \times 2} = \begin{bmatrix} b_{11} & b_{12} \\ b_{21} & b_{22} \end{bmatrix}$$

is the matrix

$$(AB)_{2 \times 2} = \begin{bmatrix} a_{11}b_{11} + a_{12}b_{21} & a_{11}b_{12} + a_{12}b_{22} \\ a_{21}b_{11} + a_{22}b_{21} & a_{21}b_{12} + a_{22}b_{22} \end{bmatrix}.$$

7.2 Matrix Multiplication

The following schematics show how to find the entry in the first row and first column of AB:

$$\begin{bmatrix} a_{11} & a_{12} \\ a_{21} & a_{22} \end{bmatrix} \begin{bmatrix} b_{11} & b_{12} \\ b_{21} & b_{22} \end{bmatrix} = \begin{bmatrix} a_{11}b_{11} + a_{12}b_{21} \end{bmatrix},$$

or

$$\begin{bmatrix} a_{11} & a_{12} \\ a_{21} & a_{22} \end{bmatrix} \begin{bmatrix} b_{11} & b_{12} \\ b_{21} & b_{22} \end{bmatrix} = \begin{bmatrix} a_{11}b_{11} + a_{12}b_{21} \end{bmatrix}.$$

Example If $A = \begin{bmatrix} 1 & 2 \\ -1 & 3 \end{bmatrix}$ and $B = \begin{bmatrix} 2 & 1 \\ 1 & 1 \end{bmatrix}$, find AB and BA.

Solution
$$AB = \begin{bmatrix} 1 & 2 \\ -1 & 3 \end{bmatrix} \begin{bmatrix} 2 & 1 \\ 1 & 1 \end{bmatrix} = \begin{bmatrix} 2+2 & 1+2 \\ -2+3 & -1+3 \end{bmatrix} = \begin{bmatrix} 4 & 3 \\ 1 & 2 \end{bmatrix}$$

$$BA = \begin{bmatrix} 2 & 1 \\ 1 & 1 \end{bmatrix} \begin{bmatrix} 1 & 2 \\ -1 & 3 \end{bmatrix} = \begin{bmatrix} 2-1 & 4+3 \\ 1-1 & 2+3 \end{bmatrix} = \begin{bmatrix} 1 & 7 \\ 0 & 5 \end{bmatrix}$$

PROPERTIES OF PRODUCTS

The above example shows very clearly that the multiplication of matrices, in general, is *not commutative*, that is,

$$AB \neq BA.$$

Thus, when discussing products of matrices, we must specify the *order* in which the matrices are to be considered as factors. For the product AB, we say that A is *right-multiplied* by B, and that B is *left-multiplied* by A.

Note that the definition of the product of two matrices, A and B, requires that the matrix A have the same number of *columns* as B has *rows*; the result, AB, then has the same number of rows as A and the same number of columns as B. Such matrices A and B are said to be **conformable** for multiplication. If

$$A_{m \times p} \quad \text{and} \quad B_{p \times n}$$

are multiplied together in the order

$$A_{m \times p} B_{p \times n},$$

the resulting matrix is

$$C_{m \times n}.$$

That is to say, the dimension of C is determined by

$$A_{m \times \boxed{p}} B_{\boxed{p} \times n}.$$

The fact that two matrices are conformable in the order AB, however, does **not** mean they necessarily are conformable in the order BA.

Example If $A = \begin{bmatrix} 3 & 1 \\ 1 & 0 \\ 2 & 1 \end{bmatrix}$ and $B = \begin{bmatrix} 1 & -1 \\ 2 & 1 \end{bmatrix}$, find AB.

Solution Since A is a 3×2 matrix, and B is a 2×2 matrix, they are conformable for multiplication in the order AB. We have

$$AB = \begin{bmatrix} 3 & 1 \\ 1 & 0 \\ 2 & 1 \end{bmatrix} \begin{bmatrix} 1 & -1 \\ 2 & 1 \end{bmatrix} = \begin{bmatrix} 3+2 & -3+1 \\ 1+0 & -1+0 \\ 2+2 & -2+1 \end{bmatrix} = \begin{bmatrix} 5 & -2 \\ 1 & -1 \\ 4 & -1 \end{bmatrix}.$$

Note that the matrices A and B in the example above are not conformable in the order BA.

In much of the matrix work in this book, we shall focus our attention on matrices having the same number of rows as columns. For brevity, a matrix of order $n \times n$ is often called a **square matrix** of order n. Although many of the ideas we shall discuss are applicable to conformable matrices of any order, we shall apply the notions only to square matrices.

If A, B, and C are $n \times n$ square matrices, then

(1) $\qquad\qquad\qquad (AB)C = A(BC).$

If A is a square matrix, then A^2, A^3, etc., denote AA, $(AA)A$, etc.

If A, B, and C are $n \times n$ square matrices, then

(2) $\qquad\qquad\qquad A(B+C) = AB + AC$

and

(3) $\qquad\qquad\qquad (B+C)A = BA + CA.$

The proofs of (1), (2), and (3) involve some complicated symbolism and are omitted here, but you will be asked to show their validity for the case of 2×2 matrices in the exercises (see Problems 27–29). Observe that, because matrix multiplication is not, in general, commutative, we must establish both the left-hand and the right-hand distributive property.

The **principal diagonal** of a square matrix is the ordered set of entries a_{ij}, extending from the upper left-hand corner to the lower right-hand corner of the matrix. Thus, the principal diagonal contains a_{11}, a_{22}, a_{33}, etc.

7.2 Matrix Multiplication

For example, the principal diagonal of

$$\begin{bmatrix} 1 & 3 & -1 \\ 5 & 2 & 3 \\ 6 & 4 & 0 \end{bmatrix}$$

consists of 1, 2, and 0, in that order.

> A **diagonal matrix** is a square matrix in which all entries not in the principal diagonal are 0.

Thus,

$$\begin{bmatrix} 4 & 0 \\ 0 & 2 \end{bmatrix} \quad \text{and} \quad \begin{bmatrix} 1 & 0 & 0 \\ 0 & 1 & 0 \\ 0 & 0 & 0 \end{bmatrix}$$

are diagonal matrices.

> The diagonal matrix having 1's for entries on the principal diagonal is denoted by $I_{n \times n}$.

For example,

$$I_{2 \times 2} = \begin{bmatrix} 1 & 0 \\ 0 & 1 \end{bmatrix} \quad \text{and} \quad I_{4 \times 4} = \begin{bmatrix} 1 & 0 & 0 & 0 \\ 0 & 1 & 0 & 0 \\ 0 & 0 & 1 & 0 \\ 0 & 0 & 0 & 1 \end{bmatrix}$$

The following properties are consequences of the definitions we have adopted.

> For each matrix $A_{n \times n}$,
>
> (4) $\qquad A_{n \times n} I_{n \times n} = I_{n \times n} A_{n \times n} = A_{n \times n}.$
>
> Furthermore, $I_{n \times n}$ is the unique matrix having this property for all matrices $A_{n \times n}$.

Accordingly, $I_{n \times n}$ is the **identity element for multiplication** in the set of $n \times n$ square matrices. The verification of (4) for the illustrative case $n = 2$, is left as an exercise (see Problem 26).

INTEREST Suppose $500 is invested in three different savings accounts paying, respectively, 4%, 5%, and 6% annual interest compounded annually. After 1 year the interest obtained from each account is

account 1   account 2   account 3
500 × (0.04) = $20   500 × (0.05) = $25   500 × (0.06) = $30.

These amounts can be represented by a row matrix

$$[20 \quad 25 \quad 30].$$

Let us suppose further that the interest is withdrawn from each account at the end of the year. If the money remains in the first account for 5 years, in the second account for 10 years, and in the third account for 25 years, then the total interest obtained is given by the product

$$[20 \quad 25 \quad 30] \begin{bmatrix} 5 \\ 10 \\ 25 \end{bmatrix} = (20)(5) + (25)(10) + (30)(25)$$

$$= 100 + 250 + 750$$

$$= \$1100.$$

Alternatively, the number of years could be written as the 3 × 3 matrix

$$\begin{bmatrix} 5 & 0 & 0 \\ 0 & 10 & 0 \\ 0 & 0 & 25 \end{bmatrix},$$

so that the *return* from each account after the given number of years is

$$[500 \quad 500 \quad 500] + [20 \quad 25 \quad 30] \begin{bmatrix} 5 & 0 & 0 \\ 0 & 10 & 0 \\ 0 & 0 & 25 \end{bmatrix}$$

$$= [500 \quad 500 \quad 500] + [100 \quad 250 \quad 750]$$

$$= [600 \quad 750 \quad 1250].$$

Of course, the *total* return on the investments would be the sum of the three entries in the foregoing matrix.

Now if the interest is not withdrawn, then at the end of 1 year the respective accounts contain

$$500 + 500(0.04) = 500(1.04) = \$520$$

$$500 + 500(0.05) = 500(1.05) = \$525$$

$$500 + 500(0.06) = 500(1.06) = \$530,$$

or,

$$[500 \quad 500 \quad 500] \begin{bmatrix} 1.04 & 0 & 0 \\ 0 & 1.05 & 0 \\ 0 & 0 & 1.06 \end{bmatrix} = [520 \quad 525 \quad 530].$$

7.2 Matrix Multiplication

If the interest is compounded, then at the end of the second year we would have in each account

$$[520 \quad 525 \quad 530] \begin{bmatrix} 1.04 & 0.00 & 0 \\ 0 & 1.05 & 0 \\ 0 & 0 & 1.06 \end{bmatrix}$$

$$= \left([500 \quad 500 \quad 500] \begin{bmatrix} 1.04 & 0 & 0 \\ 0 & 1.05 & 0 \\ 0 & 0 & 1.06 \end{bmatrix} \right) \begin{bmatrix} 1.04 & 0 & 0 \\ 0 & 1.05 & 0 \\ 0 & 0 & 1.06 \end{bmatrix}$$

$$= [500 \quad 500 \quad 500] \begin{bmatrix} 1.04 & 0 & 0 \\ 0 & 1.05 & 0 \\ 0 & 0 & 106 \end{bmatrix}^2 \quad \text{(By (1))}$$

$$= [500 \quad 500 \quad 500] \begin{bmatrix} 1.082 & 0 & 0 \\ 0 & 1.103 & 0 \\ 0 & 0 & 1.124 \end{bmatrix} \quad \text{(Entries rounded to three decimal places.)}$$

$$= [541 \quad 551.5 \quad 562]$$

In general, the amount accrued in each account after a period of m years would be the final entries in the matrix

$$[500 \quad 500 \quad 500] \begin{bmatrix} 1.04 & 0 & 0 \\ 0 & 1.05 & 0 \\ 0 & 0 & 1.06 \end{bmatrix}^m.$$

It is easy to verify that this expression is equivalent to

$$[500 \quad 500 \quad 500] \begin{bmatrix} (1.04)^m & 0 & 0 \\ 0 & (1.05)^m & 0 \\ 0 & 0 & (1.06)^m \end{bmatrix}.$$

ORDER OF MULTIPLICATION The following result relates the order in which matrices can be multiplied by real numbers and by other matrices.

If A and B are $n \times n$ square matrices, and a is a real number, then
$$a(AB) = (aA)B = A(aB).$$

EXERCISE 7.2

Write each product as a single matrix.

Example **a.** $3\begin{bmatrix} 2 & 1 \\ -1 & 3 \\ 2 & 0 \end{bmatrix}$

b. $\begin{bmatrix} 3 & 1 & -1 \\ 0 & -1 & 2 \end{bmatrix} \cdot \begin{bmatrix} 1 & -1 \\ 0 & 2 \\ 1 & 0 \end{bmatrix}$

Solution **a.** $\begin{bmatrix} 6 & 3 \\ -3 & 9 \\ 6 & 0 \end{bmatrix}$

b. $\begin{bmatrix} 3+0-1 & -3+2+0 \\ 0+0+2 & 0-2+0 \end{bmatrix} = \begin{bmatrix} 2 & -1 \\ 2 & -2 \end{bmatrix}$

1. $-5\begin{bmatrix} 0 & 1 & -1 \\ 3 & -1 & 2 \end{bmatrix}$

2. $2\begin{bmatrix} 2 & 1 & 3 & -2 \\ 4 & 2 & 0 & -1 \\ 0 & 0 & -1 & 2 \end{bmatrix}$

3. $\begin{bmatrix} 1 & -2 \end{bmatrix} \cdot \begin{bmatrix} 3 \\ 2 \end{bmatrix}$

4. $\begin{bmatrix} 3 & -2 & 2 \end{bmatrix} \cdot \begin{bmatrix} 1 \\ 0 \\ -2 \end{bmatrix}$

5. $\begin{bmatrix} 3 & -1 \\ 2 & 1 \end{bmatrix} \cdot \begin{bmatrix} 1 & -4 \\ 2 & 1 \end{bmatrix}$

6. $\begin{bmatrix} 1 & -5 \\ 0 & 2 \end{bmatrix} \cdot \begin{bmatrix} 3 & 1 \\ -1 & 2 \end{bmatrix}$

7. $\begin{bmatrix} 4 & -5 \\ 7 & 3 \end{bmatrix} \cdot \begin{bmatrix} 5 & -1 \\ -2 & 7 \end{bmatrix}$

8. $\begin{bmatrix} 1 & -2 \\ -3 & 1 \end{bmatrix} \cdot \begin{bmatrix} 5 & 1 \\ 0 & 2 \end{bmatrix}$

9. $\begin{bmatrix} -3 & 1 & 0 \\ 2 & 1 & 1 \end{bmatrix} \cdot \begin{bmatrix} 2 & 0 \\ 1 & -1 \\ 3 & 0 \end{bmatrix}$

10. $\begin{bmatrix} 1 & -1 & 0 \\ 2 & 1 & 3 \end{bmatrix} \cdot \begin{bmatrix} 4 & -1 \\ 2 & 0 \\ 1 & 1 \end{bmatrix}$

11. $\begin{bmatrix} -1 & 0 & 1 \\ 2 & 1 & 0 \\ 1 & 0 & 0 \end{bmatrix} \cdot \begin{bmatrix} 0 & 1 & 3 \\ 1 & 0 & 2 \\ -1 & 1 & 1 \end{bmatrix}$

12. $\begin{bmatrix} 2 & -3 & 1 \\ 0 & 1 & -1 \\ 2 & 0 & 0 \end{bmatrix} \cdot \begin{bmatrix} 1 & 0 & 0 \\ 0 & 1 & 0 \\ 0 & 0 & 1 \end{bmatrix}$

13. $\begin{bmatrix} 2 & -2 & -1 \\ 1 & 1 & -2 \\ 1 & 0 & -1 \end{bmatrix} \cdot \begin{bmatrix} -1 & -2 & 5 \\ -1 & -1 & 3 \\ -1 & -2 & 4 \end{bmatrix}$

14. $\begin{bmatrix} -1 & -2 & 5 \\ -1 & -1 & 3 \\ -1 & -2 & 4 \end{bmatrix} \cdot \begin{bmatrix} 2 & -2 & -1 \\ 1 & 1 & -2 \\ 1 & 0 & -1 \end{bmatrix}$

Let $A = \begin{bmatrix} 1 & -2 \\ 1 & 0 \end{bmatrix}$ and $B = \begin{bmatrix} -1 & 2 \\ -1 & 1 \end{bmatrix}$. *Compute each of the following products.*

15. AB

16. BA

17. $(AB)A$

18. $(BA)B$

19. $A^t B$

20. AB^t

7.2 Matrix Multiplication

Find a matrix X satisfying each matrix equation.

21. $3X + \begin{bmatrix} 1 & 0 \\ 2 & 1 \end{bmatrix} = \begin{bmatrix} -2 & 3 \\ -1 & -2 \end{bmatrix}$

22. $2X + 3\begin{bmatrix} 1 & 1 \\ 0 & 1 \end{bmatrix} = \begin{bmatrix} 7 & -1 \\ 3 & -5 \end{bmatrix}$

23. $X + 2I = \begin{bmatrix} 3 & -1 \\ 1 & 2 \end{bmatrix}$

24. $3X - 2I = \begin{bmatrix} 7 & 3 \\ 6 & 4 \end{bmatrix}$

25. Show that if $A = \begin{bmatrix} -1 & 2 \\ 0 & 1 \end{bmatrix}$ and $B = \begin{bmatrix} 1 & 0 \\ -1 & 2 \end{bmatrix}$, then

  a. $(A + B)(A + B) \neq A^2 + 2AB + B^2$,
  b. $(A + B)(A - B) \neq A^2 - B^2$.

26. **a.** Show that for each matrix $A_{2 \times 2}$,
$$A_{2 \times 2} \cdot I_{2 \times 2} = I_{2 \times 2} \cdot A_{2 \times 2} = A_{2 \times 2}.$$
  b. Show that if, for a given matrix $B_{2 \times 2}$ and for *all* $A_{2 \times 2}$,
$$A_{2 \times 2} \cdot B_{2 \times 2} = B_{2 \times 2} \cdot A_{2 \times 2} = A_{2 \times 2},$$
then $B_{2 \times 2} = I_{2 \times 2}$.

27. Show that
$$\left(\begin{bmatrix} a_{11} & a_{12} \\ a_{21} & a_{22} \end{bmatrix} \cdot \begin{bmatrix} b_{11} & b_{12} \\ b_{21} & b_{22} \end{bmatrix}\right) \cdot \begin{bmatrix} c_{11} & c_{12} \\ c_{21} & c_{22} \end{bmatrix} = \begin{bmatrix} a_{11} & a_{12} \\ a_{21} & a_{22} \end{bmatrix} \cdot \left(\begin{bmatrix} b_{11} & b_{12} \\ b_{21} & b_{22} \end{bmatrix} \cdot \begin{bmatrix} c_{11} & c_{12} \\ c_{21} & c_{22} \end{bmatrix}\right).$$

28. Show that
$$\begin{bmatrix} a_{11} & a_{12} \\ a_{21} & a_{22} \end{bmatrix} \cdot \left(\begin{bmatrix} b_{11} & b_{12} \\ b_{21} & b_{22} \end{bmatrix} + \begin{bmatrix} c_{11} & c_{12} \\ c_{21} & c_{22} \end{bmatrix}\right)$$
$$= \begin{bmatrix} a_{11} & a_{12} \\ a_{21} & a_{22} \end{bmatrix} \cdot \begin{bmatrix} b_{11} & b_{12} \\ b_{21} & b_{22} \end{bmatrix} + \begin{bmatrix} a_{11} & a_{12} \\ a_{21} & a_{22} \end{bmatrix} \cdot \begin{bmatrix} c_{11} & c_{12} \\ c_{21} & c_{22} \end{bmatrix}.$$

29. Show that
$$\left(\begin{bmatrix} b_{11} & b_{12} \\ b_{21} & b_{22} \end{bmatrix} + \begin{bmatrix} c_{11} & c_{12} \\ c_{21} & c_{22} \end{bmatrix}\right) \cdot \begin{bmatrix} a_{11} & a_{12} \\ a_{21} & a_{22} \end{bmatrix}$$
$$= \begin{bmatrix} b_{11} & b_{12} \\ b_{21} & b_{22} \end{bmatrix} \cdot \begin{bmatrix} a_{11} & a_{12} \\ a_{21} & a_{22} \end{bmatrix} + \begin{bmatrix} c_{11} & c_{12} \\ c_{21} & c_{22} \end{bmatrix} \cdot \begin{bmatrix} a_{11} & a_{12} \\ a_{21} & a_{22} \end{bmatrix}.$$

30. Show that $\begin{bmatrix} 0 & a \\ a & 0 \end{bmatrix}^2 = a^2 I.$

31. Show that $(A_{2 \times 2} \cdot B_{2 \times 2})^t = B^t_{2 \times 2} \cdot A^t_{2 \times 2}.$

32. In each case, find a 2×2 matrix $A \neq I$ that satisfies the equation.

  a. $A^2 = A$ **b.** $A^2 = I$ **c.** $A^2 = 0$

*In some business contexts an $m \times n$ matrix A is called a **game**. The **expectation** of the game is $E = PAQ$, where P and Q are, respectively, row and column matrices called **strategies**. The elements of P and Q are probabilities (see Chapter 10). In Problems 33–36, compute the expectation of the game for the given strategies.*

33. $A = \begin{bmatrix} -1 & -1 \\ 1 & -1 \end{bmatrix}$; $P = \begin{bmatrix} 1 & 1 \\ 2 & 2 \end{bmatrix}$; $Q = \begin{bmatrix} \frac{1}{2} \\ \frac{1}{2} \end{bmatrix}$

34. $A = \begin{bmatrix} -1 & 1 \\ 1 & -1 \end{bmatrix}$; $P = [x \quad 1-x]$; $Q = \begin{bmatrix} y \\ 1-y \end{bmatrix}$

35. $A = \begin{bmatrix} 1 & 0 & -2 \\ 2 & 1 & 0 \\ 3 & 1 & -1 \end{bmatrix}$; $P = \begin{bmatrix} \frac{1}{4} & \frac{1}{4} & \frac{1}{2} \end{bmatrix}$; $Q = \begin{bmatrix} \frac{1}{4} \\ \frac{1}{2} \\ \frac{1}{4} \end{bmatrix}$

36. $A = \begin{bmatrix} -2 & 1 & 0 \\ 1 & 4 & 3 \\ 0 & -1 & 1 \end{bmatrix}$, $P = \begin{bmatrix} \frac{1}{8} & \frac{4}{4} & \frac{1}{8} \end{bmatrix}$, $Q = \begin{bmatrix} \frac{3}{4} \\ \frac{1}{4} \\ \frac{1}{4} \end{bmatrix}$

Applications

37. *Biology* Two persons X and Y have infectious hepatitus. X and Y have the possibility of contact, and thus of passing on the disease, with four persons P_1, P_2, P_3, and P_4. Let a 2×4 matrix be given as follows:

$$A = \begin{matrix} & \begin{matrix} P_1 & P_2 & P_3 & P_4 \end{matrix} \\ \begin{matrix} X \\ Y \end{matrix} & \begin{bmatrix} 1 & 0 & 1 & 1 \\ 0 & 1 & 0 & 1 \end{bmatrix} \end{matrix}.$$

If person X (or Y) has contact with any of the four persons, enter 1 in the row labeled X (or Y) in the appropriate column. If X (or Y) has no contact with a particular person, enter 0. Define the contacts between P_1, P_2, P_3, and P_4 with four additional persons P_5, P_6, P_7, and P_8 by

$$B = \begin{matrix} & \begin{matrix} P_5 & P_6 & P_7 & P_8 \end{matrix} \\ \begin{matrix} P_1 \\ P_2 \\ P_3 \\ P_4 \end{matrix} & \begin{bmatrix} 0 & 1 & 0 & 0 \\ 1 & 0 & 0 & 1 \\ 1 & 1 & 0 & 1 \\ 0 & 1 & 1 & 1 \end{bmatrix} \end{matrix}.$$

Compute the product AB and interpret the entries.

38. *Business* The revenue, in thousands of dollars, for 3 successive weeks for five stores in a grocery chain are represented by the entries in the following matrices R_1, R_2, and R_3:

$$R_1 = [100 \quad 150 \quad 210 \quad 125 \quad 190], \qquad R_2 = 2R_1, \qquad R_3 = R_1.$$

7.2 Matrix Multiplication

Over the same period the costs can be represented by
$$C_1 = [40 \quad 60 \quad 80 \quad 50 \quad 70], \quad C_2 = 1.5C_1, \quad C_3 = C_1.$$
Compute the matrix $4R_1 - 3.5C_1$ and interpret the entries.

39. Amounts of $500, $1000, and $1500 are invested in three types of savings accounts paying, respectively, 5%, 6%, and 7% annual interest compounded annually. At the end of each year the interest is withdrawn from the accounts. By means of matrix addition and multiplication, give a representation of the return from each account, if the principals remain in their respective accounts 10 years, 20 years, and 30 years.

40. A person deposits $1000 and $1500 into bank savings accounts paying, respectively, 5% and 5.5% annual interest, and also deposits $2000 in a savings and loan account which pays 6% annual interest. Assume that interest at each institution is compounded annually. By means of matrix multiplication, find the amounts accrued in each account if the interest is compounded for 2 years. For 3 years.

41. A retail store purchases two brands of stereophonic equipment consisting of amplifiers, tuners, and speakers from wholesale outlets. Because of limited quantities, the store must purchase these supplies from three wholesale dealers. Matrix A gives the wholesale price of each piece of equipment in dollars. Matrix B represents the number of units of each piece of equipment purchased. (For example, $b_{11} = 1$ means that one amplifier, one tuner, and one set of speakers of brand #1 is purchased from wholesale dealer #1.)

$$A = \begin{array}{c} \text{Amplifiers} \\ \text{Tuners} \\ \text{Speakers} \end{array} \begin{bmatrix} \text{Price of} & \text{Price of} \\ \text{brand \#1} & \text{brand \#2} \\ 200 & 100 \\ 200 & 150 \\ 400 & 300 \end{bmatrix}$$

$$B = \begin{array}{c} \text{Units of brand \#1} \\ \text{Units of brand \#2} \end{array} \begin{bmatrix} \text{Wholesale} & \text{Wholesale} & \text{Wholesale} \\ \text{dealer \#1} & \text{dealer \#2} & \text{dealer \#3} \\ 1 & 2 & 3 \\ 2 & 4 & 2 \end{bmatrix}$$

If
$$C = \begin{bmatrix} \text{State} & \text{City} \\ \text{sales tax} & \text{sales tax} \\ 0.06 & 0.01 \\ 0.06 & 0.01 \\ 0.06 & 0.01 \end{bmatrix},$$
find the matrix
$$P = (AB)C,$$
and interpret the significance of its entries.

42. A television station does a weekly comparison of three supermarkets for the costs of five basic food items. In a particular week, the following matrix gives the per pound price for each item:

$$\begin{array}{c c} & \begin{array}{c c c} \text{Store 1} & \text{Store 2} & \text{Store 3} \end{array} \\ \begin{array}{r} \text{Vegetables} \\ \text{Meat} \\ \text{Bread} \\ \text{Cheese} \\ \text{Fruit} \end{array} & \begin{bmatrix} 0.39 & 0.41 & 0.38 \\ 1.50 & 1.29 & 1.35 \\ 0.72 & 0.68 & 0.70 \\ 1.00 & 0.92 & 0.98 \\ 0.50 & 0.58 & 0.52 \end{bmatrix} \end{array}$$

The number of pounds purchased of each item is given by the matrix

$$[2 \quad 3 \quad 1 \quad 2 \quad 4].$$

By an appropriate matrix multiplication, compare the total costs at the three stores.

43. A company owns five tire stores. The inventory of tires in store S_1 is given by

$$\begin{array}{c c} & \begin{array}{c c c} \text{Brand } X & \text{Brand } Y & \text{Brand } Z \end{array} \\ \begin{array}{r} \text{Belted tires} \\ \text{Radials} \\ \text{Steel-belted radials} \\ \text{Regular tires} \end{array} & \begin{bmatrix} 100 & 50 & 40 \\ 80 & 20 & 50 \\ 200 & 60 & 20 \\ 100 & 100 & 100 \end{bmatrix} \end{array}$$

Stores S_2 and S_3 each have 3 times as many tires as S_1; store S_4 has 1/2 as many tires as store S_1; and store S_5 has 2 times as many tires as S_1. Find the matrix that gives the total inventory of tires the company owns.

44. *Physics* The velocities of cars X, Y, and Z in kilometers per hour are given in the 3×1 matrix

$$A = \begin{bmatrix} 50 \\ 80 \\ 120 \end{bmatrix}.$$

The number of hours each car travels is given in the 1×3 matrix

$$B = [3 \quad 4 \quad 6].$$

Compute the products AB and BA and interpret the entries in each.

45. *Social Science* A college has 3000 students enrolled at the start of a particular academic year. A breakdown by classes is given by the matrix

$$\begin{array}{c c c c c} \text{Year} & \text{Freshmen} & \text{Sophomores} & \text{Juniors} & \text{Seniors} \\ \text{Number of students} & [\quad 1100 & 800 & 600 & 500 \quad]. \end{array}$$

It is projected that the percentage of dropouts per class, in any given year, is given by

$$\begin{bmatrix} 0.20 \\ 0.15 \\ 0.05 \\ 0.03 \end{bmatrix}$$

That is to say, 20% of the freshman class is expected to leave school before the completion of the year, and so on. By matrix multiplication, determine the projected total number of dropouts for that particular year.

7.3

SOLUTION OF LINEAR SYSTEMS USING ROW-EQUIVALENT MATRICES

ELEMENTARY TRANSFORMATIONS

An **elementary transformation** of a matrix $A_{n \times m}$ is a transformation that can be obtained in any one of the following ways:

1. Multiply the entries of any row of $A_{n \times m}$ by a nonzero constant.
2. Interchange any two rows of $A_{n \times m}$.
3. Replace any row of $A_{n \times m}$ by the sum of the row and a constant times any other row.

Notice that the inverse of an elementary transformation is an elementary transformation. That is, you can undo an elementary transformation by means of an elementary transformation. For example, if A is transformed into B by interchanging two rows, then you can regain A from B by again interchanging the same two rows. It follows that if B results from performing a *succession* of elementary transformations on A, then A can similarly be obtained from B by performing the inverse operations in reverse order. The elementary transformations of a matrix are also called **elementary row operations**.

> If B is a matrix resulting from a succession of a finite number of elementary transformations on a matrix A, then A and B are **row-equivalent** matrices. This is expressed by writing $A \sim B$ or $B \sim A$.

Example Show that $\begin{bmatrix} 1 & -2 & -1 \\ 1 & 0 & 2 \\ -4 & 3 & 1 \end{bmatrix} \sim \begin{bmatrix} 3 & -6 & -3 \\ 1 & 0 & 2 \\ -4 & 3 & 1 \end{bmatrix}$.

Solution Multiplying each entry of the first row of the left-hand matrix by 3, we obtain the right-hand matrix.

Example Show that $\begin{bmatrix} 1 & -2 & -1 \\ 1 & 0 & 2 \\ -4 & 3 & 1 \end{bmatrix} \sim \begin{bmatrix} -4 & 3 & 1 \\ 1 & 0 & 2 \\ 1 & -2 & -1 \end{bmatrix}$.

Solution Interchanging the first and third row of the left-hand matrix, we obtain the right-hand matrix.

Sometimes it is convenient to make several elementary transformations on the same matrix.

Example Show that $\begin{bmatrix} 1 & -2 & -1 \\ 1 & 0 & 2 \\ -4 & 3 & 1 \end{bmatrix} \sim \begin{bmatrix} 1 & -2 & -1 \\ 0 & 2 & 3 \\ 0 & -5 & -3 \end{bmatrix}$.

Solution Multiplying the entries of the first row of the left-hand matrix by -1 and adding these products to the corresponding entries of the second row, and then multiplying the entries of the first row by 4 and adding these products to the corresponding entries of the third row, we obtain the right-hand matrix.

Some $n \times n$ matrices are row-equivalent to $I_{n \times n}$. These are the matrices that in Section 7.5 will be defined to be **nonsingular**.

Example Show that $\begin{bmatrix} 1 & 2 & 1 \\ -1 & 1 & 0 \\ 1 & 0 & 1 \end{bmatrix} \sim \begin{bmatrix} 1 & 0 & 0 \\ 0 & 1 & 0 \\ 0 & 0 & 1 \end{bmatrix}$.

Solution We first make appropriate transformations to obtain 0 elements in each entry (except for the principal diagonal) in columns 1, 2, and 3. Each reference to a row indicates the row of the preceding matrix.

$$\begin{bmatrix} 1 & 2 & 1 \\ -1 & 1 & 0 \\ 1 & 0 & 1 \end{bmatrix} \sim \begin{bmatrix} 1 & 2 & 1 \\ 0 & 3 & 1 \\ 0 & -2 & 0 \end{bmatrix} \begin{array}{l} \text{Row 2 + Row 1} \\ \text{Row 3} + (-1 \times \text{Row 1}) \end{array}$$

$$\sim \begin{bmatrix} 1 & 0 & 1 \\ 0 & 3 & 1 \\ 0 & 0 & \frac{2}{3} \end{bmatrix} \begin{array}{l} \text{Row 1 + Row 3} \\ \\ \text{Row 3} + \left(\frac{2}{3} \times \text{Row 2}\right) \end{array}$$

$$\sim \begin{bmatrix} 1 & 0 & 0 \\ 0 & 3 & 0 \\ 0 & 0 & \frac{2}{3} \end{bmatrix} \begin{array}{l} \text{Row 1} + \left(-\frac{3}{2} \times \text{Row 3}\right) \\ \text{Row 2} + \left(-\frac{3}{2} \times \text{Row 3}\right) \end{array}$$

7.3 Solution of Linear Systems Using Row-Equivalent Matrices

$$\sim \begin{bmatrix} 1 & 0 & 0 \\ 0 & 1 & 0 \\ 0 & 0 & 1 \end{bmatrix} \begin{array}{l} \\ \frac{1}{3} \times \text{Row 2} \\ \\ \frac{3}{2} \times \text{Row 3} \end{array}$$

MATRIX SOLUTION OF LINEAR SYSTEMS

In a linear system of the form

$$a_{11}x + a_{12}y + a_{13}z = c_1$$
$$a_{21}x + a_{22}y + a_{23}z = c_2$$
$$a_{31}x + a_{32}y + a_{33}z = c_3$$

the matrices

$$\begin{bmatrix} a_{11} & a_{12} & a_{13} \\ a_{21} & a_{22} & a_{23} \\ a_{31} & a_{32} & a_{33} \end{bmatrix} \quad \text{and} \quad \begin{bmatrix} a_{11} & a_{12} & a_{13} & \vdots & c_1 \\ a_{21} & a_{22} & a_{23} & \vdots & c_2 \\ a_{31} & a_{32} & a_{33} & \vdots & c_3 \end{bmatrix}$$

are called the **coefficient matrix** and the **augmented matrix**, respectively. Similar definitions hold for a system of n linear equations.

Starting with the augmented matrix of a linear system, and generating a sequence of row-equivalent matrices, we can obtain a matrix from which the solution set of the system is evident simply by inspection. The validity of the method, which is illustrated by the example below, follows from the fact that performing elementary row transformations on the augmented matrix of a system corresponds in this context to forming equivalent systems of equations.

Example Solve

$$x + 2y - 3z = -4$$
$$2x - y + z = 3$$
$$3x + 2y + z = 10.$$

Solution The augmented matrix of the system is

$$\begin{bmatrix} 1 & 2 & -3 & \vdots & -4 \\ 2 & -1 & 1 & \vdots & 3 \\ 3 & 2 & 1 & \vdots & 10 \end{bmatrix}$$

The system of equations corresponding to each successive matrix is shown on the right of the matrix in the following solution.

$$\begin{array}{l} \\ \text{Row 2} + (-2 \times \text{Row 1}) \\ \text{Row 3} + (-3 \times \text{Row 1}) \end{array} \begin{bmatrix} 1 & 2 & -3 & \vdots & -4 \\ 0 & -5 & 7 & \vdots & 11 \\ 0 & -4 & 10 & \vdots & 22 \end{bmatrix} \quad \begin{array}{rl} x + 2y - 3z = & -4 \\ 0x - 5y + 7z = & 11 \\ 0x - 4y + 10z = & 22 \end{array}$$

$\text{Row } 1 + \left(\frac{2}{5} \times \text{Row } 2\right)$ $\begin{bmatrix} 1 & 0 & -\frac{1}{5} & \vdots & \frac{2}{5} \\ 0 & -5 & 7 & \vdots & 11 \\ 0 & 0 & \frac{22}{5} & \vdots & \frac{66}{5} \end{bmatrix}$ $\begin{array}{rrrr} x + 0y - \frac{1}{5}z = & \frac{2}{5} \\ 0x - 5y + 7z = & 11 \\ 0x + 0y + \frac{22}{5}z = & \frac{66}{5} \end{array}$

$\text{Row } 3 + \left(-\frac{4}{5} \times \text{Row } 2\right)$

$5 \times \text{Row } 1$ $\begin{bmatrix} 5 & 0 & -1 & \vdots & 2 \\ 0 & -5 & 7 & \vdots & 11 \\ 0 & 0 & 1 & \vdots & 3 \end{bmatrix}$ $\begin{array}{rrrr} 5x + 0y - z = & 2 \\ 0x - 5y + 7z = & 11 \\ 0x + 0y + z = & 3 \end{array}$

$\frac{5}{22} \times \text{Row } 3$

$\begin{array}{l} \text{Row } 1 + \text{Row } 3 \\ \text{Row } 2 + (-7 \times \text{Row } 3) \end{array}$ $\begin{bmatrix} 5 & 0 & 0 & \vdots & 5 \\ 0 & -5 & 0 & \vdots & -10 \\ 0 & 0 & 1 & \vdots & 3 \end{bmatrix}$ $\begin{array}{rrrr} 5x + 0y + 0z = & 5 \\ 0x - 5y + 0z = & -10 \\ 0x + 0y + z = & 3 \end{array}$

$\frac{1}{5} \times \text{Row } 1$ $\begin{bmatrix} 1 & 0 & 0 & \vdots & 1 \\ 0 & 1 & 0 & \vdots & 2 \\ 0 & 0 & 1 & \vdots & 3 \end{bmatrix}$ $\begin{array}{rrrr} x + 0y + 0z = & 1 \\ 0x + y + 0z = & 2 \\ 0x + 0y + z = & 3 \end{array}$

$-\frac{1}{5} \times \text{Row } 2$

The last system is equivalent to

$$x = 1$$
$$y = 2$$
$$z = 3.$$

From this, the solution set $\{(1, 2, 3)\}$ for the given system is evident by inspection.

For any given $n \times n$ linear system with a nonsingular coefficient matrix, there are many sequences of row operations that will transform the augmented matrix of a system to one of the form

$$\begin{bmatrix} 1 & 0 & 0 & \cdots & 0 & \vdots & x_1 \\ 0 & 1 & 0 & & 0 & \vdots & \cdot \\ 0 & 0 & 1 & & 0 & \vdots & \cdot \\ \vdots & & & & \vdots & \vdots & \cdot \\ 0 & 0 & 0 & \cdots & 1 & \vdots & x_n \end{bmatrix},$$

from which the solution set $\{(x_1, \ldots, x_n)\}$ of the original system is evident by inspection. Finding the most efficient sequence depends on experience and insight.

7.3 Solution of Linear Systems Using Low-Equivalent Matrices

EXERCISE 7.3

Use row transformations of the augmented matrix to solve each system of equations. (Each coefficient matrix is nonsingular and is row-equivalent to $I_{n \times n}$.)

1. $x - 2y = 4$
   $x + 3y = -1$

2. $x + y = -1$
   $x - 4y = -14$

3. $3x - 2y = 13$
   $4x - y = 19$

4. $4x - 3y = 16$
   $2x + y = 8$

5. $x - 2y = 6$
   $3x + y = 25$

6. $x - y = -8$
   $x + 2y = 9$

7. $x + y - z = 0$
   $2x - y + z = -6$
   $x + 2y - 3z = 2$

8. $2x - y + 3z = 1$
   $x + 2y - z = -1$
   $3x + y + z = 2$

9. $2x - y = 0$
   $3y + z = 7$
   $2x + 3z = 1$

10. $3x - z = 7$
    $2x + y = 6$
    $3y - z = 7$

11. $2x - 5y + 3z = -1$
    $-3x - y + 2z = 11$
    $-2x + 7y + 5z = 9$

12. $2x + y + z = 4$
    $3x - z = 3$
    $2x + 3z = 13$

Show that each product is a matrix that is row-equivalent to $\begin{bmatrix} a & b \\ c & d \end{bmatrix}$ if $k \neq 0$.

13. $\begin{bmatrix} k & 0 \\ 0 & 1 \end{bmatrix} \begin{bmatrix} a & b \\ c & d \end{bmatrix}$

14. $\begin{bmatrix} 1 & 0 \\ 0 & k \end{bmatrix} \begin{bmatrix} a & b \\ c & d \end{bmatrix}$

15. $\begin{bmatrix} 0 & 1 \\ 1 & 0 \end{bmatrix} \begin{bmatrix} a & b \\ c & d \end{bmatrix}$

16. $\begin{bmatrix} 1 & k \\ 0 & 1 \end{bmatrix} \begin{bmatrix} a & b \\ c & d \end{bmatrix}$

17. $\begin{bmatrix} 1 & 0 \\ k & 1 \end{bmatrix} \begin{bmatrix} a & b \\ c & d \end{bmatrix}$

18. $\begin{bmatrix} 0 & k \\ 1 & 0 \end{bmatrix} \begin{bmatrix} a & b \\ c & d \end{bmatrix}$

Application

19. *Business* A company has 100 employees divided into three categories A, B, and C. As the following table shows, each employee makes a different contribution to a retirement fund. Following negotiation of a new contract, the monthly contribution of the employees increases by the indicated percentage. The total monthly contribution of $4450 by all employees then increases to $5270 because of the new contract. Use the concept of an augmented matrix to determine the number of employees in each category.

	A	B	C	Total monthly contribution
Monthly contribution to pension fund per employee	$20	$30	$50	$4450
Percent increase per month per employee	10%	10%	20%	$5270

7.4

DETERMINANTS

Associated with each square matrix A having real-number entries is a real number called the **determinant** of A and denoted by δA or $\delta(A)$ (read "the determinant of A"). Thus, we have a function, δ (Greek letter delta), with domain the set of all square matrices having real-number entries, and with range the set of all real numbers; $\delta(A_{n \times n})$ is called a determinant of **order** n.

Let us examine δ over the set of 2×2 matrices.

> The determinant of the matrix
> $$\begin{bmatrix} a_{11} & a_{12} \\ a_{21} & a_{22} \end{bmatrix}$$
> is the number $a_{11}a_{22} - a_{12}a_{21}$.

The determinant of a square matrix is customarily displayed in the same form as the matrix, but with vertical bars instead of brackets. Thus,

$$\delta \begin{bmatrix} a_{11} & a_{12} \\ a_{21} & a_{22} \end{bmatrix} = \begin{vmatrix} a_{11} & a_{12} \\ a_{21} & a_{22} \end{vmatrix} = a_{11}a_{22} - a_{12}a_{21}.$$

Example $\delta \begin{bmatrix} 3 & 1 \\ -2 & 3 \end{bmatrix} = \begin{vmatrix} 3 & 1 \\ -2 & 3 \end{vmatrix} = (3)(3) - (1)(-2) = 9 + 2 = 11$

Turning next to 3×3 matrices, we have the following definition.

> The determinant of the matrix
> $$\begin{bmatrix} a_{11} & a_{12} & a_{13} \\ a_{21} & a_{22} & a_{23} \\ a_{31} & a_{32} & a_{33} \end{bmatrix}, \quad \text{denoted by} \quad \begin{vmatrix} a_{11} & a_{12} & a_{13} \\ a_{21} & a_{22} & a_{23} \\ a_{31} & a_{32} & a_{33} \end{vmatrix},$$
> is the number
> (1) $a_{11}a_{22}a_{33} - a_{11}a_{23}a_{32} + a_{12}a_{23}a_{31} - a_{12}a_{21}a_{33} + a_{13}a_{21}a_{32} - a_{13}a_{22}a_{31}.$

An inspection of the subscripts of the factors of the products involved in this determinant (as well as those of the determinant of a 2×2 matrix) will show that each product is formed by taking one entry from each row and one entry from each column, with the restriction that no two factors can be entries in the same row or

7.4 Determinants

column. The determinant consists of the sum of $\pm$ all such products as are possible. We shall see in the following discussion how the algebraic sign of a product is determined. However, it is evident from (1) that the determinant of a matrix with real-number entries is a real number. We shall refer to this real number as the **value** of the determinant. Every determinant of order n can be written as a sum; this is called **expanding** the determinant.

> The **minor** M_{ij} of the element a_{ij} in a given determinant is the determinant that remains after the ith row and jth column in the given determinant have been deleted.

For example, in the determinant

(2)
$$\begin{vmatrix} a_{11} & a_{12} & a_{13} \\ a_{21} & a_{22} & a_{23} \\ a_{31} & a_{32} & a_{33} \end{vmatrix},$$

the minor of a_{11} is

$$M_{11} = \begin{vmatrix} a_{22} & a_{23} \\ a_{32} & a_{33} \end{vmatrix},$$

the minor of a_{23} is

$$M_{23} = \begin{vmatrix} a_{11} & a_{12} \\ a_{31} & a_{32} \end{vmatrix},$$

and the minor of a_{31} is

$$M_{31} = \begin{vmatrix} a_{12} & a_{13} \\ a_{22} & a_{23} \end{vmatrix}.$$

> The **cofactor** A_{ij} of the element a_{ij} is the minor of a_{ij} if $i+j$ is an even integer, and the negative of the minor of a_{ij} if $i+j$ is an odd integer.

For example, in the determinant (2), the cofactor of a_{11} is

$$A_{11} = \begin{vmatrix} a_{22} & a_{23} \\ a_{32} & a_{33} \end{vmatrix},$$

because $1 + 1$ is 2, an even integer; the cofactor of a_{23} is

$$A_{23} = - \begin{vmatrix} a_{11} & a_{12} \\ a_{31} & a_{32} \end{vmatrix},$$

because 2 + 3 is 5, an odd integer; the cofactor of a_{31} is

$$A_{31} = \begin{vmatrix} a_{12} & a_{13} \\ a_{22} & a_{23} \end{vmatrix},$$

because 3 + 1 is 4, an even integer; etc.

The following sign array is a convenient means of determining whether the cofactor of a given element equals the minor or the negative of the minor.

$$\begin{matrix} + & - & + & \cdots & (-)^{n+1} \\ - & + & - & \cdots & \\ + & - & + & \cdots & \\ \cdot & \cdot & \cdot & \cdots & \\ \cdot & \cdot & \cdot & \cdots & \\ \cdot & \cdot & \cdot & \cdots & \\ (-)^{n+1} & \cdot & \cdot & \cdots & + \end{matrix}$$

EXPANSION BY COFACTORS

With the definition of a cofactor in mind, let us look again at the determinant of the matrix

$$A = \begin{bmatrix} a_{11} & a_{12} & a_{13} \\ a_{21} & a_{22} & a_{23} \\ a_{31} & a_{32} & a_{33} \end{bmatrix}.$$

From (1) we know that the value of this determinant is

$$\delta(A) = a_{11}a_{22}a_{33} - a_{11}a_{23}a_{32} + a_{12}a_{23}a_{31} - a_{12}a_{21}a_{33} + a_{13}a_{21}a_{32} - a_{13}a_{22}a_{31}.$$

By suitably factoring pairs of terms in the right-hand member, we obtain

$$\delta(A) = a_{11}(a_{22}a_{33} - a_{23}a_{32}) + a_{12}(a_{23}a_{31} - a_{21}a_{33}) + a_{13}(a_{21}a_{32} - a_{22}a_{31}).$$

If the binomial factor in the middle term is rewritten $-(a_{21}a_{33} - a_{23}a_{31})$, we have

$$\delta(A) = a_{11}(a_{22}a_{33} - a_{23}a_{32}) + a_{12}[-(a_{21}a_{33} - a_{23}a_{31})] + a_{13}(a_{21}a_{32} - a_{22}a_{31}),$$

which is equal to

$$a_{11}\begin{vmatrix} a_{22} & a_{23} \\ a_{32} & a_{33} \end{vmatrix} + a_{12}\left(-\begin{vmatrix} a_{21} & a_{23} \\ a_{31} & a_{33} \end{vmatrix}\right) + a_{13}\begin{vmatrix} a_{21} & a_{22} \\ a_{31} & a_{32} \end{vmatrix}$$

Accordingly, we have

$$\delta(A) = a_{11}A_{11} + a_{12}A_{12} + a_{13}A_{13}.$$

Thus, the determinant

$$\delta(A) = \begin{vmatrix} a_{11} & a_{12} & a_{13} \\ a_{21} & a_{22} & a_{23} \\ a_{31} & a_{32} & a_{33} \end{vmatrix}$$

7.4 Determinants

is equal to the sum formed by multiplying each entry in the first row by its cofactor and then adding these products.

Similar methods can be used to show that this determinant is also equal to the sum formed by multiplying each entry in *any* row (or column) by its cofactor and then adding the products.

The following definition enables us to expand a determinant of any order.

> *The determinant of the square matrix*
> $$\begin{bmatrix} a_{11} & a_{12} & \cdots & a_{1n} \\ a_{21} & a_{22} & \cdots & a_{2n} \\ \vdots & \vdots & & \vdots \\ a_{n1} & a_{n2} & \cdots & a_{nn} \end{bmatrix}$$
> *is the sum of the n products formed by multiplying each entry in any single row (or any single column) by its cofactor.*

When this definition is used to rewrite a determinant, the determinant is said to be **expanded** about whatever row (or column) is chosen.

Example If $A = \begin{bmatrix} 3 & 2 & 1 \\ 0 & 1 & -2 \\ 1 & 3 & 4 \end{bmatrix}$, find $\delta(A)$ by expansion about the first column.

Solution Noting that $a_{11} = 3$, $a_{21} = 0$, and $a_{31} = 1$, we have

$$\delta(A) = 3 \begin{vmatrix} 1 & -2 \\ 3 & 4 \end{vmatrix} + 0 \left(- \begin{vmatrix} 2 & 1 \\ 3 & 4 \end{vmatrix} \right) + 1 \begin{vmatrix} 2 & 1 \\ 1 & -2 \end{vmatrix}$$

$$= 3(10) + 0 + (-5) = 25.$$

EXERCISE 7.4

Let A be the matrix

$$\begin{bmatrix} 2 & 1 & -2 & 0 \\ 1 & 0 & 3 & -1 \\ -2 & 1 & 2 & 2 \\ 1 & -1 & 3 & 1 \end{bmatrix}.$$

Determine the minor M_{ij} and cofactor A_{ij} (in determinant form) of each entry.

1. a_{11}
2. a_{13}
3. a_{23}
4. a_{41}
5. a_{31}
6. a_{33}
7. a_{44}
8. a_{14}

Evaluate each determinant.

Examples **a.** $\begin{vmatrix} 2 & -3 \\ 1 & 4 \end{vmatrix}$

b. $\begin{vmatrix} 1 & 2 & 0 \\ 3 & -1 & 4 \\ -2 & 1 & 3 \end{vmatrix}$

Solutions **a.** $\begin{vmatrix} 2 & -3 \\ 1 & 4 \end{vmatrix} = (2)(4) - (-3)(1) = 11$

b. Expand about any row or column; the first row is used here:

$$\begin{vmatrix} 1 & 2 & 0 \\ 3 & -1 & 4 \\ -2 & 1 & 3 \end{vmatrix} = 1 \begin{vmatrix} -1 & 4 \\ 1 & 3 \end{vmatrix} - 2 \begin{vmatrix} 3 & 4 \\ -2 & 3 \end{vmatrix} + 0 \begin{vmatrix} 3 & -1 \\ -2 & 1 \end{vmatrix}$$

$$= 1[(-1)(3) - (4)(1)] - 2[(3)(3) - (4)(-2)] + 0$$

$$= -41.$$

9. $\begin{vmatrix} 1 & 0 \\ 2 & 1 \end{vmatrix}$
10. $\begin{vmatrix} 3 & -2 \\ 4 & 1 \end{vmatrix}$
11. $\begin{vmatrix} -3 & -1 \\ 3 & 1 \end{vmatrix}$
12. $\begin{vmatrix} -1 & 6 \\ 0 & -2 \end{vmatrix}$

13. $\begin{vmatrix} -4 & 2 \\ 1 & 7 \end{vmatrix}$
14. $\begin{vmatrix} 8 & -1 \\ 3 & 2 \end{vmatrix}$
15. $\begin{vmatrix} x & -1 \\ 1 & 3 \end{vmatrix}$
16. $\begin{vmatrix} -3 & x \\ -1 & 3 \end{vmatrix}$

17. $\begin{vmatrix} 2 & 0 & 1 \\ 1 & 1 & 2 \\ -1 & 0 & 1 \end{vmatrix}$
18. $\begin{vmatrix} 1 & 3 & 1 \\ -1 & 2 & 1 \\ 0 & 2 & 0 \end{vmatrix}$
19. $\begin{vmatrix} 1 & 2 & 3 \\ 3 & -1 & 2 \\ 2 & 0 & 2 \end{vmatrix}$

20. $\begin{vmatrix} 1 & 0 & 0 \\ 0 & 1 & 2 \\ 0 & 3 & 4 \end{vmatrix}$
21. $\begin{vmatrix} -1 & 0 & 2 \\ -2 & 1 & 0 \\ 0 & 1 & -3 \end{vmatrix}$
22. $\begin{vmatrix} 2 & 1 & 4 \\ 3 & 2 & 6 \\ 5 & -3 & 10 \end{vmatrix}$

23. $\begin{vmatrix} a & b & 1 \\ a & b & 1 \\ 1 & 1 & 1 \end{vmatrix}$
24. $\begin{vmatrix} a & a & a \\ 1 & 2 & 3 \\ 4 & 5 & 6 \end{vmatrix}$
25. $\begin{vmatrix} x & 0 & 0 \\ 0 & x & 0 \\ 0 & 0 & x \end{vmatrix}$
26. $\begin{vmatrix} 0 & 0 & x \\ 0 & x & 0 \\ x & 0 & 0 \end{vmatrix}$

Solve for x.

27. $\begin{vmatrix} x & -1 \\ 2 & 3 \end{vmatrix} = 17$

28. $\begin{vmatrix} 1 & -5 \\ x & 3 \end{vmatrix} = -7$

7.5 The Inverse of a Square Matrix

29. $\begin{vmatrix} x & 0 & 0 \\ 2 & 1 & 3 \\ 0 & 1 & 4 \end{vmatrix} = 3$

30. $\begin{vmatrix} x^2 & x & 1 \\ 0 & 2 & 1 \\ 3 & 1 & 4 \end{vmatrix} = 28$

Expand by cofactors and verify.

31. $\begin{vmatrix} 0 & 1 & 0 & 0 \\ 1 & 0 & 3 & 2 \\ 5 & -1 & 2 & 1 \\ 1 & 0 & 1 & 1 \end{vmatrix} = 5$

32. $\begin{vmatrix} 1 & 2 & 0 & -1 \\ 1 & 0 & -1 & 2 \\ 0 & 1 & 1 & 1 \\ 2 & -1 & 0 & 1 \end{vmatrix} = 17$

33. The determinant of an $n \times n$ matrix is the sum of a certain number of products. What is this number for $n = 2$? For $n = 3$? For $n = 4$?

34. Generalize the results of Problem 33. Make a conjecture about the number of such products in the determinant of an $n \times n$ matrix.

35. Show that for any 2×2 matrix A, $\delta(aA) = a^2 \delta(A)$.

36. Show that for any 2×2 matrix A, $\delta(A^t) = \delta(A)$.

37. Show that for any 2×2 matrices A and B, $\delta(AB) = \delta(A) \cdot \delta(B)$.

7.5
THE INVERSE OF A SQUARE MATRIX

In the real number system, every element a, except 0, has a multiplicative inverse $a^{-1} = 1/a$ with the property that $a \cdot a^{-1} = a^{-1} \cdot a = a \cdot (1/a) = (1/a) \cdot a = 1$. The question should (and does) arise, "Does every square matrix A have a multiplicative inverse A^{-1}?" We shall see that not every nonzero square matrix has a multiplicative inverse; furthermore, when a matrix A does have a multiplicative inverse A^{-1}, it is *not* $1/A$.

> *For a given square matrix A of order n, if there is a square matrix B of order n such that*
> $$AB = BA = I,$$
> *where I is the multiplicative identity of order n, then B is the **multiplicative inverse** of A.*

Symbolically, if such a matrix B exists, we shall write $B = A^{-1}$. That is, A^{-1} satisfies $AA^{-1} = A^{-1}A = I$.

INVERSE OF A 2 × 2 MATRIX

To answer the question about the existence of a multiplicative inverse for a matrix, we shall begin by considering the simple case of 2 × 2 matrices. If we let

$$A = \begin{bmatrix} a_{11} & a_{12} \\ a_{21} & a_{22} \end{bmatrix},$$

then we must see whether there exists a 2 × 2 matrix A^{-1} such that $AA^{-1} = I$. If so, let $A^{-1} = \begin{bmatrix} b & c \\ d & e \end{bmatrix}$. We wish to have

$$\begin{bmatrix} a_{11} & a_{12} \\ a_{21} & a_{22} \end{bmatrix} \begin{bmatrix} b & c \\ d & e \end{bmatrix} = \begin{bmatrix} 1 & 0 \\ 0 & 1 \end{bmatrix}.$$

This leads to

$$\begin{bmatrix} a_{11}b + a_{12}d & a_{11}c + a_{12}e \\ a_{21}b + a_{22}d & a_{21}c + a_{22}e \end{bmatrix} = \begin{bmatrix} 1 & 0 \\ 0 & 1 \end{bmatrix},$$

which is true if and only if

(1) $\quad\quad a_{11}b + a_{12}d = 1, \quad\quad a_{11}c + a_{12}e = 0,$
$\quad\quad\quad a_{21}b + a_{22}d = 0, \quad\quad a_{21}c + a_{22}e = 1.$

Solving these equations for b, c, d, and e, we have

$$(a_{11}a_{22} - a_{12}a_{21})b = a_{22}, \quad\quad (a_{11}a_{22} - a_{12}a_{21})c = -a_{12},$$
$$(a_{11}a_{22} - a_{12}a_{21})d = -a_{21}, \quad\quad (a_{11}a_{22} - a_{12}a_{21})e = a_{11},$$

from which

(2) $\quad\quad b = \dfrac{a_{22}}{a_{11}a_{22} - a_{12}a_{21}}, \quad\quad c = \dfrac{-a_{12}}{a_{11}a_{22} - a_{12}a_{21}},$

$\quad\quad\quad d = \dfrac{-a_{21}}{a_{11}a_{22} - a_{12}a_{21}}, \quad\quad e = \dfrac{a_{11}}{a_{11}a_{22} - a_{12}a_{21}},$

provided $a_{11}a_{22} - a_{12}a_{21} \neq 0$. Now, the denominator of each of these fractions is just $\delta(A)$, so that

(3) $\quad\quad A^{-1} = \begin{bmatrix} b & c \\ d & e \end{bmatrix} = \begin{bmatrix} \dfrac{a_{22}}{\delta(A)} & \dfrac{-a_{12}}{\delta(A)} \\ \dfrac{-a_{21}}{\delta(A)} & \dfrac{a_{11}}{\delta(A)} \end{bmatrix} = \dfrac{1}{\delta(A)} \begin{bmatrix} a_{22} & -a_{12} \\ -a_{21} & a_{11} \end{bmatrix}.$

By direct multiplication, it can be verified not only that

$$AA^{-1} = I,$$

but also (surprisingly, since matrix multiplication is not always commutative) that

$$A^{-1}A = I.$$

7.5 The Inverse of a Square Matrix

Thus, to write the inverse of a 2×2 square matrix A for which $\delta(A) \neq 0$, we interchange the entries on the principal diagonal, replace each of the other two entries with its negative, and multiply the result by $1/\delta(A)$.

Example If $A = \begin{bmatrix} 1 & 3 \\ 2 & -1 \end{bmatrix}$, find A^{-1}.

Solution We first observe that $\delta(A) = (1)(-1) - (3)(2) = -7$. Hence,

$$A^{-1} = -\frac{1}{7}\begin{bmatrix} -1 & -3 \\ -2 & 1 \end{bmatrix} = \begin{bmatrix} \frac{1}{7} & \frac{3}{7} \\ \frac{2}{7} & -\frac{1}{7} \end{bmatrix}.$$

It is a good idea always to check the result when finding A^{-1}, because there is much room for making mistakes in the process of determining the inverse. In the present example, we have

$$A^{-1}A = -\frac{1}{7}\begin{bmatrix} -1 & -3 \\ -2 & 1 \end{bmatrix}\begin{bmatrix} 1 & 3 \\ 2 & -1 \end{bmatrix} = -\frac{1}{7}\begin{bmatrix} -7 & 0 \\ 0 & -7 \end{bmatrix} = \begin{bmatrix} 1 & 0 \\ 0 & 1 \end{bmatrix}.$$

MATRICES WITH NO INVERSE We have now arrived at a position where we can answer the question, "Does every 2×2 square matrix A have an inverse?" The answer is "No," for if $\delta(A)$ is 0, then the above equations (2) for b, c, d, and e would have no solution.

Example The matrix $\begin{bmatrix} 3 & 5 \\ 6 & 10 \end{bmatrix}$ has no inverse because

$$\delta(A) = 3(10) - 6(5) = 0.$$

Formula (3) is just a special case of the following general result.

If

$$A = \begin{bmatrix} a_{11} & a_{12} & \cdots & a_{1n} \\ a_{21} & a_{22} & \cdots & a_{2n} \\ \vdots & \vdots & & \vdots \\ a_{n1} & a_{n2} & \cdots & a_{nn} \end{bmatrix},$$

and if $\delta(A) \neq 0$, then

(4)
$$A^{-1} = \frac{1}{\delta(A)}\begin{bmatrix} A_{11} & A_{21} & \cdots & A_{n1} \\ A_{12} & A_{22} & \cdots & A_{n2} \\ \vdots & \vdots & & \vdots \\ A_{1n} & A_{2n} & \cdots & A_{nn} \end{bmatrix},$$

where A_{ij} is the cofactor of a_{ij} in A. If $\delta(A) = 0$, then A has no inverse.

Square matrices A for which $\delta(A) = 0$ are called **singular matrices**. Those for which $\delta(A) \neq 0$ are **nonsingular**. Thus, by (4), A has an inverse if and only if A is nonsingular.

INVERSE OF AN $n \times n$ MATRIX Observe that A^{-1} is the matrix having as its entries the cofactors of the entries in A multiplied by $1/\delta(A)$, but that the cofactors of the *row* entries in A are the *column* entries in A^{-1}. One way to obtain A^{-1} is to replace each entry in A with its cofactor and multiply the *transpose* of the resulting matrix by $1/\delta(A)$. That is to say,

$$A^{-1} = \frac{1}{\delta(A)} \begin{bmatrix} A_{11} & A_{12} & \cdots & A_{1n} \\ A_{21} & A_{22} & \cdots & A_{2n} \\ \vdots & \vdots & & \vdots \\ A_{n1} & A_{n2} & \cdots & A_{nn} \end{bmatrix}^t$$

Example If $A = \begin{bmatrix} 1 & 0 & 1 \\ 2 & 1 & 0 \\ 1 & -1 & 1 \end{bmatrix}$, find A^{-1}.

Solution We first observe that $\delta(A) = -2$, and since $\delta(A)$ is not zero, A has an inverse. Next, replacing each entry in A with its cofactor, we obtain the matrix

$$\begin{bmatrix} 1 & -2 & -3 \\ -1 & 0 & 1 \\ -1 & 2 & 1 \end{bmatrix},$$

whose transpose is

$$\begin{bmatrix} 1 & -1 & -1 \\ -2 & 0 & 2 \\ -3 & 1 & 1 \end{bmatrix},$$

so that

$$A^{-1} = -\frac{1}{2} \begin{bmatrix} 1 & -1 & -1 \\ -2 & 0 & 2 \\ -3 & 1 & 1 \end{bmatrix}$$

As a check, we have

$$A^{-1}A = -\frac{1}{2} \begin{bmatrix} 1 & -1 & -1 \\ -2 & 0 & 2 \\ -3 & 1 & 1 \end{bmatrix} \begin{bmatrix} 1 & 0 & 1 \\ 2 & 1 & 0 \\ 1 & -1 & 1 \end{bmatrix}$$

$$= -\frac{1}{2} \begin{bmatrix} -2 & 0 & 0 \\ 0 & -2 & 0 \\ 0 & 0 & -2 \end{bmatrix} = \begin{bmatrix} 1 & 0 & 0 \\ 0 & 1 & 0 \\ 0 & 0 & 1 \end{bmatrix}$$

7.5 The Inverse of a Square Matrix

Equation (4) is applicable to $n \times n$ square matrices, although, clearly, the process of actually determining A^{-1} becomes very laborious for matrices much larger than 3×3. An alternative technique for calculating A^{-1} is given by the following result which we state here without proof.

> An $n \times n$ matrix A is nonsingular if and only if it is row-equivalent to $I_{n \times n}$ and the sequence of elementary row operations that transforms A into I is the same sequence of elementary row operations that transforms I into A^{-1}.

We first form the augmented matrix $[A \mid I]$ and if possible transform A into I. Simultaneously then, I is transformed into A^{-1} to give the result $[I \mid A^{-1}]$.

Example Use elementary row operations to find A^{-1} for

$$A = \begin{bmatrix} 1 & -2 & 3 \\ 1 & 1 & 2 \\ -1 & 2 & -1 \end{bmatrix}.$$

Solution

$$\begin{bmatrix} 1 & -2 & 3 & \vdots & 1 & 0 & 0 \\ 1 & 1 & 2 & \vdots & 0 & 1 & 0 \\ -1 & 2 & -1 & \vdots & 0 & 0 & 1 \end{bmatrix}$$

$$\sim \begin{bmatrix} 1 & -2 & 3 & \vdots & 1 & 0 & 0 \\ 0 & 3 & -1 & \vdots & -1 & 1 & 0 \\ 0 & 0 & 2 & \vdots & 1 & 0 & 1 \end{bmatrix} \begin{array}{l} \text{Row } 2 + (-1 \times \text{Row } 1) \\ \text{Row } 3 + \text{Row } 1 \end{array}$$

$$\sim \begin{bmatrix} 1 & 0 & \frac{7}{3} & \vdots & \frac{1}{3} & \frac{2}{3} & 0 \\ 0 & 1 & -\frac{1}{3} & \vdots & -\frac{1}{3} & \frac{1}{3} & 0 \\ 0 & 0 & 2 & \vdots & 1 & 0 & 1 \end{bmatrix} \begin{array}{l} \text{Row } 1 + \left(\frac{2}{3} \times \text{Row } 2\right) \\ \frac{1}{3} \times \text{Row } 2 \end{array}$$

$$\sim \begin{bmatrix} 1 & 0 & 0 & \vdots & -\frac{5}{6} & \frac{2}{3} & -\frac{7}{6} \\ 0 & 1 & 0 & \vdots & -\frac{1}{6} & \frac{1}{3} & \frac{1}{6} \\ 0 & 0 & 1 & \vdots & \frac{1}{2} & 0 & \frac{1}{2} \end{bmatrix} \begin{array}{l} \text{Row } 1 + \left(-\frac{7}{6} \times \text{Row } 3\right) \\ \text{Row } 2 + \left(\frac{1}{6} \times \text{Row } 3\right) \\ \frac{1}{2} \times \text{Row } 3 \end{array}$$

It can be easily verified that

$$A^{-1} = \begin{bmatrix} -\frac{5}{6} & \frac{2}{3} & -\frac{7}{6} \\ -\frac{1}{6} & \frac{1}{3} & \frac{1}{6} \\ \frac{1}{2} & 0 & \frac{1}{2} \end{bmatrix} = \frac{1}{6} \begin{bmatrix} -5 & 4 & -7 \\ -1 & 2 & 1 \\ 3 & 0 & 3 \end{bmatrix}$$

The use of elementary row operations to find A^{-1} is particularly useful when A is a large matrix.

PROPERTIES OF MATRICES AND THEIR INVERSES

There are a number of useful properties associated with matrices and their inverses. The following result gives one example.

> If A and B are $n \times n$ nonsingular square matrices, then AB has an inverse, namely
> (5) $\qquad (AB)^{-1} = B^{-1}A^{-1}.$

To obtain (5) we right-multiply AB by $B^{-1}A^{-1}$, and apply the associative law for the multiplication of matrices.

$$AB \cdot B^{-1}A^{-1} = A \cdot I \cdot A^{-1} = A \cdot A^{-1} = I.$$

Moreover, if we left-multiply AB by $B^{-1}A^{-1}$, we have

$$B^{-1}A^{-1} \cdot AB = B^{-1} \cdot I \cdot B = B^{-1} \cdot B = I.$$

Thus, since $(AB)(B^{-1}A^{-1}) = (B^{-1}A^{-1})(AB) = I$, by the definition of the inverse of a matrix, we have

$$(AB)^{-1} = B^{-1}A^{-1}.$$

Equation (5) can be used to find the inverse of products of any number of nonsingular matrices. For example, if there are three factors A, B, and C in a product.

$$(ABC)^{-1} = [(AB)C]^{-1} = C^{-1}(AB)^{-1} = C^{-1}B^{-1}A^{-1}.$$

7.5 The Inverse of a Square Matrix

EXERCISE 7.5

In Problems 1–12 use (4) to find the inverse of each matrix if one exists.

Example $B = \begin{bmatrix} 1 & 0 & -1 \\ 1 & 3 & 1 \\ 0 & 1 & 2 \end{bmatrix}$

Solution The determinant $\delta(B)$ is given by

$$\delta \begin{bmatrix} 1 & 0 & -1 \\ 1 & 3 & 1 \\ 0 & 1 & 2 \end{bmatrix} = 1(5) - 0 - 1(1) = 4.$$

Replacing each entry of B with its cofactor gives

$$\begin{bmatrix} 5 & -2 & 1 \\ -1 & 2 & -1 \\ 3 & -2 & 3 \end{bmatrix};$$

$$\begin{bmatrix} 5 & -2 & 1 \\ -1 & 2 & -1 \\ 3 & -2 & 3 \end{bmatrix}^t = \begin{bmatrix} 5 & -1 & 3 \\ -2 & 2 & -2 \\ 1 & -1 & 3 \end{bmatrix};$$

$$B^{-1} = \frac{1}{\delta(B)} \begin{bmatrix} \text{each } b_{ij} \text{ of} \\ B \text{ replaced} \\ \text{by } B_{ij} \end{bmatrix}^t = \frac{1}{4} \begin{bmatrix} 5 & -1 & 3 \\ -2 & 2 & -2 \\ 1 & -1 & 3 \end{bmatrix}$$

$$= \begin{bmatrix} \frac{5}{4} & -\frac{1}{4} & \frac{3}{4} \\ -\frac{2}{4} & \frac{2}{4} & -\frac{2}{4} \\ \frac{1}{4} & -\frac{1}{4} & \frac{3}{4} \end{bmatrix}$$

1. $\begin{bmatrix} 1 & 2 \\ 1 & 3 \end{bmatrix}$
2. $\begin{bmatrix} 3 & 1 \\ 2 & -1 \end{bmatrix}$
3. $\begin{bmatrix} 2 & -3 \\ 1 & 1 \end{bmatrix}$

4. $\begin{bmatrix} -9 & 5 \\ -4 & 2 \end{bmatrix}$
5. $\begin{bmatrix} -2 & -6 \\ -3 & -9 \end{bmatrix}$
6. $\begin{bmatrix} 21 & 7 \\ 9 & 3 \end{bmatrix}$

7. $\begin{bmatrix} 1 & -1 & 2 \\ 2 & 1 & 3 \\ 0 & 0 & 2 \end{bmatrix}$
8. $\begin{bmatrix} 0 & 4 & 2 \\ 1 & 0 & 2 \\ 0 & -1 & 1 \end{bmatrix}$
9. $\begin{bmatrix} 2 & -1 & 1 \\ 3 & 0 & 1 \\ 2 & 2 & 1 \end{bmatrix}$

10. $\begin{bmatrix} 1 & 2 & 1 \\ 0 & 2 & 1 \\ -2 & 2 & 3 \end{bmatrix}$
11. $\begin{bmatrix} 2 & 1 & 1 \\ 1 & 0 & 2 \\ 4 & 2 & 2 \end{bmatrix}$
12. $\begin{bmatrix} -3 & 1 & -6 \\ 2 & 1 & 4 \\ 2 & 0 & 4 \end{bmatrix}$

In Problems 13–24 use elementary row operations to find the inverse of each matrix if one exists.

Example $A = \begin{bmatrix} 6 & 3 \\ -2 & -1 \end{bmatrix}$

Solution $\begin{bmatrix} 6 & 3 & \vdots & 1 & 0 \\ -2 & -1 & \vdots & 0 & 1 \end{bmatrix} \sim \begin{bmatrix} 0 & 0 & \vdots & 1 & 3 \\ -2 & -1 & \vdots & 0 & 1 \end{bmatrix}$ Row 1 + (3 × Row 2)

Since A cannot possibly be made equivalent to $I_{2 \times 2}$, no inverse exists.

Example $A = \begin{bmatrix} 2 & 5 \\ 1 & 3 \end{bmatrix}$

Solution $\begin{bmatrix} 2 & 5 & \vdots & 1 & 0 \\ 1 & 3 & \vdots & 0 & 1 \end{bmatrix} \begin{bmatrix} 1 & 2 & \vdots & 1 & -1 \\ 1 & 3 & \vdots & 0 & 1 \end{bmatrix}$ Row 1 + (−1 × Row 2)

$\sim \begin{bmatrix} 1 & 2 & \vdots & 1 & -1 \\ 0 & 1 & \vdots & -1 & 2 \end{bmatrix}$ Row 2 + (−1 × Row 1)

$\sim \begin{bmatrix} 1 & 0 & \vdots & 3 & -5 \\ 0 & 1 & \vdots & -1 & 2 \end{bmatrix}$ Row 1 + (−2 × Row 2)

Thus, $A^{-1} = \begin{bmatrix} 3 & -5 \\ -1 & 2 \end{bmatrix}$.

13. $\begin{bmatrix} 5 & 7 \\ 3 & 4 \end{bmatrix}$
14. $\begin{bmatrix} 5 & -4 \\ 4 & -3 \end{bmatrix}$
15. $\begin{bmatrix} 7 & 4 \\ -4 & -2 \end{bmatrix}$

16. $\begin{bmatrix} 1 & 2 & 1 \\ 0 & 2 & 1 \\ -2 & 2 & 3 \end{bmatrix}$
17. $\begin{bmatrix} 2 & 1 & 1 \\ 1 & 0 & 2 \\ 4 & 2 & 2 \end{bmatrix}$
18. $\begin{bmatrix} -3 & 1 & -6 \\ 2 & 1 & 4 \\ 2 & 0 & 4 \end{bmatrix}$

19. $\begin{bmatrix} 1 & 2 & -3 \\ 3 & -1 & 0 \\ 5 & 3 & -6 \end{bmatrix}$
20. $\begin{bmatrix} 2 & 4 & -1 \\ 1 & 6 & 2 \\ 5 & 14 & 0 \end{bmatrix}$
21. $\begin{bmatrix} 2 & -1 & -5 \\ 1 & 3 & 4 \\ 0 & 1 & 2 \end{bmatrix}$

22. $\begin{bmatrix} 2 & 1 & -8 \\ 1 & 1 & -2 \\ 1 & 2 & 3 \end{bmatrix}$
23. $\begin{bmatrix} 0 & 0 & 1 \\ 0 & 1 & 0 \\ 1 & 0 & 0 \end{bmatrix}$
24. $\begin{bmatrix} 1 & 0 & 1 \\ 0 & 1 & 0 \\ 1 & 0 & 0 \end{bmatrix}$

25. Verify that

$$\left(\begin{bmatrix} 2 & 3 \\ 1 & -1 \end{bmatrix} \cdot \begin{bmatrix} 0 & 1 \\ 3 & 1 \end{bmatrix} \right)^{-1} = \begin{bmatrix} 0 & 1 \\ 3 & 1 \end{bmatrix}^{-1} \cdot \begin{bmatrix} 2 & 3 \\ 1 & -1 \end{bmatrix}^{-1}.$$

26. Verify that

$$\left(\begin{bmatrix} 1 & 2 \\ -1 & 0 \end{bmatrix} \cdot \begin{bmatrix} 1 & 1 \\ 2 & 0 \end{bmatrix} \cdot \begin{bmatrix} 2 & -1 \\ 0 & 1 \end{bmatrix} \right)^{-1} = \begin{bmatrix} 2 & -1 \\ 0 & 1 \end{bmatrix}^{-1} \cdot \begin{bmatrix} 1 & 1 \\ 2 & 0 \end{bmatrix}^{-1} \cdot \begin{bmatrix} 1 & 2 \\ -1 & 0 \end{bmatrix}^{-1}.$$

27. Verify that
$$\left(\begin{bmatrix} 3 & 0 & 1 \\ 2 & 1 & 0 \\ 0 & 1 & 2 \end{bmatrix} \begin{bmatrix} 2 & 1 & 0 \\ 1 & 1 & 2 \\ 0 & 1 & 0 \end{bmatrix}\right)^{-1} = \begin{bmatrix} 2 & 1 & 0 \\ 1 & 1 & 2 \\ 0 & 1 & 0 \end{bmatrix}^{-1} \begin{bmatrix} 3 & 0 & 1 \\ 2 & 1 & 0 \\ 0 & 1 & 2 \end{bmatrix}^{-1}$$

28. Show that $[A^t]^{-1} = [A^{-1}]^t$ for each nonsingular 2×2 matrix.

29. Show that $\delta(A^{-1}) = 1/\delta(A)$ for each nonsingular 2×2 matrix.

30. If $A^{-1} = \begin{bmatrix} 3 & -1 \\ -2 & 2 \end{bmatrix}$, what is A?

7.6

SOLUTION OF LINEAR SYSTEMS USING INVERSES OF MATRICES

In Section 7.3 we solved linear systems using row-equivalent matrices. The solution for a linear system can also be found by using the inverse of a matrix.

We first verify the matrix product equation

$$\begin{bmatrix} a_{11} & a_{12} & \cdots & a_{1n} \\ \vdots & \vdots & & \vdots \\ a_{n1} & a_{n2} & \cdots & a_{nn} \end{bmatrix} \begin{bmatrix} x_1 \\ \vdots \\ x_n \end{bmatrix} = \begin{bmatrix} a_{11}x_1 + a_{12}x_2 + \cdots + a_{1n}x_n \\ \vdots \\ a_{n1}x_1 + a_{n2}x_2 + \cdots + a_{nn}x_n \end{bmatrix},$$

and hence note that the linear system

(1)
$$\begin{aligned} a_{11}x_1 + a_{12}x_2 + \cdots + a_{1n}x_n &= c_1 \\ a_{21}x_1 + a_{22}x_2 + \cdots + a_{2n}x_n &= c_2 \\ \vdots \quad \vdots \quad \vdots \quad \vdots& \\ a_{n1}x_1 + a_{n2}x_2 + \cdots + a_{nn}x_n &= c_n \end{aligned}$$

can be written as the matrix equation

$$\begin{bmatrix} a_{11} & a_{12} & \cdots & a_{1n} \\ \vdots & \vdots & & \vdots \\ a_{n1} & a_{n2} & \cdots & a_{nn} \end{bmatrix} \begin{bmatrix} x_1 \\ \vdots \\ x_n \end{bmatrix} = \begin{bmatrix} c_1 \\ \vdots \\ c_n \end{bmatrix},$$

where the first factor in the left-hand member is the coefficient matrix for the system. In more concise notation, this latter equation can be written

(2) $$AX = B,$$

where A is an $n \times n$ square matrix, and X and B are $n \times 1$ column matrices.

SOLUTION OF SYSTEMS

If A in (2) is nonsingular, we can left-multiply both members of this equation by A^{-1} to obtain the equivalent matrices

(3)
$$A^{-1}AX = A^{-1}B$$
$$IX = A^{-1}B$$
$$X = A^{-1}B,$$

where $A^{-1}B$ is an $n \times 1$ column matrix. Since X and $A^{-1}B$ are equal, each entry in X is equal to the corresponding entry in $A^{-1}B$, and hence these latter entries constitute the components of the solution of the given linear system. If A is a singular matrix, then, of course, it has no inverse, and either the system has no solution or the solution is not unique.

Example Use matrices to find the solution set of

$$2x + y + z = 1$$
$$x - 2y - 3z = 1$$
$$3x + 2y + 4z = 5.$$

Solution We first write this as a matrix equation of the form $AX = B$.

$$\begin{bmatrix} 2 & 1 & 1 \\ 1 & -2 & -3 \\ 3 & 2 & 4 \end{bmatrix} \begin{bmatrix} x \\ y \\ z \end{bmatrix} = \begin{bmatrix} 1 \\ 1 \\ 5 \end{bmatrix}$$

We next determine $\delta(A)$, obtaining

$$\delta \begin{bmatrix} 2 & 1 & 1 \\ 1 & -2 & -3 \\ 3 & 2 & 4 \end{bmatrix} = 2(-2) - 1(13) + 1(8) = -9.$$

Observe that A is nonsingular, and then find A^{-1} by the methods of Section 7.5.

$$A^{-1} = \begin{bmatrix} 2 & 1 & 1 \\ 1 & -2 & -3 \\ 3 & 2 & 4 \end{bmatrix}^{-1} = -\frac{1}{9} \begin{bmatrix} -2 & -2 & -1 \\ -13 & 5 & 7 \\ 8 & -1 & -5 \end{bmatrix}$$

As a matter of routine, we check the latter by verifying that $A^{-1}A = I$.

$$A^{-1}A = -\frac{1}{9} \begin{bmatrix} -2 & -2 & -1 \\ -13 & 5 & 7 \\ 8 & -1 & -5 \end{bmatrix} \begin{bmatrix} 2 & 1 & 1 \\ 1 & -2 & -3 \\ 3 & 2 & 4 \end{bmatrix}$$

$$= -\frac{1}{9} \begin{bmatrix} -9 & 0 & 0 \\ 0 & -9 & 0 \\ 0 & 0 & -9 \end{bmatrix}$$

$$= \begin{bmatrix} 1 & 0 & 0 \\ 0 & 1 & 0 \\ 0 & 0 & 1 \end{bmatrix}$$

7.6 Solution of Linear Systems Using Inverses of Matrices

Now, since $X = A^{-1}B$, we have

$$\begin{bmatrix} x \\ y \\ z \end{bmatrix} = -\frac{1}{9} \begin{bmatrix} -2 & -2 & -1 \\ -13 & 5 & 7 \\ 8 & -1 & -5 \end{bmatrix} \begin{bmatrix} 1 \\ 1 \\ 5 \end{bmatrix} = -\frac{1}{9} \begin{bmatrix} -9 \\ 27 \\ -18 \end{bmatrix} = \begin{bmatrix} 1 \\ -3 \\ 2 \end{bmatrix}$$

Hence, $x = 1$, $y = -3$, and $z = 2$, and the solution set of the system is $\{(1, -3, 2)\}$.

The computation of A^{-1} is laborious when A is a square matrix containing many rows and columns. The foregoing method is not always the easiest to use in solving systems, but it is most valuable for theoretical developments.

A LEONTIEF MODEL   In economics a so-called **open Leontief model*** is characterized by a system of equations in the form

(4) $$X = AX + B$$

where

$$X = \begin{bmatrix} x_1 \\ x_2 \\ \vdots \\ x_n \end{bmatrix}$$

and B is an $n \times 1$-column matrix of constants and A is an $n \times n$ matrix of constants known as an *input–output* matrix. The fundamental idea of an open Leontief model is that whatever is produced by the industrial sector is absorbed by both the industrial and consumer sectors. That is, consumption equals production. For three industries, each producing a single item C_1, C_2, and C_3, an input–output matrix could be as follows:

$$\begin{array}{c} \\ \text{input} \end{array} \begin{array}{c} \\ C_1 \\ C_2 \\ C_3 \end{array} \overset{\displaystyle\text{Output}}{\overset{\displaystyle C_1 \quad C_2 \quad C_3}{\begin{bmatrix} \frac{1}{5} & \frac{1}{2} & \frac{1}{3} \\ \frac{1}{8} & \frac{1}{3} & \frac{1}{2} \\ \frac{3}{8} & 0 & \frac{1}{5} \end{bmatrix}}}$$

The entries in a column, say the first, mean one-fifth unit, of item C_1, one-eighth unit of item C_2, and three-eighths unit of C_3 are used as inputs to produce one unit of item

---

*Named for Wassily Leontief, Nobel Laureate in economics in 1973.

C_1 as output. In other words, the fractional entries represent the proportionate amounts of the total production of, say item C_1, attributable to C_1, C_2, and C_3. For example, if it takes 40 units of C_1, 25 units of C_2, and 75 units of C_3 to make 200 units of C_1, then the entries in the first column are $40/200 = 1/5$, $25/200 = 1/8$, and $75/200 = 3/8$.

The matrix B in (4) is called a *demand vector*. By solving for X we determine the production outputs needed from the industries in order to satisfy that demand. Observe that (4) can also be written as

$$X - AX = B$$

or

$$(I - A)X = B.$$

If the matrix $I - A$ is nonsingular it follows that

(5) $$X = (I - A)^{-1}B.$$

Example Solve (4) when the input–output and demand matrices are

$$A = \begin{bmatrix} \frac{1}{2} & \frac{1}{4} & 0 \\ \frac{1}{4} & \frac{1}{2} & \frac{1}{2} \\ 0 & 0 & \frac{1}{4} \end{bmatrix}, \quad B = \begin{bmatrix} 15 \\ 3 \\ 27 \end{bmatrix},$$

respectively.

Solution First we find

$$I - A = \begin{bmatrix} 1 & 0 & 0 \\ 0 & 1 & 0 \\ 0 & 0 & 1 \end{bmatrix} - \begin{bmatrix} \frac{1}{2} & \frac{1}{4} & 0 \\ \frac{1}{4} & \frac{1}{2} & \frac{1}{2} \\ 0 & 0 & \frac{1}{4} \end{bmatrix}$$

$$= \begin{bmatrix} \frac{1}{2} & -\frac{1}{4} & 0 \\ -\frac{1}{4} & \frac{1}{2} & -\frac{1}{2} \\ 0 & 0 & \frac{3}{4} \end{bmatrix}$$

and then $\delta(I - A) = 9/64$. From (4) of Section 7.5 it follows that

7.6 Solution of Linear Systems Using Inverses of Matrices

$$(I-A)^{-1} = \frac{64}{9}\begin{bmatrix} \frac{3}{8} & \frac{3}{16} & \frac{1}{8} \\ \frac{3}{16} & \frac{3}{8} & \frac{1}{4} \\ 0 & 0 & \frac{3}{16} \end{bmatrix} = \begin{bmatrix} \frac{8}{3} & \frac{4}{3} & \frac{8}{9} \\ \frac{4}{3} & \frac{8}{3} & \frac{16}{9} \\ 0 & 0 & \frac{4}{3} \end{bmatrix}.$$

Hence by (5),

$$X = \begin{bmatrix} \frac{8}{3} & \frac{4}{3} & \frac{8}{9} \\ \frac{4}{3} & \frac{8}{3} & \frac{16}{9} \\ 0 & 0 & \frac{4}{3} \end{bmatrix} \begin{bmatrix} 15 \\ 3 \\ 27 \end{bmatrix} = \begin{bmatrix} 68 \\ 76 \\ 36 \end{bmatrix}.$$

That is, $x_1 = 68$ units, $x_2 = 76$ units, $x_3 = 36$ units.

EXERCISE 7.6

Find the solution set of the given system by means of matrices. If the system has no unique solution, so state.

1. $2x - 3y = -1$
   $x + 4y = 5$

2. $3x - 4y = -2$
   $x - 2y = 0$

3. $3x - 4y = -2$
   $6x + 12y = 36$

4. $2x - 4y = 7$
   $x - 2y = 1$

5. $2x - 3y = 0$
   $2x + y = 16$

6. $2x + 3y = 3$
   $3x - 4y = 0$

7. $x + y = 2$
   $2x - z = 1$
   $2y - 3z = -1$

8. $2x - 6y + 3z = -12$
   $3x - 2y + 5z = -4$
   $4x + 5y - 2z = 10$

9. $x - 2y + z = -1$
   $3x + y - 2z = 4$
   $y - z = 1$

10. $2x + 5z = 9$
    $4x + 3y = -1$
    $3y - 4z = -13$

11. $2x + 2y + z = 1$
    $x - y + 6z = 21$
    $3x + 2y - z = -4$

12. $4x + 8y + z = -6$
    $2x - 3y + 2z = 0$
    $x + 7y - 3z = -8$

13. $x + y + z = 0$
    $2x - y - 4z = 15$
    $x - 2y - z = 7$

14. $x + y - 2z = 3$
    $3x - y + z = 5$
    $3x + 3y - 6z = 9$

Applications *Determine X by finding $(I - A)^{-1}$ using the given input–output matrix and demand vector.*

15. $A = \begin{bmatrix} \frac{1}{2} & 0 \\ \frac{1}{4} & \frac{1}{2} \end{bmatrix}$; $B = \begin{bmatrix} 20 \\ 30 \end{bmatrix}$

16. $A = \begin{bmatrix} 0.1 & 0.5 \\ 0.3 & 0.4 \end{bmatrix}$; $B = \begin{bmatrix} 100 \\ 600 \end{bmatrix}$

17. $A = \begin{bmatrix} \frac{1}{2} & \frac{1}{8} & \frac{1}{4} \\ 0 & \frac{1}{2} & 0 \\ \frac{1}{8} & 0 & \frac{1}{8} \end{bmatrix}$; $B = \begin{bmatrix} 13 \\ 52 \\ 26 \end{bmatrix}$

18. $A = \begin{bmatrix} \frac{1}{10} & \frac{1}{10} & \frac{5}{10} \\ \frac{3}{10} & \frac{4}{10} & \frac{1}{10} \\ \frac{1}{10} & \frac{1}{10} & \frac{3}{10} \end{bmatrix}$; $B = \begin{bmatrix} 1 \\ 1 \\ 1 \end{bmatrix}$

19. *Business* A small manufacturing company makes three items, A, B, and C, each of which is constructed out of materials I, II, and III. The number of units of material used in each item is given in the array

$$\begin{array}{c} \\ \text{I} \\ \text{II} \\ \text{III} \end{array} \begin{array}{ccc} A & B & C \\ \begin{bmatrix} 3 & 2 & 1 \\ 2 & 4 & 5 \\ 1 & 1 & 3 \end{bmatrix} \end{array}.$$

In any particular week the company uses the following number of units of each type of material:

$$\begin{array}{c} \text{I} \\ \text{II} \\ \text{III} \end{array} \begin{bmatrix} 39 \\ 70 \\ 27 \end{bmatrix}$$

Determine $X = \begin{bmatrix} x \\ y \\ z \end{bmatrix}$, where x, y, and z denote, respectively, the number of items A, B, and C produced each week.

20. A certain economic model is characterized by the equations

$$Y = C + J$$
$$C = a + bY$$
$$J = c + dY$$

7.7 Cramer's Rule

where a, b, c, d are constants and C, J, Y represent monetary values. If $X = \begin{bmatrix} C \\ J \\ Y \end{bmatrix}$ write the given system in form (2). Solve the system by (3) provided $b + d \neq 1$.

7.7
CRAMER'S RULE

In Section 7.6, we obtained the solution set of the linear system (1) on page 258 with nonsingular coefficient matrix by first expressing the system in the matrix form $AX = B$ and then left-multiplying both members of the equation by A^{-1} to obtain

$$A^{-1}AX = X = A^{-1}B.$$

If, now, this technique is viewed in terms of determinants, we arrive at a general solution for such systems.

If the coefficient matrix A is nonsingular, then its inverse, A^{-1}, is

$$A^{-1} = \frac{1}{\delta(A)} \begin{bmatrix} A_{11} & A_{21} & \cdots & A_{n1} \\ \vdots & \vdots & & \vdots \\ A_{1n} & A_{2n} & \cdots & A_{nn} \end{bmatrix}$$

Now, since $B = \begin{bmatrix} c_1 \\ c_2 \\ \vdots \\ c_n \end{bmatrix}$, we have

$$X = A^{-1}B = \frac{1}{\delta(A)} \begin{bmatrix} c_1 A_{11} + c_2 A_{21} + \cdots + c_n A_{n1} \\ c_1 A_{12} + c_2 A_{22} + \cdots + c_n A_{n2} \\ \vdots & \vdots & & \vdots \\ c_1 A_{1n} + c_2 A_{2n} + \cdots + c_n A_{nn} \end{bmatrix}.$$

Each entry in $X = A^{-1}B$ can be seen to be of the form

$$\frac{c_1 A_{1j} + c_2 A_{2j} + \cdots + c_n A_{nj}}{\delta(A)}.$$

But $c_1 A_{1j} + c_2 A_{2j} + \cdots + c_n A_{nj}$ is just the expansion of the determinant

$$\begin{array}{c} j\text{th column} \\ \downarrow \\ \begin{vmatrix} a_{11} & a_{12} & \cdots & c_1 & \cdots & a_{1n} \\ a_{21} & a_{22} & \cdots & c_2 & \cdots & a_{2n} \\ \vdots & \vdots & & \vdots & & \vdots \\ a_{n1} & a_{n2} & \cdots & c_n & \cdots & a_{nn} \end{vmatrix} \end{array}$$

about the jth column, which has entries $c_1, c_2, c_3, \ldots, c_n$ in place of $a_{1j}, a_{2j}, \ldots, a_{nj}$. Thus, each entry x_j in the matrix

$$X = \begin{bmatrix} x_1 \\ x_2 \\ \vdots \\ x_n \end{bmatrix} = A^{-1}B$$

is

$$x_j = \frac{\delta(A_j)}{\delta(A)} = \frac{\begin{vmatrix} a_{11} & a_{12} & \cdots & c_1 & \cdots & a_{1n} \\ a_{21} & a_{22} & \cdots & c_2 & \cdots & a_{2n} \\ \vdots & \vdots & & \vdots & & \vdots \\ a_{n1} & a_{n2} & \cdots & c_n & \cdots & a_{nn} \end{vmatrix}}{\begin{vmatrix} a_{11} & a_{12} & \cdots & & & a_{1n} \\ a_{21} & a_{22} & \cdots & & & a_{2n} \\ \vdots & \vdots & & & & \vdots \\ a_{n1} & a_{n2} & \cdots & & & a_{nn} \end{vmatrix}}.$$

with the jth column indicated.

APPLICATION OF CRAMER'S RULE

This relationship expresses **Cramer's rule**, which is the assertion that if the determinant of the coefficient matrix of an $n \times n$ linear system *is not* zero, then the equations are consistent (the system has a solution) and have a unique solution which can be found as follows.

To find x_j in solving the matrix equation $AX = B$:

1. Write the determinant of the coefficient matrix for the system.

2. Replace each entry in the jth column of the coefficient matrix A with the corresponding entry from the column matrix B, and find the determinant of the resulting matrix.

3. Divide the result in Step 2 by the result in Step 1.

Hence, in a nonsingular 3×3 system:

$$x = \frac{\delta(A_x)}{\delta(A)}, \quad y = \frac{\delta(A_y)}{\delta(A)}, \quad z = \frac{\delta(A_z)}{\delta(A)}.$$

Example Use Cramer's rule to solve the system

$$\begin{aligned} -4x + 2y - 9z &= 2 \\ 3x + 4y + z &= 5 \\ x - 3y + 2z &= 8. \end{aligned}$$

7.7 Cramer's Rule

Solution By inspection,

$$\delta(A) = \begin{vmatrix} -4 & 2 & -9 \\ 3 & 4 & 1 \\ 1 & -3 & 2 \end{vmatrix}$$
$$= -4(11) - 2(5) - 9(-13)$$
$$= -44 - 10 + 117 = 63.$$

Replacing the entries in the first column of A with corresponding constants 2, 5, and 8, we have

$$\delta(A_x) = \begin{vmatrix} 2 & 2 & -9 \\ 5 & 4 & 1 \\ 8 & -3 & 2 \end{vmatrix}$$
$$= 2(11) - 2(2) - 9(-47)$$
$$= 22 - 4 + 423 = 441.$$

Hence, by Cramer's rule,

$$x = \frac{\delta(A_x)}{\delta(A)} = \frac{441}{63} = 7.$$

Similarly, by replacing the entries of the second and third columns of A with the corresponding constants 2, 5, and 8, we have

$$\delta(A_y) = \begin{vmatrix} -4 & 2 & -9 \\ 3 & 5 & 1 \\ 1 & 8 & 2 \end{vmatrix} \quad \text{and} \quad \delta(A_z) = \begin{vmatrix} -4 & 2 & 2 \\ 3 & 4 & 5 \\ 1 & -3 & 8 \end{vmatrix}.$$

Now,

$$\delta(A_y) = -4(2) - 2(5) - 9(19)$$
$$= -8 - 10 - 171 = -189,$$

and

$$\delta(A_z) = -4(47) - 2(19) + 2(-13)$$
$$= -188 - 38 - 26 = -252,$$

so that

$$y = \frac{\delta(A_y)}{\delta(A)} = \frac{-189}{63} = -3,$$

and

$$z = \frac{\delta(A_z)}{\delta(A)} = \frac{-252}{63} = -4.$$

The solution set of the system is $\{(7, -3, -4)\}$.

TEST FOR CONSISTENCY If $\delta(A) = 0$ for a linear system, then the system either has infinitely many members in its solution set (the equations are consistent and one of them can be obtained from the others by linear combinations) or has an empty solution set (the equations are inconsistent). The distinction can be determined as follows. Consider the matrix of coefficients

$$\begin{bmatrix} a_{11} & \cdots & a_{1n} \\ \vdots & & \vdots \\ a_{n1} & \cdots & a_{nn} \end{bmatrix}$$

and the augmented matrix

$$\begin{bmatrix} a_{11} & \cdots & a_{1n} & c_1 \\ \vdots & & \vdots & \vdots \\ a_{n1} & \cdots & a_{nn} & c_n \end{bmatrix},$$

and in each find a determinant (obtained by striking out certain rows and columns) of order as great as possible with value not zero. The order of such a nonvanishing determinant is called the **rank** of the matrix. The rank of the augmented matrix is either the same as, or 1 greater than, that of the matrix of coefficients. The equations are consistent if and only if the two ranks are the same.

For example, the coefficient matrix C for the system

$$\begin{aligned} x + 2y + 3z &= 2 \\ 2x + 4y + 2z &= -1 \\ x + 2y - 2z &= 5 \end{aligned}$$

is given by

$$C = \begin{bmatrix} 1 & 2 & 3 \\ 2 & 4 & 2 \\ 1 & 2 & -2 \end{bmatrix},$$

while the augmented matrix C_A for the system is given by (after deleting the vertical line)

$$C_A = \begin{bmatrix} 1 & 2 & 3 & 2 \\ 2 & 4 & 2 & -1 \\ 1 & 2 & -2 & 5 \end{bmatrix}.$$

Since $\delta(C) = 0$, the rank of C is not 3, and we therefore know that the system does not have a unique solution. If the first column and third row are deleted, the remaining determinant

$$\begin{vmatrix} 2 & 3 \\ 4 & 2 \end{vmatrix}$$

is not zero (check this), so C has rank 2.

Now, if the first column of C_A is deleted, the remaining entries form the determinant

7.7 Cramer's Rule

$$\delta \begin{bmatrix} 2 & 3 & 2 \\ 4 & 2 & -1 \\ 2 & -2 & 5 \end{bmatrix},$$

which is also not zero (check this). Hence, C_A has rank 3. Therefore, the equations in the system are inconsistent.

EXERCISE 7.7

Find the solution set of each system by Cramer's rule. If $\delta(A) = 0$, use the ranks of the coefficient matrix and the augmented matrix to determine whether the equations in the system are consistent.

1. $x - y = 2$
   $x + 4y = 5$

2. $x + y = 4$
   $x - 2y = 0$

3. $3x - 4y = -2$
   $x + y = 6$

4. $2x - 4y = 7$
   $x - 2y = 1$

5. $\frac{1}{3}x - \frac{1}{2}y = 0$
   $\frac{1}{2}x + \frac{1}{4}y = 4$

6. $\frac{2}{3}x + y = 1$
   $x - \frac{4}{3}y = 0$

7. $x - 2y = 6$
   $\frac{2}{3}x - \frac{4}{3}y = 6$

8. $\frac{1}{2}x + y = 3$
   $-\frac{1}{4}x - y = -3$

9. $x - 3y = 1$
   $y = 1$

10. $2x - 3y = 12$
    $x = 4$

11. $ax + by = 1$
    $bx + ay = 1$

12. $x + y = a$
    $x - y = b$

13. $x - 2y + z = -1$
    $3x + y - 2z = 4$
    $y - z = 1$

14. $2x + 5z = 9$
    $4x + 3y = -1$
    $3y - 4z = -13$

15. $2x + 2y + z = 1$
    $x - y + 6z = 21$
    $3x + 2y - z = -4$

16. $4x + 8y + z = -6$
    $2x - 3y + 2z = 0$
    $x + 7y - 3z = -8$

17. $x + y + z = 0$
    $2x - y - 4z = 15$
    $x - 2y - z = 7$

18. $x + y - 2z = 2$
    $3x - y + z = 5$
    $3x + 3y - 6z = 6$

19. $x - 2y - 2z = 3$
    $2x - 4y + 4z = 1$
    $3x - 3y - 3z = 4$

20. $3x - 2y + 5z = 6$
    $4x - 4y + 3z = 0$
    $5x - 4y + z = -5$

21. $x - 4z = -1$
    $3x + 3y = 2$
    $3x + 4z = 5$

22. $2x - \frac{2}{3}y + z = 2$
    $\frac{1}{2}x - \frac{1}{3}y - \frac{1}{4}z = 0$
    $4x + 5y - 3z = -1$

23. $x + y + z = 0$
    $w + 2y - z = 4$
    $2w - y + 2z = 3$
    $-2w + 2y - z = -2$

24. $x + y + z = 0$
    $x + z + w = 0$
    $x + y + w = 0$
    $y + z + w = 0$

25. For the system
$$a_1 x + b_1 y + c_1 = 0$$
$$a_2 x + b_2 y + c_2 = 0,$$
show that if both $\delta(A_y) = 0$ and $\delta(A_x) = 0$, and if c_1 and c_2 are not both zero, then $\delta(A) = 0$, and the equations are consistent. [*Hint*: Show that the first two determinant equations imply that $a_1 c_2 = a_2 c_1$ and $b_1 c_2 = b_2 c_1$, and that the rest follows from the formation of a proportion with these equations.]

26. Show that if $\delta(A) = 0$ and $\delta(A_x) = 0$, and if a_1 and a_2 are not both zero, then $\delta(A_y) = 0$, where $\delta(A)$ is the determinant of the coefficient matrix of the system in Problem 25.

Application

27. *Biology* The following table shows the percentage U.S. recommended daily allowances (U.S. RDA) vitamin content per ounce of three food groups X, Y, and Z.

	X	Y	Z
Vitamin A	9	5	4
Vitamin C	24	10	5
Thiamine	3	5	0

Use Cramer's rule to determine how many ounces of each food group would have to be consumed each day to achieve 100% of the daily recommended allowance of vitamin A, 200% of the daily recommended allowance of vitamin C, and 30% of the daily recommended allowance of thiamine.

CHAPTER TEST

Write each sum or difference as a single matrix.

1. $\begin{bmatrix} 4 & -7 \\ 2 & 1 \end{bmatrix} + \begin{bmatrix} -3 & 6 \\ -1 & 0 \end{bmatrix}$

2. $\begin{bmatrix} 3 & -1 & 7 \\ 6 & 2 & 5 \end{bmatrix} + \begin{bmatrix} -1 & 6 & -9 \\ 8 & -3 & 7 \end{bmatrix}$

3. $\begin{bmatrix} 2 & -4 & 3 \\ 6 & 1 & 7 \\ 2 & 8 & 0 \end{bmatrix} - \begin{bmatrix} 4 & -1 & 2 \\ 3 & 8 & 1 \\ 7 & 6 & -5 \end{bmatrix}$

4. $\begin{bmatrix} -11 & 2 & -6 \\ 7 & 1 & 2 \\ -3 & 4 & 8 \end{bmatrix} - \begin{bmatrix} -3 & 5 & 0 \\ 1 & 4 & 2 \\ 6 & -1 & 3 \end{bmatrix}$

Write each product as a single matrix.

5. $-7 \begin{bmatrix} 3 & -1 \\ 2 & 0 \\ 1 & 1 \end{bmatrix}$

6. $\begin{bmatrix} 3 & -1 & 2 \end{bmatrix} \begin{bmatrix} 4 \\ -1 \\ 0 \end{bmatrix}$

7. $\begin{bmatrix} 3 & -1 \\ 6 & 5 \end{bmatrix} \cdot \begin{bmatrix} -4 & 2 \\ 1 & 3 \end{bmatrix}$

8. $\begin{bmatrix} -1 & 7 & 6 \\ 3 & 1 & 2 \\ 1 & 0 & 1 \end{bmatrix} \cdot \begin{bmatrix} 1 & -1 & 2 \\ 1 & 0 & 3 \\ 2 & 1 & 1 \end{bmatrix}$

Use row transformation on the augmented matrix to solve each system of equations.

9. $2x - y = 5$
$x + 3y = -1$

10. $2x - y + z = 4$
$x + 3y - z = 4$
$x + 2y + z = 5$

Evaluate each determinant.

11. $\begin{vmatrix} -3 & 0 \\ 2 & 1 \end{vmatrix}$

12. $\begin{vmatrix} 1 & 5 \\ -1 & 2 \end{vmatrix}$

13. $\begin{vmatrix} 3 & 1 & 0 \\ 2 & 0 & 1 \\ 1 & 2 & -1 \end{vmatrix}$

14. $\begin{vmatrix} 3 & 1 & -2 \\ -1 & 2 & 1 \\ 1 & -2 & 1 \end{vmatrix}$

Find the inverse of each nonsingular matrix.

15. $\begin{bmatrix} -4 & 2 \\ 11 & 3 \end{bmatrix}$

16. $\begin{bmatrix} 1 & -1 & 2 \\ 3 & 1 & 0 \\ 2 & 1 & 1 \end{bmatrix}$

Use matrices to solve each system.

17. $\begin{aligned} x - y &= -3 \\ 2x + 3y &= -1 \end{aligned}$

18. $\begin{aligned} 2x + z &= 7 \\ y + 2z &= 1 \\ 3x + y + z &= 9 \end{aligned}$

Find the production output X for the open Leontief model with input–output matrix A and demand vector B.

19. $A = \begin{bmatrix} 0.3 & 0.1 \\ 0.1 & 0.3 \end{bmatrix}; \quad B = \begin{bmatrix} 60 \\ 180 \end{bmatrix}$

20. $A = \begin{bmatrix} \frac{1}{2} & 0 & \frac{1}{5} \\ \frac{1}{5} & \frac{3}{5} & 0 \\ \frac{2}{5} & \frac{1}{5} & \frac{1}{5} \end{bmatrix}; \quad B = \begin{bmatrix} 18 \\ 6 \\ 12 \end{bmatrix}$

Use Cramer's rule to solve each system.

21. $\begin{aligned} 3x - y &= -5 \\ x + 2y &= -6 \end{aligned}$

22. $\begin{aligned} x - y + 2z &= 3 \\ 2x + y - z &= 3 \\ x - 2y + 2z &= 4 \end{aligned}$

PART III

8 INEQUALITIES AND LINEAR PROGRAMMING
9 SEQUENCES AND SERIES
10 PROBABILITY
11 MATHEMATICS OF FINANCE

8
INEQUALITIES AND LINEAR PROGRAMMING

8.1
GRAPHS OF FIRST-DEGREE RELATIONS

ASSOCIATED EQUATION OF AN INEQUALITY

A linear inequality of form

$$Ax + By + C \leq 0 \quad \text{or} \quad Ax + By + C < 0,$$

A and B not both zero defines the relation

$$\{(x, y) | Ax + By + C \leq 0\} \quad \text{or} \quad \{(x, y) | Ax + By + C < 0\}.$$

Such relations can be graphed in the plane, but the graph will be a region of the plane rather than a straight line. For example, consider the inequality

(1) $$2x + y - 3 < 0.$$

When (1) is rewritten in the equivalent form

(2) $$y < -2x + 3,$$

we see that the solutions (x, y) are such that for each x, y is less than $-2x + 3$. The graph of the equation

(3) $$y = -2x + 3$$

is simply a straight line, as illustrated in Figure 8.1, part a. To graph the relation defined by (1), we need only observe that any point below this line has a y-coordinate that satisfies (2), and consequently the solution set of (2) corresponds to the entire region below the line. The region is indicated by shading in Figure 8.1, part b. The dashed line indicates that the line itself is not included in the graph. If the inequality had been

$$2x + y - 3 \leq 0,$$

275

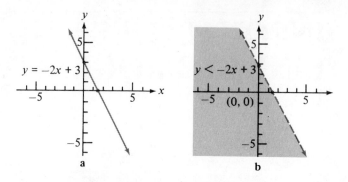

Figure 8.1

the line would be part of the graph and would be shown as a solid line. In general, the graphs of the members of

$$\{(x, y) \mid Ax + By + C < 0\} \quad \text{or} \quad \{(x, y) \mid Ax + By + C > 0\}$$

are the points in a **half-plane** on one side of the graph of the associated equation

$$Ax + By + C = 0,$$

depending on the constants and inequality symbols involved.

DETERMINATION OF THE GRAPH OF AN INEQUALITY

To determine which half-plane to shade in constructing graphs of first-degree inequalities, one can select any point in either half-plane and test its coordinates in the defining sentence to see whether the selected point lies in the graph. If the coordinates satisfy the sentence, then the half-plane containing the selected point is shaded; if not, the opposite half-plane is shaded. A very convenient point to use in this process is the origin, provided the origin is not contained in the graph of the associated equation. Thus, in the above example, the replacement of x and y by 0 in

$$2x + y - 3 \leq 0$$

results in

$$2(0) + (0) - 3 \leq 0,$$

which is true, and hence the half-plane containing the origin is shaded.

Inequalities do not ordinarily define functions, according to our definition in Section 4.1, because it usually is not true that each element of the domain is associated with a unique element in the range.

8.1 Graphs of First-Degree Relations

EXERCISE 8.1

Graph each inequality in the plane.

Example $2x + y \geq 4$

Solution Solve the defining inequality explicitly for y.

$$y \geq 4 - 2x$$

Graph the equality $y = 4 - 2x$.

Observe that the origin is not part of the graph of the inequality, because (0, 0) is not a member of the given relation; that is, $0 \not\geq 4 - 2(0)$. Hence, the region above the graph of the equation $y = 4 - 2x$ is shaded.

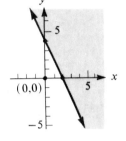

The line is included in the graph.

1. $y < x$
2. $y > x$
3. $y \leq x + 2$
4. $y \geq x - 2$
5. $x + y < 5$
6. $2x + y < 2$
7. $x - y < 3$
8. $x - 2y < 5$
9. $x \leq 2y - 4$
10. $2x \leq y + 1$
11. $3 \geq 2x - 2y$
12. $0 \geq x + y$

Example $x > 2$

Solution Graph the equality $x = 2$.

Shade the region to the right of the graph of $x = 2$, that is, the set of all points such that $x > 2$.

The line is excluded from the graph.

13. $x > 0$
14. $y < 0$
15. $x < 0$
16. $x < -2$
17. $-1 < x < 5$
18. $0 \leq y \leq 1$
19. $|x| < 3$
20. $|y| > 1$
21. $|x| + |y| \leq 1$
22. $|x| + |y| \geq 1$

[*Hint*: Consider the graphs in each quadrant separately:

$x, y \geq 0 \qquad x \leq 0, y \geq 0 \qquad x, y \leq 0 \qquad x \geq 0, y \leq 0.$]

8.2

SYSTEMS OF LINEAR INEQUALITIES

GRAPHS OF SYSTEMS OF INEQUALITIES We have observed that the graph of the solution set of an inequality in two variables might be a region in the plane. The graph of the solution set of a system of linear inequalities in two variables, which consists of the *intersection* of the graphs of the inequalities in the system, might also be a plane region.

Example Graph the solution set of the system

$$x + 2y \leq 6$$
$$2x - 3y \geq 12.$$

Solution Graphing each inequality by the method discussed in Section 8.1, we obtain the figure shown, where the doubly shaded region is the graph of the solution set of the system. We have graphed the intersection of two relations:

$$\{(x, y) | x + 2y \leq 6\} \cap \{(x, y) | 2x - 3y \geq 12\}.$$

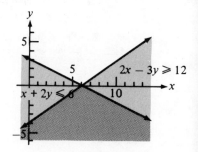

Example Graph the solution set of the system

$$y > 2$$
$$x > -2$$
$$x + y > 1.$$

Solution The triply shaded region in the figure constitutes the graph of the solution set of the system. We have graphed

$$\{(x, y) | y > 2\} \cap \{(x, y) | x > -2\}$$
$$\cap \{(x, y) | x + y > 1\}.$$

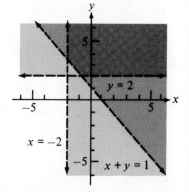

Example Food group X contains 40 calories per ounce and food group Y contains 20 calories per ounce. A person desires to limit the daily caloric intake from these two foods to no more than 600 calories. Show graphically the various combinations of quantities of the two food groups that would make this limitation possible.

8.2 Systems of Linear Inequalities

Solution Let x and y represent the number of ounces of foods X and Y, respectively, consumed each day. It is understood that we must have $x \geq 0$ and $y \geq 0$. The number of calories obtained by eating both food groups is then $40x + 20y$. The graph of the solution set of

$$40x + 20y \leq 600$$
$$x \geq 0, \quad y \geq 0$$

is the doubly shaded region in the accompanying figure.

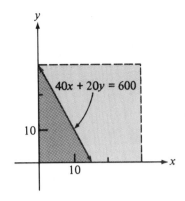

POSTAL REGULATIONS The following example translates a Government postal regulation into a system of linear inequalities.

Example A postal bulletin dated November 16, 1978, states that "All first-class items weighing one ounce or less and all single-piece third-class items weighing two ounces or less" are subject to an extra mailing fee, or surcharge, when

1. the height is greater than $6\frac{1}{8}$ inches, or
2. the length is greater than $11\frac{1}{2}$ inches, or
3. the length is less than 1.3 times the height, or
4. the length is greater than 2.5 times the height.

The bulletin goes on to say that "The standards ... may appear difficult to understand and administer." Use a system of linear inequalities to describe the sizes of envelopes for first- and third-class letters (satisfying the weight requirements) that are *not* subject to a surcharge.

Solution Suppose an envelope has length x and height y as shown in the first figure. Then for a surcharge we must have either

(1) $\qquad y > 6\frac{1}{8}$,

or

(2) $\qquad x > 11\frac{1}{2}$,

or

(3) $\qquad x < 1.3y$,

or

(4) $\qquad x > 2.5y$,

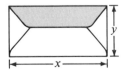

where $x > 0$, $y > 0$. The inequalities in (3) and (4) can be written as $y > \frac{10}{13}x$ and $y < \frac{2}{5}x$, respectively. Thus, in order *not* to have a surcharge (x, y) must be a solution of the system

$$y \leq 6\tfrac{1}{8}$$
$$x \leq 11\tfrac{1}{2}$$
$$y \leq \frac{10}{13}x$$
$$y \geq \frac{2}{5}x$$
$$x > 0, \qquad y > 0.$$

The shaded region given in the second figure corresponds to envelope sizes not subject to the surcharge. The standard business envelope is $9\tfrac{1}{2}$ inches long and $4\tfrac{1}{8}$ inches high. Inspection of the figure shows that the ordered pair $(9\tfrac{1}{2}, 4\tfrac{1}{8})$ is in the shaded region.

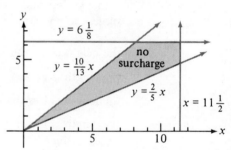

EXERCISE 8.2

By double or triple shading, indicate the region in the plane that represents the solution set of each system.

Example $x \geq -1$
$2y \geq 3x + 8$
$2y \leq x + 4$

Solution   Graph the two relations,

$$\{(x, y) | x \geq -1\} \cap \{(x, y) | 2y \geq 3x + 8\}$$

and

$$\{(x, y) | 2y \leq x + 4\}$$

8.2 Systems of Linear Inequalities

and then take their intersection. We see from the figure that the two sets have no points in common. The solution set is $\emptyset$.

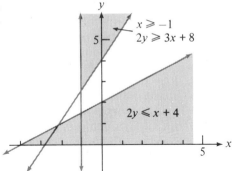

1. $y \geq x + 1$
   $y \geq 5 - x$

2. $y \geq 4$
   $x \geq 2$

3. $y - x \geq 0$
   $y + x \geq 0$

4. $2y - x \geq 1$
   $x < -3$

5. $y + 3x < 6$
   $y > 2$

6. $x > 3$
   $x + y \geq 5$

7. $x \leq 3$
   $x + y < 4$

8. $y \leq 2$
   $2y + x < 3$

9. $x - 3y < 6$
   $y + x \geq 1$

10. $2x + y \geq 4$
    $x - y \geq -2$

11. $y \leq 3$
    $x \leq 2$
    $y < x$

12. $y \geq 2$
    $x \geq 2$

13. $|y| \geq 2$
    $|x| \geq 2$

14. $-1 \leq x - y \leq 1$
    $x + y \geq 1$
    $x + y \leq -1$

15. $y - x < -2$
    $y > x + 3$

16. $-1 < x < 1$
    $-1 < y < 1$
    $y > x$

17. $x + y \geq 4$
    $3x + y \leq 12$
    $x + 3y \leq 12$

18. $x \geq 0$
    $y \geq 0$
    $x + 2y \leq 10$
    $x + y \leq 8$

Applications

19. *Biology* A patient must take at least 20 units of medicine X and at least 30 units of medicine Y each day. The combined dosage of the two medicines must not exceed 100 units each day. Summarize these requirements by linear inequalities and graph the solution set.

20. A person consumes two food groups X and Y each day. Food X contains 2 grams of protein per ounce and food Y contains 5 grams of protein per ounce. The person wishes to obtain at least 50 grams of protein each day. Summarize this requirement by linear inequalities and graph the solution set.

21. A patient must take at least twice as much medicine Y as medicine X each day. The combined dosage of the two medicines must not exceed 50 units each day. Summarize these requirements by linear inequalities and graph the solution set.

22. *Business* An automobile manufacturing company decides to limit the combined production of cars X and Y to at least 100,000 but not more than 300,000. Also, the company

wants the number of cars X to be at least 3 times the number of cars Y. Summarize these requirements by linear inequalities and graph the solution set.

23. A small mill makes wheat and rye flour. Each ton of wheat must spend three-fourths of an hour in coarse-grind milling machines and one-half hour in fine-grind milling machines. Each ton of rye must spend one-fourth of an hour in the coarse-grind milling machines and one-half hour in the fine-grind milling machines. Because of maintenance requirements, the coarse milling machines can run at most 120 hours per week and the fine milling machines can run at most 140 hours per week. Let x and y denote the number of tons of wheat and rye, respectively, that are processed each week. By means of linear inequalities summarize all the possible numbers of tons of wheat and rye the mill can process and satisfy the weekly hour requirements. Graph the solution set.

24. Use the example on page 279 to find the sizes of envelopes that can be sent as first- and third-class letters (subject to weight restrictions) without surcharge if the length of the envelope is

   a. $6\frac{1}{2}$ inches   **b.** $9\frac{1}{2}$ inches

8.3

CONVEX SETS; POLYGONAL REGIONS

In Section 8.2, we saw that the graph of the solution set of a system of linear inequalities in two variables is simply the common intersection of a number of open or closed half-planes. Any such intersection is an example of a **convex set**.

> A set $\mathscr{S}$ of points is a convex set if and only if, for each two points P and Q in $\mathscr{S}$, the line segment PQ lies entirely in $\mathscr{S}$.

Figure 8.2 shows three sets of points in a plane. Convex sets are pictured in parts a and b. The set in part c is not convex because a portion of the line segment PQ does

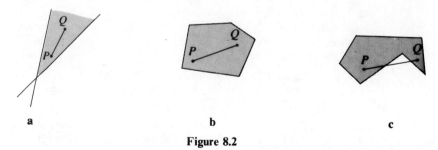

a   b   c

Figure 8.2

8.3 Convex Sets: Polygonal Regions

not lie in the set. If the boundary of a convex set is a (closed) polygon, then the boundary is called a **convex polygon**, and the region enclosed by the polygon (including the boundary) is called a **closed convex polygonal set**. In Figure 8.2, part a, we say the region is an **unbounded convex set**.

We have two immediate results that are intuitively true; their proofs are omitted.

> *Any half-plane is a convex set.*

> *The intersection of two convex sets is a convex set.*

As suggested by Figure 8.3, this latter result extends to any number of convex sets. If the intersection of the graphs of a system of linear inequalities constitutes a (closed) polygonal set $\mathcal{P}$, then we can locate the set $\mathcal{P}$ in the plane by graphing the equations associated with the given inequalities and identifying the vertices of $\mathcal{P}$.

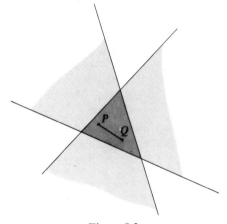

Figure 8.3

Example Locate and identify by shading the polygonal set specified by the system

$$x + y \leq 5$$
$$x - y \leq 2$$
$$x - y \geq -2$$
$$x \geq 0$$
$$y \geq 0$$

Find the vertices of the set.

Solution Graph the associated equations

$$x + y = 5$$
$$x - y = 2$$
$$x - y = -2$$
$$x = 0$$
$$y = 0$$

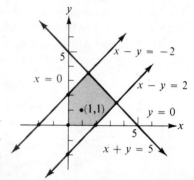

as shown. Note which half-plane is determined by each of the five inequalities, and shade the intersection of the five half-planes. The intersection is the pentagonal region shown. To check the result, note that (1, 1) is in the region and its coordinates satisfy each given inequality. The vertices are (0, 0), (0, 2), (3/2, 7/2), (7/2, 3/2), and (2, 0). These are obtained by finding the points of intersection of the equations representing the respective sides.

We should not conclude from this example that, simply because the five equations associated with the set of inequalities determine a convex pentagon, the pentagon is necessarily the graph of the system of inequalities. If the inequality $y \geq 0$ is replaced by $y \leq 0$, then the graphs of the equations remain unchanged, but the graph of the system of inequalities is the triangular region shown in Figure 8.4, part a. On the other hand, if the inequality $x + y \leq 5$ is replaced by $x + y \geq 5$, then the intersection of the graphs becomes the (infinite) rectangular region shown in Figure 8.4, part b. Finally, we observe that if both of these changes are made, that is, if $x + y \geq 5$ replaces $x + y \leq 5$ and $y \leq 0$ replaces $y \geq 0$, then the intersection of the graphs of the inequalities in the system is $\emptyset$.

A linear inequality involving three variables such as $3x + 4y + z \leq 12$ would represent a **half-space** in three dimensions. Any system of such inequalities would be a convex set which might be a closed polyhedral set.

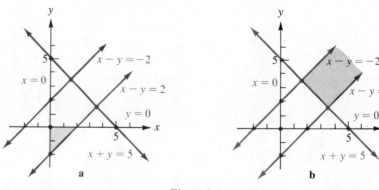

Figure 8.4

**NON-
NEGATIVE
VARIABLES**

We shall be concerned subsequently only with systems of linear inequalities in which all the variables are assumed to be nonnegative. In the case of two variables, the restriction $x \geq 0$, $y \geq 0$ will either yield a convex set which is a subset of the first quadrant or a region in the first quadrant, two boundaries of which are portions of the nonnegative coordinate axes.

EXERCISE 8.3

Graph the convex polygonal set defined by the given system of inequalities. Find the vertices of the set.

1. $x - y \leq 2$
   $x - y \geq -2$
   $x + y \geq 2$
   $x + y \leq 6$

2. $0 \leq x \leq 5$
   $0 \leq y \leq 5$
   $x + y \leq 6$

3. $0 \leq x \leq 5$
   $0 \leq y \leq 4$
   $x + 2y \leq 10$
   $2x + y \leq 10$

4. $0 \leq x$
   $0 \leq y \leq 4$
   $x - 2y \leq 6$
   $2x + y \leq 12$
   $2x - y \leq 10$

5. $0 \leq x \leq 5$
   $y \geq x - 3$
   $x + y \leq 9$
   $3y \leq 2x + 12$

6. $0 \leq x$
   $0 \leq y \leq 6$
   $x + 2y \leq 13$
   $2x + y \leq 11$
   $3x + y \leq 15$

7. Repeat Problem 1 with $x - y \leq 2$ replaced by $x - y \geq 2$. Is the graph a closed polygonal set?

8. Repeat Problem 3 with $x + 2y \leq 10$ replaced by $x + 2y \geq 10$. Is the graph a closed polygonal set?

9. Repeat Problem 4 with $0 \leq y \leq 4$ replaced by $y \leq 4$. Is the graph a closed polygonal set?

10. Repeat Problem 6 without the condition $x \geq 0$. Is the graph a closed polygonal set?

11. Show by example that the union of two convex sets need not be convex.

8.4

LINEAR PROGRAMMING

For each ordered pair of real numbers (x, y) the linear expression $F = ax + by$ has only one value. Hence, F is a function of two variables x and y. In particular, if the linear function F is defined only over a convex set $\mathscr{S}$, then the fundamental problem of linear programming is to find its maximum and minimum values over the set. We shall call $F = ax + by$ the **objective function**.

It can be shown that the following result holds for the values of F over a closed polygonal subset $\mathscr{P}$ of the Cartesian plane.

> If $\mathscr{P}$ is a closed convex polygonal set, then the function $F = ax + by$ takes on its maximum and minimum values over $\mathscr{P}$ at one of the vertices of $\mathscr{P}$.

Thus, if $\mathscr{P}$ is the closed convex polygonal set with vertices (1, 2), (3, 1), (5, 3), (4, 5), and (3, 6), as pictured in Figure 8.5, then the objective function $F = 3x - 2y$ has a value for every ordered pair (x, y) in or on the boundary of $\mathscr{P}$. In particular, at the vertices we have the values shown in the table.

Vertex	(1, 2)	(3, 1)	(5, 3)	(4, 5)	(3, 6)
Value of $3x - 2y$	-1	7	9	2	-3

By inspection, the maximum of these values is **9** and the minimum is **−3**. We conclude that 9 is the greatest value of $F = 3x - 2y$ over the entire set $\mathscr{P}$, and the least value of F over $\mathscr{P}$ is -3.

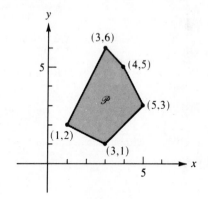

Figure 8.5

Example A company manufactures two kinds of electric shavers, one that uses a cord and another that is cordless. The company can make up to 500 cord models and 400 cordless models per day, but it can make a total of only 600 shavers per day. It takes 2 man-hours to manufacture the cord model and 3 man-hours to manufacture the cordless model, and the company has available at most 1400 man-hours per day. If there is a profit of $2.50 on a cord shaver and $3.50 on a cordless shaver, how many of each kind should the company make per day in order to realize the greatest profit?

8.4 Linear Programming

Example Let x = number of cord shavers made in 1 day.
Let y = number of cordless shavers made in 1 day.

We wish to maximize the linear profit function

$$F = 2.5x + 3.5y$$

subject to the following constraints on x and y:

$$0 \le x \le 500$$
$$0 \le y \le 400$$
$$x + y \le 600$$
$$2x + 3y \le 1400.$$

Solving the associated equations for these inequalities in pairs yields the vertices of the polygonal region shown in the figure. The values of the objective function at the vertices are listed in the table. Clearly, the most profitable numbers of shavers are 400 cord models and 200 cordless models.

Vertex	Profit
(0, 0)	$ 0
(0, 400)	$1400
(100, 400)	$1650
(400, 200)	**$1700**
(500, 100)	$1600
(500, 0)	$1250

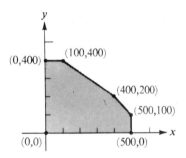

The point which maximizes or minimizes the objective function F does not have to be unique.

Example Find the maximum and minimum of $F = 4x + 6y$ over the convex set defined by

$$2x + 3y \le 21$$
$$x - 4y \le -6$$
$$-3x + y \le -4$$
$$x \ge 0$$
$$y \ge 0.$$

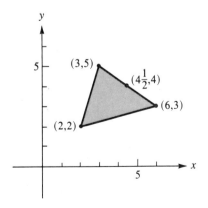

Solution The vertices are (3, 5), (2, 2), and (6, 3). The minimum is $F = 20$ at (2, 2). But the maximum of F is 42, which occurs at every point along the line segment joining (3, 5) and (6, 3); for example, at the arbitrary point $(4\frac{1}{2}, 4)$ we have $F = 42$.

A linear function $F = ax + by$ defined over an unbounded convex set does not necessarily have either a maximum or a minimum. We state the following result without proof.

> A function $F = ax + by$ defined over an unbounded convex set may not have a maximum or a minimum. If F has a maximum or minimum it will occur at a vertex of the set.

In the next example we assume (actually it can be proved) that the objective function has a minimum subject to the given constraints.

Example A person on a special diet is required to consume at least 240 units of protein and at least 200 units of carbohydrate each day. The number of units of protein, carbohydrate, and fat that are contained in two foods X and Y, per ounce, is shown in the following table.

	Food X	Food Y
Protein	60	40
Carbohydrate	40	50
Fat	10	25

How many ounces of foods X and Y can be eaten each day to satisfy the protein and carbohydrate requirements while minimizing the intake of fat?

Solution Let $x =$ number of ounces of food X consumed each day.
Let $y =$ number of ounces of food Y consumed each day.

From the table we see that the constraints on x and y are

$$x \geq 0, \quad y \geq 0$$
$$60x + 40y \geq 240$$
$$40x + 50y \geq 200.$$

We wish to minimize the total fat intake from these two foods each day:

$$F = 10x + 25y.$$

Now graph the linear inequalities and find the vertices.

We conclude that by consuming 5 ounces of food X alone we can satisfy the protein and carbohydrate requirement while obtaining a minimum of 50 units of fat.

8.4 Linear Programming

Vertex	Fat
(0, 6)	150
$\left(\dfrac{20}{7}, \dfrac{12}{7}\right)$	$\dfrac{500}{7}$
(5, 0)	50

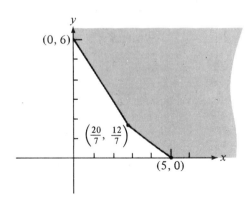

In the preceding example the objective function $F = 10x + 25y$ has no maximum. Along the two edges of the set, $y = 0$, $x > 5$, and $x = 0$, $y > 6$, F can be increased without bound.

EXERCISE 8.4

Find the maximum and minimum values of the given objective function over the closed convex polygonal set cited.

1. $F = x + 3y$; the set in Problem 1, Exercise 8.3
2. $F = 3x + 5y$; the set in Problem 2, Exercise 8.3
3. $F = 3x + 5y$; the set in Problem 3, Exercise 8.3
4. $F = x + 2y$; the set in Problem 4, Exercise 8.3
5. $F = 5x - 4y$; the set in Problem 5, Exercise 8.3
6. $F = 10x - 2y$; the set in Problem 6, Exercise 8.3
7. Determine whether

$$F = 6x - 2y$$

has a maximum and a minimum subject to

$$y \leq 3x$$
$$3x + 2y \leq 27$$
$$x \geq 0$$
$$3 \leq y \leq 6.$$

8. Determine whether

$$F = -x - 2y$$

has a maximum and a minimum subject to

$$5x + 3y \geq 18$$
$$x + 3y \geq 6$$
$$x \geq 0$$
$$y \geq 0.$$

Applications

9. *Biology* Rework the example on page 288 when the objective function describing the fat content of the diet is $F - 4.9x + 3.5y$.

10. A patient on a special diet is advised to supplement the diet by taking at least 600 units of iron, at least 1400 units of vitamin C, and at least 800 units of thiamine each day. These substances can be obtained from two commercial brands of multiple-vitamin tablets X and Y. The following table shows the content of each tablet. How many of each tablet can be taken each day to fulfill the requirement but yet at the same time minimize the intake of vitamin A?

	X	Y
Iron	100	100
Vitamin C	300	200
Thiamine	100	200
Vitamin A	60	50

For Problems 11–14 use Tables A and B, which show the number of units of labor, machinery, materials, and overhead necessary for a manufacturer to produce one unit of each of two products x and y. The tables also show the total available units of each of these items.

Table A

	Units needed for 1 unit of product x	Units needed for 1 unit of product y	Total units available
Labor	1	4	40
Machinery	1	3	31
Materials	5	2	60
Overhead	1	1	15

11. *Business* Using Table A, find the maximum profit if each unit of product x earns a profit of \$50 and each unit of product y earns a profit of \$60.

12. Using Table A, find the maximum profit if each unit of product x earns a profit of \$60 and each unit of product y earns a profit of \$50.

8.5 The Simplex Method

Table B

	Units needed for 1 unit of product x	Units needed for 1 unit of product y	Total units available
Labor	2	7	63
Machinery	5	1	50
Materials	1	2	21
Overhead	1	1	14

13. Using Table B, find the maximum profit if each unit of product x earns a profit of $50 and each unit of product y earns a profit of $60.

14. Using Table B, find the maximum profit if each unit of product x earns a profit of $60 and each unit of product y earns a profit of $50.

15. An electronics company makes two kinds of electronic ranges, a standard model that earns a $50 profit and a deluxe model that earns a $60 profit. The company has machinery capable of producing any number of deluxe ranges up to 400 per month and any number of standard ranges up to 500 per month, but it has enough man-hours available to produce only 600 ranges of both kinds in a month. How many of each kind of range should the company produce to realize a maximum profit?

16. A pharmacy has 300 ounces of a drug it can use to make two different kinds of medicines. From each ounce of the drug, 15 bottles of medicine A or 25 bottles of medicine B can be produced. The pharmacy must keep all the bottles in its own storage room, which has a capacity of 6000 bottles in all. At least 600 bottles of medicine A and 1250 bottles of medicine B must be retained in inventory and the rest will be used. If medicine A sells for $2.75 per bottle and medicine B sells for $2.00 per bottle, how many ounces of the drug should be devoted to each medicine to maximize the total return?

8.5

THE SIMPLEX METHOD

If a linear programming problem involves many constraints or more than two variables, then the geometric procedure of Section 8.4 is not very satisfactory. While the results of the preceding section remain true for linear expressions such as $F = ax + by + cz$ defined over a closed three-dimensional polyhedral convex set, finding all vertices and evaluating the objective function at these points can be a tedious job. We need a more systematic approach to the problem.

In this section we shall consider a procedure for maximizing a linear function of any number of variables by considering certain operations on an augmented matrix.

SLACK VARIABLES Recall from Section 1.2 (Problems 55–58), that for real numbers a and b, $a \leq b$ if and only if there is a number $d \geq 0$ such that $a + d = b$. Similarly, for systems of inequalities such as

$$7x + 3y \leq 21$$
$$x + 2y \leq 10$$
$$x \geq 0$$
$$y \geq 0,$$

the numbers $u \geq 0$ and $v \geq 0$ are defined as those numbers which take up the slack in the inequalities. That is,

$$7x + 3y + u \phantom{{}+v} = 21$$
$$x + 2y \phantom{{}+u} + v = 10.$$

The nonnegative numbers u and v are called **slack variables**.

By defining their coefficients as zero, we shall assume that an objective function contains as many slack variables as there are equations in the constraints.

Example Write the following problem in terms of slack variables:

Maximize

$$F = 5x + 7y$$

subject to

$$x + 2y \leq 12$$
$$x + y \leq 7$$
$$3x + 2y \leq 18$$
$$x \geq 0$$
$$y \geq 0.$$

Solution We use u, v, and w as slack variables.

$$x + 2y + u + 0v + 0w = 12$$
$$x + y + 0u + v + 0w = 7$$
$$3x + 2y + 0u + 0v + w = 18$$
$$F - 5x - 7y + 0u + 0v + 0w = 0$$

$$x \geq 0 \qquad y \geq 0 \qquad u \geq 0 \qquad v \geq 0 \qquad w \geq 0.$$

8.5 The Simplex Method

The augmented matrix of coefficients of the variables is called the initial **simplex tableau**. For the above example the tableau is

$$\begin{bmatrix} 1 & 2 & 1 & 0 & 0 & \vdots & 12 \\ 1 & 1 & 0 & 1 & 0 & \vdots & 7 \\ 3 & 2 & 0 & 0 & 1 & \vdots & 18 \\ -5 & -7 & 0 & 0 & 0 & \vdots & 0 \end{bmatrix},$$

where the last line means $F - 5x - 7y = 0$, or $F = 5x + 7y$.

THE SIMPLEX METHOD

We shall now describe row operations on the simplex tableau which will lead to the maximum of F.

1. Form the initial simplex tableau.
2. Examine the last row of the tableau and choose the entry which is the most negative. This entry determines the **pivotal column**.
3. Take each positive entry in the pivotal column and form the ratio of the entry to the right of the partition line to the corresponding entry of the pivotal column. The smallest ratio determines the **pivotal row**.
4. The intersection of the pivotal column with the pivotal row determines the **pivotal entry**.
5. Use the pivotal entry and elementary transformations to make all other entries in the pivotal column zero.
6. Examine the last row. If all entries are nonnegative, then we are finished. If there are negative entries in the last row, then repeat Steps 2–5 as many times as needed to make the last row entirely nonnegative.

Example Use the simplex method to solve the problem in the preceding example.

Solution The second column in the first matrix is the pivotal column. The ratios of the entries in the last column to the positive entries in the pivotal column are 6, 7, and 9. Thus, the pivotal entry is the number in the first row, second column. We now perform Step 5.

$$\begin{bmatrix} 1 & ② & 1 & 0 & 0 & | & 12 \\ 1 & 1 & 0 & 1 & 0 & | & 7 \\ 3 & 2 & 0 & 0 & 1 & | & 18 \\ -5 & -7 & 0 & 0 & 0 & | & 0 \end{bmatrix}$$

$$\sim \begin{bmatrix} \frac{1}{2} & 1 & \frac{1}{2} & 0 & 0 & | & 6 \\ \frac{1}{2} & 0 & -\frac{1}{2} & 1 & 0 & | & 1 \\ 2 & 0 & -1 & 0 & 1 & | & 6 \\ -\frac{3}{2} & 0 & \frac{7}{2} & 0 & 0 & | & 42 \end{bmatrix} \begin{array}{l} \frac{1}{2} \times \text{Row 1} \\ \left(-\frac{1}{2} \times \text{Row 1}\right) + \text{Row 2} \\ (-1 \times \text{Row 1}) + \text{Row 3} \\ \left(\frac{7}{2} \times \text{Row 1}\right) + \text{Row 4} \end{array}$$

Since the last row has a negative entry, we determine the pivotal entry once again and repeat the row operations.

$$\begin{bmatrix} \frac{1}{2} & 1 & \frac{1}{2} & 0 & 0 & | & 6 \\ \left(\frac{1}{2}\right) & 0 & -\frac{1}{2} & 1 & 0 & | & 1 \\ 2 & 0 & -1 & 0 & 1 & | & 6 \\ -\frac{3}{2} & 0 & \frac{7}{2} & 0 & 0 & | & 42 \end{bmatrix}$$

$$\sim \begin{bmatrix} 0 & 1 & 1 & -1 & 0 & | & 5 \\ 1 & 0 & -1 & 2 & 0 & | & 2 \\ 0 & 0 & 1 & -4 & 1 & | & 2 \\ 0 & 0 & 2 & 3 & 0 & | & 45 \end{bmatrix} \begin{array}{l} (-1 \times \text{Row 2}) + \text{Row 1} \\ 2 \times \text{Row 2} \\ (-4 \times \text{Row 2}) + \text{Row 3} \\ (3 \times \text{Row 2}) + \text{Row 4} \end{array}$$

We shall now set the variables corresponding to the *nonzero entries* in the last row equal to zero. Thus, if we set $u = 0$ and $v = 0$, then necessarily from the first row, $y = 5$; from the second row, $x = 2$; and from the third row, $w = 2$. The entry in the last row and last column is the maximum of F. We therefore have $F = 45$ at $(2, 5)$ as the solution.

INTERPRE-TATION OF THE PROCEDURE

To see why this procedure works, let us examine the various steps in the last example.

Since we are to maximimize $F = 5x + 7y$, we see that $x = 0$, $y = 0$ satisfy the constraints and give the value $F = 0$. This obviously is not the maximum, since F can be increased by increasing either x or y. The largest increase in F occurs when y is increased. This determination is the same as finding the most negative number in the last row of the initial tableau. Now, y cannot be increased without

8.5 The Simplex Method

bound because of the constraints. If x is held at zero, then from the original system

$$x + 2y + u = 12$$
$$x + y + v = 7$$
$$3x + 2y + w = 18$$
$$F - 5x - 7y = 0$$

we see that the first equation implies y can be increased up to 6 without disturbing the nonnegativity of u. The second equation implies y can be increased up to 7 without affecting the nonnegativity of v. Similarly, from the third equation we conclude w remains nonnegative provided y does not exceed 9. Thus, to maintain the nonnegativity of the slack variables u, v, and w, y can be at most the smaller of the three numbers $\{6, 7, 9\}$. In effect we are at one corner of the polygonal convex set defined by the constraints, namely $(0, 6)$. The corresponding value of F is 42.

Now the second tableau means

$$\frac{1}{2}x + y + \frac{1}{2}u = 6$$
$$\frac{1}{2}x - \frac{1}{2}u + v = 1$$
$$2x - u + w = 6$$
$$F - \frac{3}{2}x + \frac{7}{2}u = 42.$$

The last row is $F = 3/2\, x - 7/2\, u + 42$. If $x = 0$, $u = 0$, then $F = 42$. The question is: Can F be increased further? Any change in u would decrease F, whereas a change in x would increase F. Again, the first three equations in the above system dictate how far x can be increased. If we hold $u = 0$, then from the first equation we see that x can be advanced up to 12 without making y negative; the second and third equations imply that x can increase as far as 2 and 3, respectively, while keeping v and w nonnegative. Hence, x can be increased only up to the smaller of the three numbers $\{12, 2, 3\}$. When $u = 0$ and $x = 2$, then $y = 5$. We are at a new corner of the convex set with $F = 45$ as the value of the objective function. The last tableau gives this information

$$y + u - v = 5$$
$$x - u + 2v = 2$$
$$u - 4v + w = 2$$
$$F + 2u + 3v = 45.$$

Here the last row is $F = 45 - 2u - 3v$. Clearly, F can only decrease if u and v are increased. Hence, the maximum value of F must occur when $u = 0$ and $v = 0$. The four rows of the tableau give $y = 5$, $x = 2$, $w = 2$, and $F = 45$.

EXERCISE 8.5

In the following problems use the simplex method to find the maximum value of the objective function.

Example $F = 3x + 2y + z$

$2x + y + 2z \leq 4$
$2x + 2y + 3z \leq 6$
$x \geq 0 \quad y \geq 0 \quad z \geq 0$

Solution Use slack variables $u \geq 0$ and $v \geq 0$.

$$2x + y + 2z + u + 0v = 4$$
$$2x + 2y + 3z + 0u + v = 6$$
$$F - 3x - 2y - z + 0u + 0v = 0$$

Form the simplex tableau and perform Steps 2–6.

$$\begin{bmatrix} ② & 1 & 2 & 1 & 0 & \vdots & 4 \\ 2 & 2 & 3 & 0 & 1 & \vdots & 6 \\ -3 & -2 & -1 & 0 & 0 & \vdots & 0 \end{bmatrix}$$

$$\sim \begin{bmatrix} 1 & \frac{1}{2} & 1 & \frac{1}{2} & 0 & \vdots & 2 \\ 0 & ① & 1 & -1 & 1 & \vdots & 2 \\ 0 & -\frac{1}{2} & 2 & \frac{3}{2} & 0 & \vdots & 6 \end{bmatrix} \begin{matrix} \frac{1}{2} \times \text{Row 1} \\ (-1 \times \text{Row 1}) + \text{Row 2} \\ \left(\frac{3}{2} \times \text{Row 1}\right) + \text{Row 3} \end{matrix}$$

$$\sim \begin{bmatrix} 1 & 0 & \frac{1}{2} & 1 & -\frac{1}{2} & \vdots & 1 \\ 0 & 1 & 1 & -1 & 1 & \vdots & 2 \\ 0 & 0 & \frac{5}{2} & 1 & \frac{1}{2} & \vdots & 7 \end{bmatrix} \begin{matrix} \left(-\frac{1}{2} \times \text{Row 2}\right) + \text{Row 1} \\ \\ \left(\frac{1}{2} \times \text{Row 2}\right) + \text{Row 3} \end{matrix}$$

Set $z = 0$, $u = 0$, and $v = 0$. The maximum of F is 7 at $x = 1$, $y = 2$, and $z = 0$.

1. $F = 3x + 4y$
   $x + 2y \leq 30$
   $2x + y \leq 30$
   $x \geq 0 \quad y \geq 0$

2. $F = 2x + 8y$
   $x + 2y \leq 5$
   $5x + 3y \leq 18$
   $x \geq 0 \quad y \geq 0$

8.5 The Simplex Method

3. $F = 50x + 40y$
   $4x + 5y \leq 60$
   $20x + 7y \leq 120$
   $x \geq 0 \quad y \geq 0$

4. $F = 20x + 50y$
   $-2x + 5y \leq 20$
   $6x + 5y \leq 60$
   $x \geq 0 \quad y \geq 0$

5. $F = 600x + 400y$
   $x + 4y \leq 36$
   $2x + y \leq 16$
   $4x + y \leq 28$
   $x \geq 0 \quad y \geq 0$

6. $F = 40x + 30y$
   $-x + 5y \leq 15$
   $x + y \leq 9$
   $8x - 3y \leq 50$
   $x \geq 0 \quad y \geq 0$

7. $F = 2x + 3y + z$
   $3x + 4y + 9z \leq 18$
   $x + 2y + 4z \leq 8$
   $x \geq 0 \quad y \geq 0 \quad z \geq 0$

8. $F = 5x + 16y + 20z$
   $x + 2y + 7z \leq 20$
   $x + 8y + 2z \leq 10$
   $x \geq 0 \quad y \geq 0 \quad z \geq 0$

Applications

9. *Biology* A fish and game commission wishes to restock a depleted game refuge with three species of animals X, Y, and Z. The entries in the following table

	X	Y	Z
A	1	3	2
B	4	2	4
C	2	2	3

   represent the units of daily consumption of three types of foods A, B, and C, inherent to the refuge, by each animal of a species. It is desired that the daily consumption of the foods not exceed 160 units of A, 320 units of B, and 200 units of C. How should the refuge be restocked to maximize the total number of animals subject to the food constraints? [*Hint:* In this case the pivotal column can be any column. Use the first column in the initial simplex tableau.]

10. *Business* A small tool manufacturing company has two forges F_1 and F_2, each of which, because of maintenance, can run at most 20 hours per day. The company makes two types of tools, A and B. Tool A must spend 1 hour in forge F_1 and 3 hours in forge F_2. Tool B must spend 2 hours in forge F_1 and 1 hour in forge F_2. The company makes a profit of $20 on tool A and $10 on tool B. Determine the number of each type of tool the company must make in order to maximize its daily profit.

11. Suppose the company described in Problem 10 realizes that in order to make a better product a third forge F_3 must be added to the production process of both types of tools. If the forge F_3 can run at most 12.5 hours per day, and if tools A and B must, respectively, spend 1.75 and 1 hour in this forge, then what daily production level will maximize the profit?

12. A butcher has 100 units of fat, 240 units of chuck meat, and 360 units of steak from which two types of ground beef are made: lean and extra lean. In order to make 1 pound of the lean mixture the butcher uses 2 units of fat, 4 units of chuck, and 2 units of steak. To make 1 pound of the extra lean, 2 units of fat, 6 units of chuck, and 12 units of steak are used. If

the revenue from 1 pound of lean ground is 90 cents and the revenue from 1 pound of extra lean ground is 150 cents, how many pounds of each should be made in order to maximize the revenue?

13. An airline which flies between Los Angeles and New York has available each week a fleet of 707's, 747's, and DC-10's, but the combined number of available planes does not exceed 18. The combined total of 707's and DC-10's is at least as many as the available 747's, but there are at most twice as many DC-10's as 707's. If a 707 can carry 200 passengers, a 747 can carry 400 passengers, and a DC-10 can carry 300 passengers, how many of each plane should be used per week in order to maximize the number of passengers?

14. Suppose the profit for the ground beef mixtures described in Problem 12 is 80 cents for 1 pound of lean and 70 cents for 1 pound of extra lean. Does the number of pounds of each kind that maximizes revenue also maximize profit?

8.6

THE DUAL PROBLEM

In Section 8.5 we dealt only with maximization problems in which the constraints were all of the form $\leq$. The constraints for minimization problems are usually of the form $\geq$. While the simplex procedure can be modified to handle this type of problem directly, it is interesting to note that corresponding to every minimization problem there is a maximization problem called the **dual** problem.

MATRIX NOTATION Before formally defining the dual problem we observe that a maximization problem can be written in the form:

Maximize
$$F = M^t X$$
subject to
$$AX \leq B \qquad (X \geq 0),$$

where M, B, and X are column matrices (then M^t is a row matrix) and A is the matrix of coefficients in the constraint equations.

Example Write in matrix form:

Maximize
$$F = 4x_1 + 5x_2$$
subject to
$$3x_1 + 2x_2 \leq 6$$
$$4x_1 + x_2 \leq 4$$
$$x_1 \geq 0 \qquad x_2 \geq 0.$$

8.6 The Dual Problem

Solution

$$F = \begin{bmatrix} 4 & 5 \end{bmatrix} \begin{bmatrix} x_1 \\ x_2 \end{bmatrix}$$

$$\begin{bmatrix} 3 & 2 \\ 4 & 1 \end{bmatrix} \begin{bmatrix} x_1 \\ x_2 \end{bmatrix} \leq \begin{bmatrix} 6 \\ 4 \end{bmatrix} \quad \left(\begin{bmatrix} x_1 \\ x_2 \end{bmatrix} \geq \begin{bmatrix} 0 \\ 0 \end{bmatrix} \right)$$

Using matrix notation we have the following definition.

If the problem is to

maximize $\quad F = M^t X$

subject to $\quad AX \leq B \quad (X \geq 0)$,

then to minimize $\quad G = B^t Y$

subject to $\quad A^t Y \geq M \quad (Y \geq 0)$

*is called the **dual** of the original problem. The first problem is called the **primal**.*

EITHER CAN BE THE DUAL It should be observed that we could just as well start with the minimization problem as the primal, so that its dual would then be the maximization problem. Though we shall not prove it, if both problems possess solutions, then the minimum of G is the same as the maximum of F. We are primarily interested in converting a minimization problem into a maximization problem. The dual problem can be solved either by the geometric procedure of Section 8.4 or by the simplex method of the preceding section.

Example Write the dual of the following:

Minimize

$$G = 3y_1 + 2y_2$$

subject to

$$y_1 + 10y_2 \geq 20$$
$$2y_1 + y_2 \geq 8$$
$$4y_1 + y_2 \geq 12$$
$$y_1 \geq 0 \qquad y_2 \geq 0.$$

Solution First write the problem in matrix form:

$$G = \begin{bmatrix} 3 & 2 \end{bmatrix} \begin{bmatrix} y_1 \\ y_2 \end{bmatrix}$$

$$\begin{bmatrix} 1 & 10 \\ 2 & 1 \\ 4 & 1 \end{bmatrix} \begin{bmatrix} y_1 \\ y_2 \end{bmatrix} \geq \begin{bmatrix} 20 \\ 8 \\ 12 \end{bmatrix} \quad \left(\begin{bmatrix} y_1 \\ y_2 \end{bmatrix} \geq \begin{bmatrix} 0 \\ 0 \end{bmatrix} \right),$$

and read off

$$B = \begin{bmatrix} 3 \\ 2 \end{bmatrix} \quad A = \begin{bmatrix} 1 & 2 & 4 \\ 10 & 1 & 1 \end{bmatrix} \quad M^t = [20 \quad 8 \quad 12].$$

Form the dual:

Maximize

$$F = [20 \quad 8 \quad 12] \begin{bmatrix} x_1 \\ x_2 \\ x_3 \end{bmatrix}$$

subject to

$$\begin{bmatrix} 1 & 2 & 4 \\ 10 & 1 & 1 \end{bmatrix} \begin{bmatrix} x_1 \\ x_2 \\ x_3 \end{bmatrix} \le \begin{bmatrix} 3 \\ 2 \end{bmatrix} \quad \left(\begin{bmatrix} x_1 \\ x_2 \\ x_3 \end{bmatrix} \ge \begin{bmatrix} 0 \\ 0 \\ 0 \end{bmatrix} \right),$$

that is,

$$F = 20x_1 + 8x_2 + 12x_3$$

and

$$x_1 + 2x_2 + 4x_3 \le 3$$
$$10x_1 + x_2 + x_3 \le 2$$
$$x_1 \ge 0 \qquad x_2 \ge 0 \qquad x_3 \ge 0.$$

Example Solve:

Minimize

$$G = 5y_1 + 3y_2$$

subject to

$$y_1 + y_2 \ge 5$$
$$2y_1 + y_2 \ge 8$$
$$y_1 \ge 0 \qquad y_2 \ge 0.$$

Solution We first find the dual problem:

Maximize

$$F = 5x_1 + 8x_2$$

subject to

$$x_1 + 2x_2 \le 5$$
$$x_1 + x_2 \le 3$$
$$x_1 \ge 0 \qquad x_2 \ge 0.$$

8.6 The Dual Problem

Form the simplex tableau and proceed as in Section 8.5.

$$\begin{bmatrix} 1 & ② & 1 & 0 & \vdots & 5 \\ 1 & 1 & 0 & 1 & \vdots & 3 \\ -5 & -8 & 0 & 0 & \vdots & 0 \end{bmatrix}$$

$$\sim \begin{bmatrix} \frac{1}{2} & 1 & \frac{1}{2} & 0 & \vdots & \frac{5}{2} \\ ⓐ\frac{1}{2} & 0 & -\frac{1}{2} & 1 & \vdots & \frac{1}{2} \\ -1 & 0 & 4 & 0 & \vdots & 20 \end{bmatrix} \begin{matrix} \frac{1}{2} \times \text{Row 1} \\ \left(-\frac{1}{2} \times \text{Row 1}\right) + \text{Row 2} \\ (4 \times \text{Row 1}) + \text{Row 3} \end{matrix}$$

$$\sim \begin{bmatrix} 0 & 1 & 1 & -1 & \vdots & 2 \\ 1 & 0 & -1 & 2 & \vdots & 1 \\ 0 & 0 & 3 & 2 & \vdots & 21 \end{bmatrix} \begin{matrix} (-1 \times \text{Row 2}) + \text{Row 1} \\ 2 \times \text{Row 2} \\ (2 \times \text{Row 2}) + \text{Row 3} \end{matrix}$$

The maximum of F is 21 at $x_1 = 1$, $x_2 = 2$; and the minimum of G is also 21. To find the corresponding values of y_1 and y_2 we read off the entries in the last two columns of the last row to the left of the vertical partition line. Thus, $y_1 = 3$ and $y_2 = 2$.

If the primal contains three variables, then the values of y_1, y_2, and y_3 which minimize G are found in the last row beginning three columns to the left of the vertical line in the final augmented matrix of the dual problem.

EXERCISE 8.6

In Problems 1–4 write the dual of the given problem. Assume all variables are nonnegative.

1. Minimize
$$G = 9y_1 + 27y_2$$
subject to
$$9y_1 + 2y_2 \geq 18$$
$$y_1 + 2y_2 \geq 10.$$

2. Minimize
$$G = 15y_1 + 12y_2 + 20y_3$$
subject to
$$y_1 + y_2 + y_3 \geq 5$$
$$y_1 + 2y_2 + y_3 \geq 5$$
$$3y_1 + y_2 + y_3 \geq 5.$$

3. Maximize

$$F = 10x_1 + 10x_2 + 12x_3$$

subject to

$$x_1 + x_2 + 3x_3 \le 75$$
$$x_1 + 3x_2 + x_3 \le 45$$
$$3x_1 + 2x_2 + x_3 \le 90.$$

4. Maximize

$$F = 4x_1 + 8x_2$$

subject to

$$5x_1 + 3x_2 \le 15$$
$$3x_1 + 4x_2 \le 12$$
$$3x_1 + 5x_2 \le 15.$$

5. Use the dual to solve the following:

Minimize

$$G = 20y_1 + 6y_2$$

subject to

$$4y_1 + 2y_2 \ge 9$$
$$5y_1 + y_2 \ge 6$$
$$y_1 \ge 0 \qquad y_2 \ge 0.$$

6. Use the dual to solve the following:

Minimize

$$G = y_1 + 0.6y_2 + 0.8y_3$$

subject to

$$1.5y_1 + 0.5y_2 + y_3 \ge 5$$
$$0.5y_1 + 0.5y_2 + 2y_3 \ge 6$$
$$y_1 + 1.5y_2 + 0.5y_3 \ge 6.5$$
$$y_1 \ge 0 \qquad y_2 \ge 0 \qquad y_3 \ge 0.$$

Applications

7. *Biology* A pond is stocked with three species of fish X, Y, Z. The accompanying table indicates the number of units of nutrient I, nutrient II, and pollutants consumed by each fish per day. To maintain the biological balance of the pond it is necessary that the fish consume at least 22 units of nutrient I and at least 40 units of nutrient II each day. It is desired to minimize the daily intake of pollutants by the fish.

	X	Y	Z
Nutrient I	1	2	3
Nutrient II	1	5	1
Pollutants	12	45	24

a. Set up the objective function and constraints describing this minimization problem.
b. Form the dual of the problem.
c. Find the minimum value of the objective function by the geometric method of Section 8.4.

8. Use the simplex method to solve the dual in Problem 7. Determine how many fish should be stocked in the pond to minimize the daily intake of pollutants.

9. *Business* A company owns two factories, A and B, each of which produces three types of television sets. Each hour, factory A can make 10 units of high-quality sets, 20 units of medium-quality sets, and 50 units of low-quality sets. Each hour, factory B can make 20 units of high-quality sets, 10 units of medium-quality sets, and 10 units of low-quality sets. In order to meet the demand each day, the company needs at least 100 units of high-quality sets, 100 units of medium-quality sets, and 150 units of low-quality sets. It costs the company $900 per hour to run factory A and $300 per hour to run factory B. By setting

8.6 The Dual Problem

up and solving the dual problem, determine how many hours each factory should be run in order to keep the company's daily cost to a minimum.

10. An insurance company uses two computers, and IBC 490 and a CDM 500. Each hour, the IBC can process 8 units (1 unit = 1000) of medical claims, 1 unit of life insurance claims, and 2 units of car insurance claims. Each hour, the CDM can process 2 units of medical claims, 1 unit of life insurance claims, and 7 units of car insurance claims. The company finds it necessary to process at least 16 units of the medical claims, at least 5 units of the life insurance claims, and at least 20 units of the car insurance claims per day. If it costs the company 10 units (1 unit = $100) an hour to run the IBC and 20 units an hour to run the CDM, at most how many hours should each computer be run each day in order to keep the company's cost per day at a minimum? What is the minimum cost? Is there a maximum cost per day?

CHAPTER TEST

Graph the given inequality.

1. $2x - y < 6$
2. $2y + x > 0$
3. $-2 \leq y < 3$
4. $3 < x \leq 5$

By double shading, indicate the region representing the solution set of each system.

5. $y < 2 - x$
   $y < 2 + x$

6. $y + 3 > 0$
   $y < x - 3$

Graph the polygonal set defined by each system of inequalities.

7. $0 \leq x \leq 3$
   $0 \leq y \leq 4$
   $x + y \leq 5$

8. $0 \leq x \leq 3$
   $0 \leq y$
   $y \leq 2 + x$
   $x + y \leq 4$

9. Find the maximum and minimum values of $F = 2x + y$ over the set in Problem 7.

10. Find the maximum and minimum values of $F = 3x + 2y$ over the set in Problem 8.

11. Using Table A on page 290, find the maximum profit if each unit of product x earns a profit of $80 and each unit of product y earns a profit of $60.

12. Using Table B on page 291, find the maximum profit if each unit of product x earns a profit of $10 and each unit of product y earns a profit of $20.

13. Use the simplex method to find the maximum of F subject to the given constraints.

$$F = 20x + 30y$$
$$2x + y \leq 12$$
$$x + 2y \leq 12$$
$$x \geq 0 \qquad y \geq 0$$

Chapter Test

14. Write the dual of the following problem:

    Minimize
    $$G = y_1 + y_2$$

    subject to
    $$3y_1 + 4y_2 \geq 12$$
    $$y_1 + 3y_2 \geq 6$$
    $$y_1 \geq 0 \quad y_2 \geq 0.$$

15. Use the dual problem to find the minimum value of $G = 12y_1 + 11y_2$

    subject to
    $$y_1 + y_2 \geq 6$$
    $$2y_1 + y_2 \geq 10$$
    $$y_1 \geq 0 \quad y_2 \geq 0.$$

9
SEQUENCES AND SERIES

9.1
SEQUENCES

Suppose $1 is deposited into a savings account which pays interest at a rate of r per year. The following scheme indicates how $1 increases in worth through compounding of interest.

Amount: $\underbrace{1+r}_{}, \underbrace{(1+r)r + (1+r)}_{}, \underbrace{[(1+r)r + (1+r)]r + (1+r)r + (1+r)}_{}, \ldots$

After year: 1   2   3

By factoring, the amounts can be written as

$$(1+r), \quad (1+r)^2, \quad (1+r)^3, \quad \ldots$$

In other words, after the nth year the amount accrued in the account is given by the formula $s(n) = (1+r)^n$. We note that for each positive integer n there is a single value of $s(n)$. Thus, the formula $s(n) = (1+r)^n$ describes a function with domain the set N of positive integers.

In Section 7.2 we considered a matrix of the form

$$\begin{bmatrix} 1.04 & 0 & 0 \\ 0 & 1.05 & 0 \\ 0 & 0 & 1.06 \end{bmatrix}^n,$$

where n was a positive integer. By letting n take on consecutive positive integer values, we generate a **sequence** of matrices,

$$\begin{bmatrix} 1.04 & 0 & 0 \\ 0 & 1.05 & 0 \\ 0 & 0 & 1.06 \end{bmatrix}, \begin{bmatrix} 1.04 & 0 & 0 \\ 0 & 1.05 & 0 \\ 0 & 0 & 1.06 \end{bmatrix}^2, \begin{bmatrix} 1.04 & 0 & 0 \\ 0 & 1.05 & 0 \\ 0 & 0 & 1.06 \end{bmatrix}^3, \ldots,$$

9.1 Sequences

or

$$\begin{bmatrix} 1.04 & 0 & 0 \\ 0 & 1.05 & 0 \\ 0 & 0 & 1.06 \end{bmatrix}, \begin{bmatrix} 1.082 & 0 & 0 \\ 0 & 1.103 & 0 \\ 0 & 0 & 1.124 \end{bmatrix}, \ldots .$$

We summarize the foregoing concepts.

> *A sequence is a function having as its domain the set N of positive integers or a subset of successive elements of N.*

For example, the function defined by

$$s(n) = n + 3 \qquad (n = 1, 2, 3, \ldots)$$

is a sequence. The elements in the range of such a function considered in order are

$$s(1), \; s(2), \; s(3), \; s(4), \; \ldots .$$

For simplicity, we shall refer to the listing of the elements of the range, corresponding to the natural order of the elements of N, as the sequence. Thus, the sequence for the above formula is

$$4, \; 5, \; 6, \; 7, \; \ldots .$$

The domain of a sequence need not be the entire set of positive integers. For example,

$$s(n) = 3n^2 + 1 \qquad (n = 1, 2, 3, 4, 5)$$

is equivalent to

$$\begin{aligned} s(1) &= 3(1)^2 + 1 = 4, \\ s(2) &= 3(2)^2 + 1 = 13, \\ s(3) &= 3(3)^2 + 1 = 28, \\ s(4) &= 3(4)^2 + 1 = 49, \\ s(5) &= 3(5)^2 + 1 = 76, \end{aligned}$$

or

$$4, \; 13, \; 28, \; 49, \; 76.$$

Whether the domain of a sequence is a finite subset of the set of positive integers or the positive integers themselves should be clear from the context.

Example Find the first five terms of the sequence defined by

$$s(n) = \frac{3}{2n-1} \qquad (n = 1, 2, 3, \ldots).$$

Find the twenty-fifth term.

Solution The first five terms are

$$s(1) = \frac{3}{1}, \quad s(2) = \frac{3}{3}, \quad s(3) = \frac{3}{5}, \quad s(4) = \frac{3}{7}, \quad s(5) = \frac{3}{9},$$

or simply,

$$3, \; 1, \; \frac{3}{5}, \; \frac{3}{7}, \; \frac{1}{3}, \; \ldots \; .$$

The twenty-fifth term is

$$s(25) = \frac{3}{2(25) - 1} = \frac{3}{49}.$$

Given several terms in a sequence, we are often able to construct an expression for a general term of a sequence to which they belong. Such a sequence is not unique. Thus, if the first three terms in a sequence are 2, 4, 8, ..., we may *surmise* that the general term is $s(n) = 2^n$. Note, however, that the sequences for both

$$s(n) = 2^n$$

and

$$s(n) = 2^n + (n-1)(n-2)(n-3)$$

start with 2, 4, 8, but that the two sequences differ for terms following the third term.

SEQUENCE NOTATION The notation ordinarily used for the terms in a sequence is not function notation as such. It is customary to denote a term in a sequence by means of a subscript. Thus, the sequence $s(1), s(2), s(3), s(4), \ldots$ would appear as $s_1, s_2, s_3, s_4, \ldots$.

ARITHMETIC PROGRESSIONS Let us next consider two special kinds of sequences that have many applications. The first kind can be defined as follows:

9.1 Sequences

> An **arithmetic progression** is a sequence defined by equations of the form
> $$s_1 = a, \quad s_{n+1} = s_n + d,$$
> where a and d are real numbers.

Since each term in such a sequence is obtained from the preceding term by adding d, d is called the **common difference**. Thus, 3, 7, 11, 15, ... is an arithmetic progression, in which $s_1 = 3$ and $d = 4$. Of course, the positive integers themselves, 1, 2, 3, ..., form an arithmetic progression with $s_1 = 1$ and $d = 1$.

For each arithmetic progression, the general term is established by the following.

> The nth term in the sequence defined by
> $$s_1 = a, \quad s_{n+1} = s_n + d,$$
> is
> (1) $$s_n = a + (n-1)d.$$

Example Suppose $300 is initially deposited into a checking account and that 11 consecutive deposits of $50 are made. Assuming no money is withdrawn, how much is the account worth after the eleventh deposit?

Solution In (1) we use $a = 300$, $n = 12$, and $d = 50$. Thus,
$$s_{12} = 300 + (12-1)50$$
$$= \$850.$$

Note that we used $n = 12$ and not $n = 11$, since there are exactly twelve terms in the sequence
$$300, \; 350, \; 400, \; \ldots, \; 850.$$

GEOMETRIC PROGRESSIONS

The second kind of sequence we shall consider can be defined as follows:

> A **geometric progression** is a sequence defined by equations of the form
> $$s_1 = a, \quad s_{n+1} = rs_n,$$
> where $a, r \neq 0$.

Thus, 3, 9, 27, 81, ... is a geometric progression in which each term except the first is obtained by multiplying the preceding term by 3. Since the effect of multiplying the terms in this way is to produce a fixed ratio s_{n+1}/s_n between any two successive terms, the multiplier, r, is called the **common ratio**. Sequences defined by the formulas $s_n = 2^n$ and $s_n = (1+x)^n$ give geometric progressions in which the common ratios are, respectively, $r = 2$ and $r = (1+x)$.

If n takes on the values 1, 2, 3, ... in the defining equation of a geometric progression, $s_{n+1} = rs_n$, we obtain

$$n = 1, \quad s_2 = rs_1 = ra = ar,$$
$$n = 2, \quad s_3 = rs_2 = ar^2,$$
$$n = 3, \quad s_4 = rs_3 = ar^3,$$
$$\vdots \qquad\qquad \vdots$$

It is easily seen that the nth general term of the sequence is given by $s_n = ar^{n-1}$. The results are summarized as follows.

The nth term in the sequence defined by

$$s_1 = a, \quad s_{n+1} = rs_n,$$

where $a, r \neq 0$, is

$$s_n = ar^{n-1}.$$

Example To clear out the merchandise on hand, a produce market offers a case discount on apples as follows: the first case costs $15, the second case is discounted by 75 cents, the third case is discounted by an additional 75 cents, or $1.50 over the first case price, and so on up to a limiting purchase of 20 cases. What is the price of the tenth case?

Solution The prices

$$\$15, \$14.25, \$13.50, \ldots$$

form an arithmetic progression with $d = -0.75$. From (1) we can write the price p_n of the nth case as

$$p_n = 15 - (n-1)0.75.$$

The price of the tenth case is then

$$p_{10} = 15 - 9(0.75)$$
$$= \$8.25.$$

9.1 Sequences

EXERCISE 9.1

Find the first four terms in the sequence with the general term as given.

Examples **a.** $s_n = \dfrac{n(n+1)}{2}$ **b.** $s_n = (-1)^n 2^n$

Solutions **a.** $s_1 = \dfrac{1(1+1)}{2} = 1$ **b.** $s_1 = (-1)^1 2^1 = -2$

$s_2 = \dfrac{2(2+1)}{2} = 3$ $s_2 = (-1)^2 2^2 = 4$

$s_3 = \dfrac{3(3+1)}{2} = 6$ $s_3 = (-1)^3 2^3 = -8$

$s_4 = \dfrac{4(4+1)}{2} = 10$ $s_4 = (-1)^4 2^4 = 16$

1, 3, 6, 10 -2, 4, -8, 16

1. $s_n = n - 5$ **2.** $s_n = 2n - 3$ **3.** $s_n = \dfrac{n^2 - 2}{2}$

4. $s_n = \dfrac{3}{n^2 + 1}$ **5.** $s_n = 1 + \dfrac{1}{n}$ **6.** $s_n = \dfrac{n}{2n - 1}$

7. $s_n = \dfrac{n(n-1)}{2}$ **8.** $s_n = \dfrac{5}{n(n+1)}$ **9.** $s_n = (-1)^n$

10. $s_n = \left(1 + \dfrac{1}{n}\right)^n$ **11.** $s_n = \dfrac{(-1)^n(n-2)}{n}$ **12.** $s_n = (-1)^{n-1} 3^{n+1}$

Write the next three terms in each arithmetic progression.

Examples **a.** 5, 9, ... **b.** x, $x - a$, ...

Solutions **a.** Find the common difference and then continue the sequence. **b.** Find the common difference and then continue the sequence.

$d = 9 - 5 = 4$ $d = (x - a) - x = -a$
13, 17, 21 $x - 2a$, $x - 3a$, $x - 4a$

13. 3, 7, ... **14.** -6, -1, ... **15.** x, $x + 1$, ...

16. a, $a + 5$, ... **17.** $2x + 1$, $2x + 4$, ... **18.** $3a$, $5a$, ...

Write the next four terms in each geometric progression.

Examples   **a.** 3, 6, ...   **b.** x, 2, ...

Solutions   **a.** Find the common ratio, and then continue the sequence.   **b.** Find the common ratio, and then continue the sequence.

$$r = \frac{6}{3} = 2$$

$$r = \frac{2}{x} \quad (x \neq 0)$$

12, 24, 48, 96

$$\frac{4}{x}, \frac{8}{x^2}, \frac{16}{x^3}, \frac{32}{x^4}$$

19. 2, 8, ...   **20.** 4, 8, ...   **21.** $\frac{2}{3}, \frac{4}{3}, \ldots$

22. $\frac{1}{2}, -\frac{3}{2}, \ldots$ **23.** $\frac{a}{x}, -1, \ldots$ **24.** $\frac{a}{b}, \frac{a}{bc}, \ldots$

Example   Find the general term and the fourteenth term of the arithmetic progression $-6, -1, \ldots$.

Solution   Find the common difference.

$$d = -1 - (-6) = 5$$

Use $s_n = a + (n-1)d$.

$$s_n = -6 + (n-1)5 = 5n - 11$$
$$s_{14} = 5(14) - 11 = 59$$

25. Find the general term and the seventh term in the arithmetic progression 7, 11, ...
26. Find the general term and the twelfth term in the arithmetic progression 2, 5/2, ...
27. Find the general term and the twentieth term in the arithmetic progression 3, -2, ...
28. Find the general term and the ninth term in the arithmetic progression 3/4, 2, ...

Example   Find the general term and the ninth term of the geometric progression $-24, 12, \ldots$.

Solution   Find the common ratio.

$$r = \frac{12}{-24} = -\frac{1}{2}$$

Use $s_n = ar^{n-1}$.

$$s_n = -24\left(-\frac{1}{2}\right)^{n-1}$$

$$s_9 = -24\left(-\frac{1}{2}\right)^8 = -\frac{3}{32}$$

9.1 Sequences

29. Find the general term and the sixth term in the geometric progression 48, 96, ...
30. Find the general term and the eighth term in the geometric progression -3, $3/2$, ...
31. Find the general term and the seventh term in the geometric progression $-1/3$, 1, ...
32. Find the general term and the ninth term in the geometric progression -81, 27, ...

Example Find the first term in an arithmetic progression in which the third term is 7 and the eleventh term is 55.

Solution Find the common difference by considering an arithmetic progression with first term 7 and with ninth term 55. Use $s_n = a + (n - 1)d$.

$$s_9 = 7 + (9 - 1)d$$
$$55 = 7 + 8d$$
$$d = 6$$

Use the difference to find the first term in an arithmetic progression in which the third term is 7. Again use $s_n = a + (n - 1)d$.

$$s_3 = a + (3 - 1)6$$
$$7 = a + 12$$
$$a = -5$$

33. If the third term in an arithmetic progression is 7 and the eighth term is 17, find the common difference. What are the first and twentieth terms?
34. If the fifth term of an arithmetic progression is -16 and the twentieth term is -46, what is the twelfth term?
35. Which term in the arithmetic progression 4, 1, ... is -77?
36. What is the twelfth term in an arithmetic progression in which the second term is x and the third term is y?
37. Find the first term of a geometric progression with fifth term 48 and ratio 2.
38. Find two different values for x so that $-3/2$, x, $-8/27$ will be a geometric progression.

Applications
39. *Biology* The initial population of a town is 20,000. It is observed that the population increases by 30% after each year. What is the population after n years?
40. Use a calculator in Problem 39 to determine the population of the town after 4 years. After 15 years.
41. Each person has two parents. How many great-great-great grandparents will each person have?
42. *Business* A company has a profit of $10,000 in its first year of business. It is expected that profits will continue to increase at a rate of $3000 a year. What profit can the company expect in its twentieth year of operation?
43. A company found that it lost $500 on the first sale of item X. On each subsequent sale the profit per unit increased by $50. Find a formula that gives the profit P_n per unit after n sales. [*Hint*: Treat a loss as a negative number.]

44. In Problem 43 what is the profit per unit after eight sales? After 51 sales? Determine the values of n for which $P_n \geq 0$.

45. A winery offers a case discount of $1.50 on the second case purchased, an additional $1.50 on the third case, and so on. If the first case of wine costs $45, what is the price of the eighth case? Determine the limitation on the number of sales to avoid the possibility of free cases.

46. A machine costs $25,000 when new and has a salvage value of $1000 at the end of its estimated useful life of 20 years. Assuming straight-line depreciation, find a formula that gives the value v_n of the machine after n years ($1 \leq n \leq 20$). [*Hint*: See Problem 46 of Section 4.3.]

47. In the **declining balance method** of depreciation an item's initial cost is decreased by the same percentage each year. A machine costs $25,000 when new and depreciates at an annual rate of 20%. Find a formula that gives the value v_n of the machine after n years.

48. A student decides to place 1 cent on a first square of a checkerboard, 2 cents on a second square, 4 cents on a third square, 8 cents on a fourth square, and so on. Will the student have enough money to place on the last square of the checkerboard?

49. *Physics* A ball is dropped from a height of 8 meters. On each rebound the ball returns to a height of one-half its preceding height. How high does the ball go on the fourth rebound?

50. A body falling from a great height will travel 16 feet during the first second, 48 feet during the second second, 80 feet during the third second, and so on. How far will the body fall during the eleventh second?

9.2

SERIES

Associated with any sequence is a **series**.

> *A series is the indicated sum of the terms in a sequence.*

For example, with the finite sequence

$$4, \quad 7, \quad 10, \quad \ldots, \quad 3n + 1,$$

for a given positive integer n, there is associated the finite series

$$S_n = 4 + 7 + 10 + \cdots + (3n + 1).$$

Similarly, with the finite sequence

$$x, \quad x^2, \quad x^3, \quad x^4, \quad \ldots, \quad x^n,$$

there is associated the finite series

$$S_n = x + x^2 + x^3 + x^4 + \cdots + x^n.$$

9.2 Series

Since the terms in the series are the same as those in the sequence, we can refer to the first term or the second term or the general term of a series in the same manner as we do for a sequence.

SUM OF THE FIRST n TERMS OF AN ARITHMETIC PROGRESSION

Consider the series S_n of the first n terms of the general arithmetic progression,

(1) $$S_n = a + (a + d) + (a + 2d) + \cdots + [a + (n - 1)d],$$

and then consider the same series written as

(2) $$S_n = s_n + (s_n - d) + (s_n - 2d) + \cdots + [s_n - (n - 1)d],$$

where the terms are displayed in reverse order. Adding (1) and (2) term-by-term, we have

$$S_n + S_n = (a + s_n) + (a + s_n) + (a + s_n) + \cdots + (a + s_n),$$

where the term $(a + s_n)$ occurs n times. Then

$$2S_n = n(a + s_n)$$

(3) $$S_n = \frac{n}{2}(a + s_n).$$

If (3) is rewritten as

$$S_n = n\left(\frac{a + s_n}{2}\right),$$

we observe that the sum is given by the product of the number of terms in the series and the average of the first and last terms.

An alternative form for (3) is obtained by substituting $a + (n - 1)d$ for s_n in (3) to obtain

$$S_n = \frac{n}{2}(a + [a + (n - 1)d])$$

(3a) $$S_n = \frac{n}{2}[2a + (n - 1)d],$$

where the sum is now expressed in terms of a, n, and d.

Example Show that the sum of the first n positive integers is $n(n + 1)/2$.

Solution The sequence of positive integers

$$1, 2, 3, 4, 5, \ldots, n$$

forms an arithmetic progression with the first term $a = 1$, the common difference $d = 1$, and the last term the same as the number of terms, namely, n. Thus, by (3a) we have

$$S_n = \frac{n}{2}[2 + (n-1)1]$$
$$= \frac{n}{2}(n+1).$$

For example, the sum of the first 1000 positive integers is $1000(1001)/2 = 500{,}500$.

SUM OF THE FIRST n TERMS OF A GEOMETRIC PROGRESSION

To find an explicit representation for the sum of a given number of terms in a geometric progression in terms of a, r, and n, we employ a device somewhat similar to the one used in finding the sum in an arithmetic progression. Consider the geometric series (4) containing n terms, and the series (5) obtained by multiplying both members of (4) by r:

(4) $\qquad S_n = a + ar + ar^2 + ar^3 + \cdots + ar^{n-2} + ar^{n-1},$
(5) $\qquad rS_n = ar + ar^2 + ar^3 + ar^4 + \cdots + ar^{n-1} + ar^n.$

When we subtract (5) from (4), all terms in the right-hand members except the first term in (4) and the last term in (5) vanish, yielding

$$S_n - rS_n = a - ar^n.$$

Factoring S_n from the left-hand member gives

$$(1-r)S_n = a - ar^n$$

(6) $\qquad S_n = \dfrac{a - ar^n}{1 - r},$

if $r \neq 1$, and we have a formula for the sum of the first n terms of a geometric progression.

An alternative expression for (6) can be obtained by first writing

$$S_n = \frac{a - r(ar^{n-1})}{1 - r},$$

and then, since $s_n = ar^{n-1}$, expressing this as

(7) $\qquad S_n = \dfrac{a - rs_n}{1 - r},$

where the sum is now given in terms of a, s_n, and r.

Example A patient takes 15 milligrams of a drug each day, and of the amount accumulated 80% is excreted each day by bodily functions. How much of the drug has accumulated in 6 days (measured immediately after the sixth dosage)?

Solution After the patient takes the first dosage, there are 15 milligrams of the drug accumulated. One day later $15 - (0.8)15 = (0.2)15$ milligrams remain of the initial dosage so that measurements immediately after the second dosage show the patient has accumulated $15 + (0.2)15$ milligrams of the drug. After the third dosage there are

$$15 + (0.2)[15 + (0.2)15] = 15 + (0.2)15 + (0.2)^2 15 \text{ milligrams,}$$

and so on. Thus, after the sixth dosage there are

$$15 + (0.2)15 + (0.2)^2 15 + (0.2)^3 15 + (0.2)^4 15 + (0.2)^5 15 \text{ milligrams.}$$

From (6) the last line can be written

$$S_6 = \frac{15 - 15(0.2)^6}{1 - 0.2} = \frac{15[1 - (0.2)^6]}{0.8}$$

$$= 18.7 \text{ milligrams.}$$

Example When a series of deposits P are made at equally spaced intervals of time, the savings plan is called an **annuity** (see Section 11.2). If r is the yearly rate of interest compounded continuously, then after m deposits the amount accrued is given by

$$S = P + Pe^r + Pe^{2r} + \cdots + Pe^{(m-1)r}.$$

Simplify the expression for S.

Solution The above is a sum of a geometric progression with $a = P$ and common ratio e^r. From equation (6) we have

$$S = \frac{P - Pe^{mr}}{1 - e^r} = P\left(\frac{e^{mr} - 1}{e^r - 1}\right).$$

SIGMA NOTATION A series for which the general term is known can be represented in a very convenient, compact way by means of what is called **sigma**, or **summation**, **notation**. The Greek letter $\sum$ (sigma) is used to denote a sum. For example,

$$S_n = 4 + 7 + 10 + \cdots + (3n + 1)$$

can be written

$$S_n = \sum_{j=1}^{n} (3j + 1),$$

where we understand that S_n is the series having terms obtained by replacing j in the expression $3j + 1$ with the numbers $1, 2, 3, \ldots, n$, successively. Similarly,

$$S = \sum_{j=3}^{6} j^2$$

appears in expanded form as

$$S = 3^2 + 4^2 + 5^2 + 6^2,$$

where the first value for j is 3 and the last is 6.

The variable used in conjunction with summation notation is called the **index of summation**, and the set of integers over which we sum (in this case, {3, 4, 5, 6}) is called the **range of summation**.

NOTATION FOR AN INFINITE SUM

To indicate that a series has an *infinite* number of terms, we cannot use the notation S_n for the sum, because there is no value to substitute for n. We therefore adopt the notation

(8) $$S_\infty = \sum_{j=1}^{\infty} \frac{1}{2^j}$$

to indicate that there is no last term in a series. In expanded form, the infinite series (8) is given by

$$S_\infty = \frac{1}{2} + \frac{1}{4} + \frac{1}{8} + \cdots.$$

The meaning, if any, of such an infinite sum will be discussed in Section 9.3.

EXERCISE 9.2

Write each series in expanded form.

Examples   a. $\sum_{j=2}^{5} (j^2 + 1)$ b. $\sum_{k=1}^{\infty} (-1)^k 2^{k+1}$

Solutions   a. $j = 2,\ 2^2 + 1 = 5,$
$\ j = 3,\ 3^2 + 1 = 10,$
$\ j = 4,\ 4^2 + 1 = 17,$
$\ j = 5,\ 5^2 + 1 = 26.$
$\ \sum_{j=2}^{5} (j^2 + 1) = 5 + 10 + 17 + 26$

b. $k = 1,\ (-1)^1 2^{1+1} = (-1)(4) = -4,$
$\ k = 2,\ (-1)^2 2^{2+1} = (1)(8) = 8,$
$\ k = 3,\ (-1)^3 2^{3+1} = (-1)(16) = -16.$
$\ \vdots$
$\ \sum_{k=1}^{\infty} (-1)^k 2^{k+1} = -4 + 8 - 16 + \cdots$

1. $\sum_{j=1}^{4} j^2$
2. $\sum_{j=1}^{4} (3j - 2)$
3. $\sum_{j=1}^{3} \frac{(-1)^j}{2^j}$
4. $\sum_{k=3}^{5} \frac{(-1)^{k+1}}{k-2}$
5. $\sum_{k=0}^{\infty} \frac{1}{2^k}$
6. $\sum_{k=0}^{\infty} \frac{k}{1+k}$

9.2 Series

Write each series in sigma notation.

Examples **a.** $5 + 8 + 11 + 14$ **b.** $x^2 + x^4 + x^6$ **c.** $\dfrac{3}{5} + \dfrac{5}{7} + \dfrac{7}{9} + \cdots$

Solutions Find an expression for the general term and write in sigma notation.

a. $3j + 2$ **b.** x^{2j} **c.** $\dfrac{2j + 1}{2j + 3}$

$$\sum_{j=1}^{4}(3j+2) \qquad \sum_{j=1}^{3}x^{2j} \qquad \sum_{j=1}^{\infty}\dfrac{2j+1}{2j+3}$$

7. $x + x^3 + x^5 + x^7$

8. $x^3 + x^5 + x^7 + x^9 + x^{11}$

9. $1 + 4 + 9 + 16 + 25$

10. $\dfrac{1}{3} + \dfrac{1}{9} + \dfrac{1}{27} + \dfrac{1}{81}$

11. $1 \cdot 2 + 2 \cdot 3 + 3 \cdot 4 + 4 \cdot 5 + \cdots$

12. $\dfrac{1}{2} + \dfrac{2}{3} + \dfrac{3}{4} + \dfrac{4}{5} + \cdots$

13. $\dfrac{2}{1} + \dfrac{3}{2} + \dfrac{4}{3} + \dfrac{5}{4} + \cdots$

14. $\dfrac{1}{1} + \dfrac{2}{3} + \dfrac{3}{5} + \dfrac{4}{7} + \cdots$

Find each sum.

Example $\displaystyle\sum_{j=1}^{12}(4j+1)$

Solution Write the first two or three terms in expanded form.

$$5 + 9 + 13 + \cdots$$

This is an arithmetic series. The first term is 5 and the common difference is 4. Therefore, we can use $S_n = \dfrac{n}{2}[2a + (n-1)d]$ to obtain

$$S_{12} = \dfrac{12}{2}[2(5) + (12-1)4] = 324.$$

15. $\displaystyle\sum_{j=1}^{7}(2j+1)$ **16.** $\displaystyle\sum_{j=1}^{21}(3j-2)$ **17.** $\displaystyle\sum_{j=3}^{15}(7j-1)$

18. $\displaystyle\sum_{j=10}^{20}(2j-3)$ **19.** $\displaystyle\sum_{k=1}^{8}\left(\dfrac{1}{2}k-3\right)$ **20.** $\displaystyle\sum_{k=1}^{100}k$

Example $\displaystyle\sum_{j=2}^{5}\left(\dfrac{1}{3}\right)^j$

Solution Write the first two terms in expanded form.

$$\left(\frac{1}{3}\right)^2 + \left(\frac{1}{3}\right)^3 + \cdots$$

This is a geometric series in which the first term is 1/9, the common ratio is 1/3, and $n = 4$.
Therefore we can use $S_n = \dfrac{a - ar^n}{1 - r}$ to obtain

$$S_4 = \frac{\frac{1}{9} - \frac{1}{9}\left(\frac{1}{3}\right)^4}{1 - \frac{1}{3}} = \frac{40}{243}.$$

21. $\sum\limits_{j=1}^{6} 3^j$

22. $\sum\limits_{j=1}^{4} (-2)^j$

23. $\sum\limits_{k=3}^{7} \left(\frac{1}{2}\right)^{k-2}$

24. $\sum\limits_{j=3}^{12} 2^{j-5}$

25. $\sum\limits_{j=1}^{6} \left(\frac{1}{3}\right)^j$

26. $\sum\limits_{j=1}^{4} (3 + 2^j)$

27. Find the sum of all even integers n for $13 < n < 29$.

28. Find the sum of all integral multiples of 7 between 8 and 110.

29. Find $\sum\limits_{j=1}^{n} \left(\frac{1}{2}\right)^j$ for $n = 2, 3, 4, 5$. What value do you think $\sum\limits_{j=1}^{n} \left(\frac{1}{2}\right)^j$ approximates as n becomes larger and larger?

30. Find p if $\sum\limits_{j=1}^{5} pj = 14$.

31. Find p and q if $\sum\limits_{j=1}^{4} (pj + q) = 28$ and $\sum\limits_{j=2}^{5} (pj + q) = 44$.

32. Consider $S_n = \sum\limits_{j=1}^{n} f(j)$. Explain why this equation defines a sequence. What is the variable denoting an element in the domain? The range?

Applications

33. How many bricks will there be in a pile 1 brick thick if there are 27 bricks in the bottom row, 25 in the second row, and so forth, to the top row, which has 1 brick?

34. If there are a total of 256 bricks in a pile arranged in the manner of those in Problem 33, how many bricks are there in the third row from the bottom of the pile?

35. A student, made to stay after school, is told to add the positive integers up to 120. What is the total?

36. *Biology* How many ancestors will each person have going back to the great-great-great grandparents?

37. A patient takes 30 milligrams of a drug each day. If 70% of the amount accumulated is excreted each day, how much of the drug has accumulated as measured immediately after the patient takes the fifth dosage?

38. *Business* A company found that it lost $500 on the first sale of item *X*. On each

subsequent sale the profit per unit increased by $50. Find the total profit that the company realizes from the sale of 51 units.

39. A winery offers a case discount of $1.50 on the second case purchased, an additional $1.50 on the third case, and so on. If the first case of wine costs $45, what is the total purchase price of 20 cases?

40. *Physics* A body falling from a great height will travel 16 feet during the first second, 48 feet during the second second, 80 feet during the third second, and so on. What is the total distance that the body falls in 11 seconds?

9.3

LIMITS OF SEQUENCES AND SERIES

A SEQUENCE THAT IS STRICTLY INCREASING BUT BOUNDED

Consider the sequence defined by

(1) $$s_n = \frac{n}{n+1}.$$

If we write the range of (1) in the form

$$\frac{1}{2}, \frac{2}{3}, \frac{3}{4}, \frac{4}{5}, \ldots, \frac{n}{n+1}, \ldots,$$

then it is clear that each of the terms is greater than the preceding term; indeed, the difference of consecutive terms is

$$\frac{n+1}{n+2} - \frac{n}{n+1} = \frac{(n^2+2n+1)-(n^2+2n)}{(n+1)(n+2)} = \frac{1}{(n+1)(n+2)} > 0.$$

Such a sequence is said to be **strictly increasing**. On the other hand, it is also clear that, no matter how large a value is assigned to n, we have

$$\frac{n}{n+1} < 1,$$

because the denominator is one larger than the numerator; in fact, we have

$$1 - \frac{n}{n+1} = \frac{(n+1)-n}{n+1} = \frac{1}{n+1} > 0.$$

Thus, we have a sequence in which each term is greater than the preceding term and yet no term is equal to or greater than 1.

LIMIT OF A SEQUENCE

We note, however—and this is a very basic consideration—that the value of $n/(n + 1)$ is as close to 1 as we please if n is large enough. For example, the difference satisfies

$$1 - \frac{n}{n+1} = \frac{1}{n+1},$$

and we have

$$\frac{1}{n+1} < \frac{1}{1000},$$

provided $n + 1 > 1000$; that is, $n > 999$. If it is true that the nth term in a sequence differs from the number L by as little as we please for all sufficiently large n, we say that **the sequence approaches the number L as a limit**. The symbolism

$$\lim_{n \to \infty} s_n = L$$

(read "the limit, as n increases without bound, of s_n is L") is used to denote this. A thorough discussion of the notion of a limit is included in courses in calculus, and will not be attempted here. A few elementary ideas, however, are in order.

A sequence in which the nth term approaches a number L as $n \to \infty$ is said to be a **convergent sequence**, and the sequence is said to **converge** to L. It is not necessary for convergence that a sequence be strictly increasing. For example,

$$1, \frac{1}{2}, \frac{1}{3}, \frac{1}{4}, \ldots, \frac{1}{n}, \ldots$$

converges to zero, but each term in the sequence is less than, instead of greater than, the term that precedes it. Again, the sequence

$$-1, \frac{1}{2}, -\frac{1}{3}, \frac{1}{4}, \ldots, \frac{(-1)^n}{n}, \ldots$$

converges to zero, but is neither increasing nor decreasing. We can rephrase the concept of convergence of a sequence as follows:

A sequence $s_1, s_2, \ldots, s_n, \ldots$ converges to the number L,

$$\lim_{n \to \infty} s_n = L,$$

if and only if the absolute value of the difference between the nth term in the sequence and the number L is as small as we please for all sufficiently large n. Thus, the sequence converges to the number L if and only if

$$\lim_{n \to \infty} |L - s_n| = 0.$$

9.3 Limits of Sequences and Series

For example, the **alternating sequence** (because the signs alternate)

$$\frac{-2}{3}, \frac{4}{9}, \frac{-8}{27}, \ldots, \left(\frac{-2}{3}\right)^n, \ldots$$

converges to zero, since the absolute value of the difference between $(-2/3)^n$ and zero, that is, $|0 - (-2/3)^n|$, is as small as we please for n large enough. Thus, for $n = 10$ and $n = 25$, we have $|-(-2/3)^n| < 0.02$ and $|-(-2/3)^n| < 0.00004$, respectively. We express this by writing

$$\lim_{n \to \infty} \left(\frac{-2}{3}\right)^n = 0.$$

On the other hand, the alternating sequence

$$\frac{1}{2}, -\frac{2}{3}, \frac{3}{4}, -\frac{4}{5}, \ldots, (-1)^{n+1}\frac{n}{n+1}, \ldots$$

does not converge. As n increases, the nth term oscillates back and forth from the neighborhood of $+1$ to the neighborhood of -1, and we cannot find a number L such that $\lim_{n \to \infty} |L - s_n| = 0$. Such a sequence is said to **diverge**. Sequences such as

$$1, 2, 3, \ldots, n, \ldots,$$

or

$$2, 4, 8, \ldots, 2^n, \ldots$$

are also said to diverge.

An answer to the logical question of what we mean by "enough" when we say "for n large enough" requires a more precise definition of limit than we have given here. A course in calculus will treat this in detail.

SEQUENCE OF PARTIAL SUMS OF A SERIES

For an infinite series

$$S_\infty = \sum_{j=1}^{\infty} s_j$$

we can consider the infinite sequence of **partial sums**:

$$S_1 = s_1$$
$$S_2 = s_1 + s_2$$
$$\vdots$$
$$S_n = s_1 + s_2 + \cdots + s_n$$
$$\vdots$$

> *An infinite series*
>
> $$S_\infty = \sum_{j=1}^{\infty} s_j$$
>
> *converges if and only if* $S_1, S_2, \ldots, S_n, \ldots$, *the corresponding sequence of partial sums, converges.*

If the sequence of partial sums converges to the number L,

$$\lim_{n \to \infty} S_n = L,$$

then L is said to be the sum of the infinite series, and we write

$$S_\infty = \sum_{j=1}^{\infty} s_j = L.$$

If the sequence of partial sums diverges, then the series is said to diverge.

SUM OF AN INFINITE GEOMETRIC PROGRESSION

We recall from Section 9.2 that the sum of n terms (the nth partial sum) of a geometric progression is given, for $r \neq 1$, by

(2) $$S_n = \frac{a - ar^n}{1 - r}.$$

If $|r| < 1$, that is, if $-1 < r < 1$, then $|r|^n$ becomes smaller and smaller for increasingly large n. For example, if $r = 1/2$, then

$$r^2 = \frac{1}{4}, \quad r^3 = \frac{1}{8}, \quad r^4 = \frac{1}{16},$$

and so forth; and $(1/2)^n$ is as small as we please if n is sufficiently large. Writing (2) as

(3) $$S_n = \frac{a}{1 - r}(1 - r^n),$$

we see that the value of the factor $(1 - r^n)$ is as close as we please to 1 provided $|r| < 1$ and n is taken large enough. Since this argument shows that the sequence of partial sums (3) converges to

$$\frac{a}{1 - r},$$

we have the following result.

9.3 Limits of Sequences and Series

> *The sum of an infinite geometric progression,*
>
> $$a + ar + ar^2 + \cdots + ar^n + \cdots,$$
>
> *with* $|r| < 1$, *is*
>
> (4) $$S_\infty = \lim_{n \to \infty} S_n = \frac{a}{1-r}.$$

An interesting application of this sum arises in connection with repeating decimals—that is, decimal numerals that, after a finite number of decimal places, have endlessly repeating groups of digits. For example,

$$0.21\overline{21} \quad \text{and} \quad 0.138\overline{512}$$

are repeating decimals. The bar denotes that the numerals appearing under it are repeated endlessly. Consider the problem of expressing such a decimal fraction as an arithmetic fraction. We illustrate the process involved with the decimal $0.21\overline{21}$, which can be written as

(5) $$0.21 + 0.0021 + 0.000021 + \cdots.$$

This is a geometric progression with ratio $r = 0.01$. Since the ratio is less than 1 in absolute value, we can use (4) to find the sum of the infinite series (5). Thus,

$$S_\infty = \frac{a}{1-r} = \frac{0.21}{1 - 0.01} = \frac{21}{99} = \frac{7}{33},$$

and the given decimal fraction is equivalent to 7/33.

EXERCISE 9.3

Discuss the limiting behavior of each expression as $n \to \infty$.

Example $\dfrac{n^2 + 3}{n^2}$

Solution By writing $\dfrac{n^2 + 3}{n^2}$ as $\dfrac{n^2}{n^2} + \dfrac{3}{n^2}$ and then as $1 + \dfrac{3}{n^2}$, we observe that

$$\lim_{n \to \infty} \frac{n^2 + 3}{n^2} = \lim_{n \to \infty} \left(1 + \frac{3}{n^2}\right) = 1 + 0 = 1.$$

1. $\dfrac{1}{n}$ 2. $1 + \dfrac{1}{n^2}$ 3. $\dfrac{n+1}{n}$ 4. $\dfrac{n+3}{n^2}$

5. $2n$ 6. $(-1)^n$ 7. $\dfrac{1}{2^n}$ 8. $(-1)^n \dfrac{1}{n}$

State which of the sequences are convergent.

Example $1, \dfrac{3}{2}, \dfrac{7}{4}, \dfrac{15}{8}, \ldots, \dfrac{2^n - 1}{2^{n-1}}, \ldots$

Solution Writing the general term as $\dfrac{2^n}{2^{n-1}} - \dfrac{1}{2^{n-1}}$, or $2 - \dfrac{1}{2^{n-1}}$, we observe that

$$\lim_{n \to \infty} \dfrac{2^n - 1}{2^{n-1}} = \lim_{n \to \infty} \left(2 - \dfrac{1}{2^{n-1}}\right) = 2 - 0 = 2.$$

The sequence is convergent.

9. $\dfrac{1}{2}, \dfrac{1}{4}, \dfrac{1}{8}, \dfrac{1}{16}, \ldots, \dfrac{1}{2^n}, \ldots$ 10. $2, \dfrac{3}{2}, \dfrac{4}{3}, \dfrac{5}{4}, \ldots, \dfrac{n+1}{n}, \ldots$

11. $1, 2, 3, 4, 5, \ldots, n, \ldots$ 12. $2, 4, 6, 8, \ldots, 2n, \ldots$

13. $1, -\dfrac{1}{2}, \dfrac{1}{4}, -\dfrac{1}{8}, \ldots, (-1)^{n-1} \dfrac{1}{2^{n-1}}, \ldots$

14. $1, -1, 1, -1, \ldots, (-1)^{n+1}, \ldots$

Find the sum of each infinite geometric series. If the series has no sum, so state.

Examples a. $3 + 2 + \cdots$ b. $\dfrac{1}{81} - \dfrac{1}{54} + \cdots$

Solutions a. $r = 2/3$ b. $r = -\dfrac{1}{54} \div \dfrac{1}{81} = -\dfrac{3}{2}$

Series has a sum since $|r| < 1$.   Series does not have a sum since $|r| > 1$.
Using (4), we have

$$S_\infty = \dfrac{a}{1-r} = \dfrac{3}{1 - \dfrac{2}{3}} = 9.$$

15. $12 + 6 + \cdots$ 16. $2 + 1 + \cdots$ 17. $\dfrac{1}{36} + \dfrac{1}{30} + \cdots$

18. $\dfrac{1}{16} - \dfrac{1}{8} + \cdots$ 19. $\displaystyle\sum_{j=1}^{\infty} \left(\dfrac{2}{3}\right)^j$ 20. $\displaystyle\sum_{j=1}^{\infty} \left(-\dfrac{1}{4}\right)^j$

9.3 Limits of Sequences and Series

Find an arithmetic fraction equal to each of the given decimal numerals.

Example $0.81818\overline{1}$

Solution Rewrite as a series.

$$0.81 + 0.0081 + 0.000081 + \cdots$$

Find the common ratio.

$$r = 0.01$$

Use $S_\infty = \dfrac{a}{1 - r}$.

$$S_\infty = \frac{0.81}{1 - 0.01} = \frac{81}{99} = \frac{9}{11}$$

21. $0.3131\overline{31}$
22. $0.4545\overline{45}$
23. $2.4104\overline{10}$
24. $3.0270\overline{27}$
25. $0.12888\overline{8}$
26. $0.8333\overline{3}$

Applications

27. *Biology* A patient takes 25 milligrams of a drug each day. If 80% of the amount accumulated is excreted each day, how much of the drug will accumulate after a very long period of time (that is, $n \to \infty$)? [*Hint*: See page 316.]

28. In some circumstances, if a dosage A of a drug is administered each day, then the amount that will accumulate in a patient's body after a very long period of time is given by

$$A + Ae^{-k} + Ae^{-2k} + \cdots$$

where k is a positive constant. Simplify this last expression. [*Hint*: Recall the definition of the number e.]

29. *Business* One-fourth, or $600,000, of a company's yearly earnings is reinvested in the company. It is estimated that one-fourth of this investment will find its way back into the company as another investment, and so on. Eventually, how much reinvestment does the original $600,000 generate?

30. *Physics* The arc length through which the bob on a pendulum moves is nine-tenths of its preceding arc length. Approximately how far will the bob move before coming to rest if the first arc length is 12 inches?

31. A ball returns to two-thirds of its preceding height on each bounce. If the ball is dropped from a height of 6 feet, approximately what is the total distance the ball travels before coming to rest?

32. A force is applied to a particle moving in a straight line in such a fashion that each second it moves only one-half the distance it moved the preceding second. If the particle moves 10 centimeters the first second, approximately how far will it move before coming to rest?

9.4
THE BINOMIAL THEOREM

FACTORIAL NOTATION

There are situations in which it is necessary to write the product of several consecutive positive integers. To facilitate writing products of this type, we use a special symbol $n!$ (read "n factorial" or "factorial n"), which is defined by

$$n! = n(n-1)(n-2) \cdots (3)(2)(1).$$

Thus,
$$5! = 5 \cdot 4 \cdot 3 \cdot 2 \cdot 1 \qquad \text{(read "five factorial")},$$
and
$$8! = 8 \cdot 7 \cdot 6 \cdot 5 \cdot 4 \cdot 3 \cdot 2 \cdot 1 \qquad \text{(read "eight factorial")}.$$

Factorial notation can also be used to represent products of consecutive positive integers, beginning with integers different from 1. For example,

$$8 \cdot 7 \cdot 6 \cdot 5 = \frac{8!}{4!},$$

because

$$\frac{8!}{4!} = \frac{8 \cdot 7 \cdot 6 \cdot 5 \cdot 4 \cdot 3 \cdot 2 \cdot 1}{4 \cdot 3 \cdot 2 \cdot 1} = 8 \cdot 7 \cdot 6 \cdot 5.$$

Since
$$n! = n(n-1)(n-2)(n-3) \cdots 5 \cdot 4 \cdot 3 \cdot 2 \cdot 1$$
and
$$(n-1)! = (n-1)(n-2)(n-3) \cdots 5 \cdot 4 \cdot 3 \cdot 2 \cdot 1,$$

for $n > 1$ we can write the recursive relationship

(1) $$n! = n(n-1)!.$$

For example,
$$7! = 7 \cdot 6!$$
$$27! = 27 \cdot 26!$$
$$(n+2)! = (n+2)(n+1)!.$$

If (1) is to hold also for $n = 1$, then we must have

$$1! = 1 \cdot (1-1)!,$$

or

$$1! = 1 \cdot 0!.$$

Therefore, for consistency, we define $0!$ by

$$0! = 1.$$

9.4 The Binomial Theorem

Another useful factorial notation occurs in the form

$$\binom{n}{r} = \frac{n!}{r!(n-r)!}.$$

Some examples are

$$\binom{5}{3} = \frac{5!}{3!(5-3)!} = \frac{5!}{3!2!} = \frac{5 \cdot 4 \cdot 3!}{3!2 \cdot 1} = 10,$$

$$\binom{5}{1} = \frac{5!}{1!(5-1)!} = \frac{5!}{4!} = \frac{5 \cdot 4!}{4!} = 5,$$

$$\binom{5}{0} = \frac{5!}{0!(5-0)!} = \frac{5!}{5!} = 1.$$

BINOMIAL EXPANSIONS The series obtained by expanding a binomial of the form

$$(a + b)^n$$

is important in many branches of mathematics. Starting with familiar examples, where n takes the values 1, 2, 3, 4, 5, we can show by direct multiplication that

$$(a + b)^1 = a + b$$
$$(a + b)^2 = a^2 + 2ab + b^2$$
$$(a + b)^3 = a^3 + 3a^2b + 3ab^2 + b^3$$
$$(a + b)^4 = a^4 + 4a^3b + 6a^2b^2 + 4ab^3 + b^4$$
$$(a + b)^5 = a^5 + 5a^4b + 10a^3b^2 + 10a^2b^3 + 5ab^4 + b^5.$$

We observe that in each case:

1. The first term is a^n.
2. The variable factors of the second term are $a^{n-1}b^1$, and the coefficient is n, which can be written in the form

$$\frac{n}{1!}.$$

3. The variable factors of the third term are $a^{n-2}b^2$, and the coefficient can be written in the form

$$\frac{n(n-1)}{2!}.$$

4. The variable factors of the fourth term are $a^{n-3}b^3$, and the coefficient can be written in the form

$$\frac{n(n-1)(n-2)}{3!}.$$

These expansions suggest the following result, known as the **binomial theorem**.

For each positive integer n,

(2) $\quad (a+b)^n = a^n + \dfrac{n}{1!} a^{n-1} b + \dfrac{n(n-1)}{2!} a^{n-2} b^2 + \dfrac{n(n-1)(n-2)}{3!} a^{n-3} b^3$

$\quad\quad + \cdots + \dfrac{n(n-1)(n-2)\cdots(n-r+2)}{(r-1)!} a^{n-r+1} b^{r-1} + \cdots + b^n,$

where r is the number of the term.

For example,

$(x-2)^4 = x^4 + \dfrac{4}{1!} x^3 (-2)^1 + \dfrac{4 \cdot 3}{2!} x^2 (-2)^2 + \dfrac{4 \cdot 3 \cdot 2}{3!} x(-2)^3 + \dfrac{4 \cdot 3 \cdot 2 \cdot 1}{4!} (-2)^4$

$= x^4 - 8x^3 + 24x^2 - 32x + 16.$

In this case, $a = x$ and $b = -2$ in the binomial expansion.

Observe that the coefficients of the terms in the binomial expansion (2) can be represented as follows.

First term: $\quad \dbinom{n}{0} = \dfrac{n!}{0! n!} = 1$

Second term: $\quad \dbinom{n}{1} = \dfrac{n!}{1!(n-1)!} = \dfrac{n \cdot (n-1)!}{1 \cdot (n-1)!} = n$

Third term: $\quad \dbinom{n}{2} = \dfrac{n!}{2!(n-2)!} = \dfrac{n(n-1)(n-2)!}{2!(n-2)!} = \dfrac{n(n-1)}{2!}$

rth term: $\quad \dbinom{n}{r-1} = \dfrac{n!}{(r-1)!(n-r+1)!}$

$\quad\quad\quad\quad\quad = \dfrac{n(n-1)(n-2) \cdots (n-r+2)(n-r+1)!}{(r-1)!(n-r+1)!}$

$\quad\quad\quad\quad\quad = \dfrac{n(n-1)(n-2) \cdots (n-r+2)}{(r-1)!}$

Hence, the binomial expansion (2) can be represented by the expression

(3) $\quad (a+b)^n = \dbinom{n}{0} a^n + \dbinom{n}{1} a^{n-1} b + \dbinom{n}{2} a^{n-2} b^2 + \dbinom{n}{3} a^{n-3} b^3$

$\quad\quad\quad\quad\quad + \cdots + \dbinom{n}{r-1} a^{n-r+1} b^{r-1} + \cdots + \dbinom{n}{n} b^n.$

For example,

9.4 The Binomial Theorem

$$(x-2)^4 = \binom{4}{0}x^4 + \binom{4}{1}x^3(-2)^1 + \binom{4}{2}x^2(-2)^2 + \binom{4}{3}x(-2)^3 + \binom{4}{4}(-2)^4.$$

Using the formula for $\binom{n}{r}$ to find the coefficients, we obtain the same result as above,

$$(x-2)^4 = x^4 - 8x^3 + 24x^2 - 32x + 16.$$

THE rth TERM IN A BINOMIAL EXPANSION

Note that the rth term in a binomial expansion is given by

(4) $\binom{n}{r-1}a^{n-r+1}b^{r-1} = \dfrac{n!}{(r-1)!(n-r+1)!}a^{n-r+1}b^{r-1}$

$= \dfrac{n(n-1)(n-2)\cdots(n-r+2)}{(r-1)!}a^{n-r+1}b^{r-1}.$

For example, by the right-hand member in the first line of (4), the seventh term of $(x-2)^{10}$ is

$$\binom{10}{6}x^4(-2)^6 = \dfrac{10!}{6!4!}x^4(-2)^6 = \dfrac{10\cdot 9\cdot 8\cdot 7\cdot 6!}{6!\cdot 4\cdot 3\cdot 2\cdot 1}x^4(64)$$

$$= 13{,}440x^4,$$

while, by the second line of (4), we have

$$\dfrac{10\cdot 9\cdot 8\cdot 7\cdot 6\cdot 5}{6\cdot 5\cdot 4\cdot 3\cdot 2\cdot 1}x^4(64) = 13{,}440x^4.$$

EXERCISE 9.4

1. Write $(2n)!$ in expanded form for $n = 4$.
2. Write $2n!$ in expanded form for $n = 4$.
3. Write $n(n-1)!$ in expanded form for $n = 6$.
4. Write $2n(2n-1)!$ in expanded form for $n = 2$.

Write in expanded form and simplify.

Examples a. $\dfrac{7!}{4!}$ b. $\dfrac{4!6!}{8!}$

Solutions   a. $\dfrac{7 \cdot 6 \cdot 5 \cdot 4!}{4!} = 210$ b. $\dfrac{4 \cdot 3 \cdot 2 \cdot 1 \cdot 6!}{8 \cdot 7 \cdot 6!} = \dfrac{3}{7}$

5. $5!$ 6. $7!$ 7. $\dfrac{9!}{7!}$ 8. $\dfrac{12!}{11!}$

9. $\dfrac{5!7!}{8!}$ 10. $\dfrac{12!8!}{16!}$ 11. $\dfrac{8!}{2!(8-2)!}$ 12. $\dfrac{10!}{4!(10-4)!}$

Write each product in factorial notation.

Examples   a. $1 \cdot 2 \cdot 3 \cdot 4 \cdot 5 \cdot 6$ b. $11 \cdot 12 \cdot 13 \cdot 14$ c. 150

Solutions   a. $6!$ b. $\dfrac{14!}{10!}$ c. $\dfrac{150!}{149!}$

13. $1 \cdot 2 \cdot 3$ 14. $1 \cdot 2 \cdot 3 \cdot 4 \cdot 5$ 15. $3 \cdot 4 \cdot 5 \cdot 6$
16. 7 17. $8 \cdot 7 \cdot 6$ 18. $28 \cdot 27 \cdot 26 \cdot 25 \cdot 24$

Write each expression in factorial notation and simplify.

Examples   a. $\binom{6}{2}$ b. $\binom{4}{4}$

Solutions   a. $\binom{6}{2} = \dfrac{6!}{2!(6-2)!} = \dfrac{6!}{2!4!}$ b. $\binom{4}{4} = \dfrac{4!}{4!(4-4)!}$

$= \dfrac{6 \cdot 5 \cdot 4!}{2 \cdot 1 \cdot 4!} = 15$ $= \dfrac{4!}{4!0!} = 1$

19. $\binom{6}{5}$ 20. $\binom{4}{2}$ 21. $\binom{3}{3}$ 22. $\binom{5}{5}$
23. $\binom{7}{0}$ 24. $\binom{2}{0}$ 25. $\binom{5}{2}$ 26. $\binom{5}{3}$

Write each expression in factored form, and show the first three factors and the last three factors.

Example $(2n + 1)!$

9.4 The Binomial Theorem

Solution $(2n + 1)(2n)(2n - 1) \cdots 3 \cdot 2 \cdot 1$

27. $n!$
28. $(n + 4)!$
29. $(3n)!$
30. $3n!$
31. $(n - 2)!$
32. $(3n - 2)!$

Simplify each expression.

Examples a. $\dfrac{(n-1)!}{(n-3)!}$ b. $\dfrac{(n-1)!(2n)!}{2n!(2n-2)!}$

Solutions a. $\dfrac{(n-1)(n-2)(n-3)!}{(n-3)!} = (n-1)(n-2)$ b. $\dfrac{(n-1)!(2n)(2n-1)(2n-2)!}{2n(n-1)!(2n-2)!} = 2n - 1$

33. $\dfrac{(n+2)!}{n!}$ 34. $\dfrac{(n+2)!}{(n-1)!}$ 35. $\dfrac{(n+1)(n+2)!}{(n+3)!}$

36. $\dfrac{(2n+4)!}{(2n+2)!}$ 37. $\dfrac{(2n)!(n-2)!}{4(2n-2)!(n)!}$ 38. $\dfrac{(2n+1)!(2n-1)!}{[(2n)!]^2}$

Expand.

Example $(a - 3b)^4$

Solution From the binomial expansion (2),

$$(a - 3b)^4 = a^4 + \frac{4}{1!}a^3(-3b) + \frac{4 \cdot 3}{2!}a^2(-3b)^2 + \frac{4 \cdot 3 \cdot 2}{3!}a(-3b)^3 + \frac{4 \cdot 3 \cdot 2 \cdot 1}{4!}(-3b)^4$$

$$= a^4 - 12a^3b + 54a^2b^2 - 108ab^3 + 81b^4.$$

Alternatively, from (3) we have

$$(a - 3b)^4 = \binom{4}{0}a^4 + \binom{4}{1}a^3(-3b) + \binom{4}{2}a^2(-3b)^2 + \binom{4}{3}a(-3b)^3 + \binom{4}{4}(-3b)^4,$$

which also simplifies to the expression obtained above.

39. $(x + 3)^5$ 40. $(2x + y)^4$ 41. $(x - 3)^4$ 42. $(2x - 1)^5$

43. $\left(2x - \dfrac{y}{2}\right)^3$ 44. $\left(\dfrac{x}{3} + 3\right)^5$ 45. $\left(\dfrac{x}{2} + 2\right)^6$ 46. $\left(\dfrac{2}{3} - a^2\right)^4$

Write the first four terms in each expansion. Do not simplify the terms.

Example $(x + 2y)^{15}$

Solution From the binomial expansion (2),

$$(x + 2y)^{15} = x^{15} + \frac{15}{1!}x^{14}(2y) + \frac{15 \cdot 14}{2!}x^{13}(2y)^2 + \frac{15 \cdot 14 \cdot 13}{3!}x^{12}(2y)^3 + \cdots,$$

or alternatively from the form (3),

$$(x + 2y)^{15} = \binom{15}{0}x^{15} + \binom{15}{1}x^{14}(2y) + \binom{15}{2}x^{13}(2y)^2 + \binom{15}{3}x^{12}(2y)^3 + \cdots.$$

47. $(x + y)^{20}$ 48. $(x - y)^{15}$ 49. $(a - 2b)^{12}$
50. $(2a - b)^{12}$ 51. $(x - \sqrt{2})^{10}$ 52. $\left(\dfrac{x}{2} + 2\right)^8$

Find each power to the nearest hundredth.

Example $(0.97)^7$

Solution Either form of the binomial expansion, (2) or (3), can be used. We shall use (2).

$$(0.97)^7 = (1 - 0.03)^7$$

$$= 1^7 + \frac{7}{1!}(1)^6(-0.03)^1 + \frac{7 \cdot 6}{2!}(1)^5(0.03)^2 + \frac{7 \cdot 6 \cdot 5}{3!}(1)^4(-0.03)^3 + \cdots$$

$$= 1 - 0.21 + 0.0189 - 0.000945 + \cdots$$

$$\approx 0.807955$$

Hence, to the nearest hundredth, $(0.97)^7 = 0.81$.

53. $(1.02)^{10}$ [*Hint:* $1.02 = 1 + 0.02$] 54. $(1.01)^{15}$

Find the specified term.

Example $(x - 2y)^{12}$; the seventh term

Solution In (4) use $n = 12$ and $r = 6$.

$$\frac{12!}{6!(12-6)!}x^6(-2y)^6 = \frac{12 \cdot 11 \cdot 10 \cdot 9 \cdot 8 \cdot 7}{6!}x^6(-2y)^6 = 59{,}136x^6y^6$$

9.4 The Binomial Theorem

55. $(a - b)^{15}$; the sixth term
56. $(x + 2)^{12}$; the fifth term
57. $(x - 2y)^{10}$; the fifth term
58. $(a^3 - b)^9$; the seventh term
59. Given that the binomial formula (2) holds for $(1 + x)^n$ where n is a negative integer:
    a. Write the first four terms of $(1 + x)^{-1}$.
    b. Find the first four terms of the quotient $1/(1 + x)$ by dividing 1 by $(1 + x)$.

    Compare the results of parts a and b.
60. Given that the binomial formula (2) holds as an infinite "sum" for $(1 + x)^n$, where n is a rational number and $|x| < 1$, find the following to two decimal places.
    a. $\sqrt{1.02}$
    b. $\sqrt{0.99}$
61. The coefficients in the binomial expansion of $(a + b)^n$ for $n = 0, 1, 2, 3, 4$ are given by **Pascal's triangle**:

    $$\begin{array}{c}1\\1\quad 1\\1\quad 2\quad 1\\1\quad 3\quad 3\quad 1\\1\quad 4\quad 6\quad 4\quad 1\end{array}$$

    Discern the pattern in the triangle and complete it for $n = 5$.
62. Show that
    $$2^n = \binom{n}{0} + \binom{n}{1} + \binom{n}{2} + \cdots + \binom{n}{n}.$$

Applications
63. *Business* If an amount of money P is invested at 4% compounded annually, the amount S present at the end of m years is given by $S = P(1 + 0.04)^m$. Find the amount S (to the nearest dollar) if $1000 was invested for 10 years.
64. In Problem 63, find the amount present at the end of 20 years.

CHAPTER TEST

Write the next three terms of each arithmetic progression.

1. 7, 10, ...

2. $a, a-2, ...$

Write the next three terms in each geometric progression.

3. $-2, 6, ...$

4. $\frac{2}{3}, 1, ...$

5. Find the general term and the seventh term of the arithmetic progression $-3, 2, ...$.

6. Find the general term and the fifth term of the geometric progression $-2, 2/3, ...$.

7. If the fourth term of an arithmetic progression is 13 and the ninth term is 33, find the seventh term.

8. Which term in a geometric progression $-2/9, 2/3, ...$ is 54?

9. Write $\sum_{k=2}^{5} k(k-1)$ in expanded form.

10. Write $x^2 + x^3 + x^4 + \cdots$ in sigma notation.

11. Find the value of $\sum_{j=3}^{9} (3j-1)$.

12. Find the value of $\sum_{j=1}^{5} \left(\frac{1}{3}\right)^j$.

13. Specify the limit of $\frac{3n^2 - 1}{n^2}$ as $n \to \infty$.

14. Find the value of the geometric series $4 - 2 + 1 - 1/2 + \cdots$.

15. Find the value of $\sum_{i=1}^{\infty} \left(\frac{1}{3}\right)^i$.

16. Find a fraction equivalent to $0.44\overline{4}$.

17. Write $n(n-3)!$ in expanded form for $n = 5$.

Chapter Test

Write each expression in expanded form and simplify.

18. $\dfrac{8!3!}{7!}$
19. $\dbinom{7}{2}$
20. $\dfrac{(n-1)!}{n!(n+1)!}$

21. Write the first four terms of the binomial expansion of $(x - 2y)^{10}$.

22. Find the eighth term in the expansion of $(x - 2y)^{10}$.

23. A person must select one of the following gifts:
    a. $100 every month for 22 years,
    b. $1 the first month, $2 the second, $3 the third, $4 the fourth, and so on, for 22 years,
    c. 1 cent the first month, 2 cents the second, 4 cents the third, 8 cents the fourth, and so on, for only 2 years.
    Determine the best gift.

10
PROBABILITY

10.1

BASIC COUNTING PRINCIPLES; PERMUTATIONS

Associated with each finite set A is a nonnegative integer n, namely the number of elements in A. Hence, we have a function from the set of all finite sets to the set of nonnegative integers. The symbolism $n(A)$ is used to denote elements in the range of this set function n. For example, if

$$A = \{5, 7, 9\}, \qquad B = \left\{\frac{1}{2}, 0, 3, -5, 7\right\}, \qquad C = \emptyset,$$

then

$$n(A) = 3, \qquad n(B) = 5, \qquad n(C) = 0.$$

COUNTING PROPERTIES   All the sets with which we shall be concerned are assumed to be finite sets. We then have the following properties I–III, called **counting properties**, for the function n.

$$\text{I} \quad n(A \cup B) = n(A) + n(B) \qquad \text{if} \quad A \cap B = \emptyset$$

Thus, if A and B are disjoint sets, then the number of elements in their union is the sum of the number of elements in A and the number of elements in B.

For example, suppose there are 5 roads from town R to town S, and 2 railroads from town R to town S. If A is the set of roads and B is the set of railroads from R to S, then $n(A) = 5$, $n(B) = 2$, and $n(A \cup B) = 5 + 2 = 7$; thus, there are 7 ways one can go from town R to town S by driving or riding on a train.

$$\text{II} \quad n(A \cup B) = n(A) + n(B) - n(A \cap B) \qquad \text{if} \quad A \cap B \neq \emptyset$$

10.1 Basic Counting Principles; Permutations

That is, if A and B overlap, then to count the number of elements in $A \cup B$, we might add the number of elements in A to the number of elements in B. But since any elements in the intersection of A and B would be counted twice in this process (once in A and once in B), we must subtract the number of such elements from the sum $n(A) + n(B)$ to obtain the number of elements in $A \cup B$, as suggested in Figure 10.1.

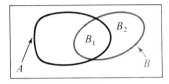

Figure 10.1

For example, suppose there are 15 unrelated girls and 17 unrelated boys in a business class, and suppose there are precisely 2 brother–sister pairs in the class. If A denotes the set of different families represented by the girls and B denotes the set of different families represented by the boys, then the number of different families represented by all the members of the class is

$$n(A) + n(B) - n(A \cap B) = 15 + 17 - 2 = 30.$$

Actually, property II is a consequence of property I. The third property,

$$\text{III} \quad n(A \times B) = n(A) \cdot n(B),$$

asserts that the number of elements in the Cartesian product of sets A and B is the product of the number of elements in A and the number of elements in B.

For example, suppose there are 5 roads from town R to town S (set A), and there are 3 roads from town S to town T (set B). Then for each element of A there are 3 elements of B, and the total possible ways one can drive from R to T via S is

$$n(A \times B) = n(A) \cdot n(B) = 5 \cdot 3 = 15.$$

PERMUTA-TIONS Given the set of digits $A = \{1, 2, 3\}$, how many different three-digit numerals can be constructed from the members of A if no member is used more than once? The answer to this question can be obtained by simply listing the different three-digit numerals, 1 2 3, 1 3 2, 2 1 3, 2 3 1, 3 1 2, 3 2 1, and counting them. Such a procedure would be quite impractical, however, if the number of members of the given set of digits were very large. Another way to arrive at the same conclusion is by applying counting property III. If we let A denote the set of possible first digits in the above numerals, then $n(A) = 3$. Since no numeral may be used more than once, and since one numeral has already been used for a first digit, there remain only two possibilities for the second digit. If B denotes the set of possible second digits after the first digit has been chosen, then $n(B) = 2$. By similar reasoning, if C is the set of possible third digits after the first two have been chosen, then $n(C) = 1$. By property III (applied twice), we find that

$$n(A \times B \times C) = [n(A) \cdot n(B)] \cdot n(C) = 3 \cdot 2 \cdot 1 = 6.$$

Each of the three-digit numerals discussed above is called a **permutation** of the elements of the set of numerals $\{1, 2, 3\}$. Formally, we have the following.

A permutation of a set A is an ordering (first, second, etc.) of the members of A.

We can now obtain the following result.

Let $P_{n,n}$ denote the number of distinct permutations of the members of a set A containing n members. Then

(1) $$P_{n,n} = n!$$

The symbol $P_{n,n}$ is read "the number of permutations of n things taken n at a time."

To justify (1), let A_1 denote the set of possible selections for the first member. Then $A_1 = A$, and $n(A_1) = n(A) = n$. Having made a first selection, let A_2 denote the set of possible second selections. Then $A_2 \subset A$, and $n(A_2) = n(A_1) - 1 = n - 1$. A continuation of this procedure, together with successive application of principle III, leads to

$$P_{n,n} = n(A_1) \cdot n(A_2) \cdots n(A_n) = n \cdot (n-1) \cdot (n-2) \cdots 1 = n!,$$

as was to be shown.

Example In how many ways can 9 women be assigned positions to form distinct softball teams?

Solution Let A denote the set of women, so that $n(A) = 9$. The total number of ways in which 9 women can be assigned 9 positions on a team, or, in other words, the number of possible permutations of the members of a 9-element set, is by (1),

$$P_{9,9} = 9! = 9 \cdot 8 \cdot 7 \cdots 1 = 362{,}880.$$

Let $P_{n,r}$ denote the number of permutations of the members, taken r at a time, of a set A containing n members; that is let $P_{n,r}$ be the number of distinct orderings of r elements when there is a set A of n elements from which to choose. Then

(2) $$P_{n,r} = n(n-1)(n-2)\cdots(n-r+1).$$

The justification of (2) follows that of (1), except that the last subset considered is such that $n(A_r) = n - (r-1) = n - r + 1$.

10.1 Basic Counting Principles; Permutations

Example In how many ways can a basketball team be formed by choosing players for the 5 positions from a set of 10 players?

Solution Let A denote the set of players, so that $n(A) = 10$. Then from (2) and the fact that a basketball team consists of 5 players, we have

$$P_{10,5} = 10 \cdot 9 \cdot 8 \cdots (10 - 5 + 1) = 10 \cdot 9 \cdot 8 \cdot 7 \cdot 6 = 30{,}240.$$

An alternative expression for $P_{n,r}$ can be obtained by observing that

$$P_{n,r} = n(n-1)(n-2)\cdots(n-r+1)$$
$$= \frac{n(n-1)(n-2)\cdots(n-r+1)(n-r)!}{(n-r)!},$$

so that

(3) $$P_{n,r} = \frac{n!}{(n-r)!}.$$

DISTINGUISHABLE PERMUTATIONS The problem of finding the number of distinguishable permutations of n objects taken n at a time, if some of the objects are identical, requires a little more careful analysis. As an example, consider the number of permutations of the letters of the word *DIVISIBLE*. We can make a distinction between the 3 I's by assigning subscripts to each so that we have 9 distinct letters,

$$D, I_1, V, I_2, S, I_3, B, L, E.$$

The number of permutations of these 9 letters is, of course, 9!. If the letters other than I_1, I_2, and I_3 are retained in the positions they occupy in a permutation of the above 9 letters, I_1, I_2, and I_3 can be permuted among themselves 3! ways. Thus, if P is the number of *distinguishable* permutations of the letters

$$D, I, V, I, S, I, B, L, E,$$

then, since for each of these there are 3! ways in which the I's can be permuted without otherwise changing the order of the other letters, it follows that

$$3! \cdot P = 9!,$$

from which

$$P = \frac{9!}{3!} = 60{,}480.$$

As another example, consider the letters of the word *MISSISSIPPI*. There would exist 11! distinguishable permutations of the letters in this word if each letter were distinct. Note, however, that the letters S and I each appear 4 times and the letter P appears 2 times. Reasoning as we did in the previous example, we see that the number P of distinguishable permutations of the letters in *MISSISSIPPI* is given by

$$4!\, 4!\, 2!\, P = 11!,$$

from which

$$P = \frac{11!}{4!\, 4!\, 2!} = 34{,}650.$$

EXERCISE 10.1

Given the following sets, find $n(A \cap B)$, $n(A \cup B)$, and $n(A \times B)$.

Example $A = \{a, b, c\}$; $B = \{c, d\}$

Solution $A \cap B = \{c\}$

Therefore,

$$n(A \cap B) = 1,$$
$$n(A \cup B) = n(A) + n(B) - n(A \cap B) = 3 + 2 - 1 = 4,$$
$$n(A \times B) = n(A) \cdot n(B) = 3 \cdot 2 = 6.$$

1. $A = \{d, e\}$; $B = \{e, f, g, h\}$
2. $A = \{e\}$; $B = \{a, b, c, d\}$
3. $A = \{1, 2, 3, 4\}$; $B = \{3, 4, 5, 6\}$
4. $A = \{1, 2\}$; $B = \{3, 4, 5\}$
5. $A = \{1, 2\}$; $B = \{1, 2\}$
6. $A = \emptyset$; $B = \{2, 3, 4\}$

Example In how many ways can 3 members of a class be assigned a grade of A, B, C, or D?

Solution Sometimes a simple diagram, such as ___, ___, ___, designating a sequence, is a helpful preliminary device. Since each of the students may receive any 1 of 4 different grades, the sequence would appear as

$$\underline{4},\, \underline{4},\, \underline{4}.$$

From counting property III, there are $4 \cdot 4 \cdot 4$, or 64, possible ways the grades may be assigned.

Example In how many different ways can 3 members of a class be assigned a grade of A, B, C, or D so that no 2 members receive the same grade?

10.1 Basic Counting Principles; Permutations

Solution Since the first student may receive any 1 of 4 different grades, the second student may then receive any 1 of 3 different grades, and the third student may then receive any 1 of 2 different grades, the sequence would appear as

$$\underline{4}, \underline{3}, \underline{2}.$$

From counting property III, there are $4 \cdot 3 \cdot 2$, or 24, possible ways the grades may be assigned. In this case, we could have obtained the same result directly from (2) since $P_{4,3} = 4 \cdot 3 \cdot 2 = 24$.

In each problem, a digit or letter may be used more than once unless stated otherwise.

7. How many different two-digit numerals can be formed from the digits 5 and 6?
8. How many different two-digit numerals can be formed from the digits 7, 8, and 9?
9. In how many different ways can 4 students be seated in a row?
10. In how many different ways can 5 students be seated in a row?
11. In how many different ways can 4 questions on a true–false test be answered?
12. In how many different ways can 5 questions on a true–false test be answered?
13. In how many ways can you write different three-digit numerals from $\{2, 3, 4, 5\}$?
14. How many different seven-digit telephone numbers can be formed from the set of digits $\{1, 2, 3, 4, 5, 6, 7, 8, 9, 0\}$?
15. In how many ways can you write different three-digit numerals using $\{2, 3, 4, 5\}$ if no digit is to be used more than once in each numeral?
16. How many different seven-digit telephone numbers can be formed from the set of digits $\{1, 2, 3, 4, 5, 6, 7, 8, 9, 0\}$ if no digit is to be used more than once in any number?
17. How many three-letter arrangements can be formed from $\{A, N, S, W, E, R\}$?
18. If no letter is to be used more than once in any arrangement, how many different three-letter arrangements can be formed from $\{A, N, S, W, E, R\}$?
19. How many four-digit numerals for positive odd integers can be formed from $\{1, 2, 3, 4, 5\}$?
20. How many four-digit numerals for positive even integers can be formed from $\{1, 2, 3, 4, 5\}$?
21. How many numerals for positive integers less than 500 can be formed from $\{3, 4, 5\}$?
22. How many numerals for positive odd integers less than 500 can be formed from $\{3, 4, 5\}$?
23. How many numerals for positive even integers less than 500 can be formed from $\{3, 4, 5\}$?
24. How many numerals for positive even integers between 400 and 500, inclusive, can be formed from $\{3, 4, 5\}$?
25. How many permutations of the elements of $\{P, R, I, M, E\}$ end in a vowel?
26. How many permutations of the elements of $\{P, R, O, D, U, C, T\}$ end in a vowel?
27. Find the number of distinguishable permutations of the letters in the word

    LIMIT.

28. Find the number of distinguishable permutations of the letters in the word

    COMBINATION.

29. Find the number of distinguishable permutations of the letters in the word

    COLORADO.

30. Find the number of distinguishable permutations of the letters in the word

    TALLAHASSEE.

31. Show that $P_{5,3} = 5(P_{4,2})$.
32. Show that $P_{5,r} = 5(P_{4,r-1})$.
33. Show that $P_{n,3} = n(P_{n-1,2})$.
34. Show that $P_{n,3} - P_{n,2} = (n-3)(P_{n,2})$.
35. Solve for n: $P_{n,5} = 5(P_{n,4})$.
36. Solve for n: $P_{n,5} = 9(P_{n-1,4})$.

Example In how many ways can 4 students be seated around a circular table?

Solution In any such arrangement (which is called a **circular permutation**), there is no first position. Each person can take 4 different initial positions without affecting the arrangement. Thus, there are

$$\frac{4!}{4} = 6$$

arrangements. [In general, there are $n!/n$, or $(n-1)!$, circular permutations of n things taken n at a time.]

37. In how many ways can 5 students be seated around a circular table?
38. In how many ways can 6 students be seated around a circular table?
39. In how many ways can 6 students be seated around a circular table if a certain two must be seated together?
40. In how many ways can 3 different keys be arranged on a key ring? [*Hint*: Arrangements should be considered identical if one can be obtained from the other by turning the ring over. In general, there are only $(1/2)(n-1)!$ distinct arrangements of n keys on a ring $(n \geq 3)$.]

Applications
41. *Biology* There are 20 male and 20 female blood donors. In how many ways can 2 donors be selected using one person of each sex?
42. A certain demonstration calls for one animal of each of three types A, B, and C. If there are 5 animals of type A, 6 animals of type B, and 7 animals of type C, in how many ways can one of each type be chosen?
43. In how many ways can 5 test tubes, each containing a different culture, be arranged on a straight rack with 5 positions?
44. Ten culture dishes are marked $1, 2, ..., 10$. In how many ways can 4 dishes be selected?
45. *Business* A company is to select its governing board from a set of 20 executives. If each board member holds a different position (such as chairman, president, and so on), how many 5-person boards can be formed?
46. There are 3 letters, in addition to 3 numbers, on a California license plate. Assuming no repetition, how many 3-letter arrangements are possible?
47. Allowing repetition of numbers and letters, how many distinct California license plates can be manufactured?

48. There are 10 ways of connecting a terminal to computer A and 16 ways in which computer A can be connected to computer B. In how many ways can the terminal be connected to computer B?

10.2 COMBINATIONS

An additional counting concept is needed before we turn our attention to probability—namely, finding the number of distinct r-element subsets of an n-element set with no reference to relative order of the elements in the subset. For example, 5 different cards can be arranged in 5! permutations, but to a poker player they represent the same hand. The set of 5 cards (with no reference to the arrangement of the cards) is called a **combination**.

> *A subset of an n-element set A is called a combination.*

The counting of combinations is related to the counting of permutations. From (2) of Section 10.1, we know that the number of distinct permutations of n elements of a set A taken r at a time is given by

$$P_{n,r} = \frac{n!}{(n-r)!}.$$

With this in mind, consider the following result concerning the number $\binom{n}{r}$ $= \frac{n!}{r!(n-r)!}$ that was introduced in Section 9.4. In this context we denote $\binom{n}{r}$ by $C_{n,r}$.

> *The number $C_{n,r}$ of distinct combinations of the members, taken r at a time, of a set containing n members is given by*
>
> (1) $$C_{n,r} = \frac{P_{n,r}}{r!} = \frac{n!}{r!(n-r)!}.$$

To derive (1) we note there are, by definition $C_{n,r}$ r-element subsets of the set A, where $n(A) = n$. Also, each of these subsets has $r!$ permutations of its members. There are therefore $C_{n,r} r!$ permutations of n elements of A taken r at a time. Thus,

$$P_{n,r} = C_{n,r} r!,$$

from which we obtain

$$C_{n,r} = \frac{P_{n,r}}{r!}, = \frac{n!/(n-r)!}{r!} = \frac{n!}{r!(n-r)!}$$

as was to be shown.

We note that since

$$P_{n,r} = n(n-1)(n-2)\cdots(n-r+1),$$

it follows that (1) has the alternative form

(2) $$C_{n,r} = \frac{P_{n,r}}{r!} = \frac{n(n-1)(n-2)\cdots(n-r+1)}{r!}$$

Example In how many ways can a committee of 5 be selected from a set of 12 persons?

Solution What we wish here is the number of 5-element subsets of a 12-element set. From (1), we have

$$C_{12,5} = \frac{12!}{5!\,7!} = \frac{12 \cdot 11 \cdot 10 \cdot 9 \cdot 8}{5 \cdot 4 \cdot 3 \cdot 2 \cdot 1} = 792.$$

EXERCISE 10.2

Example How many different amounts of money can be formed from a penny, a nickel, a dime, and a quarter?

Solution We want to find the total number of combinations that can be formed by taking the coins 1, 2, 3, and 4 at a time. From (1) we have

$$C_{4,1} = \frac{4!}{1!\,3!} = 4, \quad C_{4,2} = \frac{4!}{2!\,2!} = 6, \quad C_{4,3} = \frac{4!}{3!\,1!} = 4, \quad C_{4,4} = \frac{4!}{4!\,0!} = 1.$$

and the total number of combinations is 15. Clearly, each combination gives a different amount.

1. How many different amounts of money can be formed from a penny, a nickel, and a dime?
2. How many different amounts of money can be formed from a penny, a nickel, a dime, a quarter, and a half-dollar?
3. How many different committees of 4 persons each can be chosen from a group of 6 persons?
4. How many different committees of 4 persons each can be chosen from a group of 10 persons?

10.2 Combinations

5. In how many different ways can a set of 5 cards be selected from a standard bridge deck containing 52 cards?
6. In how many different ways can a set of 13 cards be selected from a standard bridge deck of 52 cards?
7. In how many different ways can a hand consisting of 5 spades, 5 hearts, and 3 diamonds be selected from a standard bridge deck of 52 cards?
8. In how many different ways can a hand consisting of 10 spades, 1 heart, 1 diamond, and 1 club be selected from a standard bridge deck of 52 cards?
9. In how many different ways can a hand consisting of either 5 spades, 5 hearts, 5 diamonds, or 5 clubs be selected from a standard bridge deck of 52 cards?
10. In how many different ways can a hand consisting of 3 aces and 2 cards that are not aces be selected from a standard bridge deck of 52 cards?
11. A combination of 3 balls is picked at random from a box containing 5 red, 4 white, and 3 blue balls. In how many ways can the set chosen contain at least 1 white ball?
12. In Problem 11, in how many ways can the set chosen contain at least 1 white and 1 blue ball?
13. A set of 5 distinct points lies on a circle. How many inscribed triangles can be drawn having all their vertices in this set?
14. A set of 10 distinct points lies on a circle. How many inscribed quadrilaterals can be drawn having all their vertices in this set?
15. A set of 10 distinct points lies on a circle. How many inscribed hexagons can be drawn having all their vertices in this set?
16. Given $C_{n,3} = C_{50,47}$, find n.
17. Given $C_{n,7} = C_{n,5}$, find n.
18. Show that $C_{n,r} = C_{n,n-r}$.

Applications

19. *Biology* Out of a collection of 15 mice, 10 must be selected for an experiment. In how many ways can this be done?
20. Five flasks contain a culture of virus A and 6 flasks contain a culture of virus B. Three flasks of the A virus and 4 flasks of the B virus are to be injected with an experimental serum. In how many ways can this be done?
21. *Business* A car rental agency has a fleet consisting of 10 compact cars and 10 large sedans. In how many ways can a reservation for 2 compact cars and 3 sedans be filled?
22. A board of trustees of a company consists of 10 persons. For a business meeting to be conducted, at least 6 members of the board must be present. How many member combinations are there that would enable a business meeting to be held?
23. *Social Science* A sociology class of 16 students is divided into 3 groups: 3 students in group A, 5 students in group B, and 8 students in group C. The instructor wishes to organize a discussion panel of 5 students by choosing 1 student from group A, 2 from group B, and 2 from group C. In how many ways can the panel be formed?

10.3

PROBABILITY FUNCTIONS

SAMPLE SPACES, OUTCOMES, AND EVENTS

When an experiment of some kind is undertaken, associated with the experiment is a set of possible results. For example, when a die is rolled, it will come to a stop with the number of spots on its upper face corresponding to one of the numerals 1, 2, 3, 4, 5, 6. This exhausts all possibilities. If the die has not been tampered with, exactly which one of these random results will occur cannot be precisely specified in advance of the experiment. A question of great practical importance regarding the result of casting a die, and the result of any comparable experiment in which the outcome is uncertain, is "Can we assign some kind of measure to the degree of uncertainty involved?" The answer is "Yes," but before we assign a measure, let us define some necessary terms.

> The set of all possible results of an experiment is called a **sample space** for the experiment.
>
> Each element of a sample space is called an **outcome**, or **sample point**.
>
> Any subset of a sample space is called an **event**, and is commonly denoted by the letter E.

The reason for the terminology in the last sentence is that, in conducting an experiment, one may be interested in sets of outcomes rather than in individual outcomes. In the tossing of a die, for example, if the sample space is taken as $\{1, 2, 3, 4, 5, 6\}$, then the event that an outcome (a numeral) denotes an even integer is the set $\{2, 4, 6\}$, which is a subset of the sample space. The event that an outcome denotes an odd integer is the set $\{1, 3, 5\}$. These two events are complements of each other, and are examples of **complementary events**, that is, two events whose intersection is $\emptyset$ and whose union is the entire sample space.

PROBABILITY

We can define a *probability function* P on a sample space S by means of either a priori or a posteriori considerations, as follows.

A priori considerations involve physical, geometric, and other inherent properties of the experiment in question. They involve no sampling of outcomes. One way to assign a probability to an event is simply to use the ratio of the number of outcomes in the event to the number of possible outcomes. Thus, when a die is cast and we admit as outcomes the die's stopping with any of its 6 different faces uppermost, then without making any trial throws we would assign the value 1/6 as the probability of each of the 6 possible outcomes.

More generally, we have the following.

10.3 Probability Functions

> *If E is any subset containing n(E) members (outcomes) of a sample space containing n(S) equally likely outcomes, then the **probability** of the occurrence of E, P(E), is given by*
>
> (1) $$P(E) = \frac{n(E)}{n(S)}.$$

Example If 2 dice are cast, what is the (a priori) probability that the sum of the number of dots showing on the top faces of the dice is less than 6?

Solution For our sample spaces, let us consider A the set of possible outcomes for one die and B the set for the other. Then

$$n(A) = 6 \quad \text{and} \quad n(B) = 6.$$

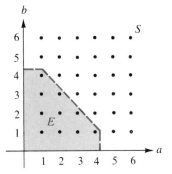

The possible outcomes for both would be $S = A \times B$, the Cartesian product of A and B, and $n(A \times B) = n(S) = 36$. Each outcome here is an ordered pair (a, b), where a is the numeral on the upper face of the first die, and b is the numeral on the second die. The event we seek is $\{(a, b) \mid a + b < 6\}$. The lattice in the figure shows the situation schematically. Since

$$E = \{(1, 1), (1, 2), (1, 3), (1, 4), (2, 1), (2, 2), (2, 3), (3, 1), (3, 2), (4, 1)\},$$

we have $n(E) = 10$, and

$$P(E) = \frac{n(E)}{n(S)} = \frac{10}{36} = \frac{5}{18}.$$

Care must be taken in interpreting the meaning of "the probability of an event." In the above example, 5/18 does not assure us, for instance, that E will occur 5 times out of 18 casts, or, indeed, that even 1 cast out of 18 will produce the event described. What it does imply, however, is that if you cast the dice a very great number of times, then you can *expect* the sum on the exposed faces to be less than 6 about 5/18 of the time.

A posteriori considerations involve testing the experiment a certain number of times. Mortality tables give probability functions of this sort. Actually, (1) can still be used in defining such probability functions, provided we interpret the function $n(E)$ as being the number of times the event E occurred in the test and $n(S)$ as being the total number of times the experiment was performed in the test.

Although diagrams are often useful in illustrating an event and a sample space, the number of outcomes $n(E)$ in the event and the number of possible outcomes $n(S)$ in the sample space ordinarily are computed directly.

Example If 3 marbles are drawn at random from an urn containing 6 white marbles and 4 blue marbles, what is the probability that the 3 marbles are all white?

Solution Since the number of ways of drawing 3 white marbles from 6 white marbles is $C_{6,3}$ and the number of ways of drawing 3 marbles from the total number of marbles is $C_{10,3}$, we have

$$P(E) = \frac{n(E)}{n(S)} = \frac{C_{6,3}}{C_{10,3}} = \frac{\frac{6!}{3!3!}}{\frac{10!}{3!7!}} = \frac{1}{6}.$$

Example In the preceding example, what is the probability that of the 3 marbles drawn 2 are white and 1 is blue?

Solution The number of ways that 2 white marbles and 1 blue marble can be drawn is

$$C_{6,2} \cdot C_{4,1} = 60.$$

Hence, in this case we have

$$P(E) = \frac{C_{6,2} \cdot C_{4,1}}{C_{10,3}} = \frac{60}{120} = \frac{1}{2}.$$

EXERCISE 10.3

1. A die is cast. List the outcomes in the sample space. List the outcomes in the event E that the number on the upper face of the die is greater than 2. Determine $P(E)$.

2. A coin is tossed. List the outcomes in the sample space. List the outcomes in the event E that a head appears. Determine $P(E)$.

3. Two coins are tossed. List the outcomes in the sample space. List the outcomes in the event E that both coins show the same face. Determine $P(E)$.

4. A die is cast and a coin is tossed. List the outcomes in the sample space. List the outcomes in the event E that the coin shows a head and the die shows a numeral greater than 4. Determine $P(E)$.

10.3 Probability Functions

In Problems 5–10, consider an experiment in which two dice are cast. Determine the probability that the specified event occurs.

5. The sum of the numbers of dots shown is 7.
6. The sum of the numbers of dots shown is 8.
7. The sum of the numbers of dots shown is 3.
8. The sum of the numbers of dots shown is 12.
9. At least one of the numbers of dots shown is less than 3.
10. Both dice show the same number of dots.

In Problems 11–14, consider an experiment in which a single card is drawn at random from a standard pack of 52 cards. Determine the probability that the specified event occurs.

11. The card is the king of hearts.
12. The card is an ace.
13. The card is a black face card.
14. The card is a 5, 6, 7, or 8.

In Problems 15–18, consider an experiment in which 2 cards are drawn at random from a standard pack of 52 cards. Determine the probability that the specified event occurs.

15. Both cards are spades.
16. Both cards are red.
17. Both cards are face cards.
18. Both cards are aces.

In Problems 19–22, consider an experiment in which 2 marbles are drawn at random from an urn containing 8 red, 6 blue, 4 green, and 2 white marbles. Determine the probability that the specified event occurs.

19. Both marbles are white.
20. Both marbles are green.
21. Both marbles are blue.
22. Both marbles are red.

23. Four marbles are drawn at random from an urn containing 5 red marbles and 5 white marbles. What is the probability that 2 of the marbles are red and 2 are white?
24. Four marbles are drawn at random from an urn containing 5 red marbles, 3 white marbles, and 2 black marbles. What is the probability that 2 of the marbles are white, 1 is red, and 1 is black?

Applications

25. *Biology* A husband and wife wish to have 3 children, one at a time over a number of years. Determine the sample space describing the possible combinations of sexes that can occur in 3 successive single childbirths. (We assume here that the birth of each sex is equally likely.)
26. Use the information in Problem 25 to determine the probability of the couple having, in order, a boy, followed by a girl, followed by another boy.
27. In a set of 8 mice, 2 are known to be diseased. If 3 mice are chosen at random for an experiment, what is the probability that one of the chosen mice is diseased?

28. *Business* A grievance committee consists of 10 union members: 7 men and 3 women. If 4 are chosen at random to form a subcommittee, what is the probability that all the subcommittee members will be men?

29. In Problem 28, what is the probability that the subcommittee will consist of 2 men and 2 women?

30. A company has 5 draftsmen and 10 machinists with the same seniority. It is determined that 8 of these employees, chosen at random, must be laid off at a certain time. What is the probability that all 8 of these employees will be machinists.

31. What is the probability that all 3 numbers on a California license plate will be the same positive even integer?

32. What is the probability that a California license plate will be 111 ZZZ? (see Problem 47, Exercise 10.1).

33. *Social Science* In a state gubernatorial election there are 2 candidates for governor (Democrat and Republican) and 3 candidates for lieutenant governor (Democrat, Republican, and Peace and Freedom). It is projected that each candidate is equally likely to win. What is the probability of a Democratic governor and a Republican lieutenant governor being elected?

34. The following table shows the results of a preelection poll concerning the attitudes of 1500 declared voters toward candidate X.

	Will vote in favor of candidate X	Will vote against candidate X
Persons under 25	100	500
Persons over 25	600	300

What is the probability that

a. someone who will vote in favor of candidate X is over 25?

b. someone under 25 will vote in favor of candidate X?

c. someone who will vote against candidate X is under 25?

d. someone will vote in favor of candidate X?

e. someone over 25 will vote against candidate X?

f. someone who will vote will also be under 25?

10.4

PROBABILITY OF THE UNION OF EVENTS

PROBABILITY OF THE COMPLEMENT OF AN EVENT

If we denote the complement of an event E in a sample space S by E', then $P(E')$ denotes the probability of the occurrence of E'. Since an outcome in S must lie in either E or E', but not both, and since $P(S) = 1$, it follows that

$$P(E) + P(E') = 1,$$

or

$$P(E') = 1 - P(E).$$

We can often use this fact to simplify the computation of a probability.

Example If 2 marbles are drawn at random from an urn containing 6 white, 2 red, and 5 green marbles, what is the probability that not both are white, that is, that at least 1 is not white?

Solution Rather than consider all the possible pairs of marbles in which not both are white, we simply compute the probability of the event E that they *are* both white and subtract the result from 1. We have

$$P(E) = \frac{C_{6,2}}{C_{13,2}} = \frac{15}{78} = \frac{5}{26}.$$

Then

$$P(E') = 1 - \frac{5}{26} = \frac{21}{26},$$

and this is just the probability that not both marbles are white.

The ratio of the probability of an event to the probability of its complement is given a special name.

> The **odds** that an experiment with sample space S will result in an event E are given by
>
> (1) $$\frac{P(E)}{P(E')} = \frac{P(E)}{1 - P(E)} \qquad (P(E') \neq 0).$$

Example Find the odds that the sum of the number determined by a single cast of 2 dice will be less than 6.

Solution From the example on page 349, the probability of the event is 5/18. Hence,

Then, by (1),
$$P(E') = 1 - P(E) = 1 - \frac{5}{18} = \frac{13}{18}.$$

$$\frac{P(E)}{P(E')} = \frac{\frac{5}{18}}{\frac{13}{18}} = \frac{5}{13},$$

and the odds that the sum of the numbers will be less than 6 are 5/13, or 5 to 13.

Example A textbook editor feels that the odds that a calculus text will succeed are 4 to 10. Determine the probability $P(E)$ that the text will succeed.

Solution From Definition 10.7, we have
$$\frac{P(E)}{1 - P(E)} = \frac{4}{10}.$$

Solving for $P(E)$ gives
$$P(E) = 0.4 - 0.4 P(E)$$
$$1.4 P(E) = 0.4$$
$$P(E) = \frac{0.4}{1.4}$$
$$= \frac{2}{7}.$$

PROBABILITY OF THE UNION OF EVENTS

By counting principle II on page 338, if E_1 and E_2 are events in a sample space S, then
$$n(E_1 \cup E_2) = n(E_1) + n(E_2) - n(E_1 \cap E_2).$$
This leads directly to the following theorem.

> If S is a sample space, and E_1 and E_2 are any events in S, then
> (2) $P(E_1 \text{ or } E_2) = P(E_1 \cup E_2) = P(E_1) + P(E_2) - P(E_1 \cap E_2).$

Example If 2 cards are drawn from a standard deck of playing cards, what is the probability that either both are red or both are jacks?

Solution A deck of cards contains 52 cards, and an outcome here consists of 2 cards. Hence, the number of elements of the sample space is the number of ways (combinations) one can draw 2 cards from 52, which is $\binom{52}{2}$. Let E_1 be the event that both are red. Since there

10.4 Probability of the Union of Events

are 26 red cards in a deck, the number of outcomes (combinations) in the event E_1 is

$$n(E_1) = \binom{26}{2}.$$

Let E_2 be the event that both cards are jacks. Then, because there are 4 jacks,

$$n(E_2) = \binom{4}{2}.$$

Since there is only 1 red pair of jacks, $n(E_1 \cap E_2) = 1$. We then have from (2),

$$P(E_1 \cup E_2) = P(E_1) + P(E_2) - P(E_1 \cap E_2)$$

$$= \frac{C_{26,2}}{C_{52,2}} + \frac{C_{4,2}}{C_{52,2}} - \frac{1}{C_{52,2}}$$

$$= \frac{C_{26,2} + C_{4,2} - 1}{C_{52,2}}$$

$$= \frac{\frac{26 \cdot 25}{1 \cdot 2} + \frac{4 \cdot 3}{1 \cdot 2} - 1}{\frac{52 \cdot 51}{1 \cdot 2}} = \frac{325 + 6 - 1}{1326} = \frac{330}{1326} = \frac{165}{663} = \frac{55}{221}.$$

PROBABILITY OF THE UNION OF DISJOINT EVENTS

Of course, if E_1 and E_2 are **disjoint**, then $E_1 \cap E_2 = \emptyset$, so that $P(E_1 \cap E_2) = 0$, and then (2) reduces to

(3) $$P(E_1 \cup E_2) = P(E_1) + P(E_2).$$

Disjoint events are said to be **mutually exclusive**.

Example A card is drawn at random from a standard deck of 52 cards. What is the probability that the card is either a face card (jack, queen, or king) or a 4?

Solution Let E_1 be the event that the card is a 4. There are 4 such cards in a deck, so that $n(E_1) = 4$. Let E_2 be the event that the card is a jack, queen, or king. There are 12 such cards in a deck. Hence, $n(E_2) = 12$. Since the sample space is just the entire deck, $n(S) = 52$, and since E_1 and E_2 are mutually exclusive,

$$P(E_1 \cup E_2) = P(E_1) + P(E_2)$$

$$= \frac{4}{52} + \frac{12}{52} = \frac{16}{52} = \frac{4}{13}.$$

Therefore, the probability is 4/13.

EXERCISE 10.4

Two dice are cast. Let E be the event that both dice show the same numeral. Let F be the event that the sum of the numbers thrown is greater than 8. Find each probability.

1. $P(E)$
2. $P(F)$
3. $P(E \cup F)$
4. $P(E')$
5. $P(F')$
6. $P(E' \cup F')$

A box contains 5 red, 4 white, and 3 blue marbles; 2 marbles are drawn from the box. Let RR be the event that both marbles are red, WW that both marbles are white, BB that both marbles are blue, and RW, RB, BW that a red and a white, a red and a blue, and a blue and a white are drawn, respectively. Find each probability.

7. $P(RR)$
8. $P(BB)$
9. $P(WW)$
10. $P(RW)$
11. $P(RB)$
12. $P(BW)$

13. What is the probability that neither is white?
14. What is the probability that neither is blue?
15. What is the probability that at least 1 is red?
16. What is the probability that either 1 is red or else both are white?
17. What is the probability that by drawing a single card from a standard deck of 52 cards, one will get a 2, 3, or 4?
18. What is the probability that if 2 cards are drawn from a standard deck of 52 cards they will be of the same suit? Different suits?

If the probability of the event E that a person will receive k dollars is P(E), then the person's **mathematical expectation** *relative to this event is kP(E).*

19. A lottery offers a prize of $50, and 70 tickets are sold. What is the mathematical expectation of a person who buys 3 tickets? If each ticket costs $1, is the person's expectation greater or less than his outlay?
20. The odds that a certain horse will win the Irish Sweepstakes are 2 to 7. If you hold a ticket on this horse to pay $100,000 if he wins, what is your mathematical expectation?

If E_1, E_2, E_3, etc., are mutually exclusive events, and the return to you is k_1 if E_1 occurs, k_2 if E_2 occurs, etc., then your mathematical expection is $\sum_{i=1}^{n} k_i \, P(E_i)$.

21. One coin is selected at random from a collection containing a penny, a nickel, and a dime. What is the expectation?
22. One coin is selected at random from a collection containing a dime, a quarter, and half-dollar. What is the expectation?
23. Three $1 bills and four $5 bills are hidden from view. What is the expectation on a single selection?

10.4 Probability of the Union of Events

Applications

24. Three $1 bills, four $5 bills, and one $10 bill are hidden from view. What is the expectation on a single draw?

25. *Biology* Of a control group of 100 volunteers, it is known that 20 persons have hypertension and 10 have a cardiovascular disease. Eight of the persons who have hypertension are also known to have a cardiovascular disease. If one person is chosen at random from the group, what is the probability that this person will have at least one of the diseases?

26. In Problem 25, what is the probability that the person chosen will have neither disease?

27. A mouse is selected at random from a collection consisting of 20 white mice, 10 black mice, and 20 grey mice. What is the probability that the mouse selected will be either white or black?

28. Thirty rats are infected with a virus. Five of the rats are inoculated with serum A, 10 are inoculated with serum B, and 15 are inoculated with serum C. If 2 rats are chosen at random, what is the probability that not both are inoculated with serum A?

29. *Business* A garment manufacturer produces 1000 unit lots of a given size each day. It is determined that usually 8% of the garments in a lot are too small and 6% are too large. If an inspector selects a garment at random from a lot, what is the probability that it will be the correct size?

30. When renewing its inventory of television sets, a company knows that the odds that a shipment of the sets will arrive in time to start a sale on a certain date are 6 to 9. What is the probability that the television sets will not arrive at the expected time?

31. The odds that a business venture will show a profit are 5 to 3. Determine the probability that the business will be profitable.

32. A personnel manager of a large company knows from long experience that 10 out of every 100 persons hired will quit within one week. What are the odds that a person hired will not quit within one week?

33. *Social Science* In a poll of 40 female and 60 male college students, it was found that 90% of the female students were opposed to cheating under any circumstances and 70% of the male students were so opposed. If one student is selected at random from those polled, what is the probability that this student will be opposed to cheating?

34. In a special class of 20 students, 4 are known to have an IQ above 150, 10 an IQ between 120 and 150, and 6 an IQ between 100 and 120. If 3 students are chosen at random from the class, what is the probability that at least one of the students has an IQ above 150?

35. Candidates X and Y of the same political party are running for the Senate and the House, respectively. In a survey, it was found that 60% of the people plan to vote for candidate X and 40% plan to vote for candidate Y. Also, 20% of the people indicated that they will vote for both X and Y. On the basis of this survey, what is the probability of at least one of the candidates winning?

10.5

PROBABILITY OF THE INTERSECTION OF EVENTS

In some experiments, we may be interested in events that are not dependent on each other, in the sense that the occurrence of one may have no effect on the probability of the occurrence or nonoccurrence of the other.

INDE-PENDENT EVENTS

Consider an experiment in which 2 cards are drawn at random, one after the other, from a deck of 10 cards, 6 of which are red and 4 blue. We can inquire into the probability that the first card drawn is red and the second blue. The simplest such situation would be one in which the first card is drawn, observed, and returned to the deck, which is then shuffled thoroughly before the second card is drawn. In this case, the sample space would consist of a set of ordered pairs (x, y), where x is the result of the first draw and y is the result of the second draw. Since there are 10 possibilities in each case, the sample space would consist of $10 \times 10 = 100$ ordered pairs. In the Cartesian graph of Figure 10.2, r_i and b_i are used to designate the drawing of red and blue cards, respectively.

The events E_1 that the first card drawn is a red card and E_2 that the second card drawn is a blue card are outlined in the figure. The event E that both E_1 and E_2 occur is the intersection of E_1 and E_2; that is, $E = E_1 \cap E_2$. By inspection,

$$P(E_1) = \frac{60}{100} = \frac{3}{5}, \qquad P(E_2) = \frac{40}{100} = \frac{2}{5},$$

and

$$P(E) = P(E_1 \cap E_2) = \frac{24}{100} = \frac{6}{25}.$$

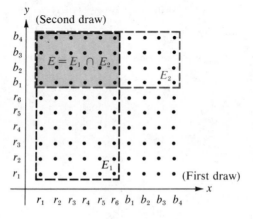

Figure 10.2

10.5 Probability of the Intersection of Events

Moreover, in this example, it is evident that

$$P(E) = P(E_1 \cap E_2) = P(E_1) \cdot P(E_2).$$

If E_1 and E_2 are events in a sample space, then E_1 and E_2 are said to be **independent events** if and only if,

(1) $\qquad P(E_1 \cap E_2) = P(E_1) \cdot P(E_2).$

If two events are not independent, then they are said to be **dependent**.

DEPENDENT EVENTS

Now consider the same experiment, except that this time the first card is not returned to the deck before the second is taken. Then there will be 10 possible first draws, but only 9 possible second draws. The sample space will therefore contain 10×9 ordered pairs, such that no ordered pair with first and second components the same remains in the set.

Figure 10.3 shows a graph of the sample space, which is the same as that in Figure 10.2 except that one diagonal is missing. Sets with graphs that have a missing diagonal are called **deleted Cartesian sets**. Again, the figure shows E_1 and E_2, the events that a red and a blue are obtained on the first and second draw, respectively. The event that both occur is $E = E_1 \cap E_2$, which is also shown in the figure. By inspection,

$$P(E_1) = \frac{54}{90} = \frac{3}{5}, \qquad P(E_2) = \frac{36}{90} = \frac{2}{5},$$

and

$$P(E) = P(E_1 \cap E_2) = \frac{24}{90} = \frac{4}{15}.$$

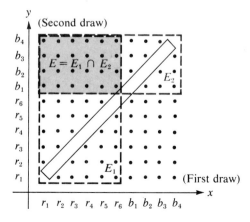

Figure 10.3

This time,
$$P(E_1 \cap E_2) \neq P(E_1) \cdot P(E_2),$$
so events are dependent.

Of course, (1) can be extended to include more than two events. For example, if E_1, E_2, and E_3 are independent events, then $P(E_1 \cap E_2 \cap E_3) = P(E_1) \cdot P(E_2) \cdot P(E_3)$. However, it should be observed that the multiplicative property $P(E_1 \cap E_2 \cap E_3) = P(E_1) \cdot P(E_2) \cdot P(E_3)$ may hold for three or more events and yet the events may be dependent (see Problem 16).

Example A die is rolled 3 times. What is the probability of throwing 3 straight sixes?

Solution The outcome of any one roll does not affect the outcome of another; thus, the events E_1, E_2, and E_3 that a six will occur on the first, second, and third roll are independent events. Since
$$P(E_1) = P(E_2) = P(E_3) = \frac{1}{6}.$$
we have
$$P(E_1 \cap E_2 \cap E_3) = P(E_1) \cdot P(E_2) \cdot P(E_3)$$
$$= \frac{1}{6} \cdot \frac{1}{6} \cdot \frac{1}{6} = \frac{1}{216}.$$

Example A family has 3 children, born in succession. What is the probability that at least one of them is a girl?

Solution We assume that the birth of either sex is equally likely. Let E_1, E_2, and E_3 denote the events of the mother having a boy on the first, second, and third births, respectively. Since the events are independent, the probability that the three successive births will all produce boys is
$$P(E_1 \cap E_2 \cap E_3) = P(E_1) \cdot P(E_2) \cdot P(E_3) = \frac{1}{2} \cdot \frac{1}{2} \cdot \frac{1}{2} = \frac{1}{8}.$$

If E_4 denotes the event of this *not* happening, that is, of the birth of at least one girl, then
$$P(E_4) = 1 - \frac{1}{8} = \frac{7}{8}.$$

Example In a class of 25 students, what is the probability that at least 2 of the students have the same birthday?

Solution Let us calculate the probability of the event E that none of the 25 students have the same birthday. If one student is chosen, the probability that he or she has a birthday is 365/365. The probability that a second student does not have the same birthday as the first is 364/365. The probability that a third student does not have either of the birthdays of the first and second student is 363/365, and so on. Since the events of the individual birthdays are independent, we have

$$P(E) = \frac{365}{365} \cdot \frac{364}{365} \cdot \frac{363}{365} \cdots \frac{342}{365} \cdot \frac{341}{365} = 0.43.$$

Thus, the probability that at least 2 students in the class have the same birthday is

$$P(E') = 1 - P(E) = 0.57.$$

EXERCISE 10.5

1. A bag contains 4 red marbles and 10 blue marbles. If 2 marbles are drawn in succession, and if the first is not replaced, what is the probability that the first is red and the second is blue? Are the 2 draws independent events?

2. A red die and a green die are cast. What is the probability of obtaining a sum greater than 9, given that the green die shows 4?

3. A bag contains 4 white and 6 red marbles. Two marbles are drawn from the bag and replaced, and 2 more marbles are then drawn from the bag. What is the probability of drawing
   a. 2 red marbles on the first draw and 2 white ones on the second draw?
   b. a total of 2 white marbles?
   c. 4 white marbles?
   d. 4 red marbles?

4. In Problem 3, what is the probability of drawing
   a. exactly 3 white marbles in the 2 draws?
   b. at least 3 white marbles in the 2 draws?
   c. exactly 2 red marbles in the 2 draws?
   d. at least 2 red marbles in the 2 draws?

5. A red and a green die are cast. Let E_1 be the event that at least 1 die shows 3 and E_2 be the event that the sum of the 2 numbers thrown is 8.
   a. Find $P(E_1)$.   b. Find $P(E_2)$.
   c. Are E_1 and E_2 independent?

6. In Problem 5, let E_1 be the event that neither die shows a result larger than 4 and let E_2 be the event that the dice do not show the same number.
   a. Find $P(E_1)$.   b. Find $P(E_2)$.
   c. Are E_1 and E_2 independent?

7. A coin is tossed 3 consecutive times. What is the probability that

a. the second toss is a head?
b. the third toss is a head?
c. both the second and third tosses are heads?
d. the first and third tosses are heads?
e. the first and third tosses are heads but the second is not?

8. In Problem 7, state whether each of the following pairs of events is independent.
   a. parts a and b   **b.** parts a and c   **c.** parts a and d
   d. parts a and e   **e.** parts d and e

9. Of 2 dice, 1 is normal, but the other has 3 faces showing 4 spots and 3 faces showing 3 spots. If a die is chosen at random and a 4 is thrown with it, what is the probability that the normal die was chosen?

10. Two identical urns contain marbles. One urn contains all red marbles, while half of the contents of the other urn are white and half are red. If a marble is drawn at random from one of the urns and found to be red, what is the probability that it was drawn from the urn containing only red marbles?

11. One box contains 3 red and 8 white marbles, and a second box contains 5 red and 2 white marbles. If 1 marble is drawn from each box, what is the probability of drawing
    a. 2 red marbles?   **b.** 2 white marbles?   **c.** 1 red and 1 white marble?

12. In Problem 11, what are the odds of drawing
    a. 2 red marbles?   **b.** 2 white marbles?   **c.** 1 red and 1 white marble?

13. A coin is tossed 4 times. What is the probability that 2 heads are tossed followed by 2 tails?

14. Argue that if E_1 and E_2 are mutually exclusive events with nonzero probabilities, then E_1 and E_2 are not independent.

15. A bag contains 4 red marbles and 2 green marbles. A die is tossed, a coin is flipped, and a marble is selected from the bag. What is the probability that consecutively a 6 is tossed, a head is flipped, and a green marble is selected?

Three events E_1, E_2, and E_3 of a sample space are independent if and only if $P(E_1 \cap E_2 \cap E_3) = P(E_1) \cdot P(E_2) \cdot P(E_3)$ and $P(E_1 \cap E_2) = P(E_1) \cdot P(E_2)$, $P(E_2 \cap E_3) = P(E_2) \cdot P(E_3)$, and $P(E_3 \cap E_1) = P(E_3) \cdot P(E_1)$.

16. A box contains marbles which are labeled 1, 2, 3, 4, 5, 6, 7, 8. One marble is selected at random from the box. Suppose E_1 consists of the event of selecting a marble labeled 1, 2, 3, 4, and events E_2 and E_3 consist of selecting marbles labeled 1, 4, 5, 6 and 1, 6, 7, 8, respectively. Show that $P(E_1 \cap E_2 \cap E_3) = P(E_1) \cdot P(E_2) \cdot P(E_3)$, but that the events are not independent.

Applications

17. *Biology* In 3 successive births, what is the probability that a boy, girl, boy will occur in that order?

18. The probability of patient A having a side reaction to a drug is 0.5 and the probability that patient B will have a side reaction to the same drug is 0.2. What is the probability that both A and B will have no side reaction to the drug?

19. For man A, who smokes, it is estimated that the probability that he will live to age 65 is 0.3,

10.6 Binomial Probability

whereas for man B, who does not smoke, the probability of living to age 65 is 0.82. What is the probability that A will live to 65 but B will not?

20. *Business* The probability that worker A will learn a job in one week is 4/5. The probability that workers B and C will learn their jobs in the same time is 5/9 and 2/3, respectively. What is the probability that all 3 will learn their jobs in one week?

21. In Problem 20, what is the probability that at least one of the workers will not learn the job in one week?

22. In Problem 20, what is the probability that worker A will learn a job in one week but neither B nor C will learn their jobs in the same time?

23. A stamping machine produces license plates. For a given day, it is observed that usually 5% of the plates are misstamped. Assuming that the successive stampings are independent events, what is the probability that 2 license plates in a row are not misstamped?

24. A company manufactures two-engine airplanes. There is a 0.004 probability that one engine may fail in flight. What is the probability of one or the other engine failing in flight? What is the probability that one or the other engine will not fail in flight? What is the probability that both engines will not fail in flight?

25. *Social Science* The probability that student A will pass a course in psychology is 5/6, that student B will pass is 3/4, and that student C will pass is 2/3. What is the probability
   a. at least 1 of the 3 will pass?
   b. at least A and C will pass?
   c. A and C will pass but B will not?
   d. At least 2 of the 3 will pass?

26. In Problem 25, are the events of parts a and c independent?

27. A day is selected at random in some fashion such that any day of the week is an equally likely choice. Let the probability be 1/30 that a day selected at random will be a rainy day.
   a. What is the probability that a rainy Wednesday will be selected?
   b. What is the probability that a dry Thursday will be selected?
   c. What is the probability that either Monday, Tuesday, or Wednesday will be selected, and that it will not rain that day?

28. In Problem 27, let E_1 be the selection of Sunday, and let E_2 be the event that it does not rain on the day selected. Are E_1 and E_2 independent?

10.6
BINOMIAL PROBABILITY

BERNOULLI TRIALS   In a single toss of a coin, the coin either comes up heads or it does not. In a single throw of a die, either a designated number between one and six appears or it does not. In sampling an item from a production line, it either is found to be defective or it is not. Experiments in which there are two possible outcomes, success E or failure E', are sometimes referred to as **Bernoulli trials**. We are interested in calculating the probability of a specific number of successes occurring within a sequence of n independent Bernoulli trials. To this end, we denote the probability of success by

$$P(E) = p$$

and the probability of failure by

$$P(E') = 1 - p = q.$$

Consider the experiment of throwing a die 4 times. Suppose we wish to determine the probability of throwing a five *exactly* 2 times out of the 4 tosses.

If ☐ ☐ ☐ ☐ denotes the 4 possible outcomes of the tosses and 5' denotes the outcome of obtaining any number other than the number five, then 2 successes can occur in the following ways:

(1) [5] [5] [5'] [5'] (2) [5] [5'] [5'] [5]
(3) [5'] [5'] [5] [5] (4) [5'] [5] [5] [5']
(5) [5] [5'] [5] [5'] (6) [5'] [5] [5'] [5]

The probability that a five will occur on one toss is $p = 1/6$ and so the probability that it will not occur is $q = 5/6$. Since the outcome of any one toss is independent of the others, the probabilities corresponding to (1), (2), (3), (4), (5), and (6) are

(1') $\dfrac{1}{6} \cdot \dfrac{1}{6} \cdot \dfrac{5}{6} \cdot \dfrac{5}{6}$ (2') $\dfrac{1}{6} \cdot \dfrac{5}{6} \cdot \dfrac{5}{6} \cdot \dfrac{1}{6}$

(3') $\dfrac{5}{6} \cdot \dfrac{5}{6} \cdot \dfrac{1}{6} \cdot \dfrac{1}{6}$ (4') $\dfrac{5}{6} \cdot \dfrac{1}{6} \cdot \dfrac{1}{6} \cdot \dfrac{5}{6}$

(5') $\dfrac{1}{6} \cdot \dfrac{5}{6} \cdot \dfrac{1}{6} \cdot \dfrac{5}{6}$ (6') $\dfrac{5}{6} \cdot \dfrac{1}{6} \cdot \dfrac{5}{6} \cdot \dfrac{1}{6}.$

The sum of (1'), (2'), (3'), (4'), (5'), and (6') would give the probability P that a five will occur exactly 2 times. But each of the 6 listed probabilities has the same numerical value, and so

(7) $$P = 6 \left(\dfrac{1}{6}\right)^2 \left(\dfrac{5}{6}\right)^2$$

$$= \dfrac{150}{1296}.$$

Observe that (7) is really a result of a combination problem. The fact that 2 fives can be distributed throughout the 4 tosses in 6 different ways is simply a consequence of our being able to fill the boxes in the sequence ☐ ☐ ☐ ☐ with a five in $C_{4,2} = 4!/2!2! = 6$ (the combination of 4 things taken 2 at a time) number of ways. Thus, (7) can be written

$$P = C_{4,2} \left(\dfrac{1}{6}\right)^2 \left(\dfrac{5}{6}\right)^2.$$

10.6 Binomial Probability

Were the problem changed to read: "Find the probability of throwing exactly 3 fives in the 4 tosses of the die," then we note that the 4 boxes can be filled with a five in $C_{4,3} = 4!/3!1! = 4$ number of ways and that 5' would appear only once in the list of outcomes. In this case, the probability would be

$$P = C_{4,3} \left(\frac{1}{6}\right)^3 \frac{5}{6}$$

$$= \frac{20}{1296}.$$

The foregoing discussion illustrates the following general result.

If p is the probability that a success will occur in a single Bernoulli trial, then the probability P that exactly r successes will occur in n independent Bernoulli trials is

(8) $$P = C_{n,r}\, p^r\, q^{n-r}$$

where $q = 1 - p$.

Recall, that $C_{n,r}$ is another symbol for the binomial coefficient $\binom{n}{r}$, hence (8) is called a **binomial probability**.

Example What is the probability of throwing exactly 3 fours in 5 tosses of a die?

Solution The probability of a four appearing on one throw of a die is $p = 1/6$ and therefore $q = 5/6$. Hence, the probability of exactly 3 fours occurring in 5 tosses is obtained from (8),

$$P = C_{5,3} \left(\frac{1}{6}\right)^3 \left(\frac{5}{6}\right)^2$$

$$= \frac{5!}{3!2!} \cdot \frac{1}{216} \cdot \frac{25}{36}$$

$$= \frac{250}{7776}.$$

Example A company estimates that one out of every 200 lightbulbs that it manufactures will be defective. If an inspector chooses 10 lightbulbs at random from a production line, what is the probability that exactly one is defective?

Solution The probability of choosing a defective bulb is $p = 1/200$ and so the probability of choosing a bulb that is not defective is $q = 199/200$. Hence, the desired probability is

$$P = C_{10,1} \left(\frac{1}{200}\right)^1 \left(\frac{199}{200}\right)^9$$

$$= \frac{10!}{1!9!} \cdot \frac{1}{200} \cdot \frac{199}{200} \cdot \frac{199}{200} \cdots \frac{199}{200}$$

$$\approx 0.05.$$

Example A coin is tossed 4 times. What is the probability that heads will occur at least twice?

Solution The probability that heads will occur exactly 2 times is

$$P_1 = C_{4,2} \left(\frac{1}{2}\right)^2 \left(\frac{1}{2}\right)^2$$

$$= \frac{4!}{2!2!} \cdot \frac{1}{4} \cdot \frac{1}{4}$$

$$= \frac{3}{8}.$$

The probability that heads will occur exactly 3 times is

$$P_2 = C_{4,3} \left(\frac{1}{2}\right)^3 \left(\frac{1}{2}\right)$$

$$= \frac{4!}{3!1!} \cdot \frac{1}{8} \cdot \frac{1}{2}$$

$$= \frac{1}{4}.$$

The probability that heads will occur exactly 4 times is

$$P_3 = C_{4,4} \left(\frac{1}{2}\right)^4 \left(\frac{1}{2}\right)^0$$

$$= \frac{4!}{4!0!} \cdot \frac{1}{16} \qquad (0! = 1)$$

$$= \frac{1}{16}.$$

Thus, the probability of heads occurring at least twice is

$$P = P_1 + P_2 + P_3$$

$$= \frac{3}{8} + \frac{1}{4} + \frac{1}{16}$$

$$= \frac{11}{16}.$$

10.6 Binomial Probability

EXERCISE 10.6

A die is thrown 4 times.

1. What is the probability of exactly one 6 occurring?
2. What is the probability of exactly 3 ones occurring?
3. What is the probability of exactly 2 threes occurring?
4. What is the probability of at least 3 ones occurring?
5. What is the probability of at least 2 threes occurring?

A coin is tossed 6 times.

6. What is the probability of exactly 4 heads occurring?
7. What is the probability of exactly 2 tails occurring?
8. What is the probability of exactly 5 tails occurring?
9. What is the probability of at least 5 heads occurring?
10. What is the probability of at least 3 tails occurring?

Applications

11. *Biology* A family consists of 5 children born in succession. Assuming that either sex is equally likely, what is the probability of the family consisting of exactly 2 boys and 3 girls?
12. In Problem 11, what is the probability of the family consisting of at least 3 boys?
13. It is estimated that a vaccine has a side effect in one out of every 10 persons innoculated. In a random sample of 6 innoculated persons, what is the probability that exactly 2 will be affected by the vaccine?
14. In classification by blood types, it is estimated that only 2% of all people have blood type B −. If 4 people are chosen at random from a large population, what is the probability that exactly one person will have B − blood? What is the probability that at least one person out of the 4 will have this blood type?
15. *Business* A salesperson has a record that indicates that she is successful in making a sale in 2 out of 3 attempts. What is the probability that she will make exactly 3 sales in the next 4 attempts?
16. In Problem 15, what is the probability of making:
    a. 3 or more sales in 4 attempts?
    b. no sales in 4 attempts?
    c. one or more sales in 4 attempts?
17. Television station KXXX in Los Angeles estimates that 40% of the area's television viewers at 5 o'clock watch the news on KXXX. In a random sample of 6 people who watch TV at 5 o'clock, what is the probability that
    a. all 6 watch KXXX?
    b. exactly 3 watch KXXX?

c. at least 3 watch KXXX?
d. none of the 6 watch KXXX?

18. A manufacturing firm produces an item in large lots. Each lot is subject to inspection by the sampling of 10 items at random. If 3 or more defective items are discovered, the entire lot is rejected. Over a long period of time, it has been observed that usually 95% of the items in a lot are nondefective. What is the probability that a lot will pass inspection? What is the probability that a lot will be rejected?

19. *Social Science* A psychology examination consists of 20 true–false questions. If a student simply guesses on each question, what is the probability that the student will obtain a score of 80% or better?

20. An aptitude test accurately predicts the subsequent job performance of 75% of those tested. What is the probability that the test will accurately predict the performance of the next 5 people taking it? What is the probability that the test will accurately predict the performances of 3 of the next 5 people taking it? What is the probability that the test will not accurately predict the performance of one of the next 5 people taking it?

10.7

CONDITIONAL PROBABILITY

We have seen in Section 10.5 that if 2 cards are drawn successively without replacement from a deck consisting of 6 red cards and 4 blue cards, then the probability that a red card is drawn followed by a blue card is

$$P(E_1 \cap E_2) = \frac{4}{15}.$$

The probability that a red card is obtained on the first draw is $P(E_1) = 6/10 = 3/5$. If we now write 4/15 as

$$\frac{4}{15} = \frac{3}{5} \cdot \frac{4}{9},$$

we have

$$P(E_1 \cap E_2) = P(E_1) \cdot \frac{4}{9},$$

where 4/9 can be interpreted as the probability of the occurrence of E_2, *given the occurrence of* E_1. This probability is denoted by the symbol $P(E_2|E_1)$.

10.7 Conditional Probability

> If E_1 and E_2 are events in a sample space, and $P(E_1) \neq 0$, then the **conditional probability** $P(E_2 | E_1)$ of E_2 given E_1 is
>
> (1) $$P(E_2 | E_1) = \frac{P(E_1 \cap E_2)}{P(E_1)}.$$

If E_1 and E_2 are independent and $P(E_1) \neq 0$, then (1) becomes

$$P(E_2 | E_1) = \frac{P(E_1 \cap E_2)}{P(E_1)} = \frac{P(E_1) \cdot P(E_2)}{P(E_1)} = P(E_2).$$

Example In an automobile assembly plant, 60 cars of model A and 40 cars of model B are made each day. Of the total cars produced on a given day, 68% passed the final inspection. The inspection was passed by 80% of the model A's and 50% of the model B's. One car was chosen at random from the collection of cars that passed the final inspection. What is the probability that the car selected was a model A? A model B?

Solution Let E_1 and E_2, respectively, denote the selection of a model B and a model A, and let E_3 represent passing the final inspection. We have

$$P(E_3) = \frac{68}{100}.$$

The probability of being a model A that passed the inspection is

$$P(E_2 \cap E_3) = \frac{(0.8)60}{100} = \frac{48}{100}.$$

Thus,

$$P(E_2 | E_3) = \frac{P(E_2 \cap E_3)}{P(E_3)}$$

$$= \frac{\frac{48}{100}}{\frac{68}{100}} = \frac{12}{17}.$$

Now the probability of being a model B that passed is

$$P(E_1 \cap E_3) = \frac{(0.5)40}{100} = \frac{20}{100},$$

and therefore,

$$P(E_1|E_3) = \frac{P(E_1 \cap E_3)}{P(E_3)}$$

$$= \frac{\frac{20}{100}}{\frac{68}{100}} = \frac{5}{17}.$$

Note that $P(E_1|E_3)$ could also be computed by $P(E_1|E_3) = 1 - P(E_2|E_3)$.

EXERCISE 10.7

1. Find $P(E_2|E_1)$ in Problem 5 of Exercise 10.5.
2. Find $P(E_2|E_1)$ in Problem 6 of Exercise 10.5.
3. What is the conditional probability of E_2 given the occurrence of E_1 in Problem 28, of Exercise 10.5?
4. Two cards are drawn at random from a deck of 52 cards. If the first card is not replaced, what is the probability that an ace is drawn by a king?
5. Two cards are drawn at random without replacement from a deck of 10 cards consisting of 6 red and 4 blue cards. What is the probability that the 2 cards are red and blue?
6. A coin is tossed 4 consecutive times. What is the conditional probability that a head occurs on the third toss if it is known that at least 1 tail occurs in the 4 tosses?
7. A deck of 10 cards consists of the following; 4 red and white cards, 2 red and green cards, 1 blue and white card, and 3 blue and green cards. A person selects 1 card at random and reveals to a second person only the fact that 1 of the 2 colors on the card is white. What probability should the second person assign to the card also being red?
8. A box contains red and blue marbles. Two marbles are taken successively out of the box without replacement. If the probability of drawing 2 red marbles in a row is 1/24 and if the probability of drawing a red marble on the first draw is 1/6, then what is the probability of drawing a red marble on the second draw?

Equation (1) can be extended to three events by the following:

$$P(E_1 \cap E_2 \cap E_3) = P(E_1) \cdot P(E_2|E_1) \cdot P(E_3|E_1 \cap E_2).$$

9. Three cards are drawn successively from a normal deck. Assuming that the cards are not replaced in the deck, what is the probability that all the cards will be hearts? All red?
10. Show that $P(E_2) \cdot P(E_1|E_2) = P(E_1) \cdot P(E_2|E_1)$.

Applications
11. *Biology* In collection A of 20 test tubes, it is known that 5 contain a specific bacteria

10.7 Conditional Probability

culture. In collection B of 40 test tubes, 10 contain the same bacteria. One test tube is chosen at random. What is the probability that it contains bacteria and is from collection B?

12. In a large group of males, it is found that 45% are overweight and 30% have hypertension. It is also determined that 25% are both overweight and have hypertension. If one man, chosen at random from the group, has hypertension, what is the probability that he is overweight?

13. In Problem 12, if the man chosen is overweight, what is the probability that he has hypertension?

14. In Problem 12, what is the probability that the man chosen either is overweight or has hypertension?

15. *Business* The probability that a company advertises is 0.9. Also, the probability that a company shows a profit of more than $100,000 per year is 0.65. If a company advertises, then the probability that it makes a yearly profit of more than $100,000 is 0.7. What is the probability that a company advertises if it is known that it makes more than $100,000 in profit each year?

16. A store receives a shipment of 1000 items of merchandise. As a rule for a shipment this size, there are 10 broken items. Two items are chosen at random without replacement.
    a. What is the probability that both items are not broken?
    b. What is the probability that both items are broken?
    c. What is the probability that the first item is not broken and the second is broken?

17. Two hundred light bulbs, incandescent (I) and fluorescent (F), are tested and are found to be either acceptable (A) or defective (D). One bulb is drawn at random from this lot. Use the information in the given table to find the probability that:

	I	F
A	75	105
D	15	5

    a. the bulb is defective.
    b. the bulb is acceptable.
    c. the bulb is fluorescent.
    d. it is an acceptable incandescent bulb.
    e. it is a defective fluorescent bulb.
    f. the bulb is defective given that it is incandescent.
    g. the bulb is incandescent given that it is defective.
    h. the bulb is acceptable given that it is fluorescent.

18. Two ranchers mix their horses to form a herd of 5000. Two thousand horses wear brand A and the remainder wear brand B. Forty percent of the brand A and 30% of the brand B horses are spotted. If one horse is chosen at random from the herd, what is the probability that it is a spotted brand A horse? What is the probability that it is a brand B horse if it is known to be spotted?

19. *Social Science* The probability that a dog will respond to a certain stimulus is 0.8. When

it responds to the stimulus there is a probability of 0.6 that it will choose dog food A and a probability of 0.3 that it will choose dog food B. What is the probability that the dog will respond to the stimulus and choose dog food A? Dog food B?

20. In a sampling of 500 students and teachers it was found that of students, 20 smoke cigarettes, 5 smoke pipes, and 200 do not smoke; whereas of teachers, 15 smoke cigarettes, 35 smoke pipes, and 225 do not smoke.
   a. If a person does not smoke, what is the probability that the person is a teacher?
   b. If a person smokes cigarettes, what is the probability that the person is a student?

21. The probability that a student will pass the final examination in a course is estimated to be 0.8. The probability that a student will fail the final examination but pass the course is 0.15. If a student knows that he has failed the final examination, what is the probability that the student has passed the course?

CHAPTER TEST

1. How many different two-digit numerals can be formed from {6, 7, 8, 9}?

2. How many different ways can 6 questions on a true-false test be answered?

3. How many four-digit numerals for positive odd integers can be formed from {3, 4, 5, 6}?

4. How many distinguishable permutations can be formed using the letters in the word *TENNIS*?

5. How many different committees of 5 persons can be formed from a group of 12 persons?

6. In how many different ways can a hand consisting of 3 spades, 5 hearts, 4 diamonds, and 1 club be selected from a standard bridge deck of 52 cards?

7. A box contains 4 red, 6 white, and 2 blue marbles. In how many ways can you select 3 marbles from the box if at least 1 of the marbles chosen is white?

In Problems 8–14, consider an experiment in which 2 cards are drawn at random from a deck of 52 cards.

8. What is the probability that both cards are clubs?

9. What is the probability that both cards are 10's?

10. What is the probability that neither card is a face card?

11. What is the probability that both cards are black?

12. What is the probability that one card is red and one is black?

13. What is the probability that one card is a face card and the other is not?

14. What is the probability that both cards are face cards or one card is a ten and the other is an ace?

15. What are the odds of randomly drawing 1 red and 1 blue marble from a box containing 4 red and 6 blue marbles?

In Problems 16 and 17, consider an experiment in which 2 cards are drawn at random from a deck of 52 cards. After the first draw, the card is replaced in the deck.

16. What is the probability that both cards are black and at least one is the ace of spades?

17. What is the probability that both cards are red and at least one is a face card?

In Problems 18 and 19, consider an experiment in which 2 cards are drawn at random from a deck of 52 cards. After the first draw, the card is not replaced in the deck.

18. What is the probability that both cards are red and one (but not the other) is a face card?

19. What is the probability that neither card is a face card and at least one is red?

20. A food-processing company sells its products in pint and quart cans. Of its daily production, 75% is in pint cans, and 2% of these are normally rejected because of flaws. If a can is selected at random from those produced on a given day, what is the probability that it will be a pint can that is also flawed?

21. In applying for a clerical job, a person must pass a typing examination. Twenty men and 40 women take the examination. Sixty percent of the men and 90% of the women pass. It is assumed that all are equally qualified for the job, so one person is selected at random from those who pass the typing examination. What is the probability that a man will be selected?

22. Consider the experiment: Draw a card from a standard deck of 52 cards, replace it, and then shuffle the deck. If this procedure is repeated 4 times, what is the probability that exactly 2 of the cards drawn are hearts? What is the probability that exactly 2 of the cards drawn are red?

11
MATHEMATICS OF FINANCE

11.1
COMPUTATION OF INTEREST

SIMPLE INTEREST When money changes hands by means of a loan, the amount paid to the lender for the use of the money is called **interest**. The interest, I, on a given amount, P, of money (the principal) for a definite number of periods t is given by the familiar formula

$$I = Prt,$$

where r is the specified rate of interest per period. The interest determined by this formula is called **simple interest**. For example, if $1000 is invested at 6% simple interest per annum for 5 years, then

$$I = 1000(0.06)(5)$$
$$= \$300.$$

The amount, S, the investor receives at the end of 5 years is

$$S = 1000 + 300$$
$$= \$1300.$$

In general, the amount received after t periods is given by

$$S = P + I$$
$$= P + Prt,$$

or,

(1) $$S = P(1 + rt).$$

Example Second mortgages on homes are often loans made at simple interest. If a bank lends $12,000 at 11% simple interest per year, then over a loan period of 15 years the borrower must repay

375

$$S = 12{,}000[1 + 0.11(15)]$$
$$= 12{,}000(2.65)$$
$$= \$31{,}800.$$

Example An investor receives a return of $7320 on an investment of $6000 for 4 years at simple interest. What is the annual rate of interest?

Solution From (1), we have
$$7320 = 6000 + 24{,}000r.$$

Thus,
$$r = \frac{1320}{24{,}000}$$
$$= 0.055.$$

The annual rate of interest is 5.5%.

COMPOUND INTEREST If, however, the interest accruing to an amount of money is periodically added to the amount, and over the next period this new total is earning interest, we say that the principal is earning **compound interest**. For example, if the sum of $1 is earning interest at a rate r per year compounded annually, the amount present, S, is

After 1 year: $S = 1 + r$,
After 2 years: $S = (1 + r) + r(1 + r) = (1 + r)^2$,
After 3 years: $S = (1 + r)^2 + r(1 + r)^2 = (1 + r)^3$,
After m years: $S = (1 + r)^m$.

For each dollar invested under such an arrangement we have $(1 + r)^m$ dollars after m years, so that P dollars invested under the same arrangement would amount to

(2) $$S = P(1 + r)^m.$$

after m years.

If the interest is compounded t times yearly, then the rate per period is r/t instead of r, where r is the stated rate per year, and the number of periods is increased to tm, so that (2) becomes

(3) $$S = P\left(1 + \frac{r}{t}\right)^{tm}.$$

Specifically, the amount of P dollars compounded **semiannually** for m years at a yearly rate of interest r will be

(3a) $$S = P\left(1 + \frac{r}{2}\right)^{2m},$$

11.1 Computation of Interest

and when compounded **quarterly**,

(3b)
$$S = P\left(1 + \frac{r}{4}\right)^{4m}.$$

The amount S is said to be the **future value** of P.

If (3) is solved for P,

(4)
$$P = S\left(1 + \frac{r}{t}\right)^{-tm},$$

then we are able to determine the amount that must be invested in order to obtain the exact amount S after m years. The initial investment P is called the **present value** of S.

While computations involving the above formulas can be handled by means of logarithms, from a practical standpoint it is more convenient to use tables. In Tables IV and V of Appendix II we have tabulated the values of $(1+i)^n$ and $(1+i)^{-n}$, respectively, where we can make the identification $i = r/t$ and $n = tm$.

Example What rate is necessary in order that $2500 compounded quarterly will amount to $4800 in 12 years?

Solution From (3b) we have

$$4800 = 2500\left(1 + \frac{r}{4}\right)^{48}.$$

We solved this equation for r in Section 5.5 and found that $r = 0.056$ or 5.6%. Now using

$$\left(1 + \frac{r}{4}\right)^{48} = 1.92$$

and Table IV with $n = 48$, we observe that 1.92 is between 1.6122 and 2.0435 so that $r/4$ is between 0.01 and 0.015. By interpolation we find $r/4 = 0.014$, so that $r = 0.056$.

Example If $10,000 is deposited in a savings account drawing 5% interest compounded semiannually, what amount is obtained at the end of 20 years?

Solution From (3a) and Table IV, we find

$$S = 10{,}000(1 + 0.025)^{40}$$
$$= 10{,}000(2.6851)$$
$$= \$26{,}851.$$

Example If $2000 is invested at 8% compounded quarterly for 10 years, find the total interest, I, earned.

Solution From (3b) and Table IV, it follows that the amount accrued in this period is

$$S = 2000(1 + 0.02)^{40}$$
$$= 2000(2.2080)$$
$$= \$4416.$$

The interest earned is the difference between the amount accrued and the amount invested. Thus,

$$I = 4416 - 2000$$
$$= \$2416.$$

In other words, the money has more than doubled in 10 years.

GIMMICKS Banks and savings and loan institutions frequently resort to eye-catching "giveaway" promotions to attract depositors. These promotions feature gifts ranging from calendars and cookbooks to color TV sets and Hawaiian vacations. Of course, for the more expensive items such as TV sets and vacations, there is a definite catch involved. A minimum amount must be left in an account for a specified time; the depositor then receives the premium in lieu of interest over this period. A news magazine recently reported that one bank offered a "free" $84,000 Rolls-Royce Silver Shadow in exchange for a principal of $160,000 left in an interest-free account for 8 years.

Example For the Rolls-Royce offer mentioned above, how much is the depositor actually paying for the car if a principal of $160,000 can draw 10% annual interest compounded semiannually for 8 years?

Solution From (3a) and Table IV we see that the future value of the $160,000 is

$$S = 160{,}000(1 + 0.05)^{16}$$
$$= 160{,}000(2.1829)$$
$$= \$349{,}264.$$

The interest, or actual cost of the car, is

$$349{,}264 - 160{,}000 = \$189{,}264.$$

It was subsequently reported that no one took advantage of the offer.

Example How much should be put into a savings account drawing 6% interest compounded quarterly in order to receive $15,000 at the end of 5 years.

Solution From (4) and Table V, we find that the present value is

$$P = 15{,}000(1 + 0.015)^{-20}$$
$$= 15{,}000(0.74247)$$
$$= \$11{,}137.05.$$

11.1 Computation of Interest

EXERCISE 11.1

Find S for the given principal, rate of simple interest, and time.

1. $4000, 8% per year, 6 years.
2. $20,000, 12% per year, 10 years.
3. $1500, 5% per 6 months, 24 months.
4. $350, 4% per quarter, 18 months.

Find the future value of the given principal for the indicated rate of compound interest and time. Use Table IV.

5. $4000, 8% per year compounded semiannually, 6 years.
6. $10,000, 10% per year compounded semiannually, 25 years.
7. $2000, 6% per year compounded annually, 10 years.
8. $2000, 6% per year compounded quarterly, 10 years.

Find the present value of the given amount for the indicated rate of compound interest and time. Use Table V.

9. $20,000, 5% per year compounded semiannually, 10 years.
10. $100,000, 10% per year compounded semiannually, 25 years.
11. $3000, 6% per year compounded quarterly, 4 years.
12. $50,000, 8% per year compounded quarterly, 11 years.

Applications

13. *Business* A man decides to lend $10,000 to a friend at simple interest. If the yearly rate of interest is 8% and if the lender wants to make a profit of exactly $4000, how many years does the friend have in which to pay back the loan?

14. What amount should one invest at $7\frac{1}{2}$% simple interest per year for 10 years in order to receive $20,000?

15. A person borrows $5000 from a friend at simple interest per year. Over a period of 10 years, the borrower agrees to pay back $11,000. What is the annual rate of simple interest?

16. In Problem 15, what is the interest that the lender earns over 10 years?

17. A person purchases an $84,000 Rolls-Royce by securing a car loan consisting of 11% simple interest per year for 8 years. What is the total amount the purchaser repays to the lender?

18. Two women, A and B, each invested $10,000 at 4% for 20 years with a bank that computed interest quarterly. If A withdrew her interest at the end of each 3-month period, but B let her investment be compounded, how much more did B earn than A over the period of 20 years?

19. If the effective rate of interest is defined to be

$$\frac{\text{Interest earned in 1 year}}{\text{Principal invested at the start of the year}},$$

determine the effective interest rate if $1000 is invested at 6% compounded quarterly.

20. A certificate of deposit valued at $1000 initially is compounded semiannually. After 16 years, the certificate is worth approximately $3000. What is the annual rate of interest?

21. At the beginning of 1960, $2000 was deposited in a savings account paying 4% compounded semiannually. Beginning in 1962 the same interest was compounded quarterly. At the start of 1970 the rate of interest was changed to 6% compounded quarterly. What was the value of the account at the start of 1975?

22. A loan of $500 is to be paid back in 2 years. Determine the interest which must be paid in addition to the principal if the yearly rate of interest is 8% compounded quarterly.

23. A long-term certificate of deposit pays 10% compounded quarterly. How much should be invested now in order to obtain a return of $50,000 in 12 years?

24. When paying off a loan early, such as a mortgage on a house, a prepayment penalty usually is assessed on the difference between the amount of the loan and its present value. A loan will amount to $30,000 in 10 years at an interest rate of 6% compounded quarterly. Determine the amount necessary to pay off the loan now if the prepayment penalty is $1\frac{1}{2}\%$.

25. A man has an annual income of $12,000 from two investments which are drawing simple interest. He has $10,000 more invested at 8% than at 6%. How much is invested at each rate?

11.2

ANNUITIES

In Section 11.1 we assumed that the principal P was not increased over a period of time except through the compounding of interest. Suppose now that the value of a savings account is increased not only by compounding interest but also by depositing an amount equal to the first deposit at the *end* of each interest drawing period. For example, if $1 is deposited monthly at the end of the month, and if i is the rate of interest per month, then the amount accrued is

After 1 payment: $S = 1$,

After 2 payments: $S = (1 + i) + 1$,

After 3 payments: $S = [(1 + i) + 1]i + [(1 + i) + 1] + 1$

$= 1 + 2i + i^2 + 1 + i + 1$

$= (1 + i)^2 + (1 + i) + 1,$

11.2 Annuities

After n payments:

(1) $$S = (1+i)^{n-1} + (1+i)^{n-2} + \cdots + (1+i) + 1.$$

Since this is a sum of a geometric progression with common ratio $1 + i$, we have from Section 9.2,

$$S = \frac{1 - (1+i)^n}{1 - (1+i)}$$

$$= \frac{(1+i)^n - 1}{i}.$$

This last quantity is sometimes denoted by the symbol $s_{\overline{n}|i}$.

Now, in general, if the same amount R is invested at n equally spaced periods of time at an interest rate of i per period, then the amount accumulated after the nth payment is

(2) $$S = R\left[\frac{(1+i)^n - 1}{i}\right].$$

This type of investment is called an **annuity**. The amount S is said to be the **future value** of the annuity.

Inspection of (1) indicates that S is simply the sum of n compound interest problems; the first deposit draws interest for $n - 1$ periods, the second for $n - 2$ periods, and so on. It should be kept in mind that the last deposit draws no interest, so formula (2) gives the value of the account immediately after the last payment is made.

Because an annuity problem involves the term $(1 + i)^n$, to solve it we could make use of the information given in Table IV, but again it is more convenient to tabulate, as in Table VI of Appendix II, the values of

$$\frac{(1+i)^n - 1}{i}.$$

Example Find the future value of an annuity consisting of $100 per month, 3/4% interest per month, for 30 months.

Solution From (2) and Table VI, we find the future value to be

$$S = 100\left[\frac{(1.0075)^{30} - 1}{0.0075}\right]$$

$$= 100(33.5029)$$

$$= \$3350.29.$$

Example An employee of a company gets paid on the first of the month, and $50 of the paycheck is deposited in a savings plan which draws interest at a rate of 1/2% per month. What is the cash value of the account at the end of 24 months?

Solution We can imagine that the first time period ends on the first of the month when the initial deposit is made. As Figure 11.1 shows, there are 24 equally spaced payments. Now, formula (2) gives the amount in the account after the twenty-fourth payment. From Table VI, we have

$$S_1 = 50 \left[\frac{(1.005)^{24} - 1}{0.005} \right]$$

$$= 50(25.4320)$$

$$= \$1271.60$$

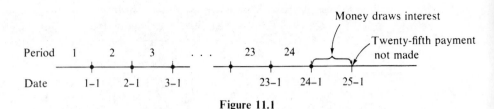

Figure 11.1

Since S_1 draws interest at $1/2\%$ for one additional month, at the end of 24 months the cash value S is

$$S = 1271.60 + (0.005)(1271.60)$$

$$= 1271.60 + 6.36$$

$$= \$1277.96.$$

Alternatively, we can imagine that 25 rather than 24 payments were made, so that after the twenty-fifth payment

$$S_2 = 50 \left[\frac{(1.005)^{25} - 1}{0.005} \right]$$

$$= 50(26.5591)$$

$$= \$1327.96.$$

Now, in actuality, the twenty-fifth payment is not made, so the amount at the end of 24 months is

$$S = 1327.96 - 50$$

$$= \$1277.96.$$

Example At the birth of their child a husband and wife decide to invest \$250 in an annuity at each birthday in order to pay for his college education. If the annuity pays 6% per year, how much have the parents saved when the child reaches age 18?

11.2 Annuities

Solution From formula (2) and Table VI, we obtain

$$S = 250 \left[\frac{(1.06)^{18} - 1}{0.06} \right]$$
$$= 250(30.9057)$$
$$= \$7726.43.$$

EXERCISE 11.2

Find the future value of the annuity consisting of the given amount, interest per period, and number of payments. Use Table VI.

1. $100, 1/2% per month, 20 months.
2. $200, 1% per month, 40 months.
3. $1500, 6% per year, 12 years.
4. $5000, 5% per year, 32 years.
5. $5000, 6% per 6 months, 10 years.
6. $10,000, 2% per quarter, 11 years.

Applications

7. *Business* A woman deposits $3000 once a year, on January 1, in a savings plan that pays 6% per year. How much will she receive after the last payment in 25 years?
8. In Problem 7, assume that all the money accumulated is withdrawn from the account immediately after the last payment. How much interest was received over the life of the annuity? [*Hint*: If no interest were received the woman would get back $3000 × 25.]
9. In Problem 7, what would be the cash value of the account at the end of 25 years?
10. What is the amount gained through interest over the life of an annuity if $100 is deposited monthly into an annuity paying 6% annual interest compounded monthly for 4 years?
11. A person puts $2000 into an annuity each year. If the annuity pays 6% interest compounded annually, in how many years will the annuity be worth at least $350,000?
12. An employee of a company gets paid on the first of the month, and $60 is witheld from her paycheck for a retirement fund. The fund pays interest at a rate of 3/4% per month. What is the value of this savings plan at the end of 4 years?
13. How much should be saved each quarter in an annuity paying 8% annual interest compounded quarterly to accrue $10,000 in 12 years from the present?
14. A wealthy parent sets up a trust account for a daughter when she is born. If the savings plan consists of depositing $5000 a year in a regular savings account paying 6% interest compounded annually, and $6000 a year in a stock option which grows at a rate of 5% a

year compounded annually, what is the total value of the trust account when the parent dies 38 years later. (Assume 38 payments of each amount are made.)

15. A loan of $8000 plus interest is to be paid in full (that is, in one payment) in 5 years. The interest on the loan is 6% compounded quarterly. What amount should be deposited each quarter in a savings plan paying 4% annual interest compounded quarterly in order to pay off the loan at the required time? (The set of equal deposits drawing interest in order to meet a future obligation consisting of a debt plus interest is called a **sinking fund**.)

16. After 25 yearly deposits of $1000 into an annuity its value is $50,000. What is the annual rate of interest?

17. An uncle sends his niece $5 on her birthday and $5 at Christmas every year. Suppose the $5 is always deposited on June 30 and December 30 in a savings account paying 4% compounded semiannually. If the gifts started at age one, what amount can the niece expect to withdraw after her twentieth birthday?

18. A person deposits $500 every 6 months in a savings account that pays 4% yearly interest compounded semiannually. What amount could be deposited at the start of this annuity in order to accrue the same amount of the annuity at the end of 10 years? (Remember, the first period *ends* when the first payment is made. This one payment, which is equivalent to the entire set of payments, is called the **present value** of the annuity.)

11.3

PRESENT VALUE OF AN ANNUITY

We have seen that an annuity over n periods of time can be considered as the sum of n compound interest problems. In turn, for each of the n payments we can determine its present value. For example, let us again consider $1 per month at an interest rate i per month for n months. At the *end* of the first month a payment of $1 is made into the annuity. However, the amount that we could invest *now* at a rate i per month in order to *receive* $1 a month from now is the present value of the first payment. From (4) of Section 11.1, this present value is $(1 + i)^{-1}$. Hence, in order to receive $1 at the end of the

first month, invest now $(1 + i)^{-1}$,

second month, invest now $(1 + i)^{-2}$,

third month, invest now $(1 + i)^{-3}$,

nth month, invest now $(1 + i)^{-n}$.

If we want to receive $1 each month for n consecutive months, we must invest now the amount

11.3 Present Value of an Annuity

(1) $$A = (1+i)^{-1} + (1+i)^{-2} + (1+i)^{-3} + \cdots + (1+i)^{-n}.$$

This is a sum of a geometric progression with common ratio $(1+i)^{-1}$, so (1) can be written as

$$A = (1+i)^{-1} \left[\frac{1-(1+i)^{-n}}{1-(1+i)^{-1}} \right]$$

$$= \frac{1-(1+i)^{-n}}{i}$$

This expression is commonly denoted as $a_{\overline{n}|i}$ and is tabulated in Table VII.

It follows that in order to receive R dollars per period for n periods we must invest now

(2) $$A = R \left[\frac{1-(1+i)^{-n}}{i} \right].$$

The amount A is said to be the **present value** of the annuity. Of course, if we receive R dollars per period, then after n periods the account will be reduced to zero.

We observe that the present value of an annuity is the sum of the present values of the payments, and that A represents the amount we could invest now—one period before the first payment is made—at compound interest with rate i per period, and receive at the end of n periods of time the same amount as the annuity.

Example Find the present value of an annuity consisting of $1000 per month, 1% interest per month, for 2 years.

Solution Two years consist of 24 periods (months). Hence, from (2) and Table VII, we find the present value to be

$$A = 1000 \left[\frac{1-(1.01)^{-24}}{0.01} \right]$$

$$= 1000(21.2434)$$

$$= \$21,243.40.$$

Example A student figures that in a year from now he will need $1500 to start college, and then $1500 a year for the following 3 years. How much could be deposited at the present time in a savings account paying 6% per year compounded annually to satisfy his requirements?

Solution From (2) and Table VII, was have

$$A = 1500 \left[\frac{1-(1.06)^{-4}}{0.06} \right]$$

$$A = 1500(3.4651)$$
$$= \$5197.65.$$

Example If $300 is deposited into a savings account each quarter, and the account pays 6% per year compounded quarterly, what is the amount accumulated after the fortieth payment?

Solution The problem as given consists of finding the *future value* of an annuity. Thus, from (2) of Section 11.2 and Table VI,

$$S = 300 \left[\frac{(1.015)^{40} - 1}{0.015} \right]$$
$$= 300(54.2679)$$
$$= \$16,280.37.$$

Alternatively, if we compute the *present value* of the annuity

$$A = 300 \left[\frac{1 - (1.015)^{-40}}{0.015} \right]$$
$$= 300(29.9158)$$
$$= \$8974.74$$

and then use the compound interest formula, (3b) of Section 11.1, we find

$$S = 8974.74(1 + 0.015)^{40}$$
$$= 8974.74(1.8140)$$
$$= \$16,280.18.$$

The slight difference in answers is due to the effects of rounding in the tables.

AMORTIZATION If we solve (2) for R, we obtain

(3) $$R = A \left[\frac{i}{1 - (1+i)^{-n}} \right].$$

This is the amount that can be paid each period at an interest rate i per period in order to pay off a debt which is due now. The set of payments made according to (3) is said to be an **amortization** of the debt. For example, the paying off of a mortgage on a home or paying off a car are amortization problems.

Example A new car is purchased for $8000 with no money down. The purchase agreement calls for 36 equal installments with a yearly interest rate of 12% compounded monthly. What are the monthly payments? What is the total amount paid for the car?

11.3 Present Value of an Annuity

Solution We use Table VII and (3).

$$R = \frac{8000}{\left[\dfrac{1 - (1.01)^{-36}}{0.01}\right]}$$

$$= \frac{8000}{30.1075}$$

$$= \$265.71$$

The total amount paid for the car is $36 \times 265.71 = \$9565.72$.

EXERCISE 11.3

Find the present value of the annuity consisting of the given amount per period, interest per period, and number of periods. Use Table VII.

1. $5000 per year, 5% per year, 10 years.
2. $10,000 per year, 6% per year, 25 years.
3. $2000 per month, 1% per month, 4 years.
4. $1500 per month, 1/2% per month, 34 months.
5. Find the future value of an annuity consisting of $1000 deposited in an account each quarter, paying 8% per year compounded quarterly, for 50 payments. Compare the answer with that obtained using the concept of present value and compound interest.
6. Using (2) of this section, rework Problem 18 of Exercise 11.2.

Applications

7. *Business* A house is purchased for $45,000 by paying $15,000 down and assuming a loan at 6% annual interest. If the balance is amortized over 28 years, what is the annual payment on the loan?
8. Beginning a month from now a man plans to lease a car at $240 a month for the next 2 years. To cover the monthly payments he decides to deposit a lump sum into a savings plan that pays 1/2% monthly interest. How much should be deposited now in order to meet the payments?
9. At 3% per period, the present value of an annuity is $6166. If $400 is received per period, after how many periods will the account be depleted?
10. A person borrows $1000 at an annual rate of interest which is compounded monthly. If the debt is amortized by paying back $104.17 each month for 10 consecutive months, what is the annual rate of interest?
11. What is the total interest paid on the loan described in Problem 10?
12. A credit card company charges interest at an annual rate of 12% or 1% a month on the

unpaid balance. A person decides to pay off a balance of $500 in 12 equal payments. What is the total amount the person pays to deplete the debt?

13. Each year for 35 consecutive years $1200 is deposited in a savings account paying 5% interest compounded annually. Beginning 2 years after the thirty-fifth payment the depositor wants to withdraw $10,000 each year as retirement income. How many years can the depositor expect to draw on this fund before it is exhausted? [*Hint*: Besides the annuity for 35 periods, consider compound interest for 1 year.]

14. The present value of an annuity called a **perpetuity** is defined by $A = R \cdot \lim_{n \to \infty} a_{\overline{n}|i}$, where $a_{\overline{n}|i} = [1 + (1+i)^{-n}]/i$. Show that $A = R/i$.

15. An alumnus of a college wills $40,000 to the college to endow a perpetual scholarship to be received by a deserving student in mathematics. Assume that the scholarship is received every 6 months. What is the value of the scholarship if the $40,000 is invested at 5% compounded semiannually.

11.4

CONTINUOUS COMPOUNDING OF INTEREST

It is very common for savings institutions to advertise that interest is being compounded either daily, or by the hour, or even by the minute. Of course, there is no reason to stop there, interest could as well be compounded every second, every half-second, every microsecond, and so on. That is to say, interest could be compounded **continuously**.

THE NUMBER e Before precisely defining the concept of continuous compounding of interest, we need to recall the sequence considered in Problem 10 of Section 9.1,

$$s_n = \left(1 + \frac{1}{n}\right)^n.$$

We would like to consider the behavior of s_n as the number of terms becomes large (written $n \to \infty$). It would appear at first inspection that since $1/n \to 0$ as $n \to \infty$, the terms of the sequence should approach unity. However, if we examine some representative terms,

$$s_1 = (1 + 1)^1 = 2$$
$$s_2 = \left(1 + \frac{1}{2}\right)^2 = 2.25$$

11.4 Continuous Compounding of Interest

$$s_3 = \left(1 + \frac{1}{3}\right)^3 \approx 2.37137$$

$$s_4 = \left(1 + \frac{1}{4}\right)^4 \approx 2.44141$$

$$s_5 = \left(1 + \frac{1}{5}\right)^5 \approx 2.48832$$

$$\vdots$$

$$s_{100} = \left(1 + \frac{1}{100}\right)^{100} \approx 2.70481$$

$$s_{200} = \left(1 + \frac{1}{200}\right)^{200} \approx 2.71152$$

$$s_{500} = \left(1 + \frac{1}{500}\right)^{500} \approx 2.71557$$

$$s_{1000} = \left(1 + \frac{1}{1000}\right)^{1000} \approx 2.71692$$

$$s_{10,000} = \left(1 + \frac{1}{10,000}\right)^{10,000} \approx 2.71815$$

we see that the s_n are increasing. It can be shown that the terms do not become arbitrarily large, but are bounded above by the number e which was discussed in Chapter 5. In fact, the number e is the smallest possible upper bound for the terms of the sequence; we say that e is the **limit** of the sequence, or s_n converges to e. If the student possesses a scientific calculator, he or she should verify the values of s_n given above.

The number e is the limiting value of the terms of the sequence $(1 + 1/n)^n$ as the number n of terms becomes arbitrarily large. Symbolically, we write

(1) $$e = \lim_{n \to \infty} \left(1 + \frac{1}{n}\right)^n.$$

To 15 decimal places $e = 2.718281828459045$.

The limit of a sequence such as $(1 + 0.06/n)^n$ can be written in terms of e by means of an algebraic substitution. Since

$$\left(1 + \frac{0.06}{n}\right)^n = \left(1 + \frac{0.06}{n}\right)^{(n/0.06)(0.06)}$$

if we now let $1/t = 0.06/n$, we note that as $n \to \infty$ so must $t \to \infty$, and

$$\lim_{n \to \infty} \left(1 + \frac{0.06}{n}\right)^n = \lim_{n \to \infty} \left(1 + \frac{0.06}{n}\right)^{(n/0.06)(0.06)}$$
$$= \lim_{t \to \infty} \left(1 + \frac{1}{t}\right)^{t(0.06)}$$
$$= e^{0.06}.$$

Thus, we can generalize (1) to the following.

For any real number r,

(2) $$e^r = \lim_{t \to \infty} \left(1 + \frac{r}{t}\right)^t.$$

Perhaps now would be a good time for the reader to review the formula for compound interest, (3) of Section 11.1.

If the yearly rate of interest r is to be compounded continuously, then the future value of a principal P in m years is given by

(3) $$S = Pe^{rm}.$$

Let t be the number of times the interest is compounded yearly. To obtain (3), let $t \to \infty$. From (3) of Section 11.1 and (2) above, it follows that

$$S = P \cdot \lim_{t \to \infty} \left(1 + \frac{r}{t}\right)^{tm}$$
$$= P(e^r)^m$$
$$= Pe^{rm}.$$

The present value is then given by

(4) $$P = Se^{-rm}.$$

Values of e^x and e^{-x} are given in Table II.

One should not conclude that simply because the interest is compounded continuously, the amount accrued after m years is substantially more than that obtained, say, by compounding quarterly.

Example An amount of $5000 is deposited into an account paying 6% per year. Compare the cash value of the account after 4 years if the interest is compounded quarterly, with the value when the interest is compounded continuously.

11.4 Continuous Compounding of Interest

Solution To find the future value of $5000 when the interest is compounded quarterly, we use (3b) of Section 11.1 and Table IV.

$$S = 5000(1 + 0.015)^{16}$$
$$= 5000(1.2690)$$
$$= \$6345$$

Whereas, to find the future value of the deposit when the interest is compounded continuously, we use (3) and Table II.

$$S = 5000e^{(0.06)4}$$
$$= 5000e^{0.24}$$
$$= 5000(1.2712)$$
$$= \$6356$$

It is interesting to note that the gain by continuous compounding over a period of 4 years is only $11.

Many savings institutions compound interest on deposits daily. Inspection of the last example should convince you that the formula $S = Pe^{rm}$ provides a very good approximation to the amount accrued through daily compounding of interest.

ANNUITIES Suppose an annuity consists of depositing P dollars each year into a savings plan that pays interest at an annual rate r. If the interest is compounded continuously, then the amount accumulated is

After 1 payment: P,
After 2 payments: $P + Pe^r$,
After 3 payments: $P + (P + Pe^r)e^r = P + Pe^r + Pe^{2r}$,
After m payments: $P + Pe^r + Pe^{2r} + \cdots + Pe^{(m-1)r}$.

We have already considered this as a sum of a geometric progression on page 317. Thus, an alternative form for this sum is given by

$$S = P\frac{e^{mr} - 1}{e^r - 1},$$

where S is the future value of the annuity and is the amount that can be received immediately after the last, or mth, payment into the annuity.

EXERCISE 11.4

Find the future value of the given principal for the indicated rate of interest, compounded continuously, and time. Use Table II.

1. $1000, 8% per year, 5 years.
2. $5000, 10% per year, 10 years.
3. $2500, 5.5% per year, 20 years.
4. $10,000, 9% per year, 5 years.

Find the present value of the given amount for the indicated rate of interest, compounded continuously, and time. Use Table II.

Example $10,000, 7% per year, 20 years.

Solution From (4) and Table II we find

$$P = 10{,}000 e^{-(0.07)20}$$
$$= 10{,}000 e^{-1.4}$$
$$= 10{,}000(0.2466)$$
$$= \$2466.$$

5. $50,000, 7.5% per year, 10 years.
6. $120,000, 11% per year, 30 years.
7. $20,000, 6.25% per year, 40 years.
8. $13,000, 8% per year, 25 years.

Applications

9. *Business* If $3000 is invested at 7% continuous compound interest, how much is accrued after 5 years?
10. What amount should be invested at 6% continuous compound interest in order to receive $50,000 in 15 years?
11. An amount of $2000 is deposited into a savings plan that pays 6% annual interest. If the interest is compounded continuously, how long does it take to triple the initial deposit?
12. An initial investment of $100 is valued at $2008.60 after 40 years. What is the annual rate of continuous compound interest?
13. An investor would like to double an initial investment in 10 years. Assuming interest compounded continuously, what should be the annual rate of interest?
14. An investment of $10,000 draws interest at 8% compounded continuously. How much interest is earned in 15 years?
15. A deposit of $2000 into a savings account draws interest at $4\tfrac{1}{2}$% compounded continuously. How much interest is earned in 10 years?

11.4 Continuous Compounding of Interest

16. Determine the effective rate of annual interest if $1000 is invested at 7% compounded continuously. (See Problem 19 of Section 11.1.)

17. Compare the future value of $10,000 at 8% annual interest compounded continuously and compounded quarterly for a period of $12\frac{1}{2}$ years.

18. Compare the future value of $500 at 6% annual interest compounded continuously and compounded daily for a period of 5 years. [*Hint*: use logarithms to compute the daily compounding of interest. Assume 365 days in a year.]

19. For 20 years, a person saves $500 a year. At 6% annual interest compounded continuously, what is the future value of this amount after the twentieth payment?

20. A person wants to obtain $20,000 from a retirement fund in 30 years. If the annual rate of interest is 5% compounded continuously, how much should be saved each year? Assume 30 equal payments.

21. *Calculator Project* The formula for the future value of a principal P invested at compound interest is

$$S = P\left(1 + \frac{r}{t}\right)^{tm},$$

where r is the annual rate of interest, t is the number of periods per year, and m represents years. Use a calculator to complete the following comparative table for the future values of $P = \$500$ with $r = 0.06$ over 4 years.

Method of compounding	t	Future value S
Annually	1	
Semiannually	2	
Quarterly	4	
Monthly	12	
Weekly	52	
Daily	365	
Hourly	8760	
Continuously	$t \to \infty$	

CHAPTER TEST

1. Compare the future value of $5000 deposited in a savings account for 3 years at 6% annual interest compounded semiannually with the future value of the same amount when interest is compounded quarterly.

2. A principal of $10,000 is deposited in a savings account that pays 7% per year compounded quarterly. How long does it take to double the principal?

3. A husband and wife take out a $12,000 second mortgage on their home at 12% simple interest for 15 years.
   a. What is the interest paid to the lender over the life of the loan?
   b. How much will be repaid in 15 years?
   c. What are the monthly payments?

4. At the first of each year, $1000 is deposited in a savings account paying 6% interest compounded annually. What is the future value of the account after the fifth payment? At the end of the fifth year?

5. Find the present value of an annuity of $3000 per quarter, 3% per quarter for 25 quarters.

6. In Problem 5, use compound interest to determine the future value of the annuity.

7. A person buys a new car valued at $7000 and received a trade-in value of $500 on an old car. The finance company charges 9% interest per year or 3/4% per month. What is the monthly payment if the buyer decides to pay off the car in 30 equal installments?

8. Suppose a person receives an unexpected inheritance. What amount of this inheritance could be deposited now in a savings account paying 5% annual interest in order to cover car payments of $200 a month for 36 consecutive months?

9. A deposit of $4000 becomes $5938 in 8 years when interest is compounded semiannually. What amount could be invested at continuous compound interest in order to accrue to the same amount in the same length of time?

10. An amount of $100 a year is paid into an annuity which pays 8% annual interest compounded continuously. After a certain payment the annuity is valued at $7670. How many yearly payments were made in order to obtain this amount?

PART IV

12 DIFFERENTIAL CALCULUS
13 APPLICATIONS OF THE DERIVATIVE
14 INTEGRAL CALCULUS
15 APPLICATIONS OF THE INTEGRAL
16 MULTIVARIATE DIFFERENTIAL CALCULUS

12
DIFFERENTIAL CALCULUS

12.1

TANGENT LINE TO A GRAPH

SLOPE In Chapter 4 we saw that the slope of a straight line yields information on how fast y is changing as x varies. For example, if $y = 3x + 1$, then the slope $m = 3/1$ means that for every unit increase in x, y increases 3 units (see Figure 12.1). For $y = mx + b$, the larger $|m|$ is, the steeper the graph of the line. That is, y changes faster with respect to a change in x when $|m|$ is large.

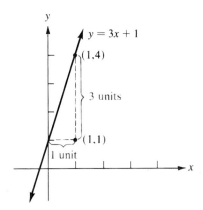

Figure 12.1

If $y = f(x)$ denotes a function, we shall see that if a straight line can be constructed **tangent** to its graph at some point $(a, f(a))$, then the slope of this line is an indicator of how fast f is changing with respect to x in a vicinity or **neighborhood** of $x = a$. The fundamental geometric problem in the study of **differential calculus** is as follows:

If the graph of a function possesses a tangent at a specified point $(a, f(a))$ on its graph, find the slope of the tangent line.

397

SLOPE OF A SECANT

To this end, let us consider the graph of any function $y = f(x)$, which, for convenience, we shall assume has domain R. Suppose h denotes a positive number, and suppose, as shown in Figure 12.2, that a **secant line** is drawn through two points $(a, f(a))$ and $(a + h, f(a + h))$. We define

$$f(a+h) - f(a) = \text{change in } f$$

and

$$(a+h) - a = h = \text{change in } x,$$

as indicated in Figure 12.2. Thus, the slope of the secant line between $(a, f(a))$ and $(a+h, f(a+h))$ is given by

$$\frac{\text{change in } f}{\text{change in } x} = \frac{f(a+h) - f(a)}{h}.$$

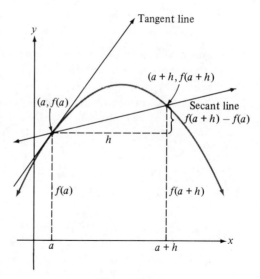

Figure 12.2

SLOPE OF TANGENT LINE

If we now let the number h become smaller and smaller (written $h \to 0$), then for each choice of h we obtain a point on the graph of $y = f(x)$ that is closer to $(a, f(a))$. Correspondingly, as shown in Figure 12.3, the secant lines L_1, L_2, L_3, and so on, become closer and closer to the tangent line L at $(a, f(a))$. It seems reasonable then to assume that the *limiting* value of the slopes of the secant lines is the **slope m of the tangent**. We write this limit as

(1) $$m = \lim_{h \to 0} \frac{f(a+h) - f(a)}{h}.$$

12.1 Tangent Line to a Graph

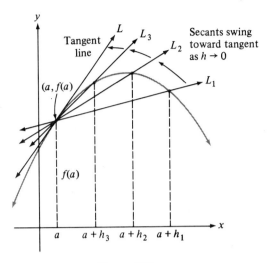

Figure 12.3

Example Find the slope of the tangent line to the graph of $y = x^2$ at $x = 2$.

Solution When $x = 2$, $f(2) = (2)^2 = 4$. Now the procedure for finding the slope of the tangent line at $(2, 4)$ consists of four steps. We find

(i) $f(2 + h) = (2 + h)^2 \ = 4 + 4h + h^2$

(ii) $f(2 + h) - f(2) \quad = 4 + 4h + h^2 - 4$
$= 4h + h^2$
$= h(4 + h)$

(iii) $\dfrac{f(2 + h) - f(2)}{h} = \dfrac{h(4 + h)}{h}$
$= 4 + h$

(iv) $\lim\limits_{h \to 0} \dfrac{f(2 + h) - f(2)}{h} = \lim\limits_{h \to 0} (4 + h)$
$= 4.$

As shown in the accompanying figure, the slope of the tangent at $(2, 4)$ is $m = 4$.

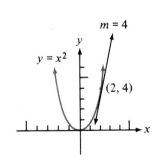

Example Find the slope of the tangent line to the graph of $y = x^2$ at $x = 3$.

Solution When $x = 3$, $f(3) = (3)^2 = 9$. In this case, we now have

(i) $f(3+h) = (3+h)^2 = 9 + 6h + h^2$

(ii) $f(3+h) - f(3) = 9 + 6h + h^2 - 9$
$= 6h + h^2$
$= h(6+h)$

(iii) $\dfrac{f(3+h) - f(3)}{h} = \dfrac{h(6+h)}{h}$
$= 6 + h$

(iv) $\lim\limits_{h \to 0} \dfrac{f(3+h) - f(3)}{h} = \lim\limits_{h \to 0} (6+h)$
$= 6.$

The accompanying figure shows that the tangent line at (3, 9) has slope $m = 6$.

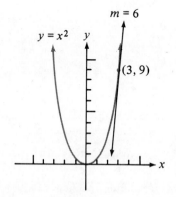

Example Find the slope of the tangent line to the graph of $y = x^2 - 7x + 6$ at $x = 2$.

Solution When $x = 2$, $f(2) = (2)^2 - 7(2) + 6 = -4$. Proceeding as in the foregoing two examples, we have

(i) $f(2+h) = (2+h)^2 - 7(2+h) + 6$
$= 4 + 4h + h^2 - 14 - 7h + 6$
$= -4 - 3h + h^2$

(ii) $f(2+h) - f(2) = -4 - 3h + h^2 - (-4)$
$= -3h + h^2$
$= h(-3 + h)$

12.1 Tangent Line to a Graph

(iii) $\dfrac{f(2+h)-f(2)}{h} = \dfrac{h(-3+h)}{h}$

$= -3+h$

(iv) $\lim\limits_{h \to 0} \dfrac{f(2+h)-f(2)}{h} = \lim\limits_{h \to 0}(-3+h)$

$= -3.$

Recall from Section 4.4 that the graph of $y = x^2 - 7x + 6$ is a parabola. The figure shows that the tangent at $(2, -4)$ has slope $m = -3$.

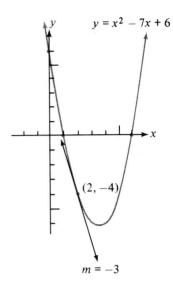

The observant reader will notice that the goal in using (1) is simply to remove h from the denominator through cancellation *before* computing the limit of the expression as $h \to 0$.

Example Find the equation of the tangent line to the graph of $y = x^2 - 7x + 6$ at $x = 2$.

Solution From the preceding example, we know that the point of tangency is $(2, -4)$ and at this point the tangent has slope -3. Using the point-slope form of the equation of a line gives

$$y - (-4) = -3(x - 2)$$

or

$$y = -3x + 2.$$

RATE OF CHANGE The **change** of a function as x changes, say from x_1 to x_2, is defined to be the difference

$$f(x_2) - f(x_1).$$

The **average rate of change** with respect to x is defined by the quotient

(2) $$\frac{f(x_2)-f(x_1)}{x_2-x_1}.$$

If we let the change in x be denoted by
$$h = x_2 - x_1,$$
then $x_2 = x_1 + h$ and (2) becomes

(3) $$\frac{f(x_1+h)-f(x_1)}{h}.$$

Geometrically, we can then interpret the average rate of change of a function (2) as the slope of the secant line between $(x_1, f(x_1))$ and $(x_2, f(x_2))$. The limit

$$\lim_{h \to 0} \frac{f(x_1+h)-f(x_1)}{h}$$

is called the **instantaneous rate of change** with respect to x of $y = f(x)$ at $(x_1, f(x_1))$.

Example Find the average rate of change with respect to x of the revenue function $R(x) = -x^2 + 400x$ as x changes from 20 to 40.

Solution We first calculate $R(20)$ and $R(40)$:

$$R(20) = -(20)^2 + 400(20) = 7600$$
$$R(40) = -(40)^2 + 400(40) = 14{,}400.$$

The change in revenue is
$$R(40) - R(20) = 14{,}400 - 7600$$
$$= 6800.$$

From (2) the average rate of change is
$$\frac{R(40)-R(20)}{40-20} = \frac{6800}{20}$$
$$= 340.$$

VELOCITY If $s(t)$ represents the distance that a body travels in time t, then the quotient

$$\frac{s(t_2)-s(t_1)}{t_2-t_1}$$

is **average velocity** between times t_1 and t_2. If $h = t_2 - t_1$ and $t_2 = t_1 + h$, then

$$\lim_{h \to 0} \frac{s(t_1+h)-s(t_1)}{h}$$

is the **instantaneous velocity** at t_1.

12.1 Tangent Line to a Graph

EXERCISE 12.1

Find the slope of the tangent line to the graph of the given function at the indicated point.

1. $y = 2x^2$; $x = 1$
2. $y = 4x^2$; $x = 2$
3. $y = -x^2$; $x = 4$
4. $y = -5x^2$; $x = 1$
5. $y = x^2 + 4$; $x = 3$
6. $y = x^2 - 2$; $x = -1$
7. $y = x^2 + x$; $x = 0$
8. $y = x^2 - 3x$; $x = 0$
9. $f(x) = -x^2 + 2x + 3$; $x = 4$
10. $f(x) = x^2 + 4x - 1$; $x = 2$
11. $f(x) = x^3$; $x = 2$
12. $f(x) = x^3 + 5$; $x = -1$
13. $f(x) = \dfrac{1}{x}$; $x = -1$
14. $f(x) = \dfrac{2}{x}$; $x = \dfrac{1}{2}$

Find the equation of the tangent line to the graph of the given function at the indicated point.

15. $y = x^2$; $x = -3$
16. $y = 2x^2 + 1$; $x = 1$
17. $y = x^2 + 5x$; $x = 1$
18. $y = -x^2 + 10x$; $x = -1$

19. Compute $\lim\limits_{h \to 0} \dfrac{f(x+h) - f(x)}{h}$ for $f(x) = 5x^2$.

20. For $f(x) = 2 - 3x^2$ compare the values of

$$\lim_{x \to 1} \dfrac{f(x) - f(1)}{x - 1} \quad \text{and} \quad \lim_{h \to 0} \dfrac{f(1+h) - f(1)}{h}.$$

Applications

21. *Business* Compute the average rate of change with respect to x of the revenue function $R(x) = -x^2 + 100x$ as x changes from 10 to 30.

22. Let revenue and cost functions be given by $R(x) = -x^2 + 400x$ and $C(x) = 300x + 1000$, respectively. Find the average rate of change with respect to x of profit as x changes from 20 to 40.

23. Find the instantaneous rate of change with respect to x of the revenue function given in Problem 21 at $(10, R(10))$.

24. Find the instantaneous rate of change with respect to x of the profit function described in Problem 22 at $(20, P(20))$.

25. *Physics* The height s above ground, in feet, of a projectile is given by $s(t) = -16t^2 + 64t$, where t represents time in seconds. Compute the average velocity between $t = 1$ and $t = 2$. Find the instantaneous velocity at $t = 1$.

26. Find the instantaneous velocity of the projectile in Problem 25 at $t = 2$. Interpret the answer.

12.2

THE DERIVATIVE

In the last section, we defined the slope of a tangent line to the graph of a function $y = f(x)$ at a specified point $(a, f(a))$ as the limit

$$m = \lim_{h \to 0} \frac{f(a+h) - f(a)}{h}.$$

Now, for *any* value of x in the domain of the function, the limit (when it exists)

$$\lim_{h \to 0} \frac{f(x+h) - f(x)}{h}$$

is itself a function of x.

Example For $y = x^2$, the four-step procedure of the preceding discussion gives:

(i) $f(x+h) = (x+h)^2 = x^2 + 2xh + h^2$

(ii) $f(x+h) - f(x)\ \ \ = x^2 + 2xh + h^2 - x^2$
$\ = 2xh + h^2$
$\ = h(2x + h)$

(iii) $\dfrac{f(x+h) - f(x)}{h} = \dfrac{h(2x+h)}{h}$

$\ = 2x + h$

(iv) $\lim\limits_{h \to 0} \dfrac{f(x+h) - f(x)}{h} = \lim\limits_{h \to 0} (2x + h)$

(1) $\ = 2x.$

We have already seen that the slope of the tangent line to the graph of $y = x^2$ at $x = 2$ is $m = 4$. Observe that the same result follows from (1) by substituting $x = 2$. Without repeating the limiting process, we see from (1) that the slopes at, say, $x = -3$ and $x = 5$, are $m = -6$ and $m = 10$, respectively. The expression $m = 2x$ is a slope function *derived* from the original function. This prompts the following important definition.

For $y = f(x)$, the **derivative** (or derived function) is defined to be

(2) $$f'(x) = \lim_{h \to 0} \frac{f(x+h) - f(x)}{h}$$

provided this limit exists.

12.2 The Derivative

NOTATION AND TERMINOLOGY

Other notations commonly used to denote the derivative are

$$\frac{dy}{dx}, \quad y', \quad \text{or} \quad D_x y.$$

In this text, if a function is specified in a manner such as $f(x) = x^2$, we write the derivative as $f'(x) = 2x$; on the other hand, if $y = x^2$, then the derivative is written $\frac{dy}{dx} = 2x$. The **value** of the derivative at $x = a$ is written as either

$$f'(a) \quad \text{or} \quad \left.\frac{dy}{dx}\right|_{x=a}.$$

The process of finding the derivative of a function is called **differentiation**. The **operation** of differentiation is often denoted by the symbol $\frac{d}{dx}$. Thus, yet another way of writing the derivative of $y = x^2$ is

$$\frac{d}{dx} x^2 = 2x.$$

Example Find the derivative of $y = 3x + 6$.

Solution

(i) $f(x+h) = 3(x+h) + 6 = 3x + 3h + 6$

(ii) $f(x+h) - f(x) = 3x + 3h + 6 - (3x + 6)$
$= 3h$

(iii) $\dfrac{f(x+h) - f(x)}{h} = \dfrac{3h}{h}$
$= 3$

(iv) $\lim\limits_{h \to 0} \dfrac{f(x+h) - f(x)}{h} = \lim\limits_{h \to 0} 3$
$= 3.$

The derivative is

$$\frac{dy}{dx} = 3.$$

In fact, it is easily shown that if $y = mx + b$, then $dy/dx = m$. For example, if $y = x$, then dy/dx is 1, or if $y = -7x + 9$, then dy/dx is -7. Of course, this makes geometric sense since a straight line and its tangent are collinear and necessarily have the same slope.

Example Find the derivative of $f(x) = x^2 - 3x$.

Solution

(i) $f(x+h) = (x+h)^2 - 3(x+h) = x^2 + 2xh + h^2 - 3x - 3h$

(ii) $f(x+h) - f(x) = x^2 + 2xh + h^2 - 3x - 3h - (x^2 - 3x)$

$= 2xh + h^2 - 3h$

$= h(2x + h - 3)$

(iii) $\dfrac{f(x+h) - f(x)}{h} = \dfrac{h(2x + h - 3)}{h}$

$= 2x + h - 3$

(iv) $\lim\limits_{h \to 0} \dfrac{f(x+h) - f(x)}{h} = \lim\limits_{h \to 0} (2x + h - 3)$

$= 2x - 3.$

The derivative is

$$f'(x) = 2x - 3.$$

Example Find the equation of the tangent line to the graph of $f(x) = x^2 - 3x$ at $x = 5$.

Solution First compute $f(5)$:

$$f(5) = (5)^2 - 3(5) = 25 - 15$$
$$= 10.$$

Thus, the point of tangency is $(5, 10)$. Second, we compute the derivative of $f(x)$. In this case, we have found $f'(x)$ in the foregoing example:

$$f'(x) = 2x - 3.$$

Evaluate the derivative at $x = 5$:

$$f'(5) = 2(5) - 3 = 10 - 3$$
$$= 7.$$

The slope of the tangent at $(5, 10)$ is 7. We now use the point-slope form to find the equation of the tangent line:

$$y - 10 = 7(x - 5)$$
$$y = 7x - 25.$$

WORD OF CAUTION The derivative of a function is *not the equation* of a tangent line, it is simply a function that, when evaluated, gives the slope of the tangent line. In addition, it would be *incorrect* to say that the equation of a tangent line at a point (x_1, y_1) is

$$y - y_1 = f'(x)(x - x_1).$$

12.2 The Derivative

Again we observe that the derivative must be evaluated *before* using it in the point-slope form of a line. Thus,

$$y - y_1 = f'(x_1)(x - x_1)$$

would be correct.

EXERCISE 12.2

Use (2) to find the derivative of the given function.

1. $y = 3x^2$
2. $y = -2x^2$
3. $y = x^2 + 1$
4. $y = 4x^2 + 5$
5. $y = x^2 - 6x$
6. $y = x^2 + 2x$
7. $y = 2x^2 + x + 1$
8. $y = -x^2 + 5x + 3$
9. $y = -x^2 + 10x - 9$
10. $y = x^2 + 7x + 14$
11. $y = x^3$
12. $y = x^3 + 4$
13. $f(x) = x^3 + x$
14. $f(x) = x^3 + x^2$
15. $f(x) = \dfrac{1}{x}$
16. $f(x) = \dfrac{1}{x+1}$
17. $f(x) = \dfrac{1}{x^2}$
18. $f(x) = \dfrac{5}{x^2}$
19. $f(x) = \dfrac{x}{x+1}$
20. $f(x) = \dfrac{x+1}{x-1}$

Use (2) to find the derivative of the given function at the indicated point.

Example $f(x) = \sqrt{x}$ at $x = 4$

Solution $f'(x) = \lim\limits_{h \to 0} \dfrac{f(x+h) - f(x)}{h}$

$= \lim\limits_{h \to 0} \dfrac{\sqrt{x+h} - \sqrt{x}}{h}$

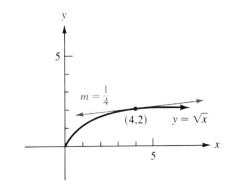

Multiply the numerator and the denominator by the conjugate factor of the numerator.

$$f'(x) = \lim_{h \to 0} \frac{(\sqrt{x+h} - \sqrt{x})}{h} \frac{(\sqrt{x+h} + \sqrt{x})}{(\sqrt{x+h} + \sqrt{x})}$$

$$= \lim_{h \to 0} \frac{h}{h(\sqrt{x+h} + \sqrt{x})}$$

$$= \lim_{h \to 0} \frac{1}{\sqrt{x+h} + \sqrt{x}}$$

$$= \frac{1}{2\sqrt{x}}.$$

At $x = 4$ we have $f'(4) = 1/4$; as shown in the figure, this is the slope of the tangent line at (4, 2).

21. $f(x) = \sqrt{x+1}$; $x = 3$
22. $f(x) = \sqrt{2x+1}$; $x = 4$

Find the equation of the tangent line to the graph of the given function at the indicated point.

23. $f(x) = \frac{1}{2}x^2 + 4$; $x = 2$
24. $f(x) = \frac{3}{2}x^2 + \frac{1}{2}$; $x = -1$

25. $f(x) = 3x^2 - x$; $x = 2$
26. $f(x) = x^2 + 9x$; $x = 0$

Without utilizing (2), determine the derivative of the given function.

27. $y = 5x - 9$
28. $y = -12x + 6$
29. $f(x) = -8x + 3$
30. $f(x) = \frac{1}{2}x - 1$

12.3

DERIVATIVES OF THE POWER FUNCTION, CONSTANTS, AND SUMS

RULES   Although the definition of the derivative, (2) of Section 12.2, is the basic building block of all differential calculus, it is clearly too inconvenient to resort to this definition each time we wish to find the derivative of a function. In this, as well as in the following two sections, we develop a list of labor-saving **rules** for obtaining derivatives.

12.3 Derivatives of the Power Function, Constants, and Sums

DERIVATIVE OF THE POWER FUNCTION

In Section 12.2 we have already seen that

$$\frac{d}{dx} x^1 = 1,$$

$$\frac{d}{dx} x^2 = 2x,$$

$$\frac{d}{dx} x^{1/2} = \frac{1}{2} x^{-1/2},$$

and from Problem 11 of Exercise 12.2 we have

$$\frac{d}{dx} x^3 = 3x^2.$$

Careful inspection of the above examples suggests the following pattern: To differentiate a power of x we must *bring down the exponent as a multiple and decrease the exponent by one*. We must summarize the general case of the derivative of the power function without proof.

I Power rule

If n is any real number, the derivative of $y = x^n$ is

$$\frac{dy}{dx} = nx^{n-1}.$$

Example Differentiate $y = x^{2/3}$.

Solution From the power rule, with $n = 2/3$, we have

$$\frac{dy}{dx} = \frac{2}{3} x^{(2/3)-1} = \frac{2}{3} x^{-1/3}.$$

Example Differentiate $y = 1/x$.

Solution By the laws of exponents, we first write the given function as $y = x^{-1}$. Thus, from rule I, with $n = -1$, we have

$$\frac{dy}{dx} = -1 \cdot x^{(-1)-1} = -x^{-2}.$$

The justification of I when n is a positive integer can be obtained from the binomial theorem; this is left as an exercise (see Problem 48).

> **II** If $y = c$, where c is a constant, then $\dfrac{dy}{dx} = 0$.

Justification of II Let $f(x) = c$ represent a constant function. From the definition of the derivative we have
$$\frac{f(x+h) - f(x)}{h} = \frac{c - c}{h} = 0$$
and hence the limit of this quotient as $h \to 0$ is necessarily 0.

Example Differentiate $y = 10$.

Solution From rule II, we have $\dfrac{dy}{dx} = 0$.

The result given in rule II simply states that the slope of a horizontal line is zero.

> **III** If $y = cf(x)$, where c is a constant, then $\dfrac{dy}{dx} = cf'(x)$.

Justification of III Let $F(x) = cf(x)$. Then
$$\frac{F(x+h) - F(x)}{h} = \frac{cf(x+h) - cf(x)}{h}$$
$$= c\left[\frac{f(x+h) - f(x)}{h}\right].$$
The limit of the quantity in the brackets as $h \to 0$ is $f'(x)$.

Example Differentiate $y = 8x^{3/2}$.

Solution From the power rule and rule III, it follows that
$$\frac{dy}{dx} = 8 \cdot \frac{3}{2} x^{(3/2) - 1}$$
$$= 12 x^{1/2}.$$

> **IV Sum rule**
>
> If $y = f(x) + g(x)$, then $\dfrac{dy}{dx} = f'(x) + g'(x)$.

12.3 Derivatives of the Power Function, Constants, and Sums

Justification of IV Let $F(x) = f(x) + g(x)$. Then

$$\frac{F(x+h) - F(x)}{h} = \frac{[f(x+h) + g(x+h)] - [f(x) + g(x)]}{h}$$

$$= \frac{[f(x+h) - f(x)] + [g(x+h) - g(x)]}{h}$$

$$= \frac{f(x+h) - f(x)}{h} + \frac{g(x+h) - g(x)}{h}.$$

The limits of the quotients in the preceding line as $h \to 0$ are $f'(x)$ and $g'(x)$, respectively.

Example Differentiate $y = 2x^{3/2} + 5x^4$.

Solution From the power rule and rules III and IV, we have

$$\frac{dy}{dx} = 2 \cdot \frac{3}{2} x^{(3/2)-1} + 5 \cdot 4x^{4-1}$$

$$= 3x^{1/2} + 20x^3.$$

Though rule IV is stated only in terms of the sum of two functions, it is equally applicable to the *difference* of two functions. In addition, rule IV can be applied to the sum (or difference) of any number of terms.

Example Find the slope of the tangent line to the graph of $y = 2x^3 - 4x^2 + 7$ at $x = 2$.

Solution From rules I, II, III, and IV, we can write

$$\frac{dy}{dx} = 2 \cdot 3x^2 - 4 \cdot 2x + 0$$

$$= 6x^2 - 8x.$$

Therefore, the slope of the tangent at $x = 2$ is

$$\left.\frac{dy}{dx}\right|_{x=2} = 6(2)^2 - 8(2)$$

$$= 8.$$

Example Find the equation of the tangent line to the graph of $y = 2x^3 - 4x^2 + 7$ at $x = 2$.

Solution The y-coordinate of the point on the graph corresponding to $x = 2$ is

$$f(2) = 2(2)^3 - 4(2)^2 + 7 = 7.$$

The point of tangency is then $(2, 7)$. From the preceding example we know that the slope of the tangent at this point is $m = 8$. Using the point-slope form of the equation of a line gives

$$y - 7 = 8(x - 2)$$

or

$$y = 8x - 9.$$

EXERCISE 12.3

Use rules I–IV to find the derivative of the given function.

1. $y = 6$
2. $y = \pi$
3. $y = x^4$
4. $y = x^8$
5. $y = x^{1/3}$
6. $y = x^{4/3}$
7. $y = \sqrt{x^3}$
8. $y = \sqrt[3]{x^2}$
9. $y = \dfrac{1}{x^2}$
10. $y = \dfrac{1}{x^3}$
11. $y = \dfrac{4}{\sqrt{x}}$
12. $y = \dfrac{6}{\sqrt[3]{x}}$
13. $y = x^{0.52}$
14. $y = x^{-0.27}$
15. $y = 2x + 1$
16. $y = -4x + 8$
17. $y = x^2 - 4$
18. $y = x^3 + 20$
19. $y = 2x^2 + 5x$
20. $y = 3x^2 - 6x$
21. $y = x^2 - 3x + 1$
22. $y = 7x^2 + 5x - 2$
23. $y = -0.1x^2 + 0.5x - 0.3$
24. $y = 0.04x^2 - 0.003x + 0.9$
25. $y = 4x^3 - 4x^2 + 6$
26. $y = -2x^3 + x^2 - 14$
27. $y = \dfrac{1}{3}x^3 - 5x^2 + 11x + 1$
28. $y = -\dfrac{2}{3}x^3 + 6x^2 + \dfrac{3}{2}x + 7$
29. $y = \dfrac{x+1}{x}$ $\left[\text{Hint: } y = 1 + \dfrac{1}{x}.\right]$
30. $y = \dfrac{x^2 + 4}{x}$
31. $y = \dfrac{2x^2 + 8}{\sqrt{x}}$
32. $y = \dfrac{\sqrt[3]{x} - \sqrt{x}}{x}$
33. $y = (3x + 1)^2$
34. $y = (x^2 - 2)^2$

12.4 Derivatives of Products and Quotients

Find the equation of the tangent line to the graph of the given function at the indicated point.

35. $y = x^2 + 4x$, $x = -1$
36. $y = 2x^2 - 3x$, $x = 2$
37. $y = x^3 - x$, $x = 2$
38. $y = x^3 + 2x + 1$, $x = 1$
39. $y = 6x^{1/3} + 1$, $x = 8$
40. $y = 2x^{3/2} + 2$, $x = 4$

Find the point(s) on the graph of the given function at which the tangent line is horizontal.

41. $y = x^2 - 6x$
42. $y = x^2 + 8x$
43. $y = x^3 - 3x^2$
44. $y = x^3 - 48x$
45. $y = \frac{1}{3}x^3 - \frac{3}{2}x^2 + 2x$
46. $y = \frac{4}{3}x^3 + 2x^2 + x$

47. Find the point on the graph of $y = x^2 + x$ at which the slope of the tangent line is 7.
48. Use the binomial theorem to justify the power rule, rule I, in the case when n is a positive integer.

12.4
DERIVATIVES OF PRODUCTS AND QUOTIENTS

In the last section, we saw that the derivative of a sum of two functions is the sum of the derivatives. However, the simplicity of this rule does not extend to the product of two functions or to the quotient of two functions. In other words, the derivative of a product is *not* the product of the derivatives, nor is the derivative of a quotient the quotient of derivatives.

V Product rule

If $y = f(x) \cdot g(x)$, then $\dfrac{dy}{dx} = f(x) \cdot g'(x) + g(x) \cdot f'(x)$.

The product rule is best remembered in words:

The derivative of a product is the first factor times the derivative of the second, plus the second factor times the derivative of the first.

Justification of V Let $F(x) = f(x)g(x)$. Then

$$\frac{F(x+h) - F(x)}{h} = \frac{f(x+h)g(x+h) - f(x)g(x)}{h}$$

$$= \frac{f(x+h)g(x+h) - \overbrace{f(x+h)g(x) + f(x+h)g(x)}^{0} - f(x)g(x)}{h}$$

$$= f(x+h) \cdot \left[\frac{g(x+h) - g(x)}{h}\right] + g(x) \cdot \left[\frac{f(x+h) - f(x)}{h}\right].$$

Taking the limit as $h \to 0$ of both members of the above equation formally gives

$$\frac{dy}{dx} = F'(x) = f(x)g'(x) + g(x)f'(x).$$

It should be noted that we have assumed that $\lim_{h \to 0} f(x+h) = f(x)$. This is always true if the function *is* **continuous**, which, as the name suggests, means the graph of the function does not have a break at x (see Section 13.1).

Example Differentiate $y = 8x^{1/2}(4x^3 - 7x^2 + 2)$.

Solution One possible way of obtaining the derivative in this case is to multiply out the terms in the given function and then use rule IV. Alternatively, by the product rule, rule V, we have

$$\frac{dy}{dx} = 8x^{1/2} \cdot \frac{d}{dx}(4x^3 - 7x^2 + 2) + (4x^3 - 7x^2 + 2) \cdot \frac{d}{dx}(8x^{1/2})$$

$$= 8x^{1/2}(12x^2 - 14x) + (4x^3 - 7x^2 + 2)4x^{-1/2}.$$

Example Differentiate $y = (x^3 - 4x^2 + 1)(x^{1/2} + 5x^3 - 10x^6)$.

Solution From the product rule, rule V, we have

$$\frac{dy}{dx} = (x^3 - 4x^2 + 1) \cdot \frac{d}{dx}(x^{1/2} + 5x^3 - 10x^6)$$

$$+ (x^{1/2} + 5x^3 - 10x^6) \cdot \frac{d}{dx}(x^3 - 4x^2 + 1).$$

It follows that

$$\frac{dy}{dx} = (x^3 - 4x^2 + 1)\left(\frac{1}{2}x^{-1/2} + 15x^2 - 60x^5\right) + (x^{1/2} + 5x^3 - 10x^6)(3x^2 - 8x).$$

The formal justification of the next rule is left as an exercise.

12.4 Derivatives of Products and Quotients

> **VI Quotient rule**
>
> If $y = \dfrac{f(x)}{g(x)}$, then $\dfrac{dy}{dx} = \dfrac{g(x)f'(x) - f(x)g'(x)}{[g(x)]^2}$, $(g(x) \neq 0)$.

In words, the quotient rule is

The denominator times the derivative of the numerator minus the numerator times the derivative of the denominator all over the denominator squared.

Example Differentiate $y = \dfrac{x}{x^2 + 1}$.

Solution From the quotient rule, rule VI,

$$\frac{dy}{dx} = \frac{(x^2 + 1) \cdot \dfrac{d}{dx} x - x \cdot \dfrac{d}{dx}(x^2 + 1)}{(x^2 + 1)^2}$$

$$= \frac{(x^2 + 1) \cdot 1 - x \cdot 2x}{(x^2 + 1)^2}$$

$$= \frac{1 - x^2}{(x^2 + 1)^2}.$$

Note: Many students use the quotient rule when it is not necessary. Although the quotient rule could be applied to the functions

$$y = \frac{x^3}{4} \quad \text{and} \quad y = \frac{8}{x^2},$$

we observe that the first function is the same as $y = \dfrac{1}{4}x^3$ and hence its derivative follows from rules I and II:

$$\frac{dy}{dx} = \frac{1}{4}(3x^2) = \frac{3}{4}x^2.$$

For the latter function, we should use the laws of exponents *before* differentiating:

$$y = 8x^{-2}.$$

The derivative then follows also from rules I and II:

$$\frac{dy}{dx} = 8(-2x^{-3}) = -16x^{-3}.$$

The quotient rule can be used in conjunction with the product rule whenever the numerator or denominator (or both) is a product of two or more factors.

Example Differentiate $y = \dfrac{(x^2 - 4x)(x^3 + 6x^2 + 5)}{7x + 3}$.

Solution We start with the quotient rule:

$$\frac{dy}{dx} = \frac{(7x+3) \cdot \dfrac{d}{dx}[(x^2 - 4x)(x^3 + 6x^2 + 5)] - (x^2 - 4x)(x^3 + 6x^2 + 5) \cdot \dfrac{d}{dx}(7x+3)}{(7x+3)^2}$$

To differentiate the numerator, we must now use the product rule:

$$\frac{dy}{dx} = \frac{(7x+3) \cdot \overbrace{[(x^2 - 4x) \cdot (3x^2 + 12x) + (x^3 + 6x^2 + 5) \cdot (2x - 4)]}^{\text{product rule applied to numerator}} - (x^2 - 4x)(x^3 + 6x^2 + 5) \cdot 7}{(7x+3)^2}.$$

The problem is complete except for simplification.

In conclusion, we note that there is no need to develop a derivative formula for the product of, say, three functions, $f(x)g(x)h(x)$, since two functions can be grouped together as one: $f(x) \cdot [g(x)h(x)]$. As the next example shows, we must use the product rule twice.

Example Differentiate $y = (2x - 6)(x^2 + 6x)(5x^3 - 4x + 1)$.

Solution Write the function as

$$y = (2x - 6) \cdot [(x^2 + 6x)(5x^3 - 4x + 1)]$$

and apply the product rule,

$$\frac{dy}{dx} = (2x - 6) \cdot \frac{d}{dx}[(x^2 + 6x)(5x^3 - 4x + 1)] + (x^2 + 6x)(5x^3 - 4x + 1) \cdot \frac{d}{dx}(2x - 6).$$

A second application of the product rule then gives

$$\frac{dy}{dx} = (2x - 6) \cdot \left[\overbrace{(x^2 + 6x) \cdot \frac{d}{dx}(5x^3 - 4x + 1) + (5x^3 - 4x + 1) \cdot \frac{d}{dx}(x^2 + 6x)}^{\text{product rule again}} \right]$$

$$+ (x^2 + 6x)(5x^3 - 4x + 1) \cdot \frac{d}{dx}(2x - 6)$$

$$= (2x - 6) \cdot [(x^2 + 6x)(15x^2 - 4) + (5x^3 - 4x + 1)(2x + 6)]$$
$$+ (x^2 + 6x)(5x^3 - 4x + 1) \cdot 2.$$

The problem is complete except for simplification.

12.4 Derivatives of Products and Quotients

EXERCISE 12.4

Use the product rule to find the derivative of the given function.

1. $y = (5x + 1)(7x - 2)$
2. $y = (10x - 8)(2x + 1)$
3. $y = x^{1/3}(3x^2 - 4x + 6)$
4. $y = 6x^{2/3}(4x^2 - 5x + 9)$
5. $y = (3x^2 + 1)(2x^2 - 4)$
6. $y = (x^2 - 2)(x^2 + 16)$
7. $y = (4x + 10)(x^2 + 11x - 7)$
8. $y = (5x - 1)(2x^2 - x + 4)$
9. $y = (\sqrt{x} + 6)(\sqrt{x} - 4)$
10. $y = (\sqrt[3]{x} + 1)(\sqrt[3]{x} + 2)$
11. $y = (8x^3 - 2x^2)(-3x^2 - 6x + 3)$
12. $y = (5x^3 + 7x^2 - 4x)(x^2 - 13x)$
13. $y = (2x^4 + 2x^{1/2} + x)(x^4 - 4x^{1/2} + 1)$
14. $y = (x^4 - 6x^{1/3} + 1)(5x^4 + 12x^{1/3} - x)$
15. $y = (x^2 + 5x)^2$
16. $y = (x^3 - x)^2$

Use the quotient rule to find the derivative of the given function.

17. $y = \dfrac{1}{2x + 1}$
18. $y = \dfrac{5}{4x - 7}$
19. $y = \dfrac{x}{x + 1}$
20. $y = \dfrac{2x}{x - 3}$
21. $y = \dfrac{x - 1}{2x^2}$
22. $y = \dfrac{2x + 1}{4x^3}$
23. $y = \dfrac{x^2}{5x^2 - 1}$
24. $y = \dfrac{x^2 + 1}{7x^2 + 2}$
25. $y = \dfrac{x^2 + 8x - 3}{x^2 - x + 1}$
26. $y = \dfrac{x^2 - 7x + 5}{x^2 + 2x - 1}$
27. $y = \dfrac{\sqrt{x}}{\sqrt{x} + 1}$
28. $y = \dfrac{\sqrt[3]{x}}{\sqrt[3]{x} - 1}$
29. $y = \dfrac{x^3 + 1}{9x^3 - 1}$
30. $y = \dfrac{x^4 - 5}{x^4 + 6}$
31. $y = \dfrac{1 + 3/x}{1 + 2/x}$
32. $y = \dfrac{5/x^2}{1 + 1/x^3}$

Use the product rule to find the derivative of the given function.

33. $y = (2x + 1)(3x + 1)(4x + 1)$
34. $y = (x - 1)(5x + 3)(7x - 2)$
35. $y = (2x^2 + 4)(x^3 - 1)(8x^4 + 9x)$
36. $y = (x^2 + 3)(10x^3 + x)(x^4 - 2)$

Use the quotient and product rules to find the derivative of the given function.

37. $y = \dfrac{(x+1)(6x-2)}{2x+3}$

38. $y = \dfrac{(3x-1)(x+1)}{x-1}$

39. $y = \dfrac{x^2}{(x^2+1)(x^3-4)}$

40. $y = \dfrac{x^3}{(4x^2+x)(5x+9)}$

Find the equation of the tangent line to the graph of the given function at the indicated point.

41. $y = (x^2+1)(3x^2-x+1)$; $x = 1$

42. $y = (2x-1)(x^2+2x+2)$; $x = 1$

43. $y = \dfrac{4x}{2x-3}$; $x = 1$

44. $y = \dfrac{x^2}{x^2+1}$; $x = -1$

12.5

THE POWER RULE FOR FUNCTIONS

The product rule provides a method for differentiating $y = [f(x)]^n$, where n is a positive integer. For example, $y = [f(x)]^2$ can be written

$$y = f(x) \cdot f(x)$$

and hence by the product rule we have

$$\dfrac{dy}{dx} = f(x) \cdot f'(x) + f(x) \cdot f'(x)$$

(1) $\qquad\qquad\qquad = 2f(x) \cdot f'(x).$

Similarly, if

$$y = [f(x)]^3$$

then

$$y = f(x) \cdot [f(x)]^2$$

and so by (1) we can write

$$\dfrac{dy}{dx} = f(x) \cdot \overbrace{[2f(x) \cdot f'(x)]}^{\text{derivative of second factor}} + [f(x)]^2 \cdot f'(x)$$

$$= 3[f(x)]^2 \cdot f'(x).$$

This suggests, but in no way proves, the following general rule for differentiating any power of a function.

12.5 The Power Rule for Functions

> **VII Power rule for functions**
>
> If n is any real number, then the derivative of $y = [f(x)]^n$ is
>
> $$\frac{dy}{dx} = n[f(x)]^{n-1} \cdot f'(x).$$

Although rule VII is very similar to the procedure of rule I, it must be remembered that, after bringing down the exponent and decreasing the power, we multiply the result by the derivative of the function *inside* the power. That is,

$$\frac{d}{dx}[\ ]^n = n[\ ]^{n-1} \cdot \frac{d}{dx}[\ ]$$

differentiate function inside power

Example Differentiate $y = \sqrt{6x + 1}$.

Solution We first write the function as $y = (6x + 1)^{1/2}$. Hence, by rule VII we have

$$\frac{dy}{dx} = \frac{1}{2}(6x + 1)^{-1/2} \cdot 6$$

derivative

$$= 3(6x + 1)^{-1/2}.$$

Example Differentiate $y = (4x^3 + x)^{10}$.

Solution By rule VII,

$$\frac{dy}{dx} = 10(4x^3 + x)^9 \cdot (12x^2 + 1).$$

Example Differentiate $y = x^3(x^2 + 1)^7$.

Solution We first use the product rule,

$$\frac{dy}{dx} = x^3 \cdot \frac{d}{dx}(x^2 + 1)^7 + (x^2 + 1)^7 \cdot \frac{d}{dx}x^2$$

followed by the power rules I and VII,

$$\frac{dy}{dx} = x^3 \cdot 7(x^2 + 1)^6 \cdot 2x + (x^2 + 1)^7 \cdot 2x$$

$$\frac{dy}{dx} = 14x^4(x^2+1)^6 + 2x(x^2+1)^7.$$
$$= (x^2+1)^6(14x^4 + 2x^3 + 2x)$$

The power rule is a special case of a general procedure for differentiating a function $g(x)$ that is combined with another function $f(x)$. To differentiate the combined function $g(f(x))$, we proceed according to the next rule.

VIII Chain rule

If $y = g(u)$ and $u = f(x)$—that is, $y = g(f(x))$—then

$$\frac{dy}{dx} = \frac{dy}{du} \cdot \frac{du}{dx} = g'(f(x)) \cdot f'(x).$$

In rule VII, we make the identifications $y = u^n = [f(x)]^n$ and

$$\frac{dy}{du} = nu^{n-1} = n[f(x)]^{n-1}$$

$$\frac{du}{dx} = f'(x).$$

Symbolically, if we denote a function $f(x)$ by [ ], then the chain rule is

$$\frac{d}{dx} g[\ ] = g'[\ ] \frac{d}{dx}[\ ]$$

differentiate function inside g

We shall not belabor further details of the chain rule at this point, but it will prove useful in Sections 12.8 and 12.9.

EXERCISE 12.5

Find the derivative of the given function.

1. $y = (x+1)^2$
2. $y = (-x+3)^2$
3. $y = \sqrt{4x-4}$
4. $y = \sqrt{2x+7}$
5. $y = \sqrt{x^2+1}$
6. $y = \sqrt{6x^2-1}$

12.6 Implicit Differentiation

7. $y = (7x^2 - x)^3$
8. $y = (2x^2 - x + 1)^4$
9. $y = (-x + 2)^{-2}$
10. $y = (x + 5)^{-3}$
11. $y = (0.1x + 9)^{0.21}$
12. $y = (-0.2x + 4)^{-0.35}$
13. $y = (-3x^3 + x^2 + 2x)^{10}$
14. $y = (6x^3 + 2x^2 - 5x - 1)^{12}$
15. $y = \sqrt[3]{6x - 1}$
16. $y = \sqrt[3]{2x + 6}$
17. $y = \sqrt[3]{(4x + 1)^2}$
18. $y = \sqrt{(x - 1)^3}$
19. $y = \dfrac{1}{(x^3 - 1)^4}$
20. $y = \dfrac{2}{(2x^4 + 1)^5}$
21. $y = \dfrac{9}{(x^2 - 1)^{2/3}}$
22. $y = \dfrac{1}{(x^3 + 3x)^{5/3}}$
23. $y = x\sqrt{x + 1}$
24. $y = x^2\sqrt{2x - 1}$
25. $y = \sqrt{(5x + 1)(2x + 3)}$
26. $y = \sqrt{(3x - 1)(6x + 2)}$
27. $y = (2x + 1)\sqrt{x^2 - 2}$
28. $y = (3x^2 + 1)\sqrt{5x + 7}$
29. $y = \dfrac{x}{\sqrt{x + 1}}$
30. $y = \dfrac{\sqrt{x + 1}}{x}$
31. $y = \dfrac{\sqrt{x^2 + 1}}{x^2 - 1}$
32. $y = \dfrac{\sqrt{2x^3 - 1}}{x^3 + 8}$
33. $y = \sqrt{\dfrac{x}{x + 1}}$
34. $y = \left(\dfrac{x - 1}{x + 1}\right)^3$
35. $y = \sqrt{2x + 1}\sqrt[3]{3x + 1}$
36. $y = (4x - 2)^3\sqrt{2x + 9}$

Find the equation of the tangent line to the graph of the given function at the indicated point.

37. $y = (2x - 1)^4$; $x = 1$
38. $y = (3x + 2)^5$; $x = -1/3$
39. $y = \sqrt{3x + 1}$; $x = 0$
40. $y = \sqrt{x^2 + 5}$; $x = 2$

12.6
IMPLICIT DIFFERENTIATION

EXPLICIT AND IMPLICIT FUNCTIONS

Up to this point in our discussion of calculus we have been concerned with functions in which the variable *y* has been given explicitly in terms of the variable *x* and constants. Such functions are referred to as **explicit functions** and written as $y = f(x)$.

For example, $y = \frac{1}{2}x^2 - 3x + 7$ is an explicit function. In the alternative form, $2y - x^2 + 6x - 14 = 0$, we say the function is defined **implicitly**. In general, an equation of the form

$$F(x, y) = 0$$

may implicitly define *one or more* functions.

Example The equation $y^2 - x = 0$ leads to

$$y^2 = x$$

or

$$y = \pm \sqrt{x}.$$

Thus, the original equation yields two different functions

$$y = \sqrt{x} \quad \text{and} \quad y = -\sqrt{x}.$$

The graphs of all three equations are given in the accompanying figure.

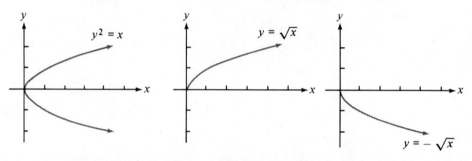

Even though most equations $F(x, y) = 0$ define functions, it is not always possible to obtain these functions in explicit form. For example, the reader should ponder the difficulty of solving $y^3 - 2xy + x^2 - 1 = 0$ for y in terms of x. However, it is always possible to determine the derivative dy/dx *without* actually solving for y. The process that we utilize for obtaining the derivative is known as **implicit differentiation**.

In the examples that follow, we shall use the power rule for functions

(1) $$\frac{d}{dx}[f(x)]^n = n[f(x)]^{n-1} \cdot f'(x).$$

By replacing $f(x)$ with y and $f'(x)$ with dy/dx, we note that (1) has the alternative form

(2) $$\frac{d}{dx} y^n = ny^{n-1} \frac{dy}{dx}.$$

12.6 Implicit Differentiation

For example, in the cases $n = 2$ and $n = 3$, (2) gives

$$\frac{d}{dx} y^2 = 2y^{2-1} \frac{dy}{dx} = 2y \frac{dy}{dx}$$

and

$$\frac{d}{dx} y^3 = 3y^{3-1} \frac{dy}{dx} = 3y^2 \frac{dy}{dx}$$

respectively.

Example Find dy/dx for $x^2 + y^2 = 1$.

Solution Implicit differentiation simply consists of differentiating both members of the equation, using (2) and other appropriate rules, and then solving for dy/dx:

$$\frac{d}{dx} x^2 + \frac{d}{dx} y^2 = \frac{d}{dx} 1$$

$$2x + 2y \frac{dy}{dx} = 0$$

$$\frac{dy}{dx} = -\frac{x}{y}.$$

Note that the answer in the previous example depends on both x and y. This reflects the fact that the slope of a tangent to the graph of the equation depends on the coordinates of a point under consideration.

Example Find dy/dx for $y^2 + x^2 y^3 - 5x = 0$.

Solution

$$\frac{d}{dx} y^2 + \underbrace{\frac{d}{dx} x^2 y^3}_{\text{product}} - 5 \frac{d}{dx} x = 0$$

$$2y \frac{dy}{dx} + \underbrace{x^2 \cdot 3y^2 \frac{dy}{dx} + y^3 \cdot 2x}_{\text{product rule}} - 5 = 0$$

$$(2y + 3x^2 y^2) \frac{dy}{dx} = 5 - 2xy^3$$

$$\frac{dy}{dx} = \frac{5 - 2xy^3}{2y + 3x^2 y^2}$$

Example Find the equation(s) of the tangent line(s) to the graph of $y^2 - x = 0$ at $x = 4$.

Solution When $x = 4$, we see that $y^2 - 4 = 0$ yields $y = 2$ and $y = -2$. Thus, as shown in the figure, there are two tangent lines, one at $(4, 2)$ and the other at $(4, -2)$. Now by implicit differentiation we obtain

$$\frac{d}{dx}y^2 - \frac{d}{dx}x = \frac{d}{dx}0$$

$$2y\frac{dy}{dx} - 1 = 0$$

$$\frac{dy}{dx} = \frac{1}{2y}.$$

Substituting $y = 2$ and $y = -2$ gives, respectively,

$$\left.\frac{dy}{dx}\right|_{(4,2)} = \frac{1}{4}$$

and

$$\left.\frac{dy}{dx}\right|_{(4,-2)} = \frac{1}{-4} = -\frac{1}{4}.$$

The equation of the tangent at $(4, 2)$ is then

$$y - 2 = \frac{1}{4}(x - 4)$$

or

$$y = \frac{1}{4}x + 1.$$

Similarly, at $(4, -2)$ we find that the tangent line is

$$y - (-2) = -\frac{1}{4}(x - 4)$$

$$y = -\frac{1}{4}x - 1.$$

EXERCISE 12.6

Find dy/dx for the given equation.

1. $2y + 4x^2 - 7 = 0$
2. $-5y + x^2 + 13 = 0$
3. $y^2 - 2x + 4 = 0$
4. $3y^2 + 12x - 9 = 0$

5. $x^2 + y^2 = 16$

6. $x^2 - y^2 = 16$

7. $3x^2 - 4y^2 - 12 = 0$

8. $\dfrac{x^2}{4} + \dfrac{y^2}{9} = 1$

9. $y^3 + y^2 + y = x + 2$

10. $y^3 - y^2 + x^2 - 2 = 0$

11. $y^3 - 3xy + 5 = 0$

12. $y^3 + xy^2 - 9 = 0$

13. $x^3 + x^2y^2 - 6x + 10 = 0$

14. $x^2y^4 - 3x^2 + 7y - 7 = 0$

15. $\sqrt{x} + \sqrt{y} = 1$

16. $\sqrt[3]{x} - \sqrt[3]{y} = 2$

17. $(1 + y^2)^2 = x^2 + 9$

18. $\sqrt{1 + 4y} - 5x^2 - 10 = 0$

19. $y^2 = \dfrac{x+1}{x-1}$

20. $x = \dfrac{y-1}{y+1}$

Find the equation(s) of the tangent line(s) to the graph of the given equation at the indicated point.

21. $y^2 - 4x - 17 = 0; \ x = 2$

22. $y^2 = x + 1; \ x = 3$

23. $x^2 + y^2 = 25; \ x = 4$

24. $x^2 + 9y^2 = 2; \ x = 1$

25. $xy = 4; \ x = 1$

26. $x^2 y = 16; \ x = 2$

27. $y^3 - xy + 5 = 0; \ (6, 1)$

28. $y + xy^2 - x + 4 = 0; \ (-2, 2)$

29. $\dfrac{1}{x} + \dfrac{1}{y} = 1; \ \left(\dfrac{1}{2}, -1\right)$

30. $\dfrac{1}{x^2} - \dfrac{1}{y^2} = 8; \ \left(\dfrac{1}{3}, 1\right)$

12.7

HIGHER-ORDER DERIVATIVES

SECOND DERIVATIVE Given a function $y = f(x)$, then its derivative $dy/dx = f'(x)$ is also a function of x. Hence, we can also take the *derivative of the derivative*. Once we have taken the second derivative we may proceed to take even higher derivatives. The **second derivative** is obtained by

$$\dfrac{d}{dx}\left(\dfrac{dy}{dx}\right)$$

and is denoted symbolically by either

$$\dfrac{d^2y}{dx^2}, \quad y'', \quad \text{or} \quad f''(x).$$

Example Find the second derivative of $y = x^2$.

Solution The first derivative is

$$\frac{dy}{dx} = 2x.$$

Hence, the derivative of this derivative—that is, the second derivative—is

$$\frac{d^2y}{dx^2} = 2.$$

Example Find the second derivative of $y = x^{3/2}$.

Solution The first derivative is

$$\frac{dy}{dx} = \frac{3}{2} x^{1/2}.$$

Applying rule I again then gives

$$\frac{d^2y}{dx^2} = \frac{3}{2}\left(\frac{1}{2}\right) x^{-1/2}$$

$$= \frac{3}{4} x^{-1/2}.$$

THIRD DERIVATIVE The **third derivative** is defined to be the derivative of the second derivative:

$$\frac{d}{dx}\left(\frac{d^2y}{dx^2}\right)$$

and is denoted by either

$$\frac{d^3y}{dx^3}, \quad y''', \quad \text{or} \quad f'''(x).$$

Higher-order derivatives are defined in the same manner. A derivative of order n is denoted by $d^n y/dx^n$.

Example Find the third derivative of $y = 8x^{1/2}$.

Solution We apply the power rule, rule I, three times:

$$\frac{dy}{dx} = 8\left(\frac{1}{2}\right) x^{-1/2} = 4x^{-1/2},$$

12.7 Higher-Order Derivatives

$$\frac{d^2y}{dx^2} = 4\left(-\frac{1}{2}\right)x^{-3/2} = -2x^{-3/2},$$

$$\frac{d^3y}{dx^3} = -2\left(-\frac{3}{2}\right)x^{-5/2} = 3x^{-5/2}.$$

Example Find the second derivative of $y = (x^2 + 1)^4$.

Solution To find the first derivative, we use the power rule for functions, rule VII,

$$\frac{dy}{dx} = 4(x^2 + 1)^3 \cdot 2x = 8x(x^2 + 1)^3.$$

Note that the first derivative is now a product of two functions. Hence, to compute the second derivative we use the product rule and the power rule for functions:

$$\frac{d^2y}{dx^2} = 8x \cdot 3(x^2 + 1)^2 \cdot 2x + (x^2 + 1)^3 \cdot 8$$

$$= 48x^2(x^2 + 1)^2 + 8(x^2 + 1)^3.$$

$$= (x^2 + 1)^2(56x^2 + 8)$$

EXERCISE 12.7

Find the second derivative of the given function.

1. $y = x^4$
2. $y = 3x^3$
3. $y = 6x^{1/2}$
4. $y = 12x^{1/3}$
5. $y = 5x^2 + x$
6. $y = x^2 - 4x + 1$
7. $y = 2x^3 - 4x + 3$
8. $y = 4x^3 + x^2 - 1$
9. $y = \dfrac{2}{x}$
10. $y = \dfrac{1}{x^2}$
11. $y = 9\sqrt[3]{x} + 4\sqrt{x}$
12. $y = 10\sqrt{x} - 6\sqrt[3]{x}$
13. $y = \dfrac{1}{x+1}$
14. $y = \dfrac{1}{2x-3}$
15. $y = (3x + 7)^4$
16. $y = (-x + 9)^3$
17. $y = \dfrac{1+x}{1-x}$
18. $y = \dfrac{1-2x}{1+4x}$

19. $y = \sqrt{x^3 + 1}$ **20.** $y = \sqrt[3]{6x^2 - 1}$

Find the third derivative of the given function.

21. $y = 6x^2 - x + 5$ **22.** $y = 3x^2 - 7x + 8$
23. $y = 36x^{1/3}$ **24.** $y = 20x^{1/2}$
25. $y = 4x^3 - 6x^2 + 9x - 1$ **26.** $y = -\frac{1}{3}x^4 + \frac{1}{2}x^2 - 7x + 20$
27. $y = \dfrac{1}{1-x}$ **28.** $y = \dfrac{1}{3x+1}$
29. $y = \dfrac{x}{x+1}$ **30.** $y = \dfrac{x+1}{x}$

Find the second derivative of the given implicit function.

Example $x^2 + y^2 = 9$

Solution The first derivative is obtained from

$$2x + 2y \frac{dy}{dx} = 0$$

or

$$\frac{dy}{dx} = -\frac{x}{y}.$$

To find the second derivative, we now use the quotient rule, rule VI,

$$\frac{d^2y}{dx^2} = -\frac{y \cdot 1 - x \cdot \frac{dy}{dx}}{y^2}.$$

Substituting $dy/dx = -x/y$ then yields

$$\frac{d^2y}{dx^2} = -\frac{y - x(-x/y)}{y^2}$$

$$= -\frac{x^2 + y^2}{y^3}.$$

Observe that since $x^2 + y^2 = 9$, the second derivative can also be written as $d^2y/dx^2 = -9/y^3$.

31. $2x^2 + 4y^2 = 1$ **32.** $x^2 - 3y^2 = 1$
33. $x^2 - 2x + y^2 = 0$ **34.** $4x^2 + y^2 - 4y = 5$

12.8
DERIVATIVE OF THE LOGARITHMIC FUNCTION

The number e introduced in Section 5.1 will play an important role in the development of the derivative of the logarithmic function $y = \log_b x$, $b > 0$, $b \neq 1$. Recall from Section 11.4 that we defined the number e by

$$e = \lim_{n \to \infty} \left(1 + \frac{1}{n}\right)^n.$$

Observe that if we let $t = 1/n$, then $n = 1/t$, and as $n \to \infty$, necessarily $t \to 0$. This leads to the alternative definition

(1) $$e = \lim_{t \to 0} (1 + t)^{1/t}.$$

We shall use this formulation immediately in justifying the next rule of differentiation.

IX If $y = \log_b x$, $b > 0$, $b \neq 1$, then

$$\frac{dy}{dx} = \frac{1}{x} \log_b e.$$

Justification of IX

$$\frac{f(x+h) - f(x)}{h} = \frac{\log_b(x+h) - \log_b x}{h}$$

$$= \frac{1}{h} \log_b \frac{x+h}{x} \qquad \text{[Property II, page 164]}$$

$$= \frac{1}{h} \log_b \left(1 + \frac{h}{x}\right)$$

$$= \frac{1}{x} \cdot \frac{x}{h} \log_b \left(1 + \frac{h}{x}\right) \qquad \left[\frac{1}{x} \cdot x = 1\right]$$

$$= \frac{1}{x} \log_b \left(1 + \frac{h}{x}\right)^{x/h} \qquad \text{[Property III, page 164]}$$

If we let $t = h/x$, then as $h \to 0$ so must $t \to 0$ for a fixed x. From (1) $\lim_{t \to 0} (1+t)^{1/t} = e$; thus, if we formally take the limit inside the logarithm,

$$\frac{dy}{dx} = \frac{1}{x} \log_b \left[\lim_{t \to 0} (1+t)^{1/t} \right]$$

$$= \frac{1}{x} \log_b e$$

Example Differentiate $y = \log_{10} x$.

Solution By rule IX, we have

$$\frac{dy}{dx} = \frac{1}{x} \log_{10} e.$$

Note that if $b = e$, then rule IX implies that the derivative of the natural logarithm $y = \log_e x$ is

(2) $$\frac{dy}{dx} = \frac{1}{x} \log_e e.$$

Since $\log_e e = 1$, (2) becomes $dy/dx = 1/x$. In calculus, the base e is used exclusively; this avoids the necessity of having to carry along the constant $\log_b e$ when differentiating a logarithmic function. Also, it is convenient to introduce a new symbol for the natural logarithm:

$$\log_e x = \ln x.$$

Very often the statement "logarithm to the base e of x" is simply read phonetically as "ell-en of x". We formally restate the result of (2) in the next rule.

X If $y = \ln x$, then

$$\frac{dy}{dx} = \frac{1}{x}.$$

Example Differentiate $y = x \ln x$.

Solution By the product rule, rule V, we have

$$\frac{dy}{dx} = x \cdot \frac{d}{dx} \ln x + \ln x \cdot \frac{d}{dx} x$$

$$= x \cdot \frac{1}{x} + 1 \cdot \ln x$$

$$= 1 + \ln x.$$

12.8 Derivative of the Logarithmic Function

To differentiate the logarithm of a function we apply the chain rule, rule VIII.

XI If $y = \ln u$, where $u = f(x)$, then
$$\frac{dy}{dx} = \frac{1}{u} \cdot \frac{du}{dx}$$
$$= \frac{1}{f(x)} \cdot f'(x).$$

In other words, to differentiate a logarithm of a function we write the reciprocal of the function times the derivative of function:

$$\frac{d}{dx} \ln [\ ] = \frac{1}{[\ ]} \cdot \frac{d}{dx} [\ ]$$

derivative of function inside logarithm

Example Differentiate $y = \ln(5x + 2)$.

Solution From rule XI, we obtain,
$$\frac{dy}{dx} = \frac{1}{5x+2} \cdot \frac{d}{dx}(5x+2)$$
$$= \frac{5}{5x+2}.$$

Example Differentiate $y = \ln(x^4 + 2x^2)$.

Solution From rule XI,
$$\frac{dy}{dx} = \frac{1}{x^4 + 2x^2} \cdot \frac{d}{dx}(x^4 + 2x^2)$$
$$= \frac{4x^3 + 4x}{x^4 + 2x^2}.$$

Example Differentiate $y = \ln x^2$.

Solution Before differentiating, we note that by the properties of logarithms we can write $\ln x^2$ as $2 \ln x$. Thus,
$$\frac{dy}{dx} = 2 \cdot \frac{1}{x}$$
$$= \frac{2}{x}.$$

Alternative Solution By applying rule XI directly, it follows that

$$\frac{dy}{dx} = \frac{1}{x^2} \cdot \frac{d}{dx} x^2$$

$$= \frac{1}{x^2} \cdot 2x$$

$$= \frac{2}{x}.$$

Example Differentiate $y = (\ln x)^2$.

Solution Before using any rule of differentiation, the reader should observe that $(\ln x)^2 \neq \ln x^2$. In this case, there is no procedure for simplifying the function prior to taking the derivative. From the power rule for functions, rule VII, we have

$$\frac{dy}{dx} = 2(\ln x)^1 \cdot \frac{d}{dx} \ln x$$

$$= 2(\ln x) \cdot \frac{1}{x}$$

$$= \frac{2 \ln x}{x}.$$

EXERCISE 12.8

Find the derivative of the given function.

1. $y = \ln 4x$
2. $y = \ln 10x$
3. $y = \ln(2x + 1)$
4. $y = \ln(6x - 5)$
5. $y = \ln(x^2 + x)$
6. $y = \ln(x^2 + 2x + 3)$
7. $y = \ln x^3$
8. $y = \ln x^4$
9. $y = 4 \ln \sqrt{x}$
10. $y = 3 \ln \sqrt[3]{x}$
11. $y = \ln(x + \sqrt{x^2 + 1})$
12. $y = \ln(x + \sqrt[3]{x})$
13. $y = (\ln x)^3$
14. $y = (\ln x)^{1/3}$
15. $y = \ln x^2(2x - 1)^3$
16. $y = \ln(3x + 3)^2(5x - 1)^3$
17. $y = \ln \dfrac{1}{x}$
18. $y = \ln \dfrac{1}{x^2}$

12.8 Derivative of the Logarithmic Function

19. $f(x) = x^2 \ln x$

20. $f(x) = (x^2 + 1) \ln 7x$

21. $f(x) = \dfrac{\ln x}{x}$

22. $f(x) = \dfrac{x^2}{\ln x}$

Use implicit differentiation to find dy/dx for the given equation.

23. $4y + x^2 - \ln y = 6$

24. $3y^2 + \ln xy + x^4 - 2 = 0$

Find the second derivative of the given function.

25. $y = \ln x$

26. $y = x \ln x$

27. $y = \ln(x^2 + 1)$

28. $y = \ln x^3 (6x + 1)$

29. Find $f'(x)$ for $f(x) = \ln(\ln x)$.

30. Explain why $g(x) = \ln kx$, $k > 0$, has the same derivative as $f(x) = \ln x$.

Find the equation of the tangent line to the graph of the given function at the indicated point.

31. $y = \ln(5x + 1)$; $x = 0$

32. $y = -\ln(2x + 1)$; $x = 0$

33. $y = \ln x^2$; $x = e$

34. $y = x \ln x$; $x = e$

*In Problems 35–40 compute the derivative by taking the logarithm of both members of the equation. The process is called **logarithmic differentiation**.*

Example $y = x^{x^2}$

Solution $\ln y = \ln x^{x^2}$

$$= x^2 \cdot \ln x$$

Take the derivative of both members

$$\frac{1}{y} \cdot \frac{dy}{dx} = x^2 \cdot \frac{1}{x} + 2x \ln x$$

$$\frac{dy}{dx} = y(x + 2x \ln x).$$

But $y = x^{x^2}$, so we obtain

$$\frac{dy}{dx} = x^{x^2}(x + 2x \ln x).$$

35. $y = x^x$

36. $y = (5x^2 + 4)^{x^2}$

37. $y = \dfrac{(x+1)^{3/2} x^{4/3}}{(x^2+1)^5}$

38. $y = \dfrac{(x^2 - x)^{1/2}(3x + 5)^{2/3}}{(4 + x^3)^2}$

39. $y = (x^4 + x^2)^{2x}$

40. $y = 5^{3x-1}$

12.9

DERIVATIVE OF THE EXPONENTIAL FUNCTION

To differentiate the exponential function

(1) $$y = e^x$$

we first use the definition of a logarithm to write (1) in the equivalent form

(2) $$x = \ln y.$$

Applying implicit differentiation to (2) then yields

$$\frac{d}{dx} x = \frac{d}{dx} \ln y$$

$$1 = \frac{1}{y} \cdot \frac{dy}{dx}$$

or

(3) $$\frac{dy}{dx} = y.$$

But $y = e^x$, and so (3) leads to the following interesting and simple result.

XII If $y = e^x$, then

$$\frac{dy}{dx} = e^x.$$

By the chain rule, rule VIII, we can immediately generalize the above rule to the following.

XIII If $y = e^u$, where $u = f(x)$, then

$$\frac{dy}{dx} = e^u \cdot \frac{du}{dx}$$

$$= e^{f(x)} \cdot f'(x).$$

Rule XIII states that the derivative of an exponential function is the exponential function multiplied by the derivative of the exponent:

12.9 Derivative of the Exponential Function

$$\frac{d}{dx} e^{[\ ]} = e^{[\ ]} \cdot \frac{d}{dx} [\ ]$$

derivative of exponent

Example Differentiate $y = e^{-x}$.

Solution By rule XIII we have

$$\frac{dy}{dx} = e^{-x} \cdot \frac{d}{dx}(-x)$$

$$= -e^{-x}.$$

Example Differentiate $y = e^{x^3}$.

Solution By rule XIII,

$$\frac{dy}{dx} = e^{x^3} \cdot \frac{d}{dx} x^3$$

$$= 3x^2 e^{x^3}.$$

Example Differentiate $y = x^2 e^{4x+3}$.

Solution By the product rule, rule V, we have

$$\frac{dy}{dx} = x^2 \cdot \frac{d}{dx} e^{4x+3} + e^{4x+3} \cdot \frac{d}{dx} x^2$$

$$= x^2 \cdot e^{4x+3} \cdot 4 + e^{4x+3} \cdot 2x$$

$$= 4x^2 e^{4x+3} + 2x e^{4x+3}.$$

Example Find the equation of the tangent line to the graph of $y = e^x$ at $x = 0$.

Solution When $x = 0$, we see that $y = e^0 = 1$. The slope of the tangent at $(0, 1)$ is then

$$\left.\frac{dy}{dx}\right|_{x=0} = e^x\big|_{x=0}$$

$$= 1.$$

The equation of the tangent line follows from the point-slope form:

$$y - 1 = 1 \cdot (x - 0)$$

$$y = x + 1.$$

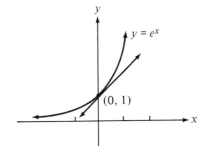

EXERCISE 12.9

Find the derivative of the given function.

1. $y = e^{4x}$
2. $y = e^{10x}$
3. $y = e^{-3x}$
4. $y = e^{-3x/2}$
5. $y = e^{2x+6}$
6. $y = e^{7x-4}$
7. $y = e^{-x^2}$
8. $y = e^{x^2+5}$
9. $y = e^{\sqrt{x}}$
10. $y = e^{\sqrt[3]{x}}$
11. $y = e^{1/x}$
12. $y = e^{1/x^2}$
13. $y = e^{x^3-x}$
14. $y = e^{4x^3+3x+1}$
15. $y = e^{(2x+1)^2}$
16. $y = e^{(-4x+2)^3}$
17. $y = \dfrac{e^{4x} - e^{-4x}}{2}$
18. $y = \dfrac{e^{3x} + e^{-3x}}{2}$
19. $y = (2x + 6)e^{9x}$
20. $y = (7x^2 - 1)e^x$
21. $y = e^{-x} \ln(2x + 4)$
22. $y = e^{3x} \ln x$
23. $y = \dfrac{x}{e^x}$
24. $y = \dfrac{e^x}{x^2}$
25. $y = \dfrac{e^x}{e^x + 1}$
26. $y = \dfrac{e^x - e^{-x}}{e^x + e^{-x}}$
27. $y = (x + 1)^3 e^{6x}$
28. $y = \sqrt{5x + 2}\, e^{-x}$
29. $y = (x^2 + 4e^{x^3})^2$
30. $y = (e^{-3x} + \ln x)^3$

Use implicit differentiation to find dy/dx for the given equation.

31. $y + x - 1 + e^{4y} = 0$
32. $e^{xy^2} = x + 8$

Find the second derivative of the given function.

33. $y = e^{-x}$
34. $y = 3e^{5x}$
35. $y = e^{x^2}$
36. $y = xe^x$
37. $y = \ln(e^x + 8)$
38. $y = \ln(e^{-x} + 1)$

Find the equation of the tangent line to the graph of the given function at the indicated point.

39. $y = 2e^{-x};\ x = 0$
40. $y = 4e^{3x};\ x = 0$

12.9 Derivative of the Exponential Function

41. $y = x^2 e^x$; $x = 1$ **42.** $y = x^3 e^{2x}$; $x = -1$

43. Use the procedure outlined in this section to find the derivative of $y = b^x$, $b > 0$, $b \neq 1$.

Use the result of Problem 43 to find the derivative of the given function.

44. $y = 10^x$ **45.** $y = 3^x$

46. Explain why the derivatives of $y = e^{\ln x}$ and $y = \ln e^x$ are the same for $x > 0$.

CHAPTER TEST

Use the definition of the derivative to find $f'(x)$ for the given function.

1. $f(x) = 3x^2 - 10x$
2. $f(x) = x^2 + 4x - 2$

Use the appropriate rule or procedure to find the derivative of the given function or equation.

3. $y = 4x + 9$
4. $y = -5x + 20$
5. $y = 6x^3 - \dfrac{1}{x}$
6. $y = 8\sqrt{x} - 3x^4$
7. $y = \sqrt{2x^2 + x}$
8. $y = (x^3 + 7x)^4$
9. $y = (x^3 + 4x^2 + 1)(5x + 2)$
10. $y = (6x^2 - 1)(x^5 - x^2)$
11. $y = \dfrac{4x + 1}{x^2}$
12. $y = \dfrac{\sqrt{x}}{x + 1}$
13. $y = \ln(7x^2 + 4)$
14. $y = \ln(x^2 + 1)^3$
15. $y = \ln(e^x + 1)$
16. $y = \ln(e^{x^2} + 1)$
17. $y = e^{4x} \cdot e^{6x}$
18. $y = \dfrac{e^{2x}}{e^{5x}}$
19. $y = e^{\sqrt{2x+8}}$
20. $y = e^{-(x^2+1)^2}$
21. $y = xe^x \ln x$
22. $y = 4xe^x \ln x^2$
23. $x - y + xy^2 - 9 = 0$
24. $y^2 + 4y = x + 3$
25. Find the second derivative of $y = (x^2 + 6)^4$.
26. Find the third derivative of $y = \dfrac{1}{(-x+1)^2}$.
27. Find the second derivative of $y = x^2 \ln x$.
28. Find $\dfrac{d^2y}{dx^2}$ for $2x^2 + y^2 = 15$.

Find the equation of the tangent line to the graph of the given function at the indicated point.

29. $y = 4x^2 - 6x + 1;\ x = 2$
30. $y = 3\ln(2x - 1);\ x = 1$

13
APPLICATIONS OF THE DERIVATIVE

13.1
THE FIRST AND SECOND DERIVATIVES AND THE SHAPE OF A GRAPH

The derivative of a function gives the slope of a tangent line to its graph. Knowledge of the behavior of the tangent is obviously related to the behavior, or *shape*, of the graph of a function. Our goal in the next two sections is to show how the first and second derivatives can be utilized to obtain a graph without the necessity for plotting a significant number of points.

CONTINUOUS FUNCTIONS In the discussion that follows, we are concerned only with **continuous functions**. For our purposes, it is sufficient to think of a continuous function as one whose graph can be drawn smoothly without lifting pencil from paper. Figure 13.1 shows the graph of the discontinuous function (not continuous) $y = 1/x$ and the continuous function $y = x^3$. It can be proved that all polynomial functions are continuous.

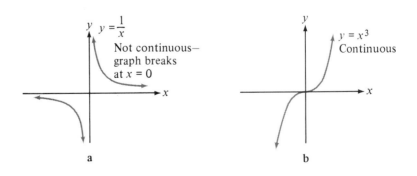

Figure 13.1

INCREASING AND DECREASING FUNCTIONS

Suppose $y = f(x)$ is a function for which we know the following two facts about the derivative $f'(x)$ near a number c on the x-axis:

(i) $f'(x) < 0$ for $x < c$, and

(ii) $f'(x) > 0$ for $x > c$.

Knowing that the slope of the tangent is negative to the left of c and positive to the right of c, we can conclude that the shape of the graph of f could *possibly* be as given in Figure 13.2, part b. Observe that $f'(x) < 0$ implies that the graph of f is going down,

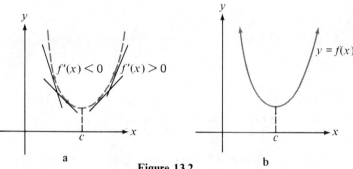

Figure 13.2

reading left to right, whereas $f'(x) > 0$ implies that the graph of the function must go up, reading left to right. Recall that when the graph goes down we say the function is **decreasing**, and when the graph goes up the function is said to be **increasing**. In general, by finding the intervals on the x-axis for which $f'(x)$ is positive or negative, we can find where $y = f(x)$ is increasing or decreasing, respectively. We state the following result without proof.

Let $y = f(x)$ be a continuous function with derivative $f'(x)$.
If $f'(x) > 0$ for all x in the interval $a < x < b$, then $y = f(x)$ is increasing on $a \leq x \leq b$.
If $f'(x) < 0$ for all x in the interval $a < x < b$, then $y = f(x)$ is decreasing on $a \leq x \leq b$.

When $f'(x) > 0$ for $x > a$ (or $x < a$), we say that f is increasing on $a \leq x < \infty$ (or $-\infty < x \leq a$). A similar comment holds when $f'(x) < 0$ for $x > a$ (or $x < a$).

Example   Given $f(x) = x^2$, find the intervals for which f is increasing. For which it is decreasing.

Solution   Compute the derivative: $f'(x) = 2x$.
Solve the inequalities $f'(x) > 0$ and $f'(x) < 0$:

$$2x > 0 \text{ implies } x > 0 \quad \text{and} \quad 2x < 0 \text{ implies } x < 0.$$

13.1 The First and Second Derivatives and the Shape of a Graph

We summarize the information in tabular form:

x	Sign of $f'(x)$	$y = f(x)$
$x > 0$	$+$	Increasing on $0 \leq x < \infty$
$x < 0$	$-$	Decreasing on $-\infty < x \leq 0$

Example Since the derivative of $f(x) = x^3$ is $f'(x) = 3x^2$ and $f'(x) > 0$ for $x > 0$ and $x < 0$, we conclude that f is increasing on the x-axis. The graph of the function, given in Figure 13.1, part b, illustrates this fact.

Example Given $f(x) = -x^2 + 6x$, find the intervals for which f is increasing. For which it is decreasing.

Solution Compute the derivative: $f'(x) = -2x + 6$.
Solve the inequalities $f'(x) > 0$ and $f'(x) < 0$:

$$-2x + 6 > 0 \text{ implies } x < 3 \quad \text{and} \quad -2x + 6 < 0 \text{ implies } x > 3.$$

x	Sign of $f'(x)$	$y = f(x)$
$x < 3$	$+$	Increasing on $-\infty < x \leq 3$
$x > 3$	$-$	Decreasing on $3 \leq x < \infty$

Example Given $f(x) = 2x^3 - 3x^2 - 12x$, find the intervals for which f is increasing. For which it is decreasing.

Solution Compute the derivative: $f'(x) = 6x^2 - 6x - 12 = 6(x + 1)(x - 2)$.
Solve the inequalities $f'(x) > 0$ and $f'(x) < 0$:
Using the procedures of Section 3.6 we find that the solution sets of the quadratic inequalities

$$6(x + 1)(x - 2) > 0 \quad \text{and} \quad 6(x + 1)(x - 2) < 0$$

are $\{x > 2 \text{ or } x < -1\}$ and $\{-1 < x < 2\}$, respectively. The results are summarized below.

x	Sign of $f'(x)$	$y = f(x)$
$x < -1$	$+$	Increasing on $-\infty < x \leq -1$
$-1 < x < 2$	$-$	Decreasing on $-1 \leq x \leq 2$
$x > 2$	$+$	Increasing on $2 \leq x < \infty$

CONCAVITY From the foregoing discussion it follows that if the second derivative $f''(x)$ is positive on some interval, then $f'(x)$ is increasing. When $f'(x)$ increases, the shape of the graph of f must necessarily be similar to that given in Figure 13.3, part a. Furthermore, if

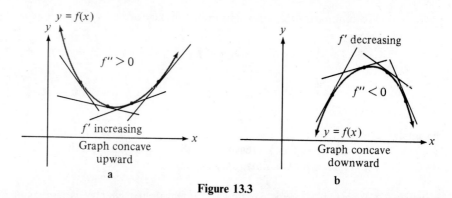

Figure 13.3

$f''(x)$ is negative on an interval, then $f'(x)$ decreases and the shape of the graph is as shown in part b of Figure 13.3. In other words, the second derivative indicates the **concavity** of a graph on an interval. When the graph on the interval lies above the tangent lines, we say that the graph is **concave upward**. When the graph on the interval lies below the tangent lines, the graph is said to be **concave downward**.

The following is a test for concavity.

Let $y = f(x)$ be a continuous function with second derivative $f''(x)$.
If $f''(x) > 0$ for all x in the interval $a < x < b$, then the graph of f is concave upward on $a < x < b$.
If $f''(x) < 0$ for all x in the interval $a < x < b$, then the graph of f is concave downward on $a < x < b$.

A point $(c, f(c))$ where the graph of $y = f(x)$ changes concavity is said to be a **point of inflection**.

Figure 13.4 illustrates a graph of a function having two points of inflection: $(b, f(b))$ and $(c, f(c))$.

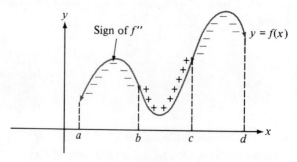

Figure 13.4

13.1 The First and Second Derivatives and the Shape of a Graph

Example The second derivative of $f(x) = x^2$ is $f''(x) = 2$. Since $f''(x) > 0$ for all values of x, the graph of the function, as shown in the figure, is concave upward on the x-axis.

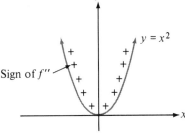

Example Given $f(x) = x^3 - 3x^2$, find the intervals for which f is concave upward. For which it is concave downward.

Solution Compute the second derivative:
$$f'(x) = 3x^2 - 6x$$
$$f''(x) = 6x - 6.$$

Solve the inequalities $f''(x) > 0$ and $f''(x) < 0$:

$6x - 6 > 0$ implies $x > 1$ and $6x - 6 < 0$ implies $x < 1$.

We again summarize the results in tabular form.

x	Sign of $f''(x)$	$y = f(x)$
$x > 1$	+	Concave upward on $1 < x < \infty$
$x < 1$	−	Concave downward on $-\infty < x < 1$

In the foregoing example, the graph of $f(x) = x^3 - 3x^2$ was shown to change concavity at a point corresponding to $x = 1$. Hence, $(1, f(1)) = (1, -2)$ is a point of inflection.

EXERCISE 13.1

Use the given figure to solve Problems 1 and 2.

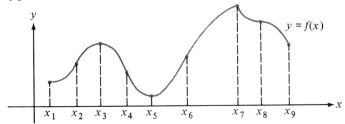

1. Find the intervals for which f is increasing. For which it is decreasing.
2. Find the intervals for which f is concave upward. For which it is concave downward. Give the points of inflection.

Use the first derivative to find the intervals for which the given function is increasing. For which it is decreasing.

3. $f(x) = 3x - 7$
4. $f(x) = -6x + 1$
5. $f(x) = -x^2 + 4x + 2$
6. $f(x) = x^2 + 6x + 1$
7. $f(x) = (x - 1)^2$
8. $f(x) = (-2x + 1)^2$
9. $f(x) = -x^3 + 5$
10. $f(x) = 4x^3 + 20$
11. $f(x) = x^3 - 3x$
12. $f(x) = x^3 - 12x$
13. $f(x) = x^3 + 12x$
14. $f(x) = 10x^3 + 30x$
15. $f(x) = 4x^3 + 6x^2$
16. $f(x) = x^3 - 4x^2 + 3$
17. $f(x) = -2x^3 - 3x^2 + 36x$
18. $f(x) = x^3 - 9x^2 + 15x$
19. $f(x) = x^4 - 2x^2$
20. $f(x) = x^4 + 8x^3$

Use the second derivative to find the intervals for which the given function is concave upward. For which it is concave downward. Find any points of inflection.

21. $f(x) = x^3 - x$
22. $f(x) = -5x^3 + 2x$
23. $f(x) = x^3 - 9x^2$
24. $f(x) = 2x^3 + 6x^2$
25. $f(x) = x^3 + 6x^2 + x$
26. $f(x) = x^3 - 15x^2 - 4x + 1$
27. $f(x) = x^4$
28. $f(x) = x^5$
29. $f(x) = x^4 - 2x^3$
30. $f(x) = x^4 - 24x^2$

Applications

31. *Business* Given the revenue and cost functions $R(x) = 100x - 0.01x^2$ and $C(x) = 50x + 100$, find the interval for which the profit function is increasing.
32. Given the cost function $C(x) = x^2 + 640x + 900$, find the interval for which the average cost function $Q(x) = C(x)/x$, $x > 0$, is decreasing.

13.2

MAXIMUM AND MINIMUM VALUES OF A FUNCTION

CRITICAL VALUES   By determining where the slope of a tangent is zero we may be able to find the turning points of a graph. These turning points are the **local maxima** or **minima** (**extrema**) of a function (see Section 4.5).

13.2 Maximum and Minimum Values of a Function

> A number c in the domain of a function $y = f(x)$ for which $f'(c) = 0$ or is undefined is said to be a **critical value**.

Critical values are the abscissas of the points on a graph where the tangent is horizontal (zero slope), vertical (undefined slope), or does not exist. Figure 13.5 shows the graphs of three continuous functions illustrating these three possibilities. We will be concerned principally with the case where $f'(x) = 0$.

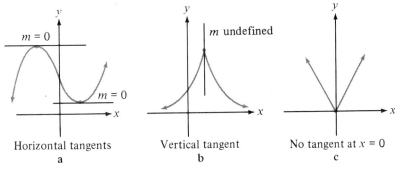

| Horizontal tangents | Vertical tangent | No tangent at $x = 0$ |
| a | b | c |

Figure 13.5

Example Find the critical values of $f(x) = x^2 + 2x + 4$.

Solution Compute the derivative: $f'(x) = 2x + 2$.
Solve the equation $f'(x) = 0$:

$$2x + 2 = 0 \text{ implies } x = -1.$$

The only critical value is -1. The tangent to the graph of the given function is horizontal at $(-1, 3)$.

Example Find the critical values of $f(x) = x^3 - 3x^2$.

Solution Compute the derivative: $f'(x) = 3x^2 - 6x$.
Solve the equation $f'(x) = 0$:

$$3x^2 - 6x = 0 \text{ is equivalent to}$$

$$3x(x - 2) = 0.$$

We see that the critical values are 0 and 2. There are horizontal tangents at $(0, 0)$ and $(2, -4)$.

Example The derivative of the function $f(x) = x^{2/3}$ is $f'(x) = \frac{2}{3} x^{-1/3}$ or $f'(x) = \frac{2}{3x^{1/3}}$. Although

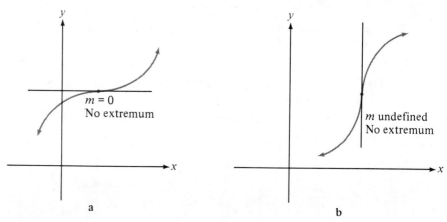

Figure 13.6

$f'(x)$ is never zero, it is seen, however, that $f'(x)$ is undefined at $x = 0$. Since $f(0) = 0$, we conclude that 0 is a critical value.

When a graph possesses a local maximum or a local minimum, these necessarily occur at points on the graph where the tangent is horizontal, vertical, or does not exist. It should be noted that the converse to this statement need not be true. Figure 13.6 shows that a graph may have a point at which the tangent is horizontal or vertical that does not correspond to a local maximum or minimum.

By finding the critical values we have a list of x values that are *possibly* the abscissas of the local extrema. In order to determine when a critical value actually corresponds to a local extremum, we have the following derivative test.

First Derivative Test

Let c be a critical value of $y = f(x)$.
If $f'(x) < 0$ for $x < c$ and $f'(x) > 0$ for $x > c$, then $(c, f(c))$ is a local minimum.
If $f'(x) > 0$ for $x < c$ and $f'(x) < 0$ for $x > c$, then $(c, f(c))$ is a local maximum.

This result follows immediately from the fact that the sign of f' indicates where f is increasing or decreasing. To the left of a local minimum (maximum) the function is decreasing (increasing), whereas to the right it is increasing (decreasing). Figure 13.7 shows two possibilities.

Also, when testing the sign of the derivative to the right and left of the critical value we should stay within a reasonably close neighborhood of the number since the function may have more than one turning point, as in Figure 13.8.

Example Find the local extrema of $f(x) = x^3 - 3x + 1$. Graph.

Solution The derivative is

13.2 Maximum and Minimum Values of a Function

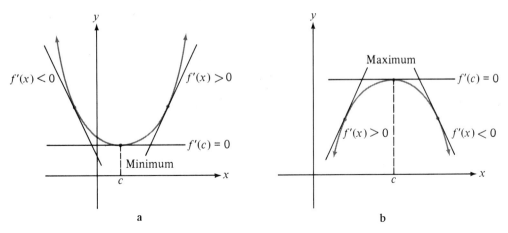

Figure 13.7

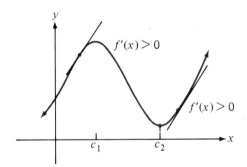

Figure 13.8

$$f'(x) = 3x^2 - 3 = 3(x^2 - 1) = 3(x + 1)(x - 1).$$

By setting $f'(x) = 0$, we find the critical values to be -1 and 1. Since

$$x < -1 \quad \text{implies} \quad f'(x) > 0,$$
$$-1 < x < 1 \quad \text{implies} \quad f'(x) < 0,$$
$$x > 1 \quad \text{implies} \quad f'(x) > 0,$$

from the first derivative test it follows that $(-1, f(-1)) = (-1, 3)$ is a local maximum and $(1, f(1)) = (1, -1)$ is a local minimum. When $x = 0$, we get the y-intercept $(0, 1)$. The graph of the function is given in the figure.

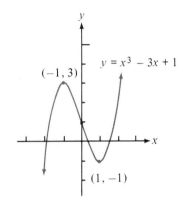

Example Find the local extrema of $f(x) = x^4 - 4x^3 + 10$. Graph.

Solution The derivative is
$$f'(x) = 4x^3 - 12x^2$$
$$= 4x^2(x-3)$$
and so the critical values are 0 and 3. Now

$x < 0$ implies $f'(x) < 0$,
$0 < x < 3$ implies $f'(x) < 0$,
$x > 3$ implies $f'(x) > 0$.

Since the function is decreasing on $-\infty < x \leq 3$ (the slope does not change sign at 0), there is no local extremum at (0, 10), although there is a horizontal tangent. The function changes from decreasing to increasing at 3, and hence $(3, -17)$ is a local minimum.

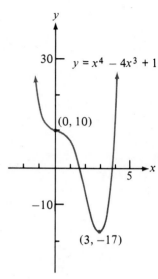

The concavity of a graph of a function can also be used to find local extrema.

Second Derivative Test

Let c be a critical value of $y = f(x)$.
If $f''(c) > 0$, then $(c, f(c))$ is a local minimum.
If $f''(c) < 0$, then $(c, f(c))$ is a local maximum.

Example Find the local extrema of $f(x) = x^3 - 3x^2$. Graph.

Solution In the second example of this section, we found the critical values to be 0 and 2. We shall test these numbers using the second derivative $f''(x) = 6x - 6$. Recall that if the second derivative is positive at a point, then the graph is concave upward in a neighborhood of the point; if the second derivative is negative, the graph is concave downward. Now
$$f''(0) = -6 < 0,$$
hence (0, 0) is a local maximum. In addition, we see that
$$f''(2) = 6 > 0$$

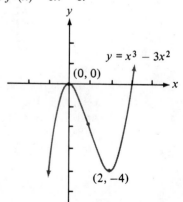

13.2 Maximum and Minimum Values of a Function

and so $(2, -4)$ is a local minimum. The graph of the function is given in the accompanying figure.

It should also be apparent in the preceding example that $f''(x) < 0$ on $-\infty < x < 1$ and $f''(x) > 0$ on $1 < x < \infty$ and so $(1, -2)$ is a point of inflection.

In conclusion, we note that the second derivative may fail to give any information about the extrema of a function. If c is a critical value for which $f''(c) = 0$, then the second derivative test provides no conclusion. In this case, the first derivative test should be used (see Problem 28).

EXERCISE 13.2

Use the first derivative test to find the local extrema for the given function. Graph.

1. $f(x) = -x^2 + 4x + 1$
2. $f(x) = x^2 - 6x + 5$
3. $f(x) = (x + 1)^2$
4. $f(x) = (2x - 3)^2$
5. $f(x) = x^3 + 2$
6. $f(x) = (x - 1)^3$
7. $f(x) = -x^3 + 12x - 4$
8. $f(x) = x^3 - 3x + 3$
9. $f(x) = x^3 - 6x^2$
10. $f(x) = -16x^3 + 12x^2 + 7$
11. $f(x) = x^3 - 3x^2 - 9x + 1$
12. $f(x) = 2x^3 - 15x^2 + 24x$
13. $f(x) = x^4 - \dfrac{4}{3}x^3$
14. $f(x) = -x^4 + 4x^3$
15. $f(x) = x^4(x - 1)^2$
16. $f(x) = 4x^5 - 5x^4$
17. $f(x) = \dfrac{4}{x^2 + 1}$
18. $f(x) = \dfrac{x}{x^2 + 1}$
19. $f(x) = xe^x$
20. $f(x) = x^2 e^x$

Use the second derivative test to find the local extrema for the given function. Graph.

21. $f(x) = x^2 + 8x$
22. $f(x) = -\dfrac{1}{2}x^2 + x + \dfrac{3}{2}$
23. $f(x) = 2x^3 - 9x^2 + 12x + 2$
24. $f(x) = x^3 - 27x$
25. $f(x) = -\dfrac{1}{3}x^3 - \dfrac{1}{2}x^2$
26. $f(x) = x^3 - 3x^2 + 1$

27. The function $f(x) = (x - 1)^{2/3}$ has a vertical tangent at its only extremum. Use the first derivative test to locate this point. Graph.

28. Given the function $f(x) = (x-1)^4 + 1$.

   a. Show that the second derivative test gives no information at the critical value of the function.

   b. Use the first derivative test to show that the function possesses a minimum at the critical value.

Applications

29. *Biology* The function $O(r) = cre^{-kr}$, where c and k are positive constants, relates the percentage r of red blood cells and the amount of oxygen O supplied by the blood. The number r is called the **hematocrit level**. Use the second derivative test to find the maximum value of O.

30. The **Reynold's number** $R(r) = c \ln r - kr$, where c and k are positive constants, is related to turbulence of the blood in an aorta of radius r. Find the maximum value of R.

31. *Business* The formula $C(x) = a\dfrac{x}{2} + (b + cx)\dfrac{q}{x}$, a, b, c, and q positive constants, gives the total yearly cost to a company for resupplying depleted inventories of a single commodity as a function of the size x of a reorder lot. Find the size of a reorder lot that will result in the least yearly cost.

32. Given the revenue and cost functions $R(x) = 6000x - 30x^2$ and $C(x) = 600x + 1000$, find the maximum profit.

33. Given the cost function $C(x) = x^2 + 640x + 900$, find the minimum average cost.

34. The value of a parcel of land t years after its purchase is given by $V(t) = 10{,}000e^{(\sqrt{t} - 0.05t)}$. What is the maximum value of V?

35. If n employees are paid $5 dollars per hour, the total cost to a company is $C = 5nt$, where t is the time necessary for the n persons to complete a job. Find the minimum cost to the company and the corresponding number of employees if it is known that $t = 2 + (3/n) + (50/n^2)$.

36. *Physics* The temperature in a controlled environment is given by $T(t) = t^3 - 3t^2 + 200$, where $t \geq 0$ is time measured in hours. Find the minimum temperature.

37. The distance above ground level of a projectile is given by $s(t) = -16t^2 + 256t$, where t is measured in seconds and s in feet. Find the maximum height attained.

38. *Social Science* Show that the normal distribution function

$$f(x) = \frac{1}{\sigma\sqrt{2\pi}} e^{-(x-\mu)^2/2\sigma^2}$$

has a critical value at $x = \mu$. Show that there is a maximum at this critical value.

13.3
ABSOLUTE EXTREMA—APPLICATIONS

We are particularly interested in functions that have **absolute extrema**.

13.3 Absolute Extrema—Applications

A function $y = f(x)$ is said to have:

an **absolute maximum** value M if $f(x) \leq M$ for all x in the domain of the function,

an **absolute minimum** value m if $f(x) \geq m$ for all x in the domain of the function.

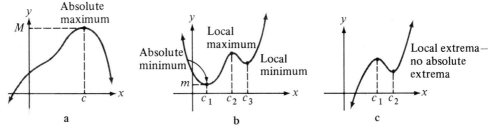

Figure 13.9

As shown in Figure 13.9, part b, a function may have an absolute extremum as well as local extrema. Note that in the cases illustrated, the extrema occur at critical values. However, when a function is defined only on a *closed* interval $a \leq x \leq b$, it can happen that the absolute extrema do not always occur at critical values.* Figure 13.10 illustrates several possibilities. For example, the revenue function $R(x) = 100x - 2x^2 = 2x(50 - x)$ only makes sense on the interval $0 \leq x \leq 50$, since $R < 0$ for $x > 50$. Now $R(0) = 0$ and $R(50) = 0$, and so 0 is the absolute minimum value of the function. Upon using a derivative test, it is seen that (25, 1250) is an absolute maximum.

We have the following general result.

(1) A continuous function defined on a closed interval $a \leq x \leq b$ always attains an absolute maximum and an absolute minimum.

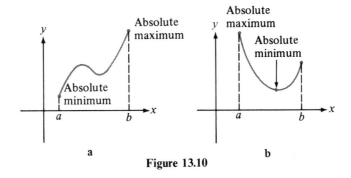

Figure 13.10

*An interval $a < x < b$ is said to be *open*.

The problems considered in this section may result in a function defined on an interval $a \leq x \leq b$ or for $x > 0$. In the former case, we must examine the endpoints of the interval, as well as the critical values, for the absolute extrema.

Example A bus company finds that each bus averages 400 fare-paying passengers per day. It is projected that for every 5 cent fare reduction each bus would have a 25% daily increase in passengers. If the fare for each passenger is originally 50 cents, how many fare reductions should the company make to maximize the daily revenue for a bus?

Solution Let x be the number of fare reductions. After x reductions, the fare would be $50 - 5x$. The increase in passengers for each bus would be $(0.25)400x$. [For example, if there were two fare reductions, each bus would have a daily passenger increase of $(0.25)400 \cdot 2 = 200$.] Therefore, the number of daily fare-paying passengers is $400 + (0.25)400x$. The revenue for one bus is

(number of daily fare paying passengers) $\cdot$ (fare),

that is,

$$R(x) = [400 + (0.25)400x] \cdot (50 - 5x)$$
$$= (400 + 100x)(50 - 5x)$$
$$= -500x^2 + 3000x + 20{,}000.$$

Observe that it is implicit that the domain of $R(x)$ is $0 \leq x \leq 10$. (Why?) The derivative is

$$R'(x) = -1000x + 3000$$

so that $R'(x) = 0$ gives the critical value 3. Since $R''(x) = -1000$ and $R''(3) < 0$, it follows from the second derivative test that $(3, R(3))$ is at least a local maximum. Now from

$$R(0) = 20{,}000 \text{ cents} = \$200$$
$$R(3) = 24{,}500 \text{ cents} = \$245$$
$$R(10) = \$0$$

it follows that $R(3)$ is the absolute maximum value of the revenue function. The fare of each bus should be $50 - 5(3) = 35$ cents.

Example A box, open at the top, is to be made from a square piece of cardboard by cutting a square out of each corner and then turning up the sides. If the cardboard is 24 inches on a side, what are the dimensions of the box that will give the maximum volume?

13.3 Absolute Extrema—Applications

Solution The accompanying figures show that we cut a square of x inches on a side from each corner. The volume V of the resulting box is given by (area of base) × height and so we have

$$V(x) = x(24 - 2x)^2.$$

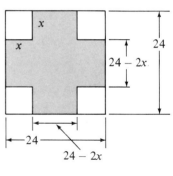

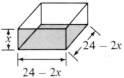

Because we require $x \geq 0$ and $24 - 2x \geq 0$, it is seen that the domain of the function must be $0 \leq x \leq 12$. But obviously 0 and 12 will not give a maximum volume since $V(0) = V(12) = 0$.

Using the product rule and simplifying then gives

$$V'(x) = -4x(24 - 2x) + (24 - 2x)^2$$
$$= (24 - 2x)(24 - 6x).$$

Hence, the only meaningful critical value is 4. Since $V'(x) > 0$ on $0 < x < 4$ and $V'(x) < 0$ on $4 < x < 12$, we conclude from the first derivative test that $(4, V(4))$ is an absolute maximum. The absolute maximum value of the volume is

$$V(4) = 1024 \text{ cubic inches.}$$

The dimensions of the box are 16 by 16 by 4 inches.

Example A rancher wishes to enclose a 1000-square-foot rectangular corral, using two different construction materials. Along two parallel sides, the material costs $4 per foot; for the other two parallel sides, the material costs $1.6 per foot. Find the dimensions of the corral that will give the minimum cost of construction.

Solution Let x represent the length of the sides that cost $4 per foot and let y represent the length of sides that cost $1.6 per foot. As can be seen from the figure, the cost function is

$$C = 4(2x) + 1.6(2y)$$
$$= 8x + 3.2y.$$

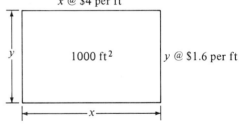

But x and y are related by the requirement $xy = 1000$ or $y = 1000/x$. Hence, the cost function can be written as

$$C(x) = 8x + 3.2\left(\frac{1000}{x}\right)$$

$$= 8x + \frac{3200}{x}.$$

The domain of this function is $x > 0$, so there are no interval endpoints to be considered.

The first derivative is

$$C'(x) = 8 - \frac{3200}{x^2}$$

and the equation $C'(x) = 0$ leads to $x^2 = 400$. Thus, 20 is the only critical point of the problem. The second derivative

$$C''(x) = \frac{6400}{x^3}$$

reveals that $C''(20) > 0$, and so $(20, C(20))$ is a minimum. Although the results given in (1) do not apply in this case, it can be verified from the graph of C that

$$C(20) = 8(20) + \frac{3200}{20}$$

$$= \$320$$

is the absolute minimum value of the cost. The dimensions of the corral are then 20 feet by 50 feet.

EXERCISE 13.3

1. Find two nonnegative numbers whose sum is 80 and whose product is a maximum.
2. Find two numbers whose difference is 60 and whose product is a minimum.
3. Find two nonnegative numbers whose sum is 1 such that the sum of their squares is a minimum.
4. Find two nonnegative numbers whose sum is 10 such that the product of the square of one and the cube of the other is a maximum.
5. The graphs of $y = x^2 - 1$ and $y = 1 - x$ intersect at -2 and 1. Find the value of x in the interval $-2 \leq x \leq 1$ for which the vertical distance between the graphs is a maximum.

13.3 Absolute Extrema—Applications

6. Find the point on the graph of $y^2 = 4x$ nearest $(6, 0)$. [*Hint*: Consider the square-of-the-distance formula. See Section 4.2.]

Applications

7. A rancher has 1000 feet of fencing. Find the dimensions of a rectangular corral that encloses a maximum area.

8. A rancher wishes to construct a rectangular corral along a straight stream with 4000 feet of fencing. No fence is to be used along the stream. Find the dimensions of the corral that encloses a maximum area.

9. A rancher decides to enclose a rectangular corral of 128,000 square feet that borders a straight stream. The fence along the three sides not along the stream will be made of wood and the fence along the stream will be wire. The cost of the wood fence is $5 per foot and the cost of the wire fence is $3 per foot. Find the dimensions of the corral that will give the minimum cost of construction.

10. A carpenter wants to frame a 128-square-foot window with metal. As shown in the figure, along the horizontal portion of the window he wants to use stainless steel costing $1.00 per foot and along the vertical portions he wants to use aluminum costing $0.25 per foot. How much of each type of metal should be used in order to keep his expenses (cost) a minimum?

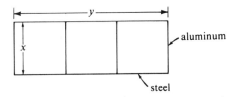

11. A charter airline charges a rate of $200 per person if the number traveling together is exactly 250. For each additional person exceeding 250 the fare for everyone is reduced by 50 cents. What number of travelers will enable the airline to take in a maximum revenue?

12. A car-leasing firm makes a profit of $80 per customer up to 50 customers. Because of the need for additional help and other increased costs, the firm loses $1 profit for each customer exceeding 50. Determine the maximum total profit.

13. When food is packaged, part of the price must reflect the cost of the packaging; for example, the cost of a tin can is directly proportional to the amount of tin used. For tomato juice sold in a cylindrical can containing 32 cubic inches, determine the dimensions of the can that will use the least amount of material.

14. A lead cylindrical container holds 100 cubic inches of radioactive material. The top of the container costs 3 times as much per unit area as the sides and bottom. Show that the dimensions that give the least total cost are a height that is four times the radius.

15. A rectangular box, open at the top, is to have a square base and a volume of 13,500 cubic inches. Find the dimensions of the box that will minimize the amount of material used.

16. A closed rectangular shipping crate is to have a square base and a volume of 32,000 cubic inches. The cost of the material used in the construction of the crate is 2 cents per square inch. Find the dimensions of the crate that will minimize the construction costs.

17. U.S. postal regulations stipulate that boxes sent by fourth-class mail must satisfy the requirement that the perimeter of one end plus the length of the box must not exceed 100

inches. A man wants to send a box with a rectangular end whose width is twice its height. Find the dimensions of the box that will give a maximum volume.

18. An entrance to a building is in the form of an arch described by the equation $y = 12 - x^2$, where x is measured in feet. A rectangular glass door is inscribed in the arch with the bottom of the door coinciding with a portion of the x-axis. Find the dimensions of the door that will give a maximum entrance capacity. [*Hint*: The entrance capacity is directly proportional to the area of the door.]

19. As shown in the figure, an oil pipeline is to connect a drilling site and a seaport. The cost per mile of laying the pipe under a one-half-mile-wide river is determined to be $\sqrt{5}$ times the cost per mile of laying the pipe under the ground. Using the information supplied in the figure, determine the value of x that will result in the minimum cost of construction.

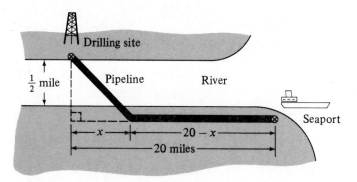

20. At noon, ship A is 60 miles north of ship B. Ship A is sailing south at 10 miles per hour and ship B is sailing east at 20 miles per hour. At what time will the distance between the ships be a minimum? [*Hint*: Use the Pythagorean theorem and the fact that distance = rate × time.]

21. A person would like to cut a wire 10 feet long into two pieces. One piece is to be bent into a square and the other into a circle. How should the wire be cut in order to maximize the sum of the areas?

22. A page with printing on it has 1-inch margins of white space all around. The area of the printed portion is 40 square inches. Determine the dimensions of the page that uses the least amount of paper.

13.4

THE DERIVATIVE AS A RATE OF CHANGE

In Section 12.1, we defined the **average rate of change** of a function with respect to x as the quotient

13.4 The Derivative as a Rate of Change

$$\frac{f(x+h)-f(x)}{h}$$

and the **instantaneous rate of change**, or simply, **rate of change**, to be the limit

$$\lim_{h\to 0}\frac{f(x+h)-f(x)}{h}.$$

That is, the rate of change of a function $y=f(x)$ is its derivative dy/dx.

It is almost impossible to study the managerial, physical, and social sciences without encountering the concept of rate. The following are some examples of rates of change.

The rate at which alcohol is absorbed into the bloodstream
The rate at which cells multiply (such as in a tumor)
The rate at which a population grows
The rate at which people learn
The rate at which rumors, or epidemics, spread throughout a community
The rate at which capital grows
The rate at which radioactive wastes decay
The rate at which a person drives a car
The rate at which the velocity of a falling body changes

Example The revenue obtained by selling x units of an item is given by $R(x) = 1000x - 5x^2$. Find the rate at which the revenue is changing when $x = 150$.

Solution The rate of change of the revenue is

$$R'(x) = 1000 - 10x.$$

The rate

$$R'(150) = 1000 - 1500$$
$$= -500 < 0$$

indicates that the revenue is decreasing at this sales level.

VELOCITY AND ACCELERATION If $s(t)$ gives distance traveled by a body as a function of time, then the rate of change of distance with respect to time is defined as **velocity**. In other words, velocity v is given by

(1) $$v(t) = \frac{ds}{dt}.$$

The rate of change of velocity with respect to time is called **acceleration** and is denoted by $a(t)$:

$$a(t) = \frac{dv}{dt}.$$

But, in view of (1), it follows that acceleration is the *second derivative* of a distance function:

$$a(t) = \frac{d^2 s}{dt^2}.$$

Example The distance, in miles, a car travels in t hours is given by $s(t) = 55t$. What are its velocity and acceleration?

Solution The velocity is the first derivative of s,

$$v(t) = 55 \text{ mi/hr}$$

and the acceleration is the second derivative of s,

$$a(t) = 0 \text{ mi/hr}^2.$$

When a projectile is fired vertically into the air from an initial height s_0 with an initial velocity v_0, the distance above ground level at any time is given by

(2) $$s(t) = -\frac{1}{2} gt^2 + v_0 t + s_0$$

where g is the constant acceleration of gravity. Time is usually measured in seconds and g is taken to be either 32 ft/sec^2, 980 cm/sec^2, or 9.8 m/sec^2. Observe that the second derivative of (2) is $a(t) = -g$.

Example A ball is thrown upward from ground level with an initial velocity of 10 m/sec. What is the velocity of the ball at $t = 1$? At $t = 2$?

Solution Identifying $s_0 = 0$, $v_0 = 10$, and $g = 9.8$, we have from (2)

$$s(t) = -4.9\, t^2 + 10t.$$

The velocity is

$$v(t) = -9.8t + 10.$$

Hence

$$v(1) = -9.8 + 10$$
$$= 0.02 \text{ m/sec}$$

and

$$v(2) = -9.8(2) + 10$$
$$= -9.6 \text{ m/sec}.$$

We can interpret $v(1) = 0.02 > 0$ to mean that the ball is going up, although it is very close to stopping, whereas $v(2) = -9.6 < 0$ means that the ball has turned around and is falling to the ground.

13.4 The Derivative as a Rate of Change

LOGISTIC FUNCTION

The so-called **logistic function**

$$f(t) = \frac{a}{1 + be^{-akt}},$$

where a, b, and k are positive constants, is often used as a mathematical model to describe limited population growth or spread of diseases. In a business setting, the logistic function can be used to describe the effect of advertising in a community.

Example One student infected with the flu returns to an isolated college that has an enrollment of 1000. After t days, the number of infected students is predicted to be

$$N(t) = \frac{1000}{1 + 999e^{-0.99t}}.$$

How fast is the disease spreading when $t = 2$? When $t = 7$? When $t = 12$?

Solution Taking the derivative of $N(t)$ and simplifying gives

$$N'(t) = \frac{989{,}010 e^{-0.99t}}{(1 + 999e^{-0.99t})^2}.$$

With the aid of a calculator, we find the rates to be

$$N'(2) = 7 \text{ students/day}, \quad N'(7) = 247.5 \text{ students/day},$$

$$N'(12) = 6.8 \text{ students/day}.$$

Consistent with our expectation that the flu spreads, we see that the derivatives are positive. At the start, the flu is spreading slowly, but as more people are infected the rate of infection increases drastically. However, it is also reasonable to assume that over a very large time (since most people will be infected) the rate at which the epidemic spreads will itself decrease. A graph of $N(t)$ is given in the accompanying figure.

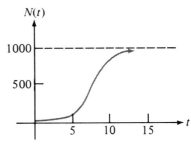

EXERCISE 13.4

1. Find the rate of change of the area of a circle with respect to its radius.

2. Show that the rate of change of the circumference of a circle with respect to its radius is constant.

3. Find the rate of change of the area of an equilateral triangle with respect to the length of its base.

4. Show that the rate of change of the volume of a sphere, $V = \frac{4}{3}\pi r^3$, with respect to its radius is the surface area $S = 4\pi r^2$.

Applications

5. *Biology* The number of milligrams of a drug present in a person's bloodstream is given by $A(t) = \frac{r}{k}(1 - e^{-kt})$, $t \geq 0$, where r and k are positive constants and t is measured in minutes. Find the rate of change of A with respect to time.

6. The number of bacteria in a culture is given by $N(t) = N_0 e^{0.4t}$, $t \geq 0$, where N_0 is the initial number of bacteria present and t is measured in hours. At what rate does the number of bacteria increase in 3 hours?

7. The graph of the function $P(t) = ae^{-be^{-ct}}$, where a, b, and c are constants, is called a **Gompertz curve** and is encountered in the study of the growth of certain populations. Find the rate of change of P with respect to t.

8. *Business* The future value of $10,000 invested at an annual rate of interest r compounded quarterly for 5 years is $S = 10,000\left(1 + \frac{r}{4}\right)^{20}$. Find the rate of change of S with respect to r when $r = 0.08$.

9. The number of units demanded of a commodity is given by $D(x) = (x - 200)^2$, $0 \leq x \leq 200$, where x represents price in dollars. At what rate is the demand changing when $x = 50$?

10. The future value of an annuity is given by $S = R\,[(1 + i)^n - 1]/i$, where R is the amount invested and i represents the interest rate per period. Assuming R and n are constant, find the rate of change of S with respect to the interest rate.

11. *Physics* A spherical balloon is filled with a gas so that in t minutes its radius, in centimeters, is as given by $r = 3t$. At what rate is the volume changing with respect to time at $t = 2$?

12. The peri   in seconds, of a simple pendulum is given by $T = 2\pi\sqrt{L/g}$, where L is measured in feet and $g = 32$ ft/sec². Find the rate of change of T with respect to L when $L = 4$.

13. The intensity of sound, in watts/cm², d centimeters from its source, is given by $I = k/d^2$, where k is a positive constant. Find the rate of change of I with respect to d when $d = 1/2$.

In Problems 14–17 each function gives distance. Find the velocity and acceleration at the indicated time.

14. $s(t) = -4.9t^2 + 24t + 4$, $t = 1$

15. $s(t) = t^4 + t^2$, $t = 2$

16. $s(t) = (1 + 2t)^{-1}$, $t = 0$

17. $s(t) = 5t + \dfrac{10}{t + 1}$, $t = 1$

18. The height above ground level, in feet, of a ball tossed up from a building 100 feet high is given by the function $s(t) = -16t^2 + 96t + 100$, where t is in seconds. What is the velocity of the ball at $t = 2$? At $t = 4$? At what time will the ball attain its maximum height?

19. In Problem 18, at what time will the falling ball be level with the height of the building? What is its velocity at this time?

20. The position of a particle moving on a coordinate line is given by $s(t) = 2t^3 - 21t^2 + 72t$, $t \geq 0$, where t is measured in minutes and s in centimeters. Find the times when the particle is not moving. What is the acceleration of the particle at each of these times? Find the time at which the acceleration is zero.

21. The position of a particle moving on a coordinate line is given by $s(t) = -t^3 + 18t^2$, $t \geq 0$, where t is measured in minutes and s in centimeters. What are the velocity and acceleration of the particle at $t = 2$? For what interval of time is the velocity increasing? Decreasing? What is the maximum velocity attained by the particle?

22. *Social Science* The number of motors installed each day by a new automobile assembly worker after x days on the job is given by $N(x) = 60 - 60e^{-0.3x}$, $x \geq 0$. Find the rate of change of N with respect to x when $x = 40$.

23. The graph of the function $f(x) = (ax + b)/(x + c)$, where a, b, and c are positive constants, is one form of learning curve. Find the rate of change of f with respect to x.

13.5

RELATED RATES

A practical problem may be described using two variables, and these variables, in turn, could be functions of time. For example, if a spherical balloon is filled with gas, then the volume V changes at a rate measured in units of volume/time and, simultaneously, the radius r of the balloon increases at a rate measured in units of length/time. We shall see that the rates at which the volume and radius change with respect to time are related since the volume V of a sphere is a function of the radius r. We say that dV/dt and dr/dt are **related rates**.

If a problem is described using two variables, and if one of these variables changes at a constant rate, then the rate at which the other variable changes is determined through the process of implicit differentiation. Before considering problems involving related rates, we make the following observation from the chain rule, rule VIII, of differentiation.

If [ ] denotes a variable dependent on time t, then

$$\frac{d}{dt}[\ ]^n = n[\ ]^{n-1}\frac{d}{dt}[\ ].$$

For example, if r depends on time, then

$$\frac{d}{dt}r^3 = 3r^2\frac{dr}{dt}.$$

Example A spherical balloon is being filled with gas at a rate of 64 ft³/min. At what rate is the radius changing when the radius is 2 feet?

Solution The problem can be analyzed in three steps:

Known: The relationship between volume and radius is
$$V = \frac{4}{3}\pi r^3.$$

Given: $\dfrac{dV}{dt} = 64$ ft³/min

Want: $\dfrac{dr}{dt}\bigg|_{r=2}$

Differentiating both members of the known formula with respect to time yields
$$\frac{dV}{dt} = \frac{4}{3}\pi \frac{d}{dt} r^3 = \frac{4}{3}\pi \left(3r^2 \frac{dr}{dt} \right)$$
$$= 4\pi r^2 \frac{dr}{dt}.$$

Since $dV/dt = 64$, we solve the last equation for dr/dt:
$$\frac{dr}{dt} = \frac{64}{4\pi r^2}$$
$$= \frac{16}{\pi r^2}.$$

At the instant corresponding to $r = 2$, the radius is increasing at the rate
$$\frac{dr}{dt}\bigg|_{r=2} = \frac{16}{\pi 4} = \frac{4}{\pi} \text{ ft/min.}$$

Example A ladder 30 feet long is leaning against a house. If the bottom of the ladder slips away from the house at a rate of 1 ft/min, at what rate does the top of the ladder slide down the side of the house when the base is 18 feet from the house?

Solution The accompanying figure shows the relationship between the variables x and y at any time. Since the ladder forms a right triangle with the house and ground, it follows from the Pythagorean theorem that
$$(30)^2 = x^2 + y^2.$$
Thus,

Known: $x^2 + y^2 = 900$

Given: $\dfrac{dx}{dt} = 1$ ft/min

Want: $\dfrac{dy}{dt}\bigg|_{x=18}$

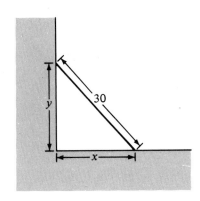

Differentiate both members of the known formula with respect to time:

$$\frac{d}{dt}x^2 + \frac{d}{dt}y^2 = \frac{d}{dt}900$$

$$2x\frac{dx}{dt} + 2y\frac{dy}{dt} = 0.$$

Using the fact that $dx/dt = 1$, we solve for dy/dt,

$$\frac{dy}{dt} = -\frac{x}{y}.$$

Now when $x = 18$, we can determine y from the known relation,

$$(18)^2 + y^2 = 90$$
$$y^2 = 900 - 324 = 576$$
$$y = \sqrt{576} = 24.$$

Hence, we obtain

$$\frac{dy}{dt}\bigg|_{x=18} = -\frac{18}{24} = -\frac{3}{4} \text{ ft/min}.$$

The fact that dy/dt is negative at this instant in time means that y is decreasing.

EXERCISE 13.5

Applications

1. A large cylindrical tank with a radius of 5 feet is filled with water at the rate of 175 ft³/min. At what rate is the height of the water increasing?

2. A cylindrical tank with a radius of 20 meters is filled in such a manner that the height of the liquid is increasing at the rate of 0.01 m/sec. At what rate does the volume increase?

3. A spherical balloon is filled at the rate of 4 cm³/sec. At what rate is the radius increasing when the radius is 3 centimeters?

4. In Problem 3, at what rate is the radius changing when the volume is $32\pi/3$ cubic centimeters?

5. The area of a circle is increasing at a rate of 10π in²/sec. At what rate is the radius increasing when the radius is 5 inches?

6. A 50-foot ladder is leaning against a vertical wall. If the bottom of the ladder moves away from the wall at a rate of 1/2 ft/min, at what rate does the top of the ladder move down the wall when the top of the ladder is 30 feet above ground?

7. A kite flies parallel to the ground at a height of 100 feet. If the kite string is paid out a rate of 1 ft/sec, at what rate is the kite moving when 200 feet of string have been paid out?

8. A parachutist is descending at a rate of 50 ft/min. An observer is stationed 300 feet from a point directly beneath the parachutist. At what rate is the distance between the observer and the parachutist decreasing when the parachutist is 400 feet off the ground?

9. A tank in the shape of an inverted cone is filled with water at a rate of 40 ft³/min. The tank is 20 feet high and has a radius of 10 feet. At what rate is the height of the water changing when it is 8 feet deep? The volume of a cone is $V = \dfrac{\pi}{3} r^2 h$. [*Hint*: Use similar triangles to find a relationship between r and h.]

10. A woman who is 5 feet tall walks away from a 20-foot-high lamppost at a rate of 3 ft/sec. At what rate is the end of her shadow moving away from the base of the lamppost? [*Hint*: Use similar triangles.]

11. *Biology* The total flow of blood in an artery of radius R in a unit time is given by Poiseuille's law: $F = \pi P R^4 / 8vl$, where P, v, and l are constants. Some drugs, such as aspirin, cause the arteries to dilate. If R increases at a rate of 0.002 cm/min, at what rate is F increasing when $R = 0.02$ centimeter?

12. *Physics* Boyle's law for gases states that at a constant temperature the volume V occupied by the gas and the pressure P exerted on it are related by the formula $PV = k$, where k is a constant. If P increases at a rate of 5 lb/in²/min, at what rate is V decreasing when $P = 100$ lb/in² and $V = 60$ in³?

13.6

FURTHER APPLICATIONS OF THE DERIVATIVE

ELASTICITY OF DEMAND If $D(x)$ represents the number of units demanded of a commodity at a price x, the quotient

13.6 Further Applications of the Derivative

(1)
$$\frac{\dfrac{D(x+h)-D(x)}{D(x)}}{\dfrac{(x+h)-x}{x}}$$

is the proportionate change in demand divided by the proportionate change in price.

Now (1) can also be written as

$$\frac{x}{D(x)} \cdot \frac{D(x+h)-D(x)}{h}.$$

If we let $h \to 0$, it follows from the definition of the derivative that

(2) $$\lim_{h \to 0} \frac{x}{D(x)} \cdot \frac{D(x+h)-D(x)}{h} = \frac{x}{D(x)} \cdot D'(x).$$

Recall, $D'(x)$ gives the slope of a tangent to the graph of the demand function. Since the demand for a commodity decreases as the price increases, we know from Section 13.1 that $D'(x)$ will be negative. For this reason, many economists will affix an arbitrary negative sign in (2) in order to obtain and discuss only nonnegative quantities. The expression

(3) $$\eta = -\frac{x}{D(x)} \cdot D'(x)$$

is called the **elasticity of demand.**

CASES Three cases are usually distinguished.

1. If $\eta = 1$, the percentage change in demand is the same as the percentage change in price.

2. If $\eta > 1$, the percentage change in demand is greater than the percentage change in price.

3. If $\eta < 1$, the percentage change in demand is less than the percentage change in price.

Cases 2 and 3 are sometimes referred to as **elastic** and **inelastic demands**, respectively.

Example Compute the elasticity of demand if $D(x) = -x^2 + 100$ and $x = 5$.

Solution The derivative of the demand function is

$$D'(x) = -2x,$$

so

$$D'(5) = -10.$$

Also,
$$D(5) = 100 - 25$$
$$= 75.$$
Hence, the elasticity of demand is
$$\eta = -\frac{5}{D(5)} \cdot D'(5)$$
$$= -\frac{5(-10)}{75}$$
$$= \frac{2}{3}$$
$$= 0.667.$$

In this case $\eta < 1$, so the demand is inelastic.

We can interpret the results of the foregoing example to mean that at $x = 5$, there is an approximate 0.667% decrease in demand for a 1% increase in price. Were the price increased by, say, 3%, the demand would decrease approximately 2% since $(\frac{2}{3})3 = 2$.

If $\eta > 1$, then for a percentage increase in price, there results a greater decrease in demand. Hence, total revenue must decrease. Similarly, if $\eta < 1$, a percentage increase in price brings with it a smaller decrease in demand, so that revenue increases. For $\eta = 1$, no change in revenue occurs.

Example The demand for a commodity is given by
$$D(x) = 4x^2 - 120x + 900, \ 0 \le x \le 15,$$
where x represents price. Find the elasticity of demand at $x = 10$.

Solution We have
$$D'(x) = 8x - 120$$
Now
$$D(10) = 4(100) - 120(10) + 900$$
$$= 100,$$
and
$$D'(10) = 8(10) - 120$$
$$= -40.$$
Therefore,
$$\eta = \frac{-10}{D(10)} \cdot D'(10)$$

$$\eta = -\frac{10(-40)}{100}$$

$$= 4.$$

The demand is elastic, so a percentage increase in price will decrease the revenue.

MARGINAL FUNCTIONS In economic analysis, such terms as "marginal revenue" or "marginal profit" usually refer to the derivative of the function. We define:

> *Marginal revenue:* $R'(x)$
> *Marginal cost:* $C'(x)$
> *Marginal profit:* $P'(x)$.

These three derivatives are sometimes denoted by MR, MC, and MP, respectively. A marginal function gives an indication of the rate at which a function, such as profit, changes for a unit increase in the amount produced x.

Let $R(x)$ be the function that gives the total revenue for selling x units of a commodity. The amount by which $R(x)$ changes when *one* more item is sold is

$$R(x+1) - R(x),$$

or if we let $h = 1$, we see from Figure 13.11 that the quotient

$$\frac{R(x+h) - R(x)}{h} = R(x+1) - R(x)$$

is the slope of the secant line through the points $(x, R(x))$ and $(x+1, R(x+1))$.

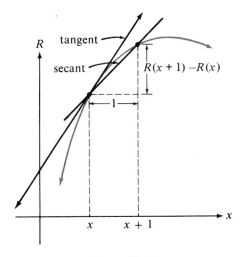

Figure 13.11

We already know that the slope of the tangent line at x is $R'(x)$. Economists take the derivative R' (slope of tangent) as an approximation to the change in revenue (slope of secant).

Example The total revenue for selling x units of a commodity is given by the function $R(x) = -2x^2 + 50x$. Find the revenue obtained in selling the fourth unit. Find the marginal revenue at $x = 3$.

Solution The actual change in revenue when the number of items sold increases from $x = 3$ to $x = 4$ is $R(4) - R(3)$. From the given function we have

$$R(4) = -2(4)^2 + 50(4)$$
$$= \$168$$

and

$$R(3) = -2(3)^2 + 50(3)$$
$$= \$132.$$

Thus,

$$R(4) - R(3) = \$36.$$

Alternatively, as another indicator of this change we can compute the marginal revenue $R'(x)$ and evaluate it at $x = 3$. That is,

$$MR = -4x + 50$$

so

$$MR(3) = -12 + 50$$
$$= \$38.$$

One obvious reason why $R'(x)$ is taken as an indicator of the change in revenues for a unit increase in selling levels is the fact that it is much easier to evaluate *one* (and hopefully, a simpler) function $R'(x)$ at *one* point than it is to evaluate $R(x)$ at *two* different points and then subtract two quantities.

Similarly, if $C(x)$ represents the total cost for producing x units of a commodity, the (approximate) change in cost for a unit increase in production is given by the marginal cost $C'(x)$. This derivative is an indication of the cost of producing the next unit.

Example The total cost for producing x units of a commodity is given by the function $C(x) = x^2 + 640x + 900$. Find the approximate cost for producing the thirty-first unit.

Solution The actual cost for producing the 31st unit is $C(31) - C(30)$. Rather than go through the arithmetic of computing this difference, we need only compute the marginal cost at $x = 30$. We have

13.6 Further Applications of the Derivative

$$MC = 2x + 640.$$

so

$$MC(30) = 2(30) + 640$$
$$= \$700.$$

Example $R(x) = -0.001x^2 + 150x$ and $C(x) = 50x + 3000$ are revenue and cost functions, respectively. Find the marginal profit.

Solution The profit is $P(x) = R(x) - C(x)$, or

$$P(x) = (-0.001x^2 + 150x) - (50x + 3000)$$
$$= -0.001x^2 + 100x - 3000.$$

The marginal profit is the derivative of the profit,

$$MP = -0.002x + 100.$$

EXERCISE 13.6

Compute the elasticity of demand for the given demand function at the indicated price.

1. $D(x) = -5x + 400$, $0 \le x \le 80$; $x = 50$
2. $D(x) = -10x + 2000$, $0 \le x \le 200$; $x = 30$
3. $D(x) = -2x^2 + 5000$, $0 \le x \le 50$; $x = 20$
4. $D(x) = 5x^2 - 300x + 4500$, $0 \le x \le 30$; $x = 10$
5. $D(x) = 100\sqrt{400 - x}$, $0 \le x \le 400$; $x = 300$
6. $D(x) = 100 + \dfrac{50}{\sqrt{x+2}}$, $x \ge 0$; $x = 7$
7. Compute the elasticity of demand for $D(x) = -x^2 + 100$, $0 \le x \le 10$; $x = 6$. If the price decreases by 8%, determine the approximate change in demand.
8. Compute the elasticity of demand for $D(x) = -x^2 + 4x + 4$, $0 \le x \le 2$; $x = 1$. If the price increases 16%, determine the approximate change in demand.
9. If $D(x) = -2x + 500$, $0 \le x \le 250$, find the prices for which the demand is inelastic.
10. The elasticity of demand η can be written as the ratio of two derivatives. Verify that

$$\eta = -\dfrac{\dfrac{d}{dx}\ln D(x)}{\dfrac{d}{dx}\ln x}.$$

11. Given the revenue function $R(x) = -x^2 + 40x$:
    a. Find the marginal revenue at $x = 6$.
    b. Compare the result of part a with $R(7) - R(6)$.

12. The total cost for making x units of a commodity is given by $C(x) = 2x^2 + 7x + 15$. Find the approximate cost of making the thirty-first unit of the commodity.

13. The total cost for making x units of a commodity is given by $C(x) = 2x^2 + 100$. Compare the marginal cost at $x = 10$ and $x = 11$ with the exact cost of making the eleventh unit.

14. In the manufacturing of a certain product, the revenue and cost functions turn out to be $R(x) = -x^2 + 1000x$ and $C(x) = 10x + 500$. Find the marginal profit.

15. Given the revenue function $R(x) = -2x^2 + 600x$ and cost function $C(x) = 2x + 40$, use the concept of marginal profit to find the approximate profit from selling the fifty-first unit.

16. Given the revenue function $R(x) = x(-x + 150)$, show that the marginal revenue is always decreasing.

17. Given the cost function $C(x) = 0.1x^3 - 9x^2 + 1000x$, determine the interval over which the marginal cost is increasing.

18. Let $R(x)$, $C(x)$, and $P(x)$ be continuous functions with derivatives $R'(x)$, $C'(x)$, and $P'(x)$, respectively. Show that the profit $P(x)$ is a maximum at the point where the marginal revenue equals the marginal cost.

Find the marginal average cost at the indicated production level.

19. $C(x) = 50x + 750$; $x = 5$

20. $C(x) = x^2 + 640x + 900$; $x = 10$

CHAPTER TEST

Find the intervals for which the given function is increasing. In which it is decreasing.

1. $f(x) = -2x^3 + 3x^2 + 72x$

2. $f(x) = 3x^4 + 6x^2$

Find the intervals for which the given function is concave upward. For which it is concave downward. Find any points of inflection.

3. $f(x) = x^3 - 6x^2 + 4x + 4$

4. $f(x) = 3x^5 - 40x^3$

Use a derivative test as an aid in graphing the given function.

5. $f(x) = 2x^4 - 16x^2 + 3$

6. $f(x) = 10 + 5xe^{-x/10}$

7. The driver of a dune buggy wishes to reach a fishing pier by traveling along a straight road parallel to the ocean and then diagonally across a beach. The buggy proceeds at a rate of 20 mi/hr along the road and 10 mi/hr over the beach. Use the information given in the figure to determine the route that should be taken to minimize the total time of the trip. [*Hint*: Time = distance/rate.]

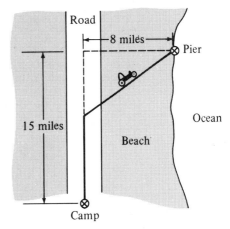

8. A wholesale company sells hand calculators to a retailer for $40 each less a discount per calculator of 20 cents times the number ordered. The wholesaler wishes to offer this discount only up to a limited order size. What size order will maximize the wholesaler's revenue?

9. The temperature of a cooling body is given by $T(t) = 70 + 230e^{-0.2t}$, $t \geq 0$, where T is measured in degrees Fahrenheit and t in minutes. Find the rate of cooling when $t = 20$.

10. A stone is tossed into a pool of still water, thereby causing a circular ripple. If the radius of the circle expands at a rate of 5 in/sec, find the rate at which the area of the circle is expanding when the area is 100 square inches.

11. If $R(x) = 43x - x^2$ and $C(x) = 3x + 100$ are revenue and cost functions, respectively, find the exact profit gained from the sale of the fifteenth unit. Use the marginal profit to find an approximation to this profit.

12. Find the elasticity of demand at $x = 1$ for the demand function $D(x) = 100/(x+1)^2$.

14
INTEGRAL CALCULUS

14.1
ANTIDERIVATIVES

Given the function $y = x^2$ it is readily seen that the derivative is $dy/dx = 2x$. But suppose we are given the derivative $dy/dx = 2x$ and asked: What is the function whose derivative is $2x$? Of course, one possible answer would be $y = x^2$, however, it should be apparent that $y = x^2 - 1$, $y = x^2 + 9$, or $y = x^2 - 4$ are also functions with derivative $2x$. In fact, all possibilities are encompassed by the statement $y = x^2 + c$, where c is any constant. We say that any member of the set of functions $\{y \mid y = x^2 + c, c \text{ arbitrary}\}$ is an **antiderivative** of $dy/dx = 2x$.

> A function $F(x)$ is said to be an antiderivative of a function $f(x)$ if $F'(x) = f(x)$.

For example, $F(x) = x$ is an antiderivative of $f(x) = 1$, since

$$F'(x) = 1 = f(x).$$

NOTATION AND TERMINOLOGY

Symbolically, when we wish to include the arbitrary constant we write the antiderivative as

$$\int f(x)\,dx = F(x) + c.$$

The notation $\int f(x)\,dx$ is also called the **indefinite integral** of $f(x)$. The process of finding an antiderivative is known as **integration**. The symbol $\int$ is called an **integral sign**.

Example Evaluate $\int x^3\,dx$.

Solution
$$\int x^3\,dx = \frac{x^4}{4} + c,$$

473

since

$$\frac{d}{dx}\left(\frac{x^4}{4}+c\right) = 4\cdot\frac{1}{4}x^3 + 0 = x^3.$$

Recall that in order to differentiate a power of x we "multiply and decrease the power by one." Hence, to antidifferentiate we must "increase the power by one and divide." We summarize the general case for powers of x by the following result.

For any real number n, except n = −1,

(1)
$$\int x^n\,dx = \frac{x^{n+1}}{n+1} + c.$$

Justification To verify that an antiderivative is correct all we need do is differentiate. Therefore, from rule I, of Section 12.3,

$$\frac{d}{dx}\left(\frac{x^{n+1}}{n+1}+c\right) = \frac{n+1}{n+1}x^{(n+1)-1} = x^n.$$

Example Evaluate $\int x^{-1/4}\,dx$.

Solution From (1) with $n = -1/4$ it follows that

$$\int x^{-1/4}\,dx = \frac{x^{-(1/4)+1}}{3/4} + c = \frac{4}{3}x^{3/4} + c.$$

We check by differentiation.

$$\frac{d}{dx}\left(\frac{4}{3}x^{3/4}+c\right) = \frac{4}{3}\cdot\frac{3}{4}x^{(3/4)-1} = x^{-1/4}.$$

Example Evaluate $\int 1\,dx$.

Solution From (1) with $n = 0$ we have

$$\int 1\,dx = \frac{x^{0+1}}{1} + c = x + c.$$

The next two properties are immediate consequences of rules III and IV of Section 12.3.

(2) $\qquad\int kf(x)\,dx = k\int f(x)\,dx$, k a constant

(3) $\qquad\int [\,f(x)+g(x)]\,dx = \int f(x)\,dx + \int g(x)\,dx.$

14.1 Antiderivatives

Example Evaluate $\int 6x^2\,dx$.

Solution From (2)
$$\int 6x^2\,dx = 6\int x^2\,dx.$$
Then from (1),
$$6\int x^2\,dx = 6\frac{x^3}{3} + c = 2x^3 + c.$$

Example Evaluate $\displaystyle\int \frac{12}{\sqrt[3]{x^2}}\,dx.$

Solution From (2) and the laws of exponents we can write
$$\int \frac{12}{\sqrt[3]{x^2}}\,dx = 12\int x^{-2/3}\,dx$$
and thus from (1),
$$12\int x^{-2/3}\,dx = 12\frac{x^{1/3}}{1/3} + c = 36x^{1/3} + c.$$

Example Evaluate $\int (5x^6 + 3x^4)\,dx$.

Solution From (2) and (3) we have
$$\int (5x^6 + 3x^4)\,dx = 5\int x^6\,dx + 3\int x^4\,dx$$
$$= 5\frac{x^7}{7} + 3\frac{x^5}{5} + c = \frac{5}{7}x^7 + \frac{3}{5}x^5 + c.$$

Property (3) extends to the sum or difference of any number of functions.

Example Evaluate $\int [2x^{1/2} - 3x^2 + 4]\,dx$.

Solution
$$\int [2x^{1/2} - 3x^2 + 4]\,dx = 2\int x^{1/2}\,dx - 3\int x^2\,dx + 4\int 1\,dx$$
$$= 2\frac{x^{3/2}}{3/2} - 3\frac{x^3}{3} + 4x + c$$
$$= \frac{4}{3}x^{3/2} - x^3 + 4x + c.$$

In the exercises that follow, readers are urged to check their answers by the process of differentiation.

EXERCISE 14.1

Evaluate the given.

1. $\int 5\,dx$
2. $\int 7\,dx$
3. $\int x^4\,dx$
4. $\int x^9\,dx$
5. $\int x^{3/2}\,dx$
6. $\int x^{7/3}\,dx$
7. $\int x^{-2/3}\,dx$
8. $\int x^{-3/2}\,dx$
9. $\int 6x^3\,dx$
10. $\int 5x^2\,dx$
11. $\int \dfrac{dx}{\sqrt{x}}$
12. $\int \dfrac{dx}{\sqrt[3]{x}}$
13. $\int \dfrac{20}{\sqrt{x^5}}\,dx$
14. $\int \dfrac{4}{\sqrt[3]{x^2}}\,dx$
15. $\int (3x^2 + 1)\,dx$
16. $\int (6x^2 + 2x)\,dx$
17. $\int (x^2 - 3x + 1)\,dx$
18. $\int (4x^2 + 5x - 6)\,dx$
19. $\int (9x^{1/2} - 8x^3)\,dx$
20. $\int (5x^{3/2} + 4x^{-1/2})\,dx$
21. $\int (x + 2)^2\,dx$
22. $\int (4x - 3)^2\,dx$
23. $\int \dfrac{x^4 + 6x^2}{x}\,dx$
24. $\int \dfrac{x^2 - 1}{x^2}\,dx$
25. $\int \dfrac{\sqrt[3]{x} + 2}{\sqrt{x}}\,dx$
26. $\int \dfrac{\sqrt[3]{x} + \sqrt{x}}{x}\,dx$

Find $y = f(x)$ from the given derivative.

27. $\dfrac{dy}{dx} = 2x + 1$
28. $\dfrac{dy}{dx} = 3x^2 + 2x$
29. $\dfrac{dy}{dx} = 4x^3 + 8x - 3$
30. $\dfrac{dy}{dx} = -10x^4 + 9x^2 - 4x + 14$

Find the function with given derivative whose graph passes through the indicated point.

Example $\dfrac{dy}{dx} = 3x^2;\ (1, 3)$

Solution Since $dy/dx = 3x^2$ we must have
$$y = \int 3x^2\,dx = x^3 + c.$$

14.1 Antiderivatives

When $x = 1$, $y = 3$, and so

$$3 = 1 + c$$

implies $c = 2$. Hence, the function is $y = x^3 + 2$.

31. $\dfrac{dy}{dx} = 4x$; $(1, 5)$

32. $\dfrac{dy}{dx} = 2x + 3$; $(-1, 3)$

33. $\dfrac{dy}{dx} = 3x^2 + 1$; $(-2, 4)$

34. $\dfrac{dy}{dx} = -x^2 + x$; $(2, 9)$

35. $\dfrac{dy}{dx} = 9\sqrt{x}$; $(4, 0)$

36. $\dfrac{dy}{dx} = \dfrac{4}{\sqrt[3]{x}} + 1$; $(1, 1)$

Applications Find the antiderivatives subject to the given conditions.

Example If the marginal cost is $MC = x + 50$ and the initial cost is $C(0) = 100$, then what is the cost?

Solution Since the marginal cost is the derivative of the cost function, the cost function must be the antiderivative of the marginal cost. That is,

$$C(x) = \int (x + 50)\,dx$$

$$= \dfrac{x^2}{2} + 50x + c.$$

Now,

$$C(0) = 0 + 0 + c = 100,$$

which gives $c = 100$. The total cost is therefore

$$C(x) = \dfrac{x^2}{2} + 50x + 100.$$

37. *Business* Marginal profit $= 300 - 0.02x$, with $P(100) = 900$.
38. Marginal revenue $= 100 - x$, with $R(0) = 0$.
39. Marginal average cost $= 0.01 - 800/x^2$, with the cost at 100 units, $C(100) = 500$.
40. Marginal cost $= 3x^2 - 12x + 4$, with the cost at 3 units, $C(3) = 11$.

Example Given that the acceleration of an object is $a(t) = -20$, find the velocity $v(t)$ and position $s(t)$ if the initial velocity and position are, respectively, $v(0) = 64$ and $s(0) = 30$.

Solution The acceleration $a(t)$ is the second derivative of position $s(t)$,

$$\dfrac{d^2s}{dt^2} = a(t),$$

and it is also the first derivative of velocity,

$$\dfrac{d}{dt}\left(\dfrac{ds}{dt}\right) = \dfrac{d}{dt}v(t) = a(t).$$

Hence,

$$v(t) = \int a(t)\, dt = -\int 20\, dt$$
$$= -20t + c_1.$$

But $v(0) = 64 = c_1$ implies
$$v(t) = -20t + 64.$$

Now,
$$\frac{ds}{dt} = v(t)$$

gives
$$s(t) = \int v(t)\, dt = \int (-20t + 64)\, dt$$
$$= -10t^2 + 64t + c_2.$$

Finally, $s(0) = 30$ implies $c_2 = 30$, so we obtain
$$s(t) = -10t^2 + 64t + 30.$$

41. *Physics* If $v(t) = 2t + 4$ and $s(1) = 10$, find $s(t)$.
42. If $a(t) = -32$, $v(0) = 64$, and $s(0) = 200$, find $s(t)$.
43. If $v(t) = -3t^2$ and $s(0) = 5$, find $s(t)$.
44. If $a(t) = t(t^2 + 3)$ and $v(0) = 30$, find $v(t)$.
45. If $a(t) = 6$, $v(0) = 2$, and $s(1) = 1$, find $s(t)$.

14.2
INTEGRATION BY SUBSTITUTION

Since
$$\frac{d}{dx} \ln x = \frac{1}{x}$$
and
$$\frac{d}{dx} e^x = e^x$$

the next two results follow immediately from the concept of an antiderivative.

(1) $$\int \frac{dx}{x} = \ln x + c$$

(2) $$\int e^x\, dx = e^x + c.$$

Observe that (1) fills the gap left in the formula for $\int x^n \, dx$ when $n = -1$. In other words,

$$\int x^n \, dx = \begin{cases} \dfrac{x^{n+1}}{n+1} + c, & n \neq -1 \\ \ln x + c, & n = -1. \end{cases}$$

POWERS OF FUNCTIONS

Suppose now we wish to evaluate

$$\int (2x + 1)^{1/2} \, dx.$$

We could think backward as follows. To obtain $(2x + 1)^{1/2}$ as the derivative, we must have differentiated $(2x + 1)^{3/2}$ since the power is decreased by 1 in the process of differentiation. But unfortunately

$$\int (2x + 1)^{1/2} \, dx = \frac{(2x + 1)^{3/2}}{\frac{3}{2}} + c = \frac{2}{3}(2x + 1)^{3/2} + c$$

is *not* correct because, by rule VII,

$$\frac{d}{dx}\left[\frac{2}{3}(2x + 1)^{3/2}\right] = \frac{2}{3}\frac{3}{2}(2x + 1)^{1/2} \cdot 2$$

$$= 2(2x + 1)^{1/2}$$

$$\neq (2x + 1)^{1/2}.$$

We must account for the extra factor of 2. The answer lies in the fact that in differentiation we *multiply* by the constant which is the derivative of the function inside the parenthesis. Hence, in integration we must *divide* by this constant. Thus,

(3) $$\int (2x + 1)^{1/2} \, dx = \int (2x + 1)^{1/2} \left(\frac{1}{2} \cdot 2\right) dx$$

$$= \frac{1}{2} \int (2x + 1)^{1/2} \cdot 2 \, dx$$

(4) $$= \frac{1}{2} \frac{(2x + 1)^{3/2}}{\frac{3}{2}} + c$$

$$= \frac{1}{3}(2x + 1)^{3/2} + c,$$

Notice that in this case

$$\frac{d}{dx}\left[\frac{1}{3}(2x + 1)^{3/2}\right] = \frac{1}{3}\frac{3}{2}(2x + 1)^{1/2} \cdot 2$$

$$= (2x + 1)^{1/2}.$$

In (3) we multiplied *and* divided by 2 (that is, multiplied by the number 1), which is the derivative of $2x + 1$. The inevitable question is: What happened to the multiple of 2 in (4)? It must be kept firmly in mind that the multiple of 2 "comes out" in the process of differentiation and hence is "absorbed" back into the function in the process of integration.

In general, the power rule for functions gives

$$\frac{d}{dx} \frac{[f(x)]^{n+1}}{n+1} = (n+1) \frac{[f(x)]^n}{n+1} f'(x)$$

$$= [f(x)]^n f'(x).$$

From the definition of an antiderivative we can write

(5) $$\int [f(x)]^n f'(x) \, dx = \frac{[f(x)]^{n+1}}{n+1} + c, \; n \neq -1.$$

DIFFERENTIAL When working with indefinite integrals and functions it is convenient to introduce simplifying *substitutions*. If $u = f(x)$ denotes a function, then the product

$$du = f'(x) \, dx$$

is called the **differential** of $f(x)$. For example, the differential of $u = x^2$ is $du = 2x \, dx$. Substituting u for $f(x)$ and du for $f'(x) \, dx$ in (5) yields the following equivalent result.

(6) $$\int u^n \, du = \frac{u^{n+1}}{n+1} + c, \; n \neq -1.$$

Example Evaluate $\int (x^2 + 1)^4 \, 2x \, dx$.

Solution First identify $u = f(x) = x^2 + 1$ and $du = f'(x) \, dx = 2x \, dx$. Hence, from (6) we find

$$\int \underbrace{(x^2 + 1)^4}_{u^4} \underbrace{2x \, dx}_{du} = \int u^4 \, du$$

$$= \frac{u^5}{5} + c$$

$$= \frac{(x^2 + 1)^5}{5} + c.$$

Remember, if you are not convinced that the answer is correct then differentiation should be used as a check. For example, in the preceding example we have

$$\frac{d}{dx} \frac{(x^2 + 1)^5}{5} + c = 5 \frac{(x^2 + 1)^4}{5} \cdot 2x = (x^2 + 1)^4 \, 2x.$$

Example Evaluate $\int \sqrt{3x + 4} \, dx$.

14.2 Integration by Substitution

Solution Since the problem is the same as $\int (3x+4)^{1/2}\,dx$, we can identify $u = f(x) = 3x + 4$ and so $du = 3\,dx$. Note that, as given, the problem lacks the constant multiple of 3. We remedy this by writing

$$\int (3x+4)^{1/2}\,dx = \frac{1}{3}\int \underbrace{(3x+4)^{1/2}}_{u^{1/2}}\,\overbrace{3\,dx}^{du}. \qquad \left(\frac{1}{3}\cdot 3 = 1\right)$$

We are now in a position to use formula (6):

$$\int (3x+4)^{1/2}\,dx = \frac{1}{3}\int u^{1/2}\,du$$

$$= \frac{1}{3}\frac{u^{3/2}}{3/2} + c$$

$$= \frac{2}{9}u^{3/2} + c$$

$$= \frac{2}{9}(3x+4)^{3/2} + c.$$

When finding an antiderivative of a power of a function, the essential item to check is whether we actually have the *du* term. If we are simply lacking a constant, as in the example above, then all we need do is multiply and divide by this appropriate constant. Do *not* attempt to multiply and divide by variables, since no variable can come outside the integral symbol. For example,

$$\int (2x^2+9)^{1/2}\,dx \ne \frac{1}{4x}\int (2x^2+9)^{1/2}\,4x\,dx.$$

Example Evaluate $\int \dfrac{(1+\ln x)^2}{x}\,dx$.

Solution Let $u = 1 + \ln x$. The differential of this function is exactly $du = \dfrac{1}{x}\,dx$. Therefore, we have from (6),

$$\int \frac{(1+\ln x)^2}{x}\,dx = \int (1+\ln x)^2 \cdot \frac{1}{x}\,dx$$

$$= \int u^2\,du$$

$$= \frac{u^3}{3} + c$$

$$= \frac{(1+\ln x)^3}{3} + c.$$

INDEFINITE INTEGRALS THAT YIELD LOGARITHMS

Recall from rule XI of Section 12.8 that

$$\frac{d}{dx}\ln f(x) = \frac{1}{f(x)} \cdot f'(x)$$

Thus, we obtain

$$\int \frac{1}{f(x)} \cdot f'(x)\,dx = \int \frac{f'(x)\,dx}{f(x)} = \ln f(x) + c.$$

With usual substitutions $u = f(x)$ and $du = f'(x)\,dx$, we obtain the next result.

(7) $$\int \frac{du}{u} = \ln u + c.$$

Example Evaluate $\int \frac{dx}{x+9}$.

Solution Suppose $u = x + 9$, then $du = 1 \cdot dx = dx$. Hence, from (7) we see immediately that

$$\int \frac{dx}{x+9} = \int \frac{du}{u}$$
$$= \ln u + c$$
$$= \ln(x+9) + c.$$

Example Evaluate $\int \frac{x\,dx}{x^2+4}$.

Solution Let $u = x^2 + 4$, and so $du = 2x\,dx$. Therefore, from (7) we obtain

$$\int \frac{x\,dx}{x^2+4} = \frac{1}{2}\int \underbrace{\frac{2x\,dx}{x^2+4}}_{u}{\Big\}\frac{du}{u}}$$
$$= \frac{1}{2}\int \frac{du}{u}$$
$$= \frac{1}{2}\ln u + c$$
$$= \frac{1}{2}\ln(x^2+4) + c.$$

EXPONENTIAL FUNCTIONS

The derivative formula

14.2 Integration by Substitution

$$\frac{d}{dx} e^{f(x)} = e^{f(x)} f'(x)$$

yields the integration formula

$$\int e^{f(x)} f'(x)\, dx = e^{f(x)} + c.$$

If $u = f(x)$ and $du = f'(x)\, dx$, we then have

(8) $$\int e^u\, du = e^u + c.$$

Example Evaluate $\int e^{5x}\, dx$.

Solution Let $u = 5x$ and $du = 5\, dx$. Therefore, from (8) we obtain

$$\int e^{5x}\, dx = \frac{1}{5} \int \underbrace{e^{5x}}_{e^u} \cdot \underbrace{5\, dx}_{du}$$

$$= \frac{1}{5} \int e^u\, du$$

$$= \frac{1}{5} e^u + c.$$

$$= \frac{1}{5} e^{5x} + c.$$

Example Evaluate $\int e^{x^3} x^2\, dx$.

Solution Let $u = x^3$, and so $du = 3x^2\, dx$. From (8) it follows that

$$\int e^{x^3} x^2\, dx = \frac{1}{3} \int e^{x^3} \cdot 3x^2\, dx$$

$$= \frac{1}{3} \int e^u\, du$$

$$= \frac{1}{3} e^u + c$$

$$= \frac{1}{3} e^{x^3} + c.$$

In conclusion, we urge the student to inspect a problem carefully before attempting to solve it. Observe that the indefinite integrals $\int \frac{x^2\, dx}{x^3 + 1}$ and $\int \frac{x^2\, dx}{(x^3 + 1)^2}$ look very similar. However, by means of the substitutions $u = x^3 + 1$ and $du = 3x^2\,dx$ we see that the first integral is of the form $\dfrac{1}{3}\displaystyle\int \dfrac{du}{u}$ whereas the second is $\dfrac{1}{3}\displaystyle\int \dfrac{du}{u^2} = \dfrac{1}{3}\displaystyle\int u^{-2}\,du$. Thus, we must use (7) to evaluate $\displaystyle\int \dfrac{x^2\,dx}{x^3 + 1}$ and (6) to evaluate $\displaystyle\int \dfrac{x^2\,dx}{(x^3 + 1)^2}$.

EXERCISE 14.2

Evaluate the given.

1. $\displaystyle\int_{}^{8} \dfrac{8}{x}\,dx$
2. $\displaystyle\int 5x^{-1}\,dx$
3. $\displaystyle\int (2x - 4x^{-1})\,dx$
4. $\displaystyle\int \left(10 + \dfrac{2}{x}\right)dx$
5. $\displaystyle\int \left(x + \dfrac{1}{x^2}\right)^2 dx$
6. $\displaystyle\int \left(x^2 + \dfrac{1}{x^3}\right)^2 dx$
7. $\displaystyle\int (x + 1)^{10}\,dx$
8. $\displaystyle\int (x + 1)^{-9}\,dx$
9. $\displaystyle\int (3x + 1)^5\,dx$
10. $\displaystyle\int (4x + 7)^3\,dx$
11. $\displaystyle\int \sqrt{x + 1}\,dx$
12. $\displaystyle\int \dfrac{dx}{\sqrt{x + 1}}$
13. $\displaystyle\int (5x + 3)^{3/2}\,dx$
14. $\displaystyle\int (-x + 9)^{1/3}\,dx$
15. $\displaystyle\int \sqrt[3]{(x + 1)^2}\,dx$
16. $\displaystyle\int \sqrt{(2x - 1)^5}\,dx$
17. $\displaystyle\int (x^2 - 4)^3\,2x\,dx$
18. $\displaystyle\int (x^2 + 7)^5\,2x\,dx$
19. $\displaystyle\int (2x^3 + 1)^4\,x^2\,dx$
20. $\displaystyle\int (3x^3 - 1)^3\,x^2\,dx$
21. $\displaystyle\int \dfrac{x}{(5x^2 + 2)^2}\,dx$
22. $\displaystyle\int \dfrac{x}{\sqrt{4x^2 + 1}}\,dx$
23. $\displaystyle\int (4x^2 + 4x + 1)^2(2x + 1)\,dx$
24. $\displaystyle\int (-3x^2 + 6x + 9)^5(x - 1)\,dx$
25. $\displaystyle\int \dfrac{x + 1}{\sqrt{x^2 + 2x + 2}}\,dx$
26. $\displaystyle\int \dfrac{4x + 2}{\sqrt{(4x^2 + 4x - 1)^3}}\,dx$
27. $\displaystyle\int \dfrac{dx}{x - 3}$
28. $\displaystyle\int \dfrac{dx}{x + 7}$
29. $\displaystyle\int \dfrac{dx}{4x + 9}$
30. $\displaystyle\int \dfrac{dx}{10 - x}$

14.3 Integration by Parts

31. $\int \dfrac{x}{x^2+1}\,dx$

32. $\int \dfrac{x}{x^2+10}\,dx$

33. $\int \dfrac{x+1}{x^2+2x+8}\,dx$

34. $\int \dfrac{4x+2}{4x^2+4x-1}\,dx$

35. $\int (x^3+5)^{-1} x^2\,dx$

36. $\int (x^3-3x+1)^{-1}(x^2-1)\,dx$

37. $\int e^{2x}\,dx$

38. $\int (e^{7x}+1)\,dx$

39. $\int e^{-x}\,dx$

40. $\int 10 e^{-5x}\,dx$

41. $\int \dfrac{e^x + e^{-x}}{2}\,dx$

42. $\int \dfrac{e^x - e^{-x}}{2}\,dx$

43. $\int e^{x^2} x\,dx$

44. $\int e^{-2x^2} x\,dx$

45. $\int e^{-x^3} x^2\,dx$

46. $\int e^{4x^3} x^2\,dx$

47. $\int e^{x^2+x}(2x+1)\,dx$

48. $\int e^{x^2+2x+3}(x+1)\,dx$

49. $\int (e^{2x}+1)^4 e^{2x}\,dx$

50. $\int (e^{-4x}+2)^9 e^{-4x}\,dx$

51. $\int \dfrac{e^x}{e^x+1}\,dx$

52. $\int \dfrac{e^{-x}}{e^{-x}+4}\,dx$

53. $\int \dfrac{\ln x}{x}\,dx$

54. $\int \dfrac{(\ln x)^2}{x}\,dx$

14.3
INTEGRATION BY PARTS

The reader should not think that the integration formulas of the last two sections will yield every possible antiderivative. For example, xe^x and $\ln x$ have antiderivatives but $\int xe^x\,dx$ is *not* of the form $\int e^u\,du$ nor can $\int \dfrac{dx}{x} = \ln x + c$ be applied to find $\int \ln x\,dx$. In the following discussion, we develop a procedure that will enable us to evaluate both $\int xe^x\,dx$ and $\int \ln x\,dx$ as well as many other indefinite integrals not covered by our previous results. This procedure is called **integration by parts** and is a direct result of the product rule for derivatives.

If $f(x)g(x)$ is a product of two functions, then by rule V

$$\frac{d}{dx}[f(x)g(x)] = f(x)g'(x) + g(x)f'(x).$$

By integrating both members of the last equation, it follows that

$$\int \frac{d}{dx}[f(x)g(x)]\,dx = \int f(x)g'(x)\,dx + \int g(x)f'(x)\,dx$$

and so

(1) $$\int f(x)g'(x)\,dx = f(x)g(x) - \int g(x)f'(x)\,dx.$$

It is common practice to express (1) in terms of differential notation. If $u = f(x)$, $v = g(x)$, $du = f'(x)dx$, and $dv = g'(x)dx$, we obtain the following equivalent result.

Integration by Parts Formula

(2) $$\int u\,dv = uv - \int v\,du.$$

Equation (2) always starts with an integration followed by a differentiation. That is,

$$\int u\,\overrightarrow{dv} = \underbrace{uv}_{\text{differentiate}} - \int v\,du.$$

We summarize the procedure.

1. Choose dv and u.
2. Integrate dv to find v.
3. Differentiate u.
4. Form the product uv.
5. Evaluate $\int v\,du$.

Example Evaluate $\int xe^x\,dx$.

Solution If we identify

$$dv = e^x\,dx$$
$$u = x$$

then

$$v = e^x$$
$$du = 1 \cdot dx = dx.$$

Substitution in (2) gives

$$\overset{u\,dv}{\int xe^x\,dx} = \overset{uv}{xe^x} - \overset{v\ du}{\int e^x\,dx}$$
$$= xe^x - e^x + c.$$

14.3 Integration by Parts

Check:

$$\frac{d}{dx}[xe^x - e^x + c] = xe^x + e^x - e^x$$

$$= xe^x.$$

Two comments are in order concerning the last example. The choice of $dv = e^x\, dx$ is one of several possible options. Had we selected $dv = x\, dx$, then $v = x^2/2$ and (2) would yield

$$\int xe^x\, dx = \frac{x^2}{2}e^x - \frac{1}{2}\int x^2 e^x\, dx.$$

The problem here is that the second integral $\int x^2 e^x\, dx$ is now actually more complicated than the original. The selection of dv must always be made with an eye to make $\int v\, du$ simpler than $\int u\, dv$. Also, note that we did not affix a constant when integrating $dv = e^x\, dx$. This constant need only be added at the end of the problem.

Example Evaluate $\int xe^{10x}\, dx$.

Solution If

$$dv = e^{10x}\, dx$$

$$u = x$$

then

$$v = \frac{1}{10}e^{10x}$$

$$du = dx.$$

Therefore, (2) gives

$$\int xe^{10x}\, dx = \frac{1}{10}xe^{10x} - \frac{1}{10}\int e^{10x}\, dx$$

$$= \frac{1}{10}xe^{10x} - \frac{1}{10}\cdot\frac{1}{10}e^{10x} + c$$

$$= \frac{1}{10}xe^{10x} - \frac{1}{100}e^{10x} + c.$$

Example Evaluate $\int \ln x\, dx$.

Solution Choosing

$$dv = dx$$

$$u = \ln x$$

leads to
$$v = x$$
$$du = \frac{dx}{x}.$$

Thus, (2) gives
$$\int \ln x \, dx = x \ln x - \int x \cdot \frac{1}{x} dx$$
$$= x \ln x - \int dx$$
$$= x \ln x - x + c.$$

Example Evaluate $\int x \sqrt{x+1} \, dx$.

Solution Let
$$dv = (x+1)^{1/2} \, dx$$
$$u = x.$$

Hence, from (6) of Section 14.2 we find
$$v = \frac{2}{3}(x+1)^{3/2}$$

and
$$du = dx.$$

Thus, (2) yields
$$\int x \sqrt{x+1} \, dx = x \cdot \frac{2}{3}(x+1)^{3/2} - \frac{2}{3} \int (x+1)^{3/2} \, dx$$
$$= \frac{2}{3} x(x+1)^{3/2} - \frac{2}{3} \cdot \frac{2}{5}(x+1)^{5/2} + c$$
$$= \frac{2}{3} x(x+1)^{3/2} - \frac{4}{15}(x+1)^{5/2} + c.$$

It may be necessary to integrate by parts more than once in a given problem.

Example Evaluate $\int x^2 e^x \, dx$.

Solution Let
$$dv = e^x \, dx$$
$$u = x^2,$$

14.3 Integration by Parts

then
$$v = e^x$$
$$du = 2x\,dx.$$

Equation (1) is then
$$\int x^2 e^x\,dx = x^2 e^x - 2\int xe^x\,dx.$$

To evaluate $\int xe^x\,dx$ we must integrate by parts again. However, we know from our first example that
$$\int xe^x\,dx = xe^x - e^x$$

Therefore,
$$\int x^2 e^x\,dx = x^2 e^x - 2(xe^x - e^x) + c$$
$$= x^2 e^x - 2xe^x + 2e^x + c.$$

EXERCISE 14.3

Integrate by parts.

1. $\int xe^{-x}\,dx$
2. $\int xe^{2x}\,dx$
3. $\int xe^{5x}\,dx$
4. $\int xe^{-3x}\,dx$
5. $\int x^2 e^{2x}\,dx$
6. $\int x^2 e^{-4x}\,dx$
7. $\int (x+1)e^x\,dx$
8. $\int (2x+3)e^{-x}\,dx$
9. $\int \ln 2x\,dx$
10. $\int \ln 4x\,dx$
11. $\int \ln x^2\,dx$
12. $\int \ln x^3\,dx$
13. $\int x \ln x\,dx$
14. $\int x^2 \ln x\,dx$
15. $\int x^{1/2} \ln x\,dx$
16. $\int \dfrac{\ln x}{\sqrt{x}}\,dx$
17. $\int (\ln x)^2\,dx$
18. $\int x(\ln x)^2\,dx$
19. $\int \ln(x+1)\,dx$ [*Hint*: $x/(x+1) = 1 - 1/(x+1)$.]
20. $\int \ln(x+5)\,dx$
21. $\int (x+1)\ln(x+1)\,dx$
22. $\int \dfrac{\ln(x+5)}{(x+5)^2}\,dx$
23. $\int x(x+2)^5\,dx$
24. $\int x(2x-1)^3\,dx$

25. $\int x\sqrt{4x+1}\,dx$
26. $\int x(x+1)^{3/2}\,dx$
27. $\int (x+1)(x+6)^4\,dx$
28. $\int (x-1)(x+1)^{10}\,dx$
29. $\int x^3 e^{x^2}\,dx$ [Hint: $x^3 e^{x^2} = x^2 \cdot xe^{x^2}$.]
30. $\int \dfrac{x^3}{e^{x^2}}\,dx$
31. Use integration by parts to evaluate $\int x^n \ln x\,dx$, $n \neq -1$.
32. Use integration by parts to evaluate $\int \dfrac{\ln x}{x}\,dx$. Choose $dv = \dfrac{dx}{x}$ and $u = \ln x$.

14.4

THE DEFINITE INTEGRAL

We review here the seven basic integration formulas of the last three sections:

(1) $\displaystyle\int x^n\,dx = \dfrac{x^{n+1}}{n+1} + c,\ n \neq -1$
(2) $\displaystyle\int u^n\,du = \dfrac{u^{n+1}}{n+1} + c,\ n \neq -1$

(3) $\displaystyle\int \dfrac{dx}{x} = \ln x + c$
(4) $\displaystyle\int \dfrac{du}{u} = \ln u + c$

(5) $\displaystyle\int e^x\,dx = e^x + c$
(6) $\displaystyle\int e^u\,du = e^u + c$

(7) $\displaystyle\int u\,dv = uv - \int v\,du.$

An antiderivative, or indefinite integral of a function, is itself another function. We are now in a position to define the concept of a **definite integral**. In contrast to the indefinite integral, the definite integral is a *real number*.

> Let f be continuous on an interval $a \leq x \leq b$ and F any antiderivative of f, that is, $F'(x) = f(x)$, then the **definite integral** of f on the interval, written $\int_a^b f(x)\,dx$, is the real number
>
> $$F(b) - F(a).$$

The numbers a and b are called **limits of integration**; a is called the **lower limit** of integration and b is the **upper limit** of integration. The function f is referred to as the **integrand**. A definite integral is the number obtained by

(antiderivative at upper limit) − (antiderivative at lower limit).

This difference is usually denoted by

$$F(x)\,\big|_a^b$$

14.4 The Definite Integral

and we write

(8)
$$\int_a^b f(x)\,dx = F(x)\,|_a^b = F(b) - F(a).$$

We note that *any* antiderivative can be used to evaluate a definite integral. In reality, the constant of antidifferentiation is superfluous since

$$F(x) + c\,|_a^b = [F(b) + c] - [F(a) + c]$$
$$= F(b) - F(a).$$

Example Evaluate $\int_1^2 x\,dx$.

Solution From (1) we know that $\int x\,dx = x^2/2$, therefore,

$$\int_1^2 x\,dx = \frac{x^2}{2}\bigg|_1^2 = \frac{2^2}{2} - \frac{1^2}{2}$$
$$= 2 - \frac{1}{2}$$
$$= \frac{3}{2}.$$

Example Evaluate $\int_0^9 \sqrt{x}\,dx$.

Solution Again from (1) we have $\int x^{1/2}\,dx = 2x^{3/2}/3$ and so the definite integral is

$$\int_0^9 \sqrt{x}\,dx = \frac{2}{3}x^{3/2}\bigg|_0^9 = \frac{2}{3}9^{3/2} - \frac{2}{3}0^{3/2}$$
$$= \frac{2}{3}\cdot 27$$
$$= 18.$$

The following properties of the definite integral are immediate consequences of (8) and the properties of the indefinite integral.

(9) $\qquad \int_a^b kf(x)\,dx = k\int_a^b f(x)\,dx$, k a constant

(10) $\qquad \int_a^b [f(x) + g(x)]\,dx = \int_a^b f(x)\,dx + \int_a^b g(x)\,dx$

(11) $\qquad \int_a^a f(x)\,dx = 0.$

Justification of (11) If F is an antiderivative of f, then
$$\int_a^a f(x)\,dx = F(x)\,|_a^a = F(a) - F(a) = 0.$$

Example Evaluate $\int_{-1}^3 (3x^2 + 1)\,dx$.

Solution Since $\int (3x^2 + 1)\,dx = 3\int x^2\,dx + \int dx = x^3 + x$ we have
$$\int_{-1}^3 (3x^2 + 1)\,dx = x^3 + x\,|_{-1}^3 = [3^3 + 3] - [(-1)^3 - 1]$$
$$= 30 - (-2)$$
$$= 32.$$

Example Evaluate $\int_0^6 \sqrt{2x+4}\,dx$.

Solution To obtain the value of $\int (2x+4)^{1/2}\,dx$, we identify $u = 2x + 4$, $du = 2dx$, and utilize (2):
$$\int (2x+4)^{1/2}\,dx = \frac{1}{2}\int (2x+4)^{1/2}\,2\,dx$$
$$= \frac{1}{2}\int u^{1/2}\,du$$
$$= \frac{1}{2}\cdot\frac{2}{3}u^{3/2}$$
$$= \frac{1}{3}(2x+4)^{3/2}.$$

Thus,
$$\int_0^6 \sqrt{2x+4}\,dx = \frac{1}{3}(2x+4)^{3/2}\,\bigg|_0^6 = \frac{1}{3}16^{3/2} - \frac{1}{3}4^{3/2}$$
$$= \frac{64}{3} - \frac{8}{3}$$
$$= \frac{56}{3}.$$

Example Evaluate $\int_0^5 \dfrac{dx}{x+1}$.

Solution With the identifications $u = x + 1$ and $du = dx$, it follows immediately from (4) that $\int dx/(x+1) = \ln(x+1)$. Hence,
$$\int_0^5 \frac{dx}{x+1} = \ln(x+1)\,\bigg|_0^5 = \ln 6 - \ln 1$$
$$= \ln 6$$

14.4 The Definite Integral

since $\ln 1 = 0$.

Example From property (11) we see $\int_3^3 \sqrt{1+x^2}\, dx = 0$. Note that in this case there is no need to seek the antiderivative of $f(x) = \sqrt{1+x^2}$.

Example Evaluate $\int_{-1}^1 xe^x\, dx$.

Solution Using integration by parts, formula (7), we have $\int xe^x\, dx = xe^x - e^x$ (see page 486). It follows that

$$\int_{-1}^1 xe^x\, dx = xe^x - e^x\big|_{-1}^1 = [e - e] - [(-1)e^{-1} - e^{-1}]$$
$$= 2e^{-1}.$$

The definite integral is more than an exercise in finding antiderivatives coupled with arithmetic. We consider applications of this concept in the following chapter. Finally, we state an additional property of the definite integral which will be immediately useful in these applications.

If c is any real number then

(12) $\qquad \int_a^b f(x)\, dx = \int_a^c f(x)\, dx + \int_c^b f(x)\, dx.$

Justification of (12) Let $F(x)$ be an antiderivative of $f(x)$. Then

$$\int_a^b f(x)\, dx = F(b) - F(a)$$
$$= F(b) - \underbrace{F(c) + F(c)}_{\text{zero}} - F(a)$$
$$= [F(c) - F(a)] + [F(b) - F(c)]$$
$$= \int_a^c f(x)\, dx + \int_c^b f(x)\, dx.$$

EXERCISE 14.4

Evaluate the given definite integral.

1. $\int_1^4 5\, dx$
2. $\int_{-1}^2 7\, dx$
3. $\int_{-1}^2 x\, dx$
4. $\int_1^3 x\, dx$
5. $\int_0^1 4x^2\, dx$
6. $\int_1^2 9x^2\, dx$

7. $\int_{1/2}^{4} \dfrac{dx}{x^2}$
8. $\int_{1/3}^{1} \dfrac{dx}{x^3}$
9. $\int_{-1}^{1} x^3\,dx$
10. $\int_{-1}^{1} 5x^4\,dx$
11. $\int_{1}^{8} x^{-1/3}\,dx$
12. $\int_{0}^{4} 3\sqrt{x}\,dx$
13. $\int_{-2}^{1} (2x+1)\,dx$
14. $\int_{-4}^{2} (4x-5)\,dx$
15. $\int_{2}^{5} (3x^2+4x-1)\,dx$
16. $\int_{3}^{4} (6x^2+8x+2)\,dx$
17. $\int_{1}^{2} \left(1+\dfrac{1}{x}\right)dx$
18. $\int_{1}^{4} \left(2x-\dfrac{2}{x}\right)dx$
19. $\int_{0}^{4} \sqrt{2x+1}\,dx$
20. $\int_{1}^{5} \sqrt{4x+5}\,dx$
21. $\int_{2}^{5} (-x+2)^3\,dx$
22. $\int_{0}^{1} (3x-1)^5\,dx$
23. $\int_{0}^{1} (x^2+1)^3 x\,dx$
24. $\int_{1}^{2} (x^2+4)^{-2} x\,dx$
25. $\int_{-1}^{1} e^x\,dx$
26. $\int_{2}^{7} e^{-x}\,dx$
27. $\int_{0}^{2} e^{2x}\,dx$
28. $\int_{-1}^{3} e^{5x}\,dx$
29. $\int_{0}^{1} e^{x^2} x\,dx$
30. $\int_{-1}^{2} e^{3x^2} x\,dx$
31. $\int_{0}^{6} \dfrac{dx}{x+2}$
32. $\int_{0}^{4} \dfrac{dx}{2x+1}$
33. $\int_{1}^{3} \dfrac{x\,dx}{x^2+1}$
34. $\int_{0}^{2} \dfrac{x^3\,dx}{x^4+1}$
35. $\int_{0}^{1} x(1+x)^3\,dx$
36. $\int_{0}^{3} x\sqrt{1+x}\,dx$
37. $\int_{1}^{2} xe^x\,dx$
38. $\int_{-2}^{1} xe^{2x}\,dx$
39. $\int_{1}^{4} x \ln x\,dx$
40. $\int_{1}^{2} x^2 \ln x\,dx$
41. $\int_{1}^{e} \dfrac{(\ln x)^2}{x}\,dx$
42. $\int_{e}^{4} \dfrac{dx}{x(\ln x)^2}$
43. $\int_{-1}^{-1}(x^3+1)^{10}\,dx$
44. $\int_{2}^{2} e^{x^2}\,dx$
45. $\int_{0}^{1} \dfrac{e^x\,dx}{e^x+1}$
46. $\int_{-2}^{1} \dfrac{e^x+1}{e^x}\,dx$
47. Verify that $\int_{1}^{4} 3x^2\,dx$ is the same as $\int_{1}^{2} 3x^2\,dx + \int_{2}^{4} 3x^2\,dx$.
48. Verify that $\displaystyle\int_{2}^{6} \dfrac{dx}{x}$ is the same as $\displaystyle\int_{2}^{3} \dfrac{dx}{x} + \int_{3}^{6} \dfrac{dx}{x}$.

CHAPTER TEST

Use an appropriate integration formula to evaluate each of the following.

1. $\int x^{1.5}\, dx$
2. $\int x^{-1.02}\, dx$
3. $\int (2x - 8)\, dx$
4. $\int (18x^2 - 6x + 5)\, dx$
5. $\int \dfrac{dx}{2x^{1/2}}$
6. $\int \dfrac{dx}{2x}$
7. $\int (2x + 1)^2\, dx$
8. $\int (x + 1)^{3/2}\, dx$
9. $\int \dfrac{x}{\sqrt{3x^2 - 4}}\, dx$
10. $\int \dfrac{6x^2}{\sqrt{2x^3 - 4}}\, dx$
11. $\int \dfrac{x^2 - 4x}{x^3 - 6x^2 + 1}\, dx$
12. $\int \dfrac{x^2 - 4x}{(x^3 - 6x^2 + 1)^{1/3}}\, dx$
13. $\int e^{-0.1x}\, dx$
14. $\int (4x - 7 + e^{9x})\, dx$
15. $\int xe^{2x}\, dx$
16. $\int xe^{2x^2}\, dx$
17. $\int \dfrac{e^{\sqrt{x}}}{\sqrt{x}}\, dx$
18. $\int \dfrac{e^{1/x}}{x^2}\, dx$
19. $\int \dfrac{(\ln x)^9}{x}\, dx$
20. $\int x^9 \ln x\, dx$
21. $\int_0^1 (2x + 1)\, dx$
22. $\int_0^1 (2x - 1)^4\, dx$
23. $\int_1^3 \dfrac{dx}{x^3}$
24. $\int_1^3 \dfrac{dx}{x}$
25. $\int_1^2 \ln 2x\, dx$
26. $\int_1^2 x \ln 2x\, dx$

15

APPLICATIONS OF THE INTEGRAL

15.1

AREA AS A DEFINITE INTEGRAL

The two basic subdivisions of calculus: differential and integral calculus, have their motivation in two geometric problems. As we saw in preceding chapters, the derivative is a means of finding the slope of a tangent line to the graph of a function $y = f(x)$. The motivation for the study of integral calculus is the problem of finding areas of geometric figures.

AREA UNDER A GRAPH

Let $y = f(x)$ be a continuous nonnegative function on an interval $a \leq x \leq b$. We shall refer to the area A shown in Figure 15.1, part b, bounded by the graph of $y = f(x)$, the vertical lines $x = a$ and $x = b$, and the x-axis as the **area under the graph**. As we shall now see, A can be found by means of definite integration.

Suppose $A(x)$ denotes the area under the graph of $y = f(x)$ from $x = a$ to a general abscissa x. If $h > 0$ is an increment in x, the corresponding change in the area A as x changes from the point x to the point $x + h$ is the difference in areas

$$A(x + h) - A(x).$$

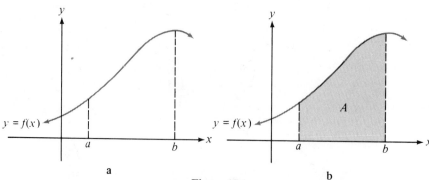

Figure 15.1

15.1 Area as a Definite Integral

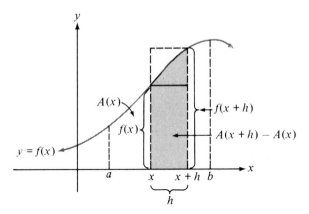

Figure 15.2

This change is indicated by the shaded region in Figure 15.2. From the figure we see that the smaller dotted rectangle has the area $f(x)h$ (height $\times$ width), whereas the larger rectangle has the area $f(x+h)h$. Since the shaded area is between these two extremes, we have

$$f(x)h \leq A(x+h) - A(x) \leq f(x+h)h.$$

If we divide by $h > 0$, then

$$f(x) \leq \frac{A(x+h) - A(x)}{h} \leq f(x+h).$$

Here we have assumed that the function $y = f(x)$ increases on the interval from x to $x+h$. If the function decreases, then the inequalities are reversed. Nevertheless, as $h \to 0$ it can be shown that

$$f(x) = \lim_{h \to 0} \frac{A(x+h) - A(x)}{h}$$

From the definition of the derivative we have

$$f(x) = \frac{dA}{dx}.$$

Since $A'(x) = f(x)$, $A(x)$ must be an antiderivative of $f(x)$. That is,

$$A(x) = \int f(x)\,dx$$
$$= F(x) + c.$$

Now to evaluate the constant c, we observe that the area A is zero when $x = a$, so that $A(a) = 0$. This implies that

$$A(a) = F(a) + c$$
$$0 = F(a) + c.$$

Thus, $c = -F(a)$. Hence,
$$A(x) = F(x) - F(a).$$
The area from $x = a$ to $x = b$ is then

(1) $$A(b) = F(b) - F(a).$$

Replacing $A(b)$ by A, it follows from (1) and the definition of the definite integral that, whenever $f(x) \geq 0$ on $a \leq x \leq b$, the area under the graph is

(2) $$A = \int_a^b f(x)\,dx.$$

Example Find the area under the graph of $y = x^2$ on (a) $0 \leq x \leq 2$, and (b) $-1 \leq x \leq 1$.

Solution (a) The area is shown in the accompanying figure. From (2) we find
$$A = \int_0^2 x^2\,dx$$
$$= \left.\frac{x^3}{3}\right|_0^2$$
$$= \frac{8}{3} \text{ square units.}$$

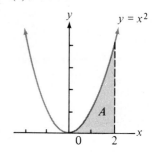

(b) The area is shown in the figure. In this case, (2) gives
$$A = \int_{-1}^1 x^2\,dx$$
$$= \left.\frac{x^3}{3}\right|_{-1}^1 = \frac{1}{3} - \left(-\frac{1}{3}\right).$$
$$= \frac{2}{3} \text{ square units.}$$

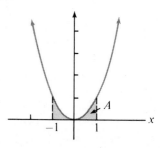

Whenever the graph of a continuous function is below the x-axis on an interval, the definite integral of the function on the interval does *not* give area. For example, as shown in Figure 15.3, the function $y = x^3$ is negative on a portion of the interval $-1 \leq x \leq 1$. The definite integral on the interval is

15.1 Area as a Definite Integral

$$\int_{-1}^{1} x^3 \, dx = \frac{x^4}{4} \Big|_{-1}^{1} = \frac{1}{4} - \frac{1}{4}$$
$$= 0.$$

Obviously, the value of the integral does not represent any area.

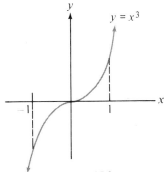

Figure 15.3

AREA AGAIN

We can, with little difficulty, find the *total* area bounded between a graph, the vertical lines $x = a$ and $x = b$, and the x-axis, even when the function is negative on the interval of consideration. For example, if the graph of $y = f(x)$ is as shown in Figure 15.4 then $\int_a^b f(x) \, dx$ is not area, since $f(x) \leq 0$ on $a \leq x \leq c$. To make the function nonnegative on the entire interval we take the absolute value of the function and use the fact that $|f(x)| = -f(x)$ if $f(x) < 0$. The total area, as shown in Figure 15.5, part b, is the sum of the areas A_1 and A_2. By (12) of Section 14.4 we can write

$$A = \int_a^b |f(x)| \, dx$$
$$= \int_a^c |f(x)| \, dx + \int_c^b |f(x)| \, dx$$
$$= -\int_a^c f(x) \, dx + \int_c^b f(x) \, dx$$
$$= A_1 + A_2.$$

In the second integral, the absolute-value signs can be ignored since the function is nonnegative on the interval $c \leq x \leq b$. Also, it should be carefully noted that A_1 is *not* the integral $\int_a^c f(x) \, dx$ but rather $A_1 = -\int_a^c f(x) \, dx$. Remember that a negative sign is

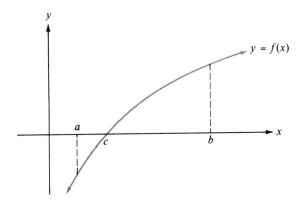

Figure 15.4

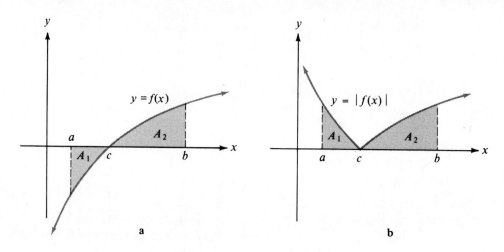

Figure 15.5

not to be confused with a negative number; here $-f(x)$ is the negative of the function, which in this case turns out to be a nonnegative quantity. In general, we have the following.

> The area bounded by a graph and the x-axis, on the interval $a \leq x \leq b$, is given by
> (3) $\qquad A = \int_a^b |f(x)|\, dx.$

Example Find the area bounded by the graph of $y = x^3$ and the x-axis on $-1 \leq x \leq 1$.

Solution The area in question is shaded in the accompanying figure. From (3) we have

$$A = \int_{-1}^{1} |x^3|\, dx$$
$$= \int_{-1}^{0} |x^3|\, dx + \int_{0}^{1} |x^3|\, dx$$
$$= -\int_{-1}^{0} x^3\, dx + \int_{0}^{1} x^3\, dx$$
$$= -\frac{1}{4}x^4 \Big|_{-1}^{0} + \frac{1}{4}x^4 \Big|_{0}^{1}$$
$$= -\left(0 - \frac{1}{4}\right) + \left(\frac{1}{4} - 0\right)$$
$$= \frac{1}{2} \text{ square unit.}$$

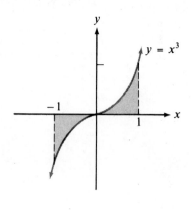

15.1 Area as a Definite Integral

With a little practice it should be apparent that all we really need do is take the negative of the function on that portion of the interval for which its graph is below the x-axis.

Example Find the area bounded by the graph of $y = 4 - x^2$ and the x-axis on $0 \leq x \leq 3$.

Solution The accompanying figure shows that $y = 4 - x^2$ is negative for $2 < x \leq 3$, and so it follows immediately from the discussion above that

$$A = \int_0^2 (4 - x^2)\,dx - \int_2^3 (4 - x^2)\,dx$$

$$= \left(4x - \frac{x^3}{3}\right)\Big|_0^2 - \left(4x - \frac{x^3}{3}\right)\Big|_2^3$$

$$= \left(8 - \frac{8}{3}\right) - (0 - 0)$$

$$\quad - (12 - 9) + \left(8 - \frac{8}{3}\right)$$

$$= \frac{23}{3} \text{ square units.}$$

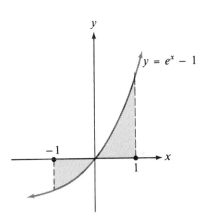

Example Find the area bounded by the graph of $y = e^x - 1$ and the x-axis on $-1 \leq x \leq 1$.

Solution The graph of $y = e^x - 1$ is given in the figure. Since the graph is below the x-axis for $-1 \leq x < 0$, the area is given by

$$A = -\int_{-1}^0 (e^x - 1)\,dx + \int_0^1 (e^x - 1)\,dx$$

$$= -(e^x - x)\Big|_{-1}^0 + (e^x - x)\Big|_0^1$$

$$= -(e^0 - 0) + (e^{-1} + 1)$$

$$\quad + (e^1 - 1) - (e^0 - 0)$$

$$= e + e^{-1} - 2$$

$$\approx 1.09 \text{ square units.}$$

EXERCISE 15.1

Find the area bounded by the graph of the given function and the x-axis on the indicated interval.

1. $y = 2, \; -3 \leq x \leq 5$
2. $y = 4, \; -1 \leq x \leq 3$
3. $y = x, \; 0 \leq x \leq 2$
4. $y = x, \; 1 \leq x \leq 4$
5. $y = x, \; -1 \leq x \leq 1$
6. $y = x, \; -1 \leq x \leq 4$
7. $y = 2x + 1, \; 0 \leq x \leq 4$
8. $y = 2x + 2, \; -1 \leq x \leq 2$
9. $y = -x + 1, \; -2 \leq x \leq 1$
10. $y = -2x + 4, \; -1 \leq x \leq 1$
11. $y = 2x - 4, \; -1 \leq x \leq 1$
12. $y = 6x - 3, \; -3 \leq x \leq 3$
13. $y = x^2, \; 0 \leq x \leq 3$
14. $y = x^2, \; -1 \leq x \leq 2$
15. $y = x^2 + 1, \; -2 \leq x \leq 1$
16. $y = x^2 + 4, \; 0 \leq x \leq 1$
17. $y = x^2 - 1, \; 1 \leq x \leq 4$
18. $y = x^2 - 1, \; -3 \leq x \leq -1$
19. $y = 1 - x^2, \; -1 \leq x \leq 1$
20. $y = 1 - x^2, \; 0 \leq x \leq 1$
21. $y = 4 - x^2, \; -3 \leq x \leq 0$
22. $y = 9 - x^2, \; -3 \leq x \leq 4$
23. $y = x^2 - 2x, \; 0 \leq x \leq 3$
24. $y = x^2 + 2x, \; -2 \leq x \leq 1$
25. $y = x^3 + 1, \; -1 \leq x \leq 2$
26. $y = 1 - x^3, \; 0 \leq x \leq 1$
27. $y = \sqrt{x}, \; 1 \leq x \leq 4$
28. $y = \sqrt{x}, \; 0 \leq x \leq 9$
29. $y = \sqrt{x} - 1, \; 0 \leq x \leq 16$
30. $y = \sqrt{x} - 1, \; 0 \leq x \leq 1$
31. $y = \sqrt[3]{x}, \; -1 \leq x \leq 1$
32. $y = x^{2/3}, \; -1 \leq x \leq 1$
33. $y = \dfrac{1}{x}, \; 1 \leq x \leq e$
34. $y = 2x^{-1}, \; 2 \leq x \leq 4$
35. $y = \dfrac{1}{x} + 1, \; 1 \leq x \leq 2$
36. $y = \dfrac{1}{x} - 1, \; \dfrac{1}{2} \leq x \leq 1$
37. $y = e^x, \; -3 \leq x \leq 1$
38. $y = e^{-x}, \; -2 \leq x \leq 2$
39. $y = e^x - 1, \; -2 \leq x \leq 1$
40. $y = 1 - e^x, \; -1 \leq x \leq 5$
41. $y = \ln x, \; 1 \leq x \leq 6$
42. $y = \ln x, \; \dfrac{1}{2} \leq x \leq 2$
43. $y = xe^x, \; 0 \leq x \leq 1$
44. $y = xe^{-x}, \; 0 \leq x \leq 2$

15.2
AREA BETWEEN TWO GRAPHS

In the last section, we used the definite integral to find the area between the graph of a function and the x-axis. For a graph such as in Figure 15.6, part a, the area is given by $A = \int_a^b f(x)\,dx$. Actually, we are finding the area between the graphs of $y = f(x)$ and $y = 0$, the latter being the equation of the x-axis. We can generalize the former results to the following rule, which enables us to find areas between two graphs such as indicated in Figure 15.6, part b.

Let $f(x)$ and $g(x)$ be continuous on $a \leq x \leq b$. If $f(x) \geq g(x)$ for all x in the interval, the area bounded by the graph of $y = f(x)$, the graph of $y = g(x)$ and the vertical lines $x = a$ and $x = b$ is given by

(1) $\qquad A = \int_a^b [f(x) - g(x)]\,dx.$

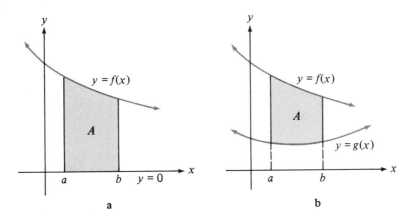

Figure 15.6

Note that if $g(x) = 0$, then we have $A = \int_a^b f(x)\,dx$, as before.

Also, $f(x) \geq g(x)$ for all x in the interval implies $f(x) - g(x) \geq 0$. Since the integral in (1) has a nonnegative integrand on the interval, it must represent area. We note that it is only necessary that $f(x) \geq g(x)$ and not $f(x) \geq 0$ and $g(x) \geq 0$ for $a \leq x \leq b$. Thus, (1) gives the area between two graphs even if, as shown in Figure 15.7, one graph passes below the x-axis. An alternative formulation of (1) is

$$A = \int_a^b [\text{upper graph ordinate} - \text{lower graph ordinate}]\,dx.$$

Example Find the area bounded by the graphs of $y = 4$ and $y = x^2$ in the first quadrant.

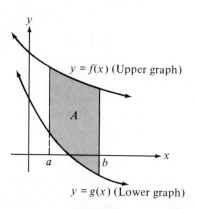

Figure 15.7

Solution The accompanying figure shows the area in question. Observe that the graphs $y = 4$ and $y = x^2$ intersect at the point corresponding to $x = 2$. Inspection of the figure also shows that the graph of $y = 4$ is the upper graph on the interval $0 \leq x \leq 2$. Hence, (1) yields

$$A = \int_0^2 [4 - x^2]\, dx$$

$$= 4x - \frac{x^3}{3}\bigg|_0^2$$

$$= \frac{16}{3} \text{ square units.}$$

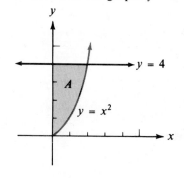

Example Find the area bounded by the graphs of $y = x^2 - x$ and $y = x + 3$ between their points of intersection.

Solution To determine where the graphs intersect, we must solve

$$x^2 - x = x + 3,$$

$$x^2 - 2x - 3 = 0,$$

$$(x - 3)(x + 1) = 0.$$

This last equation gives $x = -1$ and $x = 3$. From the figure it is apparent that $y = x + 3$ is the upper graph on the interval $-1 \leq x \leq 3$. Hence,

$$A = \int_{-1}^3 [(x + 3) - (x^2 - x)]\, dx$$

$$= \int_{-1}^3 (-x^2 + 2x + 3)\, dx$$

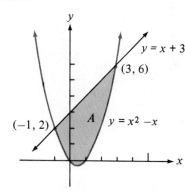

15.2 Area Between Two Graphs

$$A = -\frac{x^3}{3} + x^2 + 3x \Big|_{-1}^{3}$$

$$= (-9 + 9 + 9) - \left(\frac{1}{3} + 1 - 3\right)$$

$$= \frac{32}{3} \text{ square units.}$$

UPPER GRAPH MAY BECOME LOWER GRAPH The area between two graphs is *always* the definite integral of the upper graph ordinate minus the lower graph ordinate. So had we wanted the area between the curves given in the foregoing example, for say, $x = -2$ to $x = 3$, we must observe from Figure 15.8 that the relative position of the two graphs changes at $x = -1$. We have to break up the problem into two integrals. The area is given by

$$A = \int_{-2}^{-1} \overbrace{[(x^2 - x)}^{\substack{\text{upper graph} \\ \text{ordinate on} \\ -2 \le x < -1}} - (x + 3)] \, dx + \int_{-1}^{3} \overbrace{[(x + 3)}^{\substack{\text{upper graph} \\ \text{ordinate on} \\ -1 < x \le 3}} - (x^2 - x)] \, dx$$

$$= \int_{-2}^{-1} (x^2 - 2x - 3) \, dx + \int_{-1}^{3} (-x^2 + 2x + 3) \, dx$$

$$= \frac{x^3}{3} - x^2 - 3x \Big|_{-2}^{-1} + \left(-\frac{x^3}{3} + x^2 + 3x\right)\Big|_{-1}^{3}$$

$$= \left(-\frac{1}{3} - 1 + 3\right) - \left(-\frac{8}{3} - 4 + 6\right) + (-9 + 9 + 9) - \left(\frac{1}{3} + 1 - 3\right)$$

$$= 13 \text{ square units.}$$

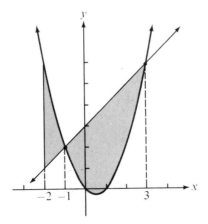

Figure 15.8

Example Find the area bounded by the graphs of $y = 1/x$ and $y = x$ on the interval $1 \le x \le e$.

Solution From the figure we see that $y = x$ is the upper graph on the interval. Hence, we obtain

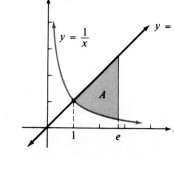

$$A = \int_1^e \left(x - \frac{1}{x}\right) dx$$

$$= \frac{1}{2}x^2 - \ln x \Big|_1^e$$

$$= \frac{1}{2}e^2 - \ln e - \frac{1}{2} + \ln 1$$

$$= \frac{1}{2}e^2 - 1 - \frac{1}{2}$$

$$= \frac{1}{2}e^2 - \frac{3}{2}$$

≈ 2.195 square units.

EXERCISE 15.2

Find the area bounded between the graphs of the given functions on the indicated interval.

1. $y = x$, $y = -x$, $1 \le x \le 3$
2. $y = x$, $y = 3x$, $0 \le x \le 2$
3. $y = x$, $y = -x + 6$, $0 \le x \le 3$
4. $y = x$, $y = -x + 6$, $3 \le x \le 4$
5. $y = x + 1$, $y = -x + 1$, $1 \le x \le 2$
6. $y = 2x + 2$, $y = 2x - 2$, $0 \le x \le 3$
7. $y = 1$, $y = x^2$, between points of intersection
8. $y = 9$, $y = x^2$, second quadrant
9. $y = 2x$, $y = x^2$, between points of intersection
10. $y = -x$, $y = x^2$, between points of intersection
11. $y = -x^2$, $y = x^2 + 1$, $-1 \le x \le 2$
12. $y = x$, $y = x^2 + 1$, $-1 \le x \le 2$
13. $y = x^2$, $y = -x^2 + 2$, first quadrant
14. $y = x^2 - 4$, $y = -x^2 + 4$, between points of intersection
15. $y = 9x$, $y = x^3$, third quadrant
16. $y = x$, $y = \sqrt{x}$, between points of intersection
17. $y = \sqrt{x}$, $y = -\sqrt{x}$, $0 \le x \le 9$
18. $y = x^3$, $y = -x^3$, $1 \le x \le 2$

19. $y = \dfrac{1}{x}$, $y = -\dfrac{1}{2}x + \dfrac{3}{2}$, between points of intersection

20. $y = \dfrac{4}{x}$, $y = x$, $2 \leq x \leq 5$

21. $y = -2x + 2$, $y = -x^2 + 5$, between points of intersection

22. $y = x + 1$, $y = x^2 + 2x + 1$, between points of intersection

23. $y = x$, $y = -x + 2$, $0 \leq x \leq 3$
24. $y = 4$, $y = x + 2$, $-1 \leq x \leq 4$
25. $y = x$, $y = x^2$, $0 \leq x \leq 2$
26. $y = x^2 - 1$, $y = -x^2 + 1$, $0 \leq x \leq 2$
27. $y = e^x$, $y = e^{-x}$, $0 \leq x \leq 3$
28. $y = e^x$, $y = e^{-x}$, $-1 \leq x \leq 2$
29. $y = 2$, $y = e^x + 1$, $-1 \leq x \leq 1$
30. $y = x + 1$, $y = e^{-x}$, $0 \leq x \leq 2$

15.3

THE AVERAGE VALUE OF A FUNCTION

In Chapter 13, we saw that the derivative means more than simply the slope of a tangent to a graph of a function. Similarly, a definite integral has many interpretations other than area under the graph of a function. The definite integral can be used to find work done by a variable force, centers of mass, volumes, moments of inertia, force exerted on a barrier such as water pressing against a dam, the output from the heart, change in profit, total revenue, the average value of a function, and many other things in the areas of science, engineering, and business. In this, and in the next section, we consider some of these applications. We begin with the notion of average value of a function.

If a and b are real numbers, then another number *between* them is the average

$$\frac{a+b}{2}.$$

For example, if $a = 3.5$ and $b = 10$, then

$$\frac{3.5 + 10}{2} = \frac{13.5}{2} = 6.75$$

is halfway between 3.5 and 10.

If a student has scores of 83%, 77%, 90%, and 88% on four examinations, then of course, the average of the scores is a number somewhere between the lowest score of 77% and the highest score of 90%. That is,

$$\frac{83 + 77 + 90 + 88}{4} = \frac{338}{4} = 84.5\%.$$

In general, the *arithmetic average* of n numbers $a_1, a_2, \ldots, a_n$ is

(1) $$\frac{a_1 + a_2 + a_3 + \cdots + a_n}{n}$$

or equivalently in terms of sigma notation (see Section 9.2),

(2) $$\frac{1}{n} \sum_{k=1}^{n} a_k.$$

In calculus, the analog of (2) is defined by means of a definite integral.

> The **average value** of a continuous function on an interval $a \leq x \leq b$ is the number
>
> (3) $$f_{av} = \frac{1}{b-a} \int_a^b f(x)\,dx.$$

Example Find the average value of $f(x) = x^2$ on the interval $1 \leq x \leq 3$.

Solution From (3) we find

$$\begin{aligned}
f_{av} &= \frac{1}{3-1} \int_1^3 x^2\,dx \\
&= \frac{1}{2} \frac{x^3}{3}\Big|_1^3 \\
&= \frac{1}{6}[27 - 1] \\
&= \frac{26}{6} \\
&= \frac{13}{3}.
\end{aligned}$$

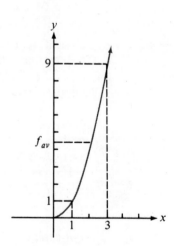

The accompanying figure shows, in this case, that $f_{av} = 13/3$ is between $f(1) = 1$ and $f(3) = 9$.

Example Suppose $P(t) = 120 + 5t + t^2$, t measured in minutes, represents the systolic blood pressure of a healthy person running a 1-mile race. What is the average blood pressure for $0 \leq t \leq 4$?

15.3 The Average Value of a Function

Solution From (3)

$$P_{av} = \frac{1}{4-0}\int_0^4 (120 + 5t + t^2)\, dt$$

$$= \frac{1}{4}\left(120t + 5\frac{t^2}{2} + \frac{t^3}{3}\right)\Big|_0^4$$

$$= \frac{1}{4}\left(480 + 40 + \frac{64}{3}\right)$$

$$= 135\frac{1}{3}.$$

GEOMETRIC INTERPRETATION If we cross-multiply both sides of (3) by the number $b - a$, then

$$f_{av}\cdot(b-a) = \int_a^b f(x)\,dx.$$

Now when $f(x) \geq 0$ on $a \leq x \leq b$, the right-hand member of the foregoing equation represents the area under the graph on the interval. Thus, as Figure 15.9 indicates, the area of the rectangle of height f_{av} and length $b - a$ is exactly the same as the area under the graph.

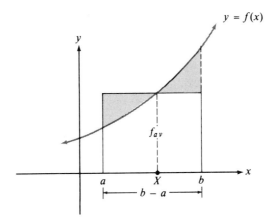

Figure 15.9

In other words, the equality of $f_{av}(b-a)$ and the integral $\int_a^b f(x)\,dx$ occurs since the area of that portion of the rectangle above the graph of $y = f(x)$ on $a \leq x \leq X$ is compensated by the area above the rectangle, but under the graph on $X \leq x \leq b$. This compensating effect is precisely why f_{av} can be considered an average.

EXERCISE 15.3

Find the average value of the given function on the indicated interval.

1. $f(x) = x$, $1 \leq x \leq 5$
2. $f(x) = 2x$, $0 \leq x \leq 6$
3. $f(x) = 2x + 1$, $1 \leq x \leq 2$
4. $f(x) = 3x + 4$, $-1 \leq x \leq 1$
5. $f(x) = 3x^2 + 2$, $0 \leq x \leq 3$
6. $f(x) = 6x^2 + 2x + 1$, $1 \leq x \leq 3$
7. $f(x) = x^3$, $-2 \leq x \leq 2$
8. $f(x) = x^3 + x$, $-1 \leq x \leq 2$
9. $f(x) = \sqrt{2x + 1}$, $0 \leq x \leq 4$
10. $f(x) = \dfrac{1}{\sqrt{x+1}}$, $3 \leq x \leq 8$
11. $f(x) = e^{3x}$, $-1 \leq x \leq 1$
12. $f(x) = 1 + e^{-x}$, $0 \leq x \leq 6$
13. $f(x) = \dfrac{1}{x}$, $\dfrac{1}{2} \leq x \leq \dfrac{5}{2}$
14. $f(x) = \dfrac{1}{x+1}$, $0 \leq x \leq \dfrac{3}{2}$
15. $f(x) = \ln x$, $1 \leq x \leq 8$
16. $f(x) = x \ln x$, $1 \leq x \leq e$

Applications

17. *Biology* The velocity of blood, in centimeters per second, in a small artery is given by $v(r) = P(R^2 - r^2)/4vl$, $0 \leq r \leq R$, where P is the blood pressure, v the viscosity of the blood, l the length of the artery, and R the radius of the artery. Find the average velocity of the blood.

18. The number of bacteria present in a culture at time $t \geq 0$ is given by $N(t) = N_0 e^{kt}$, where N_0 is the initial number of bacteria and k is a positive constant. Find the average number of bacteria present for $0 \leq t \leq T$.

19. *Business* The total cost for producing x units of an item is given by $C(x) = 100x + 20{,}000$. Find the average cost for $0 \leq x \leq 50$.

20. A company determines that its revenue obtained through the sale of x units of an item is given by $R(x) = 3x^2 + 2x + 100$. Find the average revenue for $2 \leq x \leq 4$. Compare this average value with the average $[R(2) + R(3) + R(4)]/3$.

21. *Physics* The temperature in a controlled environment is given by $T(t) = -t^2 + 5t + 100$, where $t \geq 0$ is time measured in hours. Find the average temperature for $0 \leq t \leq 5$.

15.4 FURTHER APPLICATIONS OF THE DEFINITE INTEGRAL

We conclude our study of integral calculus with several additional applications of the definite integral from the areas of business and biology.

15.4 Further Applications of the Definite Integral

CHANGE IN PROFIT Suppose the marginal revenue is $MR = f(x)$ and the marginal cost is $MC = g(x)$; then the **change in profit** for a change in the level of production from, say, $x = a$ units to $x = b$ units is defined to be

(1) $$\int_a^b [f(x) - g(x)] \, dx.$$

Recall that the marginal profit is given by

$$MP = MR - MC$$
$$= P'(x)$$
$$= R'(x) - C'(x)$$
$$= f(x) - g(x).$$

so that if we integrate both sides, we obtain

$$\int_a^b P'(x) \, dx = P(b) - P(a) = \int_a^b [f(x) - g(x)] \, dx.$$

If the integral in line (1) is positive, then the profit has increased, whereas a negative integral represents a decrease in profit.

Example If the marginal revenue and marginal cost are $f(x) = -2x + 20$ and $g(x) = 5$, respectively, find the change in profit if the production level changes from 3 to 7 units.

Solution From (1) we have

$$\int_3^7 [(-2x + 20) - 5] \, dx = \int_3^7 (-2x + 15) \, dx$$
$$= (-x^2 + 15x)\Big|_3^7$$
$$= (-49 + 105) - (-9 + 45)$$
$$= 20 \text{ units of dollars.}$$

This represents an increase in profit between the two production levels. It is also the area between the graph of $f(x)$ and the graph of $g(x)$ on the interval $3 \le x \le 7$, as can be seen from the figure.

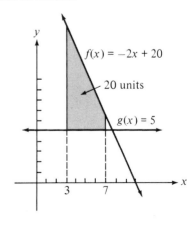

CONSUMER AND PRODUCER SURPLUSES

Let $D(x)$ and $S(x)$ represent, respectively, the demand and supply functions for a certain commodity. The variable x is the price at which $D(x)$ number of units (or $S(x)$ number of units) are demanded (or supplied). Suppose that $x = p$ is the price at which the market is in equilibrium [that is, $x = p$ is the solution of $D(x) = S(x)$; see Section 6.4] and let $x = b$ be the price of the commodity at which the demand for it is zero. In economic theory the definite integral

(2) $$CS = \int_p^b D(x)\,dx$$

is called the **consumer surplus.** When the market is in equilibrium, the consumer surplus represents, in monetary units, the combined savings realized by those consumers who would have been willing to pay an even higher price for the commodity.

Similarly, the combined savings realized by producers who would have been willing to supply a commodity at prices lower than the equilibrium price p is called the **producer surplus.** The producer surplus is also defined by a definite integral:

(3) $$PS = \int_c^p S(x)\,dx,$$

where $x = c$ is the price for which $S(x) = 0$.

Example Find the consumer and producer surpluses if $D(x) = 100 - x^2$ and $S(x) = x^2 + 10x$.

Solution We first find the equilibrium price by solving

$$100 - x^2 = x^2 + 10x$$
$$0 = 2x^2 + 10x - 100$$
$$0 = (2x - 10)(x + 10).$$

Since -10 does not make sense as a solution, the equilibrium price is $p = 5$. Now, the demand is zero if

$$100 - x^2 = 0$$
$$(10 + x)(10 - x) = 0.$$

Thus, we identify $b = 10$. Hence, from (2) the consumer surplus is given by

$$CS = \int_5^{10} (100 - x^2)\,dx$$

$$= 100x - \frac{x^3}{3}\Big|_5^{10}$$

$$= \left(1000 - \frac{1000}{3}\right) - \left(500 - \frac{125}{3}\right)$$

$$= \frac{625}{3}$$

$$= 208.33 \text{ units of dollars.}$$

15.4 Further Applications of the Definite Integral

Since the only meaningful solution to $S(x) = 0$, or

$$x^2 + 10x = 0$$
$$x(x + 10) = 0$$

is $c = 0$, it follows from (3) that the producer surplus is

$$\begin{aligned}
PS &= \int_0^5 (x^2 + 10x)\,dx \\
&= \left. \frac{x^3}{3} + 5x^2 \right|_0^5 \\
&= \frac{125}{3} + 125 \\
&= \frac{500}{3} \\
&= 166.67 \text{ units of dollars.}
\end{aligned}$$

As shown in the accompanying figures, the values of CS and PS are also the areas under the demand and supply graphs on the intervals $5 \le x \le 10$ and $0 \le x \le 5$, respectively.

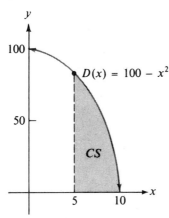
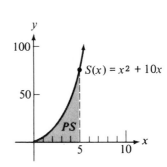

TOTAL REVENUE Let $R(t)$ be a function of time (measured in years) which describes the rate at which revenue comes in (or flows) from some source. For example, if $R(t) = 5000/(1 + t^2)$, then the rates at which revenue comes in for, say, $t = 1/2$ year and $t = 2$ years are respectively, $R(1/2) = \$4000$ per year and $R(2) = \$1000$ per year. If we compute

(4) $$\int_0^T R(t)\,dt,$$

we have the **total revenue** over a period of T years.

Example Suppose the rate at which revenue flows is given by $R(t) = 8000/\sqrt{4 + t}$ dollars/year. Determine the total revenue obtained in 12 years.

Solution From (4) we have

$$\int_0^{12} \frac{8000}{\sqrt{4+t}}\,dt = 16{,}000(4+t)^{1/2}\Big|_0^{12}$$

$$= 16{,}000(16^{1/2} - 4^{1/2})$$

$$= 16{,}000(4 - 2)$$

$$= \$32{,}000.$$

Example Rent from an apartment is $200 per month. What is the total revenue in 8 years?

Solution Intuitively we can determine the total revenue by

$$(\$200) \times (12 \text{ months}) \times (8 \text{ years}) = \$19{,}200.$$

Alternatively, it follows from (4) that

$$\int_0^8 (\text{Yearly revenue})\,dt = \int_0^8 2400\,dt$$

$$= 2400t\big|_0^8$$

$$= 2400(8 - 0)$$

$$= \$19{,}200.$$

PRESENT VALUE We can relate the idea of a continuous flow of revenue to the concept of the present value of an investment. Recall from Section 11.4 that the present value of an investment is given by

$$P = Se^{-rm},$$

where r is the annual rate of interest compounded continuously and m is time in years. The amount P represents the amount which must be invested *now* at a rate r in order to be worth S dollars m years in the future. Hence, we can say that the amount S is worth P dollars now.

Similarly, if revenue flows at a continuous and constant rate of, say, S dollars per year for a period of T years, then the **total present value** or worth of this future revenue is defined to be

(5) $$\int_0^T Se^{-rt}\,dt,$$

where r is the annual rate of interest which is compounded continuously.

Example Rent is received from an apartment over a period of 8 years at a rate of $200 a month. If the annual rate of interest is 6% compounded continuously, what is the present value of this flow of income?

Solution We have seen that the yearly revenue is a constant $2400. From (5) the present value of this flow of revenue is

15.4 Further Applications of the Definite Integral

$$\int_0^8 2400 e^{-0.06t}\, dt = 2400 \left[\frac{-1}{0.06} e^{-0.06t} \right]_0^8$$
$$= 40{,}000(1 - e^{-0.48})$$
$$= \$15{,}248.66.$$

CARDIAC OUTPUT To determine whether there is any disease present in the heart, physiologists often use a measurement of the **cardiac output**, that is, the volume of blood pumped from the heart per unit time. In the *dye dilution method*, a known amount D of dye (measured in milligrams) is inserted into the pulmonary artery. After circulating through the lungs, the pulmonary veins, and the chambers of the heart, the blood carrying the dye is pumped from the left ventricle of the heart into the aorta. A monitoring device inserted into the aorta measures the amount of dye leaving the heart over a time interval $0 \le t \le T$. T is the time at which there is no measurable amount of dye flowing from the heart. If $c(t)$ is a continuous function describing the concentration of the dye at any time (measured in milligrams per liter), it can be shown that the cardiac output, or flow rate, R is given by the following formula involving the definite integral of the concentration.

(6) $$R = \frac{D}{\int_0^T c(t)\, dt}$$

Example Five milligrams of dye are injected into the pulmonary artery. Determine the cardiac output of a person doing light exercise if it is known that $c(t) = -\frac{1}{90} t(t - 25)$ milligrams/liter and $0 \le t \le 25$.

Solution With $D = 5$ we have from (6)

$$R = \frac{5}{\int_0^{25} -\frac{1}{90} t(t-25)\, dt}$$

$$= \frac{5}{-\frac{1}{90}\left[\frac{t^3}{3} - 25\frac{t^2}{2} \right]_0^{25}}$$

$$= \frac{5}{(25)^3/540}$$

$$\approx 0.17 \text{ liters/sec}$$
$$= (0.17)\, 60$$
$$= 10.2 \text{ liters/min}.$$

The cardiac output can range from about 4.5 liters/min for a sleeping person to more than 25 liters/min for a jogger.

EXERCISE 15.4

Applications

1. *Biology* Determine the cardiac output of a person at rest if it is known that $c(t) = -\frac{1}{100}t(t-30)$ milligrams/liter and $0 \le t \le 30$. Use $D = 5$ milligrams.

2. Determine the cardiac output of a person doing moderate exercise if it is known that $c(t) = -\frac{1}{150}t(t-25)$ milligrams/liter and $0 \le t \le 25$. Use $D = 5$ milligrams.

3. The total flow of blood F inside an artery of radius R in a unit time is given by $F = \int_0^R 2\pi r v(r)\,dr$, where $v(r)$ is the velocity of the blood. Calculate F if it is known that $v(r) = P(R^2 - r^2)/4vl$, $0 \le r \le R$, P, v, and l are constants. The result is known as **Poiseuille's law**.

4. The average velocity of blood at a circular cross section of radius R and area A is given by $\bar{v} = \frac{1}{A}\int_0^R 2\pi r v(r)\,dr$, where $v(r)$ is given in Problem 3. Show that $\bar{v} = PR^2/8vl$.

5. *Business* The marginal revenue is $f(x) = 50 - 4x$ and the marginal cost is $g(x) = 20$. Determine the change in profit when the change in production level is from 1 to 3 units.

6. The marginal revenue is $f(x) = 200 - 8x$ and the marginal cost is $g(x) = 50$. Determine the change in profit when the change in production level is from 2 to 6 units.

Find the consumer and producer surpluses for the given demand and supply functions.

7. $D(x) = 200 - 4x$, $S(x) = 6x$
8. $D(x) = 12{,}000 - 24x$, $S(x) = 6x$
9. $D(x) = -3x + 12$, $S(x) = x^2 + x$
10. $D(x) = -x + 8$, $S(x) = x^2 - 4$
11. $D(x) = -x^2 + 100$, $S(x) = 4x^2 + 5x$
12. $D(x) = x^2 - 8x + 16$, $0 \le x \le 4$, $S(x) = x^2 + 8x$
13. Find the total revenue obtained in 5 years if the rate of income is 6000 dollars/year.
14. Find the total revenue obtained in 4 years if the rate of income is 100 dollars/month.
15. Find the total revenue obtained in 81 months if the rate of income is $R(t) = 1000/\sqrt{4t+9}$ dollars/year.

15.4 Further Applications of the Definite Integral

16. Find the total revenue obtained in 3 years if the rate of income is $R(t) = 500t(t^2 + 16)^{-1/2}$ dollars/year.

17. Find the present value of a 10 year flow of revenues of $5000 a year if the rate of continuous compounding is 8%.

18. Rent from a single apartment is $250 a month in a complex of 40 apartments. Determine the present value of a 5 year flow of revenues at 6% per annum compounded continuously.

19. The average life span of a certain type of fuse is 100 hours and its maximum life span is 200 hours. It is known that the probability of a fuse failing at a time between $t = a$ hours to $t = b$ hours is given by

$$\int_a^b \frac{e^{(-1/100)t}}{100} dt.$$

Suppose the fuse is still operative at the end of 25 hours. Then the probability that it will last at least an additional 25 hours, that is, that it will last for a total of at least 50 hours, is given by

$$P = \frac{\int_{50}^{200} \frac{1}{100} e^{(-1/100)t} dt}{\int_{25}^{200} \frac{1}{100} e^{(-1/100)t} dt}.$$

Determine the value of P.

20. In a certain theory of economics the value of property which is rented over an infinite period of time is given by the ratio of the yearly rent to the yearly rate of interest compounded continuously. That is to say,

$$\text{Value of property} = \frac{\text{Yearly rent}}{\text{Rate of interest}} = \frac{S}{r}.$$

Prove this by considering $\int_0^\infty Se^{-rt} dt$. [*Hint:* Define $\int_0^\infty Se^{-rt} dt = \lim_{b \to \infty} \int_0^b Se^{-rt} dt$.]

CHAPTER TEST

1. Find the area bounded by the graph of $y = -4x + 12$ and the x-axis on
    a. $0 \leq x \leq 2$,
    b. $0 \leq x \leq 4$.

2. Find the area bounded by the graph of $y = 9 - x^2$ and the x-axis on the interval $-3 \leq x \leq 3$.

3. Find the area bounded between the graphs of $y = x$ and $y = 2\sqrt{x}$
    a. between their points of intersection,
    b. on the interval $0 \leq x \leq 9$.

4. Find the area bounded between the graphs of $y = 2x + 1$ and $y = x^2 - 2$ between their points of intersection.

5. Find the average value of $f(x) = 4x^3 + 2x + 1$ on the interval $1 \leq x \leq 3$.

6. The velocity of a projectile is given by $v(t) = -32t + 128$, where t is measured in seconds. Find the average velocity for $0 \leq t \leq 4$.

7. The marginal revenue and marginal cost are $f(x) = 100 - 2x$ and $g(x) = 10$. Determine the change in profit when the production level changes from 2 to 5 units.

8. Find the consumer and producer surpluses if $D(x) = -x + 10$ and $S(x) = x^2 + 2x$.

9. Find the total revenue obtained in 13 years if the rate of income is $R(t) = 5000(2t + 1)^{-1/3}$ dollars/year.

10. Find the present value of a flow of income of $500 a month for the next 20 years if the annual rate of interest is 5% compounded continuously.

16

MULTIVARIATE DIFFERENTIAL CALCULUS

16.1

FUNCTIONS OF SEVERAL VARIABLES

A function can involve more than just one variable. For example, the area of a rectangle is $A = xy$, where x is the length and y is the width. The volume of a box depends on three variables: $V = xyz$, where x is the length, y is the width, and z is the height. Another familiar function of two variables is the volume of a cylinder, given by the formula $V = \pi r^2 h$, where r is the radius of the base and h is the height.

In this chapter we shall be primarily concerned with equations of the form $z = f(x, y)$ which describe functions of two variables. That is, when x and y are specified, there is one and only one value of z determined by $f(x, y)$. The domain of a function $z = f(x, y)$ is the set of ordered pairs (x, y) for which $f(x, y)$ is a real number.

Example If $f(x, y) = 100 - 2x^2 + xy$, find $f(-1, 3)$ and $f(2, 5)$.

Solution In the first case we simply replace x by -1 and y by 3,

$$f(-1, 3) = 100 - 2(-1)^2 + (-1)(3)$$
$$= 100 - 2 - 3$$
$$= 95,$$

and in the second case x is replaced by 2 and y by 5,

$$f(2, 5) = 100 - 2(2)^2 + (2)(5)$$
$$= 100 - 8 + 10$$
$$= 102.$$

GRAPHS We have seen that the graph of a function $y = f(x)$ is a *curve* in two dimensions. Although we shall not belabor the theory of graphing functions of two variables, we need to know that an equation $z = f(x, y)$ determines some kind of **surface** in three

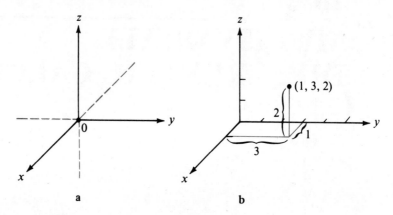

Figure 16.1

dimensions. Given three mutually perpendicular axes as shown in Figure 16.1, part a, we obtain three mutually perpendicular planes which we call the xy, yz, and xz planes. The point of intersection of these planes is labeled 0 for the origin. Any point in three dimensions can then be described as an ordered triple of numbers (x, y, z). For example, the point $(1, 3, 2)$ is illustrated in Figure 16.1, part b. We are particularly interested in the points $(x, y, 0)$ (or points in the xy plane) since the domain of a function $z = f(x, y)$ is a subset of this plane. As shown in Figure 16.2, given a point $(x, y, 0)$ in the domain of the function, we then determine a unique point $(x, y, f(x, y))$ in space. There is no geometric interpretation of a function of three variables, $w = f(x, y, z)$.

For example, portions of the graphs of $z = 9 - x^2 - y^2$ and $z = 12 - 3x - 6y$ are illustrated in Figure 16.3.

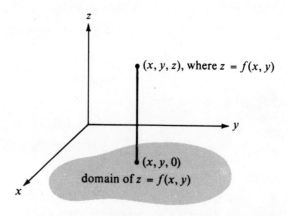

Figure 16.2

16.1 Functions of Several Variables

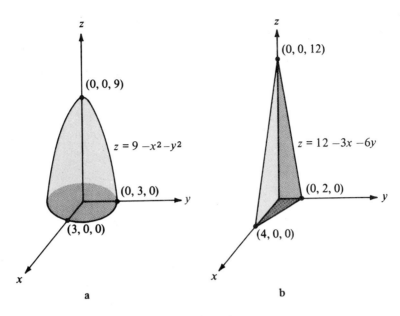

Figure 16.3

Example A hobbyist makes and sells two kinds of picture frames, each made from a different kind of wood. From long experience he knows that for every x number of frames made of one kind of wood, he will sell them at $50 - 2x$ dollars, and for every y number of frames made from the other kind of wood, he will sell them at $30 - y$ dollars. What is the total revenue function $R(x, y)$?

Solution Revenue is

$$(\text{number sold}) \times (\text{selling price}).$$

Since the total revenue is the sum of the revenues, we have

$$R(x, y) = x(50 - 2x) + y(30 - y)$$
$$= 50x - 2x^2 + 30y - y^2.$$

Example A soda can is constructed with an aluminum top and bottom and a tin lateral side. If the cost of the top is 1.5 cents per square unit, 1 cent per square unit for the bottom, and 2 cents per square unit for the side, find the cost function $C(x, y)$, where x is the radius of the can and y is the height.

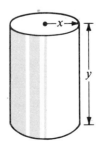

Solution   The area of the lateral side is the same as the area of a rectangle of length $2\pi x$ and width y. The cost is

$$(\text{cost per area}) \times (\text{area}).$$

Therefore, the cost of the top is

$$C_T = 1.5(\pi x^2) = 1.5\pi x^2.$$

Similarly, the cost of the bottom is

$$C_B = 1(\pi x^2) = \pi x^2.$$

The cost of the side is

$$C_S = 2(2\pi x \cdot y) = 4\pi xy.$$

The total cost (in cents) is then

$$\begin{aligned} C &= C_T + C_B + C_S \\ &= 1.5\pi x^2 + \pi x^2 + 4\pi xy \\ &= 2.5\pi x^2 + 4\pi xy. \end{aligned}$$

ENERGY CONSUMPTION   The total energy consumption per hour C of a person having a metabolic rate r and surface area S is given by the function

$$C = rS.$$

It can be shown that the surface area of a person can be expressed in terms of weight and height:

$$S = 0.2 W^{0.425} H^{0.725},$$

where W is measured in kilograms and H in meters. Thus, C can be written as a function of three variables

(1) $$C = 0.2r\, W^{0.425} H^{0.725}.$$

Example   The metabolic rate of a jogger who weighs 70 kilograms and is 1.5 meters tall is 600 cal/m²-hr. Determine the total energy consumption per hour.

Solution   From (1),

$$C = 0.2(600)(70)^{0.425}(1.5)^{0.725}.$$

With the aid of a calculator we find

$$C = 980 \text{ calories/hour}$$

The compound interest formula $S = Pe^{rm}$ is a function of the three variables P, r, and m. In physics, the voltage E across a resistor is given by the function of two variables $E = IR$, where I is current and R is the resistance.

EXERCISE 16.1

Evaluate the given function.

1. $f(x, y) = 2xy; f(2, 3), f(-4, 2)$
2. $f(x, y) = 4 + x^2y; f(1, -1), f(0, 6)$
3. $f(x, y) = x^2 - xy + y^2; f(2, 5), f(-3, -2)$
4. $f(x, y) = (x + y)^2; f(6, 4), f(3, -4)$
5. $f(x, y) = \sqrt{x + y}; f(-1, 5), f(5, 11)$
6. $f(x, y) = \sqrt[3]{2x + 3y}; f\left(\frac{1}{2}, \frac{7}{3}\right), f(0, -9)$
7. $f(x, y, z) = x^2 + y^2 + z^2; f(3, 2, -5), f(2, -2, -1)$
8. $f(x, y, z) = (x + 2y + 3z)^2; f(-1, -1, 1), f(2, 3, -2)$
9. $f(x, y, z) = \dfrac{z^2}{x^2 + y^2}; f\left(\frac{1}{2}, \frac{1}{2}, \frac{1}{4}\right), f(3, 4, 5)$
10. $f(x, y, z) = \dfrac{1}{x^2} + \dfrac{1}{y^2} + \dfrac{1}{z^2}; f(\sqrt{3}, \sqrt{3}, \sqrt{3}), f\left(\frac{1}{3}, \frac{1}{4}, \frac{1}{5}\right)$

Applications

11. *Biology* A person who is 1.6 meters tall and weighs 80 kilograms has a metabolic rate of 40 cal/m²-hr while sleeping. Use a calculator to determine the energy consumption per hour.

12. *Business* A company's cost function for production of x and y units of two different commodities is given by

$$C(x, y) = 1000 + 2x + 6y + 5xy.$$

What is the cost to the company before any of the units are produced? Evaluate $C(x + 1, y + 1)$.

13. A closed rectangular box is to be constructed from 300 square inches of cardboard. Express the volume V as a function of the length x and width y. [*Hint*: Use the fact that the total surface area is $2xy + 2yz + 2zx = 300$, where z is the height of the box.]

14. The cost of construction of a glass window framed with metal is a function of its length y and width x. Suppose that the cost of the metal for the width along the top and bottom of the window is 15 cents per foot, the cost of the sides is 20 cents per foot, and the cost of the glass is 12 cents per square foot. Find the function $C(x, y)$ which gives the total cost of construction of each window.

15. A company sells two types of television sets. It makes a profit of $40 on one type and $125 on the other. If it sells x number of the first type each day, and y number of the second type each day, what is the company's daily profit function for the television sets?

16. It costs a small manufacturing company $50 and $75 per unit, respectively, in the construction of two items A and B. If the selling prices for A and B are x and y,

respectively, what is the company's weekly profit function $P(x, y)$ if it is known that it can sell $5 + 2x + y$ units of A each week, and $12 - x + 4y$ units of B each week?

17. A company finds that its yearly revenue and cost functions for producing x and y units of two different items are, respectively,

$$R(x, y) = 500x + 2000y - 2x^2 - y^2 + 100xy,$$
$$C(x, y) = 100x + 90y + x^2 + 2y^2.$$

What is the profit at $x = 2$, $y = 3$?

18. The demand $f(x, y)$ for a certain commodity could be a function of price x and the number of units y which are available. If $z = 5 - xy + y^2$, find the price x corresponding to a demand of $z = 8$ when 3 units are available.

For a function of two variables $z = f(x, y)$, if z is held constant then the statement $f(x, y) = c$ (constant) is said to determine a **level curve** in the xy plane. In Problems 19 and 20, sketch the indicated level curves. In economic theory a sequence of level curves is called an **indifference map**.

19. $z = 3x + 4y - 12$, $z = 0$, and $z = 12$.
20. $z = y - x^2 + 1$, $z = 1$, and $z = -3$.

16.2

PARTIAL DIFFERENTIATION

In Chapter 12, we saw that the derivative dy/dx gives the slope of a tangent line to the graph of a function $y = f(x)$. This derivative can also be interpreted as the instantaneous rate of change of the function with respect to x. For a function of two variables $z = f(x, y)$, we now consider two instantaneous rates of change: one with respect to x and the other with respect to y. These respective rates of change are called the **partial derivative of the function with respect to x**, written $\dfrac{\partial z}{\partial x}$, and the **partial derivative of the function with respect to y**, written $\dfrac{\partial z}{\partial y}$.

For a given function $z = f(x, y)$, a partial derivative is obtained in the following manner.

To compute $\dfrac{\partial z}{\partial x}$ use the rules of ordinary differentiation while treating y as a constant.

16.2 Partial Differentiation

> To compute $\dfrac{\partial z}{\partial y}$ use the rules of ordinary differentiation while treating x as a constant.

ALTERNATIVE SYMBOLS   The partial derivatives of $z = f(x, y)$ with respect to x and y are also denoted by $f_x(x, y)$ and $f_y(x, y)$, respectively.

Example   For $z = 2x^3 y^2 + 2$ find (a) $\dfrac{\partial z}{\partial x}$, (b) $\dfrac{\partial z}{\partial y}$.

Solution   (a) We hold y fixed and treat constants in the usual manner. Thus,

$$\frac{\partial z}{\partial x} = 2(3x^2) y^2 + 0$$

$$= 6x^2 y^2.$$

(b) By treating x as a constant we obtain

$$\frac{\partial z}{\partial y} = 2x^3 (2y) + 0$$

$$= 4x^3 y.$$

Example   Find $\partial z/\partial y$ for $z = e^{x^2 y}$.

Solution   Since x is held constant, we treat x^2 as if it were a constant multiple of y. Thus, by rule XIII of Section 12.9,

$$\frac{\partial z}{\partial y} = e^{x^2 y} \frac{\partial}{\partial y}(x^2 y)$$

$$= e^{x^2 y}(x^2)(1)$$

$$= x^2 e^{x^2 y}.$$

Example   Find $f_y(x, y)$ for $f(x, y) = x^2 e^{3xy} \ln 4y$.

Solution   Observe that

$$f(x, y) = x^2 \underbrace{(e^{3xy} \cdot \ln 4y)}_{\text{product of two functions of } y}$$

By the product rule, rule V, we obtain

$$f_y(x,y) = x^2\left[e^{3xy} \cdot \frac{1}{4y} \cdot 4 + e^{3xy}(3x)\ln 4y\right]$$

$$= \frac{x^2}{y}e^{3xy} + 3x^3 e^{3xy} \ln 4y.$$

Example Find $\partial z/\partial x$ for $z = 4 + 2y - y^2$.

Solution Since y is held constant, and since no x appears in the expression, it follows that

$$\frac{\partial z}{\partial x} = 0.$$

Example For $z = (4x^{1/2} + xy - y^3)^5$ find (a) $\dfrac{\partial z}{\partial x}$ and (b) $\dfrac{\partial z}{\partial y}$.

Solution In each case we use the power rule for functions, rule VII.

(a) $\dfrac{\partial z}{\partial x} = 5(4x^{1/2} + xy - y^3)^4 \cdot \dfrac{\partial}{\partial x}(4x^{1/2} + xy - y^3)$

$= 5(4x^{1/2} + xy - y^3)^4 \left(4 \cdot \dfrac{1}{2}x^{-1/2} + 1 \cdot y - 0\right)$

$= (10x^{-1/2} + 5y)(4x^{1/2} + xy - y^3)^4,$

(b) $\dfrac{\partial z}{\partial y} = 5(4x^{1/2} + xy - y^3)^4 \cdot \dfrac{\partial}{\partial y}(4x^{1/2} + xy - y^3)$

$= 5(4x^{1/2} + xy - y^3)^4(0 + x \cdot 1 - 3y^2)$

$= (5x - 15y^2)(4x^{1/2} + xy - y^3)^4.$

SLOPE OF A TANGENT LINE

In Section 16.1, we saw that a function $z = f(x, y)$ determines a surface in three dimensions. The statement $x = a$ (constant) when interpreted in two dimensions represents a straight line perpendicular to the x-axis. In three dimensions, $x = a$ is the equation of a plane in which y and z are free to vary but x is held fixed; that is, it is the equation of a plane perpendicular to the x-axis. Similarly, $y = b$ (constant) is the equation of a plane perpendicular to the y-axis. As shown in Figure 16.4, the intersection of a surface $z = f(x, y)$ and the plane $x = a$ is a *curve* $\mathscr{C}_1$; the intersection of the surface and the plane $y = b$ is a *curve* $\mathscr{C}_2$. The partial derivative $\partial z/\partial y$ evaluated at (a, b), written

$$\left.\frac{\partial z}{\partial y}\right|_{(a,b)}$$

is the **slope** of the tangent line to the curve $\mathscr{C}_1$ at $(a, b, f(a, b))$. The number

$$\left.\frac{\partial z}{\partial x}\right|_{(a,b)}$$

16.2 Partial Differentiation

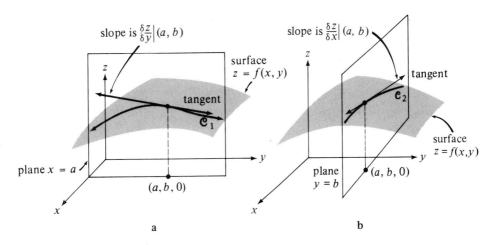

Figure 16.4

is the **slope** of the tangent line to the curve $\mathscr{C}_2$ at $(a, b, f(a, b))$.

Example Given $z = x^2 y^3$, find the slope of the tangent line in the plane $x = 2$ at $y = 1$.

Solution By specifying the plane $x = 2$ we are holding x constant in the function $z = x^2 y^3$. Hence, we compute the partial derivative with respect to y,

$$\frac{\partial z}{\partial y} = x^2 \cdot 3y^2$$

$$= 3x^2 y^2.$$

Thus, when $x = 2$ and $y = 1$, we have the slope

$$\left.\frac{\partial z}{\partial y}\right|_{(2,1)} = 3(2)^2(1)^2$$

$$= 12.$$

EXERCISE 16.2

Compute both $\partial z/\partial x$ and $\partial z/\partial y$.

1. $z = 5x + 3y - 1$
2. $z = -7x + 9y + 12$
3. $z = x^2 + y^2$
4. $z = x^2 y + 4y$
5. $z = x^2 + xy^2 + 3y^4$
6. $z = -x^3 + 4x^2 y^3 - 5y^2$

7. $z = 13x^3 + 4x^2 y^{1/2} - x + 2$

8. $z = 4x^{3/2} y^2 - 6x^2 y^{1/2} + 10y$

9. $z = 3x^2 + 4x - 1$

10. $z = 14y^2 - 7y + 6$

11. $z = (3x + 4y)^2$

12. $z = (-x + 5y)^3$

13. $z = (x^3 - 2x + y^2)^3$

14. $z = (-x^4 + 4y^2 + 2y)^5$

15. $z = e^{xy^2}$

16. $z = e^{2x^3 y^2}$

17. $z = \ln(6x + y^2)$

18. $z = \ln(x^4 + y^8)$

19. $z = xe^{x^2 y}$

20. $z = y^2 e^{2xy}$

21. $z = xy \ln(x^4 + 2y^2)$

22. $z = x^2 y^3 \ln(y^4 + 1)$

23. $z = \dfrac{\sqrt{x}}{2y - 6}$

24. $z = \dfrac{y^2}{x^2 + 1}$

25. $z = \dfrac{2x + y}{x - 3y}$

26. $z = \dfrac{x^2 - y^3}{x^3 + y^2}$

Compute the slope of the tangent line in the given plane corresponding to the indicated point.

27. $z = 2x^2 + 3y^2;\ y = 1;\ (2, 1)$

28. $z = -4x^3 + 6y^3;\ x = 4;\ (4, 2)$

29. $z = x^3 y^2 + xy + 1;\ x = 1;\ (1, 2)$

30. $z = -x^2 + 2xy^2 + 4;\ y = -1;\ (3, -1)$

Applications

31. The volume of a right circular cylinder of radius r and height h is given by $V = \pi r^2 h$. Compute $\partial V / \partial r$ and $\partial V / \partial h$.

32. **Biology** The surface area of the human body is given by $S = 0.2 W^{0.425} H^{0.725}$, where W and H represent weight and height, respectively. Compute $\partial S / \partial W$ and $\partial S / \partial H$.

33. **Business** A closed box is constructed having a volume of 60 cubic feet. The costs of the material for the top and bottom are, respectively, 5 cents per square foot and 10 cents per square foot. The cost of the sides is 1 cent per square foot. What is the cost function $C(x, y)$? Compute $\partial C / \partial x$ and $\partial C / \partial y$.

34. The cost of construction of a tin can in the shape of a right circular cylinder is a function of the area of the top and bottom as well as the area of the lateral side. If the cost of the lateral side is 1 cent per square unit and the cost of the top and bottom is 2 cents per square unit, determine the cost function C. Compute $\partial C / \partial r$ and $\partial C / \partial h$.

If $z = f(x, y)$ and $w = g(x, y)$ are demand functions for two related commodities at prices x and y, respectively, then the quantities $\partial z / \partial x$, $\partial z / \partial y$, $\partial w / \partial x$, $\partial w / \partial y$ are called **partial marginal demands**.

35. Compute the partial marginal demands if the demand functions are

$$z = 140 - 5x + 4y \qquad w = 90 + 2x - 3y.$$

36. Compute the partial marginal demands if the demand functions are

$$z = 10 e^{-2x^2 y} \qquad w = 5 e^{3x - 4y}.$$

If $z = f(x, y)$ is a demand function, where x and y are prices, then the quantities

$$\frac{x}{z} \cdot \frac{\partial z}{\partial x} \quad \text{and} \quad \frac{y}{z} \cdot \frac{\partial z}{\partial y}$$

are called **partial elasticities**. Compute the partial elasticities at the given prices.

37. $z = 60 - x^2 - 3y$ at $x = 1, y = 3$
38. $z = 500 - (x - 1)^2 - (2y + 3)^2$ at $x = 1, y = 1$
39. $z = 128 - 4x - 6y$ at $x = 6, y = 4$

16.3 HIGHER-ORDER DERIVATIVES

HIGHER AND MIXED DERIVATIVES

If $z = x^3 y^2$, the first partial derivatives are $\partial z/\partial x = 3x^2 y^2$ and $\partial z/\partial y = 2x^3 y$. In general, the first partial derivatives are themselves functions of *both* x and y. Consequently, we can compute the **second** (or higher) **partial derivative** with respect to x or y as well as the partial derivative with respect to x *of* the partial derivative with respect to y, or the partial derivative with respect to y *of* the partial derivative with respect to x. These latter types of derivatives are called **mixed partial derivatives**. For $z = f(x, y)$:

> The second partial derivative with respect to x is given by
> $$\frac{\partial^2 z}{\partial x^2} = \frac{\partial}{\partial x}\left(\frac{\partial z}{\partial x}\right).$$
>
> The second partial derivative with respect to y is given by
> $$\frac{\partial^2 z}{\partial y^2} = \frac{\partial}{\partial y}\left(\frac{\partial z}{\partial y}\right).$$
>
> The mixed partial derivatives are given by
> $$\frac{\partial^2 z}{\partial x \partial y} = \frac{\partial}{\partial x}\left(\frac{\partial z}{\partial y}\right),$$
> $$\frac{\partial^2 z}{\partial y \partial x} = \frac{\partial}{\partial y}\left(\frac{\partial z}{\partial x}\right).$$

ALTERNATIVE SYMBOLS

The second partial derivatives of $z = f(x, y)$ with respect to x and y are also denoted by $f_{xx}(x, y)$ and $f_{yy}(x, y)$, respectively. The mixed partial derivative $\partial^2 z/\partial x \partial y$ is written $f_{yx}(x, y)$. It is interesting to note that

$$\frac{\partial^2 z}{\partial x \partial y} = \frac{\partial^2 z}{\partial y \partial x}$$

for all functions that we consider in this text.

Example Compute $\partial^2 z/\partial y^2$ for $z = x^2 y^2 - y^3 + 3x^4 + 5$.

Solution First we compute $\partial z/\partial y$ by holding x constant,

$$\frac{\partial z}{\partial y} = 2x^2 y - 3y^2.$$

Therefore,

$$\frac{\partial^2 z}{\partial y^2} = \frac{\partial}{\partial y}\left(\frac{\partial z}{\partial y}\right) = 2x^2 - 6y.$$

Example Compute $\partial^2 z/\partial x^2$ for $z = \sqrt{x^2 + y^4}$.

Solution By the power rule for functions we have

$$\frac{\partial z}{\partial x} = \frac{1}{2}(x^2 + y^4)^{-1/2}(2x)$$

$$= x(x^2 + y^4)^{-1/2}.$$

To compute the second partial derivative we use the product and power rules:

$$\frac{\partial^2 z}{\partial x^2} = \frac{\partial}{\partial x}\left(\frac{\partial z}{\partial x}\right)$$

$$= x\left(-\frac{1}{2}\right)(x^2 + y^4)^{-3/2}(2x) + (x^2 + y^4)^{-1/2} \cdot 1$$

$$= -x^2(x^2 + y^4)^{-3/2} + (x^2 + y^4)^{-1/2}.$$

$$= (x^2 + y^4)^{-1/2}[-x^2(x^2 + y^4)^{-1} + 1]$$

$$= (x^2 + y^4)^{-1/2}\left(\frac{-x^2}{x^2 + y^4} + 1\right)$$

$$= (x^2 + y^4)^{-1/2}\frac{-x^2 + x^2 + y^4}{x^2 + y^4}$$

$$= (x^2 + y^4)^{-1/2} \cdot \frac{y^4}{x^2 + y^4}$$

$$= y^4(x^2 + y^4)^{-3/2}.$$

Example Compute the mixed partial $\partial^2 z/\partial x \partial y$ for $z = x^2 + 4x^3 y^2$.

16.3 Higher-Order Derivatives

Solution First we differentiate with respect to y holding x constant,

$$\frac{\partial z}{\partial y} = 0 + 4x^3 \cdot 2y = 8x^3 y$$

We then differentiate the foregoing result with respect to x holding y constant,

$$\frac{\partial^2 z}{\partial x \partial y} = \frac{\partial}{\partial x}\left(\frac{\partial z}{\partial y}\right) = 24x^2 y.$$

Example Compute the mixed partial $\partial^2 z/\partial x \partial y$ for $z = e^{xy^2}$.

Solution The partial derivative with respect to y follows from rule XIII,

$$\frac{\partial z}{\partial y} = e^{xy^2} \frac{\partial}{\partial y}(xy^2)$$

$$= e^{xy^2}(2xy)$$

$$= 2xy e^{xy^2}.$$

Therefore, by the product rule and rule XIII we obtain

$$\frac{\partial^2 z}{\partial x \partial y} = \frac{\partial}{\partial x}\left(\frac{\partial z}{\partial y}\right) = \frac{\partial}{\partial x}(2xy e^{xy^2})$$

$$= 2xy e^{xy^2}(y^2) + e^{xy^2}(2y)$$

$$= 2y e^{xy^2}(xy^2 + 1).$$

EXERCISE 16.3

Compute $\partial^2 z/\partial x^2$.

1. $z = 4x^2 - 2y + 1$
2. $z = y^3 + 7x^2 + x - 2$
3. $z = 2x^2 y - y^2$
4. $z = x^4 y^2 + 5y$
5. $z = x^3 y^2 - xy^5 + x^2$
6. $z = 3x^2 y^3 - 2x^3 y + y$
7. $z = e^{xy}$
8. $z = e^{-2xy}$
9. $z = (x^2 + y^2)^3$
10. $z = (x^3 - 2x + y^2)^2$
11. $z = x \ln(4x + y)$
12. $z = 5y \ln(x^2 + y^2)$

Compute $\partial^2 z/\partial y^2$.

13. $z = 4x^2 - \dfrac{1}{2}y^4$
14. $z = 8x^3 + 3y^5$
15. $z = x^2 y^3 + 12xy^{1/2} + 25$
16. $z = 9x^3 y^{2/3} - 4x^3 y^{-2} + 7$
17. $z = \dfrac{1}{x} + \dfrac{1}{y}$
18. $z = \dfrac{1}{x+y}$
19. $z = \dfrac{5y^2}{1+6x}$
20. $z = \dfrac{x^2+9}{3y^2}$
21. $z = 10e^{-2xy}$
22. $z = xe^{5xy}$
23. $z = \ln(x^4 + y^4)$
24. $z = y \ln xy$

Compute $\partial^2 z/\partial x \partial y$.

25. $z = x^3 y^5$
26. $z = x^4 y^{-3}$
27. $z = 5x^2 y^2 - 2xy + 8$
28. $z = 6x^{1/2} y^{3/2} + x^2 y^7 + 4x$
29. $z = e^{x^2 y}$
30. $z = xe^{xy}$
31. $z = y \ln 4x$
32. $z = x^2 \ln y^2$

Verify that $\partial^2 z/\partial x \partial y = \partial^2 z/\partial y \partial x$.

33. $z = 2x^4 y^3 - x^2 y$
34. $z = x^6 - 5x^3 y^2 + 4xy^2 + 6$

Applications

35. *Business* The Cobb–Douglas production function $z = f(x, y)$ is defined by $z = Ax^\alpha y^\beta$. The value of z is called the **efficient output** produced by inputs x and y. Show that for A, α, and β constants,

$$f_x = \frac{\alpha z}{x},$$

$$f_y = \frac{\beta z}{y},$$

$$f_{xx} = \frac{\alpha(\alpha-1)z}{x^2},$$

$$f_{yy} = \frac{\beta(\beta-1)z}{y^2},$$

and

$$f_{yx} = f_{xy} = \frac{\alpha \beta z}{xy}.$$

16.4

MAXIMA AND MINIMA

We have seen that such quantities as revenue, cost, and profit could be functions of two (or more) variables. Of course, if $z = P(x, y)$ is a profit function, we are interested in whether there are any values of x and y which maximize P. Similarly, we would like to minimize a cost function $z = C(x, y)$. In this section we shall use partial derivatives as an aid in determining if a function of two variables has any local extrema.

CRITICAL POINTS For our purposes in the subsequent discussion it suffices to take a slightly different interpretation of the concept of a critical value.

> For $z = f(x, y)$, the solutions to
> $$\frac{\partial z}{\partial x} = 0 \quad \text{and} \quad \frac{\partial z}{\partial y} = 0$$
> are called **critical points**.

The critical points correspond to the points where the function could *possibly* have a local extremum.

Example Determine the critical points for $z = x^3 + y^3 - 12x - 27y$.

Solution The first partial derivatives are
$$\frac{\partial z}{\partial x} = 3x^2 - 12,$$
$$\frac{\partial z}{\partial y} = 3y^2 - 27.$$

Thus $\partial z/\partial x = 0$ and $\partial z/\partial y = 0$ give, respectively,
$$x^2 = 4 \quad \text{and} \quad y^2 = 9.$$

The solutions are $x = \pm 2$ and $y = \pm 3$. Hence, there are four critical points: $(2, 3)$, $(-2, 3)$, $(2, -3)$, and $(-2, -3)$.

Example Determine the critical points for $z = y^2 - 4xy - 2x^2 + 16x$.

Solution The first partial derivatives are
$$\frac{\partial z}{\partial x} = -4y - 4x + 16,$$

$$\frac{\partial z}{\partial y} = 2y - 4x.$$

The simultaneous equations

$$\frac{\partial z}{\partial x} = 0 \quad \text{and} \quad \frac{\partial z}{\partial y} = 0$$

imply that we must solve

$$y + x = 4$$
$$y = 2x.$$

The last equation, when substituted in the first, gives

$$3x = 4 \quad \text{or} \quad x = \frac{4}{3}.$$

Hence, $y = 2(\frac{4}{3}) = \frac{8}{3}$. The only critical point is $(\frac{4}{3}, \frac{8}{3})$.

TEST FOR LOCAL EXTREMA Although we shall not prove it, the following result gives sufficient conditions for ascertaining local maxima and local minima. It is analogous to the second derivative test for ordinary derivatives.

Let (a, b) be a critical point of $z = f(x, y)$.

(1) If $\dfrac{\partial^2 z}{\partial x^2} \cdot \dfrac{\partial^2 z}{\partial y^2} - \left(\dfrac{\partial^2 z}{\partial x \partial y}\right)^2 > 0$ and $\dfrac{\partial^2 z}{\partial x^2} < 0$ at (a, b), then $f(a, b)$ is a local maximum value.

(2) If $\dfrac{\partial^2 z}{\partial x^2} \cdot \dfrac{\partial^2 z}{\partial y^2} - \left(\dfrac{\partial^2 z}{\partial x \partial y}\right)^2 > 0$ and $\dfrac{\partial^2 z}{\partial x^2} > 0$ at (a, b), then $f(a, b)$ is a local minimum value.

(3) If $\dfrac{\partial^2 z}{\partial x^2} \cdot \dfrac{\partial^2 z}{\partial y^2} - \left(\dfrac{\partial^2 z}{\partial x \partial y}\right)^2 < 0$ at (a, b), then $z = f(x, y)$ has no local extremum.

(4) If $\dfrac{\partial^2 z}{\partial x^2} \cdot \dfrac{\partial^2 z}{\partial y^2} - \left(\dfrac{\partial^2 z}{\partial x \partial y}\right)^2 = 0$ at (a, b), then no conclusion can be drawn concerning a local extremum.

In Figure 16.5, parts a and b illustrate, respectively, a surface with a local maximum value $f(a, b)$ and a surface with a local minimum at the origin. The above result therefore determines whether a surface is concave up or concave down in a neighborhood of a critical point. Recall that this is precisely the information obtained from the second derivative test.

16.4 Maxima and Minima

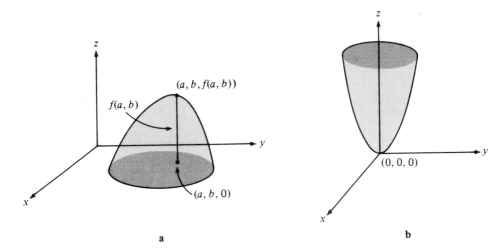

Figure 16.5

Example Find the local extrema for $z = 314 + 2x^2 + 4x + 3y^2 - 12y$.

Solution We have

$$\frac{\partial z}{\partial x} = 4x + 4 = 4(x+1) = 0,$$

$$\frac{\partial z}{\partial y} = 6y - 12 = 6(y-2) = 0$$

only at $x = -1$ and $y = 2$. That is, the critical point is $(-1, 2)$. Now

$$\frac{\partial^2 z}{\partial x\, \partial y} = 0, \qquad \frac{\partial^2 z}{\partial x^2} = 4, \qquad \frac{\partial^2 z}{\partial y^2} = 6$$

so that when evaluated at $(-1, 2)$,

$$\frac{\partial^2 z}{\partial x^2} \cdot \frac{\partial^2 z}{\partial y^2} - \left(\frac{\partial^2 z}{\partial x\, \partial y}\right)^2 = 4 \cdot 6 - 0^2 = 24 > 0.$$

Since $\partial^2 z / \partial x^2 = 4 > 0$ at $(-1, 2)$, we conclude from (2) that $f(-1, 2) = 300$ is a local minimum value.

Example A revenue function is $R(x, y) = x(50 - 3x) + y(96 - 2y)$, where x and y could denote the number of items sold of two commodities. If a corresponding cost function is given by $C(x, y) = x^2 + y^2 + 2xy - 4x + 10$, determine whether the profit P has a local maximum.

Solution Recall that profit is defined by

$$P(x, y) = R(x, y) - C(x, y).$$

Therefore,

$$P(x, y) = [x(50 - 3x) + y(96 - 2y)] - (x^2 + y^2 + 2xy - 4x + 10)$$
$$= -4x^2 - 3y^2 + 54x + 96y - 2xy - 10.$$

Now

$$\frac{\partial P}{\partial x} = -8x + 54 - 2y = 0$$

$$\frac{\partial P}{\partial y} = -6y + 96 - 2x = 0$$

means that we must solve the system of equations

$$8x + 2y = 54$$
$$2x + 6y = 96.$$

By elimination it is readily found that the solution to the system is (3, 15). The second partial derivatives are

$$\frac{\partial^2 P}{\partial x^2} = -8, \quad \frac{\partial^2 P}{\partial y^2} = -6, \quad \frac{\partial^2 P}{\partial x \partial y} = -2.$$

Thus, at (3, 15) we have

$$\frac{\partial^2 P}{\partial x^2} \cdot \frac{\partial^2 P}{\partial y^2} - \left(\frac{\partial^2 P}{\partial x \partial y}\right)^2 = (-8)(-6) - (-2)^2 = 44 > 0.$$

Since $\partial^2 P / \partial x^2 < 0$, it follows from (1) that

$$P(3, 15) = -4(9) - 3(225) + 54(3) + 96(15) - 2(3)(15) - 10$$
$$= 791 \text{ units of dollars}$$

is a local maximum profit.

EXERCISE 16.4

Find any local extrema of the given function.

1. $z = x^2 + y^2 + 4$
2. $z = 8x^2 + 4y^2$
3. $z = -x^2 - y^2 + 6x + 8y$
4. $z = 2x^2 + 3y^2 + 8x - 6y$
5. $z = 5x^2 + 5y^2 - 10x + 20y - 30$
6. $z = -2x^2 - 4y^2 + 12x - 8y + 3$

16.4 Maxima and Minima

7. $z = 4x^2 + 2y^2 - 2xy - 2x - 10y + 1$
8. $z = 5x^2 + 5y^2 + 5xy - 5x - 10y + 17$
9. $z = x^3 + 4y^3 - 3x - 12y$
10. $z = 2x^3 - y^3 - 24x + 27y + 2$
11. $z = -2x^3 - 2y^3 + 6xy + 4$
12. $z = \frac{1}{3}x^3 + \frac{1}{3}y^3 - 2xy + 9$
13. $z = (x-1)(y-2)$
14. $z = (2x+6)(y+5)$

Applications

15. Find three positive numbers x, y, and z whose sum is 18 such that their product P is a maximum. [*Hint*: Write P as a product of x and y only.]

16. Find the dimensions x, y, and z of a rectangular box with volume equal to 1 cubic foot, which has a minimal surface area S. [*Hint*: Use $1 = xyz$ to write S in terms of x and y only.]

17. *Business* A brewery makes two kinds of beer and sales are measured in x and y thousands of gallons per year. If the company's yearly cost and revenue functions are

$$C(x, y) = x^2 + y^2 + 200x + 100y + 100,$$
$$R(x, y) = 300 - x^2 - y^2 + 600x + 800y,$$

find the yearly production level that will give maximum profit.

18. A closed box is constructed that has a volume of 60 cubic feet. The costs of the material for the top and bottom are 5 cents per square foot and 10 cents per square foot, respectively. The cost of the sides is 1 cent per square foot. Determine the cost function $C(x, y)$ and find the dimensions of the box that give a minimum cost.

19. A company's yearly profit function is given by the formula

$$P(x, y) = 100x + 50y - 25x^2 + 400.$$

Is there a value of x and y that will give a maximum profit?

20. A hobbyist makes and sells two kinds of picture frames, each from a different kind of wood. From long experience he knows that for every x number of frames made of one kind of wood, he will sell them at $48 - 2x$ dollars, and for every y number of frames made from the other kind of wood, he will sell them at $30 - y$ dollars. How many of each kind should he sell in order to maximize his total revenue?

21. A drive-in restaurant sells two kinds of hamburgers. One kind is made from cheaper ground beef and the other is made from the more expensive ground chuck. The cost per hamburger is 10 cents or 20 cents, depending on whether the cheap or expensive meat is used. If x and y are the selling prices of the hamburgers made with the expensive meat and cheaper meat, respectively, the restaurant estimates that it can sell $-2x + y + 130$ of the better hamburgers and $x - 2y + 40$ of the lower-quality burgers each hour. Find the selling prices of the hamburgers that will give a maximum hourly profit. How many of each kind of hamburger does the restaurant sell each hour? [*Hint*: The profit for each kind of hamburger is $(x - 20)$ and $(y - 10)$, respectively.]

22. *Social Science* In statistics one often wants to fit an approximating curve to a set of data points $\{(x_i, y_i) | i = 1, \ldots, n\}$. A measure of the "goodness of fit" is the sum of the squares D of the distances between the ordinates on the approximating curve and the ordinate values of the data points. For a linear approximating function $y = ax + b$, called a **regression line**,

we have

$$D(a, b) = \sum_{i=1}^{n} [(ax_i + b) - y_i]^2.$$

Of course, it is desired that D be a minimum. Show that $\partial D/\partial a = 0$ and $\partial D/\partial b = 0$ when a and b satisfy

$$a = \frac{n \sum_{i=1}^{n} x_i y_i - \sum_{i=1}^{n} x_i \sum_{i=1}^{n} y_i}{n \sum_{i=1}^{n} x_i^2 - \left(\sum_{i=1}^{n} x_i\right)^2},$$

$$b = \frac{-a \sum_{i=1}^{n} x_i + \sum_{i=1}^{n} y_i}{n}$$

$$= \frac{\sum_{i=1}^{n} x_i^2 \sum_{i=1}^{n} y_i - \sum_{i=1}^{n} x_i y_i \sum_{i=1}^{n} x_i}{n \sum_{i=1}^{n} x_i^2 - \left(\sum_{i=1}^{n} x_i\right)^2}.$$

The number a is the slope of the regression line. Find the values of a and b if the data points are (1, 2), (2, 3), (3, 5), and (4, 7).

16.5

CONSTRAINED EXTREMA

CONSTRAINTS AND LAGRANGE MULTIPLIERS

In Problem 15 of Exercise 16.4 the reader was asked to find three positive numbers x, y, and z whose sum is 18 such that their product P is a maximum. The idea is to find the extremum of the function

$$P(x, y, z) = xyz$$

subject to the side condition that the positive numbers x, y, and z satisfy

$$x + y + z = 18.$$

This latter equation is called a **constraint** of the problem. We could, of course, use the constraint equation to eliminate one of the variables in the product function so that the method of the preceding section may then be applied to the problem. However, in this discussion we shall investigate an alternative procedure, called the method of **Lagrange multipliers**, for finding the extrema of functions subject to a constraining equation. Specifically, we wish to find the maximum or minimum of, say,

16.5 Constrained Extrema

$$w = f(x, y, z)$$

such that x, y, and z are connected through a constraint equation

$$g(x, y, z) = 0.$$

Or in terms of functions of two variables, find the extrema of $z = f(x, y)$ such that $g(x, y) = 0$. Although the method that we shall describe is applicable to problems involving any number of variables, we shall state the formal results in terms of functions of two variables.

Example Determine geometrically whether $z = 9 - x^2 - y^2$ subject to $x + y = 3$ has an extremum.

Solution In two dimensions the equation $x + y = 3$ determines a straight line. However, in three dimensions $x + y = 3$ can be interpreted as a plane perpendicular to the xy plane. It can be shown that $z = 9 - x^2 - y^2$ has an absolute maximum $f(0, 0) = 9$. On the other hand, the function $z = 9 - x^2 - y^2$ subject to the constraint $x + y = 3$ appears to have a new extremum for some values x_1 and y_1 satisfying $0 < x_1 < 3$, $0 < y_1 < 3$ and $x_1 + y_1 = 3$.

Geometric intuition as well as the following table should convince the reader that the new maximum occurs, in this case, at the midpoint of the line segment between $(3, 0, 0)$ and $(0, 3, 0)$. Notice that numbers such as $x = 1.6$, $y = 2.5$ cannot be used since these values do not satisfy the requirement $x + y = 3$. It appears, then, that the *constrained maximum* value is $f(1.5, 1.5) = 4.5$.

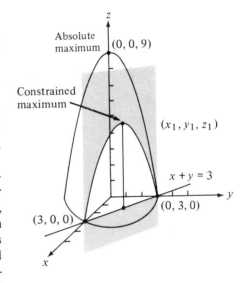

x	y	$z = 9 - x^2 - y^2$
0.5	2.5	2.5
1	2	4
1.25	1.75	4.375
1.5	1.5	4.5
1.75	1.25	4.375
2	1	4
2.5	0.5	2.5

LAGRANGE'S METHOD We state here without proof a method due to Joseph Louis Lagrange, an eighteenth century French mathematician.

> To find the extrema of $z = f(x, y)$ subject to the constraint $g(x, y) = 0$, we
> 1. Form the new function
> $$F(x, y, \lambda) = f(x, y) + \lambda g(x, y).$$
> 2. Compute the partial derivatives $\partial F/\partial x$, $\partial F/\partial y$, and $\partial F/\partial \lambda$.
> 3. Solve the system
> $$\frac{\partial F}{\partial x} = 0$$
> $$\frac{\partial F}{\partial y} = 0$$
> $$\frac{\partial F}{\partial \lambda} = 0.$$
>
> Among the solutions of this system will be the point(s) where f has an extremum.

The additional variable λ in the procedure above is called a **Lagrange multiplier**. Also, we note that the equation $\partial F/\partial \lambda = 0$ is really equivalent to the constraint equation $g(x, y) = 0$. However, we include the requirement that $\partial F/\partial \lambda = 0$ to emphasize that the constraint is to be included in the system of equations.

Example Use the method of Lagrange multipliers to find the maximum of $z = 9 - x^2 - y^2$ subject to $x + y = 3$.

Solution
1. Form the function
$$F(x, y, \lambda) = 9 - x^2 - y^2 + \lambda(x + y - 3).$$
2. Compute the partial derivatives:
$$\frac{\partial F}{\partial x} = -2x + \lambda,$$
$$\frac{\partial F}{\partial y} = -2y + \lambda,$$
$$\frac{\partial F}{\partial \lambda} = x + y - 3.$$
3. We set each of the above derivatives equal to zero and solve the resulting system of equations. If we subtract the second equation from the first in the system
$$-2x + \lambda = 0$$

16.5 Constrained Extrema

we obtain $-2x + 2y = 0$, or $x = y$. Substituting this into the equation

$$x + y - 3 = 0$$

gives

$$y + y - 3 = 0,$$

or

$$2y = 3.$$

Thus,

$$y = \frac{3}{2}, \quad x = y = \frac{3}{2}, \quad \text{and} \quad z = 9 - \left(\frac{3}{2}\right)^2 - \left(\frac{3}{2}\right)^2 = \frac{9}{2}.$$

This verifies that the constrained maximum value is $f(1.5, 1.5) = 4.5$.

Example A closed container in the form of a right circular cylinder is to have a volume of 1000 cubic inches. All sides of the container are made out of metal costing 2 cents per square inch. The lateral side is wrapped in paper that costs $\frac{1}{2}$ cent per square inch. We wish to minimize the cost function

$$C(r, h) = \overbrace{2(2\pi r^2)}^{\substack{\text{cost of} \\ \text{bottom} \\ \text{and top}}} + \overbrace{2(2\pi rh)}^{\substack{\text{cost of} \\ \text{side}}} + \overbrace{\tfrac{1}{2}(2\pi rh)}^{\substack{\text{cost of} \\ \text{paper on} \\ \text{side}}}$$

$$= 4\pi r^2 + 5\pi rh$$

subject to the constraint

$$1000 = \pi r^2 h.$$

Example Use the method of Lagrange multipliers to find the minimum value of $C(r, h) = 4\pi r^2 + 5\pi rh$ subject to $1000 = \pi r^2 h$.

Solution The steps are

1. Form the function:

$$F(r, h, \lambda) = 4\pi r^2 + 5\pi rh + \lambda(\pi r^2 h - 1000).$$

2. Compute the partial derivatives:

$$\frac{\partial F}{\partial r} = 8\pi r + 5\pi h + 2\lambda \pi rh$$

$$\frac{\partial F}{\partial h} = 5\pi r + \lambda \pi r^2$$

$$\frac{\partial F}{\partial \lambda} = \pi r^2 h - 1000.$$

3. Solve the system of equations $\partial F/\partial r = 0$, $\partial F/\partial h = 0$, $\partial F/\partial \lambda = 0$. Multiply the first equation by r and the second equation by $2h$:

$$8\pi r^2 + 5\pi rh + 2\lambda \pi r^2 h = 0,$$

$$10\pi rh + 2\lambda \pi r^2 h = 0.$$

Subtracting the second from the first gives

$$8\pi r^2 - 5\pi rh = 0,$$

or

$$\pi r(8r - 5h) = 0.$$

Since $r = 0$ does not satisfy the equation $\partial F/\partial \lambda = 0$, we take $r = \frac{5}{8}h$. Substituting this into $\pi r^2 h - 1000 = 0$ yields

$$\pi \left(\frac{5}{8}h\right)^2 h - 1000 = 0,$$

or

$$h^3 = \frac{1000 \cdot 64}{25\pi}.$$

Thus,

$$h = \frac{40}{\sqrt[3]{25\pi}}$$

$$r = \frac{5}{8}h = \frac{5}{8}\left(\frac{40}{\sqrt[3]{25\pi}}\right)$$

$$r = \frac{25}{\sqrt[3]{25\pi}}.$$

The constrained minimum cost is

$$C = 4\pi\left(\frac{25}{\sqrt[3]{25\pi}}\right)^2 + 5\pi\left(\frac{25}{\sqrt[3]{25\pi}}\right)\left(\frac{40}{\sqrt[3]{25\pi}}\right)$$

$$= \frac{25\pi}{\sqrt[3]{25\pi}}\left(\frac{100}{\sqrt[3]{25\pi}} + \frac{200}{\sqrt[3]{25\pi}}\right)$$

$$= 300\sqrt[3]{25\pi}.$$

To find the extrema of a function of three variables $w = f(x, y, z)$ subject to the constraint $g(x, y, z) = 0$, we would form the function $F(x, y, z, \lambda) = f(x, y, z) + \lambda g(x, y, z)$

16.5 Constrained Extrema

and then solve the system of four equations $\partial F/\partial x = 0, \partial F/\partial y = 0, \partial F/\partial z = 0, \partial F/\partial \lambda = 0$.

In conclusion, we note that it may not always be obvious when a function has a maximum or minimum at a solution of the system defined by step 3. For example, applying the three steps of the method to $f(x, y, z) = x^2 + y^2 + z^2$ subject to $2x - 2y - z = 5$ yields the fact that the value $f(10/9, 10/9, -5/9) = 225/81$ is a constrained extremum. However, we do not know immediately whether this value is a maximum or minimum! One way of convincing yourself as to the nature of the extremum is to compare it with the values obtained by calculating the function at other points (x, y, z) satisfying the constraint equation. In this manner, we find that $225/81$ is actually a constrained minimum value of the given function.

EXERCISE 16.5

Use the method of Lagrange multipliers to find the constrained extrema of the given function.

1. $f(x, y) = x + y$
   subject to $x^2 + y^2 = 2$.

2. $f(x, y) = xy$
   subject to $x^2 + y^2 = 2$.

3. $f(x, y) = x^2 + y^2$
   subject to $2x + y = 5$.

4. $f(x, y) = 3x^2 + 3y^2 + 3$
   subject to $x - y = 1$.

5. $f(x, y) = 2x^2 + 4y^2 - 1$
   subject to $x + 4y = 18$.

6. $f(x, y) = 4x^2 + 2y^2 + 5$
   subject to $4x^2 + y^2 = 4$.

7. $f(x, y, z) = x^2 + y^2 + z^2$
   subject to $y^2 - z^2 = 4$.

8. $f(x, y, z) = xyz$
   subject to $x^3 + y^3 + z^3 = 24$.

9. $f(x, y, z) = x^2 + y^2 + z^2$
   subject to $36x^2 + 4y^2 + 9z^2 = 36$.

10. $f(x, y, z) = x + 2y + z$
    subject to $x^2 + y^2 + z^2 = 30$.

Applications

11. Find the dimensions of the rectangular box with maximum volume that can be enclosed by the sphere $x^2 + y^2 + z^2 = 16$.

12. **Business** A **utility index** U is a function that gives a measure of satisfaction obtained from the possession of variable amounts x, y of, say, two commodities. If $U = x^{1/3} y^{2/3}$ is a utility index, find its extrema subject to $x + 6y = 18$.

CHAPTER TEST

Evaluate the functions.

1. $f(x, y) = 2x - 4y + 10$; $f(6, -5)$, $f(2, 10)$
2. $f(x, y) = x^2 - 3xy^3$; $f(-2, -1)$, $f\left(9, \dfrac{1}{3}\right)$

Compute $\partial z/\partial x$ and $\partial z/\partial y$.

3. $z = x^3 - 3x^2 y^3 + 6y^4$
4. $z = x^2 + \sqrt{xy}$
5. $z = ye^{-(x + 5y^2)}$
6. $z = \ln(x^2 + 8y^4)$

Compute $\partial^2 z/\partial x^2$ and $\partial^2 z/\partial y^2$.

7. $z = 4x^2 - 2x^4 y^5 + y^2$
8. $z = -y^3 + 2x^{-1} y^4 + x^{-3}$

Compute $\partial^2 z/\partial x \partial y$.

9. $z = (4x + 3y)^3$
10. $z = e^{x^3 + y^7}$

Find the extrema of the given function.

11. $z = x^2 + y^2 + 2x - 6y + 10$
12. $z = -x^2 - y^2 + xy - 2x - 2y + 5$

Find the constrained extrema of the given function.

13. $f(x, y) = 36 - x^2 - y^2$ subject to $3x + y - 12 = 0$
14. $f(x, y) = xy + 10$ subject to $x^2 + y^2 - 32 = 0$
15. A rectangular box, open at the top, is to be made of material that costs 1 cent per square inch. If the box is to contain 108 cubic inches, what dimensions would minimize the cost of construction?

APPENDIX I
THE FUNDAMENTAL PROPERTIES OF THE REAL NUMBER SYSTEM

1. **Properties of equality**

    $a = a$ — *Reflexive property for equality*

    If $a = b$, then $b = a$. — *Symmetric property for equality*

    If $a = b$ and $b = c$, then $a = c$. — *Transitive property for equality*

    If $a = b$, then a may be replaced by b and b by a in any mathematical statement without altering the truth or falsity of the statement. — *Substitution property for equality*

2. **Reciprocal of a nonzero number**

    For $a \neq 0$, $a^{-1} = 1/a$.

3. **Basic properties of R**

    $a + b$ is a unique element of R — *Closure for addition*

    $(a + b) + c = a + (b + c)$ — *Associative property for addition*

    There exists an element $0 \in R$ (called the **identity element for addition**) with the property — *Additive-identity*

    $$a + 0 = a \quad \text{and} \quad 0 + a = a$$

    for each $a \in R$.

    For each $a \in R$, there exists an element $^{-}a \in R$ (called the **additive inverse**, or **negative**, of a) with the property — *Additive-inverse*

    $$a + (^{-}a) = 0 \quad \text{and} \quad (^{-}a) + a = 0.$$

545

$a + b = b + a$	Commutative property for addition
$a \times b$ is a unique element of R	Closure for multiplication
$(a \times b) \times c = a \times (b \times c)$	Associative property for multiplication
$a \times (b + c) = a \times b + a \times c$ and $(b + c) \times a = b \times a + c \times a$	Distributive property
There exists an element $1 \in R$ (called the **identity element for multiplication**), $1 \neq 0$, with the property $a \times 1 = a$ and $1 \times a = a$ for each $a \in R$.	Multiplicative-identity
$a \times b = b \times a$	Commutative property for multiplication
For each element $a \in R$, $a \neq 0$, there exists an element $a^{-1} \in R$ (called the **multiplicative inverse**, or **reciprocal**, of a) with the property $a \times (a^{-1}) = 1$ and $(a^{-1}) \times a = 1$.	Multiplicative-inverse

4. **Definition of subtraction**
   $a - b = a + (-b)$, where $-b$ is the negative of b
5. **Definition of division**
   $a/b = a \times (b^{-1})$, $b \neq 0$
6. **Cancellation property of addition**
   If $a + c = b + c$, then $a = b$.
7. **Cancellation property of multiplication**
   If $c \neq 0$ and $ac = bc$, then $a = b$.
8. **A product law**
   If $ab = 0$, then either $a = 0$ or $b = 0$ or both.
9. **Laws of signs**
   I $\quad -(-a) = a$,
   II $\quad (-a) + (-b) = -(a + b)$,
   III $\quad (-a)(b) = -(ab)$,
   IV $\quad (-a)(-b) = ab$,

I The Fundamental Properties of the Real Number System

$$\text{V} \qquad \frac{-a}{b} = \frac{a}{-b} = -\frac{a}{b} \qquad (b \neq 0),$$

$$\text{VI} \qquad \frac{-a}{-b} = \frac{a}{b} \qquad (b \neq 0).$$

10. **Fundamental principle of fractions**
    $a/b = c/d$ $(b, d \neq 0)$ if and only if $ad = bc$

11. **Laws of fractions**

$$\text{I} \qquad \frac{1}{a} \cdot \frac{1}{b} = \frac{1}{ab} \qquad (a, b \neq 0),$$

$$\text{II} \qquad \frac{a}{b} \cdot \frac{c}{d} = \frac{ac}{bd} \qquad (b, d \neq 0),$$

$$\text{III} \qquad \frac{a}{c} + \frac{b}{c} = \frac{a+b}{c} \qquad (c \neq 0),$$

$$\text{IV} \qquad \frac{a}{b} + \frac{c}{d} = \frac{ad+bc}{bd} \qquad (b, d \neq 0),$$

$$\text{V} \qquad \frac{a}{b} - \frac{c}{d} = \frac{ad-bc}{bd} \qquad (b, d \neq 0),$$

$$\text{VI} \qquad 1 \div \frac{a}{b} = \frac{b}{a} \qquad (a, b \neq 0),$$

$$\text{VII} \qquad \frac{a}{b} \div \frac{c}{d} = \frac{ad}{bc} \qquad (b, c, d \neq 0).$$

APPENDIX II
TABLES

I

COMMON LOGARITHMS

x	0	1	2	3	4	5	6	7	8	9
1.0	.0000	.0043	.0086	.0128	.0170	.0212	.0253	.0294	.0334	.0374
1.1	.0414	.0453	.0492	.0531	.0569	.0607	.0645	.0682	.0719	.0755
1.2	.0792	.0828	.0864	.0899	.0934	.0969	.1004	.1038	.1072	.1106
1.3	.1139	.1173	.1206	.1239	.1271	.1303	.1335	.1367	.1399	.1430
1.4	.1461	.1492	.1523	.1553	.1584	.1614	.1644	.1673	.1703	.1732
1.5	.1761	.1790	.1818	.1847	.1875	.1903	.1931	.1959	.1987	.2014
1.6	.2041	.2068	.2095	.2122	.2148	.2175	.2201	.2227	.2253	.2279
1.7	.2304	.2330	.2355	.2380	.2405	.2430	.2455	.2480	.2504	.2529
1.8	.2553	.2577	.2601	.2625	.2648	.2672	.2695	.2718	.2742	.2765
1.9	.2788	.2810	.2833	.2856	.2878	.2900	.2923	.2945	.2967	.2989
2.0	.3010	.3032	.3054	.3075	.3096	.3118	.3139	.3160	.3181	.3201
2.1	.3222	.3243	.3263	.3284	.3304	.3324	.3345	.3365	.3385	.3404
2.2	.3424	.3444	.3464	.3483	.3502	.3522	.3541	.3560	.3579	.3598
2.3	.3617	.3636	.3655	.3674	.3692	.3711	.3729	.3747	.3766	.3784
2.4	.3802	.3820	.3838	.3856	.3874	.3892	.3909	.3927	.3945	.3962
2.5	.3979	.3997	.4014	.4031	.4048	.4065	.4082	.4099	.4116	.4133
2.6	.4150	.4166	.4183	.4200	.4216	.4232	.4249	.4265	.4281	.4298
2.7	.4314	.4330	.4346	.4362	.4378	.4393	.4409	.4425	.4440	.4456
2.8	.4472	.4487	.4502	.4518	.4533	.4548	.4564	.4579	.4594	.4609
2.9	.4624	.4639	.4654	.4669	.4683	.4698	.4713	.4728	.4742	.4757
3.0	.4771	.4786	.4800	.4814	.4829	.4843	.4857	.4871	.4886	.4900
3.1	.4914	.4928	.4942	.4955	.4969	.4983	.4997	.5011	.5024	.5038
3.2	.5051	.5065	.5079	.5092	.5105	.5119	.5132	.5145	.5159	.5172
3.3	.5185	.5198	.5211	.5224	.5237	.5250	.5263	.5276	.5289	.5302
3.4	.5315	.5328	.5340	.5353	.5366	.5378	.5391	.5403	.5416	.5428
3.5	.5441	.5453	.5465	.5478	.5490	.5502	.5514	.5527	.5539	.5551
3.6	.5563	.5575	.5587	.5599	.5611	.5623	.5635	.5647	.5658	.5670
3.7	.5682	.5694	.5705	.5717	.5729	.5740	.5752	.5763	.5775	.5786
3.8	.5798	.5809	.5821	.5832	.5843	.5855	.5866	.5877	.5888	.5899
3.9	.5911	.5922	.5933	.5944	.5955	.5966	.5977	.5988	.5999	.6010
4.0	.6021	.6031	.6042	.6053	.6064	.6075	.6085	.6096	.6107	.6117
4.1	.6128	.6138	.6149	.6160	.6170	.6180	.6191	.6201	.6212	.6222
4.2	.6232	.6243	.6253	.6263	.6274	.6284	.6294	.6304	.6314	.6325
4.3	.6335	.6345	.6355	.6365	.6375	.6385	.6395	.6405	.6415	.6425
4.4	.6435	.6444	.6454	.6464	.6474	.6484	.6493	.6503	.6513	.6522
4.5	.6532	.6542	.6551	.6561	.6571	.6580	.6590	.6599	.6609	.6618
4.6	.6628	.6637	.6646	.6656	.6665	.6675	.6684	.6693	.6702	.6712
4.7	.6721	.6730	.6739	.6749	.6758	.6767	.6776	.6785	.6794	.6803
4.8	.6812	.6821	.6830	.6839	.6848	.6857	.6866	.6875	.6884	.6893
4.9	.6902	.6911	.6920	.6928	.6937	.6946	.6955	.6964	.6972	.6981
5.0	.6990	.6998	.7007	.7016	.7024	.7033	.7042	.7050	.7059	.7067
5.1	.7076	.7084	.7093	.7101	.7110	.7118	.7126	.7135	.7143	.7152
5.2	.7160	.7168	.7177	.7185	.7193	.7202	.7210	.7218	.7226	.7235
5.3	.7243	.7251	.7259	.7267	.7275	.7284	.7292	.7300	.7308	.7316
5.4	.7324	.7332	.7340	.7348	.7356	.7364	.7372	.7380	.7388	.7396

x	0	1	2	3	4	5	6	7	8	9
5.5	.7404	.7412	.7419	.7427	.7435	.7443	.7451	.7459	.7466	.7474
5.6	.7482	.7490	.7497	.7505	.7513	.7520	.7528	.7536	.7543	.7551
5.7	.7559	.7566	.7574	.7582	.7589	.7597	.7604	.7612	.7619	.7627
5.8	.7634	.7642	.7649	.7657	.7664	.7672	.7679	.7686	.7694	.7701
5.9	.7709	.7716	.7723	.7731	.7738	.7745	.7752	.7760	.7767	.7774
6.0	.7782	.7789	.7796	.7803	.7810	.7818	.7825	.7832	.7839	.7846
6.1	.7853	.7860	.7868	.7875	.7882	.7889	.7896	.7903	.7910	.7917
6.2	.7924	.7931	.7938	.7945	.7952	.7959	.7966	.7973	.7980	.7987
6.3	.7993	.8000	.8007	.8014	.8021	.8028	.8035	.8041	.8048	.8055
6.4	.8062	.8069	.8075	.8082	.8089	.8096	.8102	.8109	.8116	.8122
6.5	.8129	.8136	.8142	.8149	.8156	.8162	.8169	.8176	.8182	.8189
6.6	.8195	.8202	.8209	.8215	.8222	.8228	.8235	.8241	.8248	.8254
6.7	.8261	.8267	.8274	.8280	.8287	.8293	.8299	.8306	.8312	.8319
6.8	.8325	.8331	.8338	.8344	.8351	.8357	.8363	.8370	.8376	.8382
6.9	.8388	.8395	.8401	.8407	.8414	.8420	.8426	.8432	.8439	.8445
7.0	.8451	.8457	.8463	.8470	.8476	.8482	.8488	.8494	.8500	.8506
7.1	.8513	.8519	.8525	.8531	.8537	.8543	.8549	.8555	.8561	.8567
7.2	.8573	.8579	.8585	.8591	.8597	.8603	.8609	.8615	.8621	.8627
7.3	.8633	.8639	.8645	.8651	.8657	.8663	.8669	.8675	.8681	.8686
7.4	.8692	.8698	.8704	.8710	.8716	.8722	.8727	.8733	.8739	.8745
7.5	.8751	.8756	.8762	.8768	.8774	.8779	.8785	.8791	.8797	.8802
7.6	.8808	.8814	.8820	.8825	.8831	.8837	.8842	.8848	.8854	.8859
7.7	.8865	.8871	.8876	.8882	.8887	.8893	.8899	.8904	.8910	.8915
7.8	.8921	.8927	.8932	.8938	.8943	.8949	.8954	.8960	.8965	.8971
7.9	.8976	.8982	.8987	.8993	.8998	.9004	.9009	.9015	.9020	.9025
8.0	.9031	.9036	.9042	.9047	.9053	.9058	.9063	.9069	.9074	.9079
8.1	.9085	.9090	.9096	.9101	.9106	.9112	.9117	.9122	.9128	.9133
8.2	.9138	.9143	.9149	.9154	.9159	.9165	.9170	.9175	.9180	.9186
8.3	.9191	.9196	.9201	.9206	.9212	.9217	.9222	.9227	.9232	.9238
8.4	.9243	.9248	.9253	.9258	.9263	.9269	.9274	.9279	.9284	.9289
8.5	.9294	.9299	.9304	.9309	.9315	.9320	.9325	.9330	.9335	.9340
8.6	.9345	.9350	.9355	.9360	.9365	.9370	.9375	.9380	.9385	.9390
8.7	.9395	.9400	.9405	.9410	.9415	.9420	.9425	.9430	.9435	.9440
8.8	.9445	.9450	.9455	.9460	.9465	.9469	.9474	.9479	.9484	.9489
8.9	.9494	.9499	.9504	.9509	.9513	.9518	.9523	.9528	.9533	.9538
9.0	.9542	.9547	.9552	.9557	.9562	.9566	.9571	.9576	.9581	.9586
9.1	.9590	.9595	.9600	.9605	.9609	.9614	.9619	.9624	.9628	.9633
9.2	.9638	.9643	.9647	.9652	.9657	.9661	.9666	.9671	.9675	.9680
9.3	.9685	.9689	.9694	.9699	.9703	.9708	.9713	.9717	.9722	.9727
9.4	.9731	.9736	.9741	.9745	.9750	.9754	.9759	.9763	.9768	.9773
9.5	.9777	.9782	.9786	.9791	.9795	.9800	.9805	.9809	.9814	.9818
9.6	.9823	.9827	.9832	.9836	.9841	.9845	.9850	.9854	.9859	.9863
9.7	.9868	.9872	.9877	.9881	.9886	.9890	.9894	.9899	.9903	.9908
9.8	.9912	.9917	.9921	.9926	.9930	.9934	.9939	.9943	.9948	.9952
9.9	.9956	.9961	.9965	.9969	.9974	.9978	.9983	.9987	.9991	.9996

II

EXPONENTIAL FUNCTIONS

x	e^x	e^{-x}	x	e^x	e^{-x}
0.00	1.0000	1.0000	1.5	4.4817	0.2231
0.01	1.0101	0.9901	1.6	4.9530	0.2019
0.02	1.0202	0.9802	1.7	5.4739	0.1827
0.03	1.0305	0.9705	1.8	6.0496	0.1653
0.04	1.0408	0.9608	1.9	6.6859	0.1496
0.05	1.0513	0.9512	2.0	7.3891	0.1353
0.06	1.0618	0.9418	2.1	8.1662	0.1225
0.07	1.0725	0.9324	2.2	9.0250	0.1108
0.08	1.0833	0.9331	2.3	9.9742	0.1003
0.09	1.0942	0.9139	2.4	11.023	0.0907
0.10	1.1052	0.9048	2.5	12.182	0.0821
0.11	1.1163	0.8958	2.6	13.464	0.0743
0.12	1.1275	0.8869	2.7	14.880	0.0672
0.13	1.1388	0.8781	2.8	16.445	0.0608
0.14	1.1503	0.8694	2.9	18.174	0.0550
0.15	1.1618	0.8607	3.0	20.086	0.0498
0.16	1.1735	0.8521	3.1	22.198	0.0450
0.17	1.1853	0.8437	3.2	24.533	0.0408
0.18	1.1972	0.8353	3.3	27.113	0.0369
0.19	1.2092	0.8270	3.4	29.964	0.0334
0.20	1.2214	0.8187	3.5	33.115	0.0302
0.21	1.2337	0.8106	3.6	36.598	0.0273
0.22	1.2461	0.8025	3.7	40.447	0.0247
0.23	1.2586	0.7945	3.8	44.701	0.0224
0.24	1.2712	0.7866	3.9	49.402	0.0202
0.25	1.2840	0.7788	4.0	54.598	0.0183
0.30	1.3499	0.7408	4.1	60.340	0.0166
0.35	1.4191	0.7047	4.2	66.686	0.0150
0.40	1.4918	0.6703	4.3	73.700	0.0136
0.45	1.5683	0.6376	4.4	81.451	0.0123
0.50	1.6487	0.6065	4.5	90.017	0.0111
0.55	1.7333	0.5769	4.6	99.484	0.0101
0.60	1.8221	0.5488	4.7	109.95	0.0091
0.65	1.9155	0.5220	4.8	121.51	0.0082
0.70	2.0138	0.4966	4.9	134.29	0.0074
0.75	2.1170	0.4724	5.0	148.41	0.0067
0.80	2.2255	0.4493	5.5	244.69	0.0041
0.85	2.3396	0.4274	6.0	403.43	0.0025
0.90	2.4596	0.4066	6.5	665.14	0.0015
0.95	2.5857	0.3867	7.0	1096.6	0.0009
1.0	2.7183	0.3679	7.5	1808.0	0.0006
1.1	3.0042	0.3329	8.0	2981.0	0.0003
1.2	3.3201	0.3012	8.5	4914.8	0.0002
1.3	3.6693	0.2725	9.0	8103.1	0.0001
1.4	4.0552	0.2466	10.0	22026	0.00005

III

NATURAL LOGARITHMS OF NUMBERS

n	$\log_e n$	n	$\log_e n$	n	$\log_e n$
	*	4.5	1.5041	9.0	2.1972
0.1	7.6974	4.6	1.5261	9.1	2.2083
0.2	8.3906	4.7	1.5476	9.2	2.2192
0.3	8.7960	4.8	1.5686	9.3	2.2300
0.4	9.0837	4.9	1.5892	9.4	2.2407
0.5	9.3069	5.0	1.6094	9.5	2.2513
0.6	9.4892	5.1	1.6292	9.6	2.2618
0.7	9.6433	5.2	1.6487	9.7	2.2721
0.8	9.7769	5.3	1.6677	9.8	2.2824
0.9	9.8946	5.4	1.6864	9.9	2.2925
1.0	0.0000	5.5	1.7047	10	2.3026
1.1	0.0953	5.6	1.7228	11	2.3979
1.2	0.1823	5.7	1.7405	12	2.4849
1.3	0.2624	5.8	1.7579	13	2.5649
1.4	0.3365	5.9	1.7750	14	2.6391
1.5	0.4055	6.0	1.7918	15	2.7081
1.6	0.4700	6.1	1.8083	16	2.7726
1.7	0.5306	6.2	1.8245	17	2.8332
1.8	0.5878	6.3	1.8405	18	2.8904
1.9	0.6419	6.4	1.8563	19	2.9444
2.0	0.6931	6.5	1.8718	20	2.9957
2.1	0.7419	6.6	1.8871	25	3.2189
2.2	0.7885	6.7	1.9021	30	3.4012
2.3	0.8329	6.8	1.9169	35	3.5553
2.4	0.8755	6.9	1.9315	40	3.6889
2.5	0.9163	7.0	1.9459	45	3.8067
2.6	0.9555	7.1	1.9601	50	3.9120
2.7	0.9933	7.2	1.9741	55	4.0073
2.8	1.0296	7.3	1.9879	60	4.0943
2.9	1.0647	7.4	2.0015	65	4.1744
3.0	1.0986	7.5	2.0149	70	4.2485
3.1	1.1314	7.6	2.0281	75	4.3175
3.2	1.1632	7.7	2.0412	80	4.3820
3.3	1.1939	7.8	2.0541	85	4.4427
3.4	1.2238	7.9	2.0669	90	4.4998
3.5	1.2528	8.0	2.0794	100	4.6052
3.6	1.2809	8.1	2.0919	110	4.7005
3.7	1.3083	8.2	2.1041	120	4.7875
3.8	1.3350	8.3	2.1163	130	4.8676
3.9	1.3610	8.4	2.1282	140	4.9416
4.0	1.3863	8.5	2.1401	150	5.0106
4.1	1.4110	8.6	2.1518	160	5.0752
4.2	1.4351	8.7	2.1633	170	5.1358
4.3	1.4586	8.8	2.1748	180	5.1930
4.4	1.4816	8.9	2.1861	190	5.2470

* Subtract 10 for $n < 1$. Thus, $\log_e 0.1 = 7.6974 - 10 = -2.3026$.

IV

COMPOUND AMOUNT OF $1

$$(1 + i)^n$$

n	1%	1½%	1¾%	2%	2½%	3%	4%	5%	6%
1	1.0100	1.0150	1.0175	1.0200	1.0250	1.0300	1.0400	1.0500	1.0600
2	1.0201	1.0302	1.0353	1.0404	1.0506	1.0609	1.0816	1.1025	1.1236
3	1.0303	1.0457	1.0534	1.0612	1.0769	1.0927	1.1249	1.1576	1.1910
4	1.0406	1.0614	1.0719	1.0824	1.1038	1.1255	1.1699	1.2155	1.2625
5	1.0510	1.0773	1.0906	1.1041	1.1314	1.1593	1.2167	1.2763	1.3382
6	1.0615	1.0934	1.1097	1.1262	1.1597	1.1941	1.2653	1.3401	1.4185
7	1.0721	1.1098	1.1291	1.1487	1.1887	1.2299	1.3159	1.4071	1.5036
8	1.0829	1.1265	1.1489	1.1717	1.2184	1.2668	1.3686	1.4775	1.5938
9	1.0937	1.1434	1.1690	1.1951	1.2489	1.3048	1.4233	1.5513	1.6895
10	1.1046	1.1605	1.1894	1.2190	1.2801	1.3439	1.4802	1.6289	1.7908
11	1.1157	1.1779	1.2103	1.2434	1.3121	1.3842	1.5395	1.7103	1.8983
12	1.1268	1.1956	1.2314	1.2682	1.3449	1.4258	1.6010	1.7959	2.0122
13	1.1381	1.2136	1.2530	1.2936	1.3785	1.4685	1.6651	1.8856	2.1329
14	1.1495	1.2318	1.2749	1.3195	1.4130	1.5126	1.7317	1.9799	2.2609
15	1.1610	1.2502	1.2972	1.3459	1.4483	1.5580	1.8009	2.0789	2.3966
16	1.1726	1.2690	1.3199	1.3728	1.4845	1.6047	1.8730	2.1829	2.5404
17	1.1843	1.2880	1.3430	1.4002	1.5216	1.6528	1.9479	2.2920	2.6928
18	1.1961	1.3073	1.3665	1.4282	1.5597	1.7024	2.0258	2.4066	2.8543
19	1.2081	1.3270	1.3904	1.4568	1.5987	1.7535	2.1068	2.5270	3.0256
20	1.2202	1.3469	1.4148	1.4859	1.6386	1.8061	2.1911	2.6533	3.2071
21	1.2324	1.3671	1.4395	1.5157	1.6796	1.8603	2.2788	2.7860	3.3996
22	1.2447	1.3876	1.4647	1.5460	1.7216	1.9161	2.3699	2.9253	3.6035
23	1.2572	1.4084	1.4904	1.5769	1.7646	1.9736	2.4647	3.0715	3.8198
24	1.2697	1.4295	1.5164	1.6084	1.8087	2.0328	2.5633	3.2251	4.0489
25	1.2824	1.4509	1.5430	1.6406	1.8539	2.0938	2.6658	3.3864	4.2919
26	1.2953	1.4727	1.5700	1.6734	1.9003	2.1566	2.7725	3.5557	4.5494
27	1.3082	1.4948	1.5975	1.7069	1.9478	2.2213	2.8834	3.7335	4.8223
28	1.3213	1.5172	1.6254	1.7410	1.9965	2.2879	2.9987	3.9201	5.1117
29	1.3345	1.5400	1.6539	1.7758	2.0464	2.3566	3.1187	4.1161	5.4184
30	1.3478	1.5631	1.6828	1.8114	2.0976	2.4273	3.2434	4.3219	5.7435
31	1.3613	1.5865	1.7122	1.8476	2.1500	2.5001	3.3731	4.5380	6.0881
32	1.3749	1.6103	1.7422	1.8845	2.2038	2.5751	3.5081	4.7649	6.4534
33	1.3887	1.6345	1.7727	1.9222	2.2589	2.6523	3.6484	5.0032	6.8406
34	1.4026	1.6590	1.8037	1.9607	2.3153	2.7319	3.7943	5.2533	7.2510
35	1.4166	1.6839	1.8353	1.9999	2.3732	2.8139	3.9461	5.5160	7.6861
36	1.4308	1.7091	1.8674	2.0399	2.4325	2.8983	4.1039	5.7918	8.1473
37	1.4451	1.7348	1.9001	2.0807	2.4933	2.9852	4.2681	6.0814	8.6361
38	1.4595	1.7608	1.9333	2.1223	2.5557	3.0748	4.4388	6.3855	9.1543
39	1.4741	1.7872	1.9672	2.1647	2.6196	3.1670	4.6164	6.7048	9.7035
40	1.4889	1.8140	2.0016	2.2080	2.6851	3.2620	4.8010	7.0400	10.2857
41	1.5038	1.8412	2.0366	2.2522	2.7522	3.3599	4.9931	7.3920	10.9029
42	1.5188	1.8688	2.0723	2.2972	2.8210	3.4607	5.1928	7.7616	11.5570
43	1.5340	1.8969	2.1085	2.3432	2.8915	3.5645	5.4005	8.1497	12.2505
44	1.5493	1.9253	2.1454	2.3901	2.9638	3.6715	5.6165	8.5572	12.9855
45	1.5648	1.9542	2.1830	2.4379	3.0379	3.7816	5.8412	8.9850	13.7646
46	1.5805	1.9835	2.2212	2.4866	3.1139	3.8950	6.0748	9.4343	14.5905
47	1.5963	2.0133	2.2600	2.5363	3.1917	4.0119	6.3178	9.9060	15.4659
48	1.6122	2.0435	2.2996	2.5871	3.2715	4.1323	6.5705	10.4013	16.3939
49	1.6283	2.0741	2.3398	2.6388	3.3533	4.2562	6.8333	10.9213	17.3775
50	1.6446	2.1052	2.3808	2.6916	3.4371	4.3839	7.1067	11.4674	18.4202

V

PRESENT VALUE OF $1

$$(1 + i)^{-n}$$

n	1%	$1\frac{1}{2}$%	$1\frac{3}{4}$%	2%	$2\frac{1}{2}$%	3%	4%	5%	6%
1	0.99010	0.98522	0.98280	0.98039	0.97561	0.97087	0.96154	0.95238	0.94340
2	0.98030	0.97066	0.96590	0.96117	0.95181	0.94260	0.92456	0.90703	0.89000
3	0.97059	0.95632	0.94929	0.94232	0.92860	0.91514	0.88900	0.86384	0.83962
4	0.96098	0.94218	0.93296	0.92385	0.90595	0.88849	0.85480	0.82270	0.79209
5	0.95147	0.92826	0.91691	0.90573	0.88385	0.86261	0.82193	0.78353	0.74726
6	0.94205	0.91454	0.90114	0.88797	0.86230	0.83748	0.79032	0.74622	0.70496
7	0.93272	0.90103	0.88564	0.87056	0.84127	0.81309	0.75992	0.71068	0.66506
8	0.92348	0.88771	0.87041	0.85349	0.82075	0.78941	0.73069	0.67684	0.62741
9	0.91434	0.87459	0.85544	0.83676	0.80073	0.76642	0.70259	0.64461	0.59190
10	0.90529	0.86167	0.84073	0.82035	0.78120	0.74409	0.67556	0.61391	0.55840
11	0.89632	0.84893	0.82627	0.80426	0.76215	0.72242	0.64958	0.58468	0.52679
12	0.88745	0.83639	0.81206	0.78849	0.74356	0.70138	0.62460	0.55684	0.49697
13	0.87866	0.82403	0.79809	0.77303	0.72542	0.68095	0.60057	0.53032	0.46884
14	0.86996	0.81185	0.78437	0.75788	0.70773	0.66112	0.57748	0.50507	0.44230
15	0.86135	0.79985	0.77088	0.74302	0.69047	0.64186	0.55527	0.48102	0.41727
16	0.85282	0.78803	0.75762	0.72845	0.67363	0.62317	0.53391	0.45811	0.39365
17	0.84438	0.77639	0.74459	0.71416	0.65720	0.60502	0.51337	0.43630	0.37136
18	0.83602	0.76491	0.73178	0.70016	0.64117	0.58740	0.49363	0.41552	0.35034
19	0.82774	0.75361	0.71919	0.68643	0.62553	0.57029	0.47464	0.39573	0.33051
20	0.81954	0.74247	0.70683	0.67297	0.61027	0.55368	0.45639	0.37689	0.31181
21	0.81143	0.73150	0.69467	0.65978	0.59539	0.53755	0.43883	0.35894	0.29416
22	0.80340	0.72069	0.68272	0.64684	0.58087	0.52189	0.42196	0.34185	0.27751
23	0.79544	0.71004	0.67098	0.63416	0.56670	0.50669	0.40573	0.32557	0.26180
24	0.78757	0.69954	0.65944	0.62172	0.55288	0.49193	0.39012	0.31007	0.24698
25	0.77977	0.68921	0.64810	0.60953	0.53940	0.47761	0.37512	0.29530	0.23300
26	0.77205	0.67902	0.63695	0.59758	0.52624	0.46370	0.36069	0.28124	0.21981
27	0.76440	0.66899	0.62600	0.58586	0.51340	0.45019	0.34682	0.26785	0.20737
28	0.75684	0.65910	0.61523	0.57438	0.50088	0.43708	0.33348	0.25509	0.19563
29	0.74934	0.64936	0.60465	0.56311	0.48866	0.42435	0.32065	0.24295	0.18456
30	0.74192	0.63976	0.59425	0.55207	0.47674	0.41199	0.30832	0.23138	0.17411
31	0.73458	0.63031	0.58403	0.54125	0.46512	0.39999	0.29646	0.22036	0.16426
32	0.72730	0.62099	0.57398	0.53063	0.45377	0.38834	0.28506	0.20987	0.15496
33	0.72010	0.61182	0.56411	0.52023	0.44270	0.37703	0.27409	0.19987	0.14619
34	0.71297	0.60277	0.55441	0.51003	0.43191	0.36605	0.26355	0.19036	0.13791
35	0.70591	0.59387	0.54487	0.50003	0.42137	0.35538	0.25342	0.18129	0.13011
36	0.69893	0.58509	0.53550	0.49022	0.41109	0.34503	0.24367	0.17266	0.12274
37	0.69201	0.57644	0.52629	0.48061	0.40107	0.33498	0.23430	0.16444	0.11579
38	0.68515	0.56792	0.51724	0.47119	0.39129	0.32523	0.22529	0.15661	0.10924
39	0.67837	0.55953	0.50834	0.46195	0.38174	0.31575	0.21662	0.14915	0.10306
40	0.67165	0.55126	0.49960	0.45289	0.37243	0.30656	0.20829	0.14205	0.09722
41	0.66500	0.54312	0.49101	0.44401	0.36335	0.29763	0.20028	0.13528	0.09172
42	0.65842	0.53509	0.48256	0.43530	0.35449	0.28896	0.19258	0.12884	0.08653
43	0.65190	0.52718	0.47426	0.42677	0.34584	0.28054	0.18517	0.12270	0.08163
44	0.64545	0.51939	0.46611	0.41840	0.33740	0.27237	0.17805	0.11686	0.07701
45	0.63906	0.51172	0.45809	0.41020	0.32917	0.26444	0.17120	0.11130	0.07265
46	0.63273	0.50415	0.45021	0.40215	0.32115	0.25674	0.16461	0.10600	0.06854
47	0.62646	0.49670	0.44247	0.39427	0.31331	0.24926	0.15828	0.10095	0.06466
48	0.62026	0.48936	0.43486	0.38654	0.30567	0.24200	0.15220	0.09614	0.06100
49	0.61412	0.48213	0.42738	0.37896	0.29822	0.23495	0.14634	0.09156	0.05755
50	0.60804	0.47501	0.42003	0.37153	0.29094	0.22811	0.14071	0.08720	0.05429

VI

AMOUNT OF AN ANNUITY OF $1 PER PERIOD

$$\frac{(1+i)^n - 1}{i}$$

n	$\frac{1}{2}\%$	$\frac{3}{4}\%$	1%	2%	3%	4%	5%	6%
1	1.0000	1.0000	1.0000	1.0000	1.0000	1.0000	1.0000	1.0000
2	2.0050	2.0075	2.0100	2.0200	2.0300	2.0400	2.0500	2.0600
3	3.0150	3.0226	3.0301	3.0604	3.0909	3.1216	3.1525	3.1836
4	4.0301	4.0452	4.0604	4.1216	4.1836	4.2465	4.3101	4.3746
5	5.0503	5.0756	5.1010	5.2040	5.3091	5.4163	5.5256	5.6371
6	6.0755	6.1136	6.1520	6.3081	6.4684	6.6330	6.8019	6.9753
7	7.1059	7.1595	7.2135	7.4343	7.6625	7.8983	8.1420	8.3938
8	8.1414	8.2132	8.2857	8.5830	8.8923	9.2142	9.5491	9.8975
9	9.1821	9.2748	9.3685	9.7546	10.1591	10.5828	11.0266	11.4913
10	10.2280	10.3443	10.4622	10.9497	11.4639	12.0061	12.5779	13.1808
11	11.2792	11.4219	11.5668	12.1687	12.8078	13.4864	14.2068	14.9716
12	12.3356	12.5076	12.6825	13.4121	14.1920	15.0258	15.9171	16.8699
13	13.3972	13.6014	13.8093	14.6803	15.6178	16.6268	17.7130	18.8821
14	14.4642	14.7034	14.9474	15.9739	17.0863	18.2919	19.5986	21.0151
15	15.5365	15.8137	16.0969	17.2934	18.5989	20.0236	21.5786	23.2760
16	16.6142	16.9323	17.2579	18.6393	20.1569	21.8245	23.6575	25.6725
17	17.6973	18.0593	18.4304	20.0121	21.7616	23.6975	25.8404	28.2129
18	18.7858	19.1947	19.6147	21.4123	23.4144	25.6454	28.1324	30.9057
19	19.8797	20.3387	20.8109	22.8406	25.1169	27.6712	30.5390	33.7600
20	20.9791	21.4912	22.0190	24.2974	26.8704	29.7781	33.0660	36.7856
21	22.0840	22.6524	23.2392	25.7833	28.6765	31.9692	35.7193	39.9927
22	23.1944	23.8223	24.4716	27.2990	30.5368	34.2480	38.5052	43.3923
23	24.3104	25.0010	25.7163	28.8450	32.4529	36.6179	41.4305	46.9958
24	25.4320	26.1885	26.9735	30.4219	34.4265	39.0826	44.5020	50.8156
25	26.5591	27.3849	28.2432	32.0303	36.4593	41.6459	47.7271	54.8645
26	27.6919	28.5903	29.5256	33.6709	38.5530	44.3117	51.1135	59.1564
27	28.8304	29.8047	30.8209	35.3443	40.7096	47.0842	54.6691	63.7058
28	29.9745	31.0282	32.1291	37.0512	42.9309	49.9676	58.4026	68.5281
29	31.1244	32.2609	33.4504	38.7922	45.2188	52.9663	62.3227	73.6398
30	32.2800	33.5029	34.7849	40.5681	47.5754	56.0849	66.4388	79.0582
31	33.4414	34.7542	36.1327	42.3794	50.0027	59.3283	70.7608	84.8017
32	34.6086	36.0148	37.4941	44.2270	52.5028	62.7015	75.2988	90.8898
33	35.7817	37.2849	38.8690	46.1116	55.0778	66.2095	80.0638	97.3432
34	36.9606	38.5646	40.2577	48.0338	57.7302	69.8579	85.0670	104.1838
35	38.1454	39.8538	41.6603	49.9945	60.4621	73.6522	90.3203	111.4348
36	39.3361	41.1527	43.0769	51.9944	63.2759	77.5983	95.8363	119.1209
37	40.5328	42.4614	44.5076	54.0343	66.1742	81.7022	101.6281	127.2681
38	41.7355	43.7798	45.9527	56.1149	69.1594	85.9703	107.7095	135.9042
39	42.9441	45.1082	47.4122	58.2372	72.2342	90.4091	114.0950	145.0585
40	44.1589	46.4465	48.8864	60.4020	75.4013	95.0255	120.7998	154.7620
41	45.3796	47.7948	50.3752	62.6100	78.6633	99.8265	127.8398	165.0477
42	46.6065	49.1533	51.8790	64.8622	82.0232	104.8196	135.2317	175.9505
43	47.8396	50.5219	53.3978	67.1595	85.4839	110.0124	142.9933	187.5076
44	49.0788	51.9009	54.9318	69.5027	89.0484	115.4129	151.1430	199.7580
45	50.3242	53.2901	56.4811	71.8927	92.7199	121.0294	159.7002	212.7435
46	51.5758	54.6898	58.0459	74.3306	96.5015	126.8706	168.6852	226.5081
47	52.8337	56.1000	59.6263	76.8172	100.3965	132.9454	178.1194	241.0986
48	54.0978	57.5207	61.2226	79.3535	104.4084	139.2632	188.0254	256.5645
49	55.3683	58.9521	62.8348	81.9406	108.5406	145.8337	198.4267	272.9584
50	56.6452	60.3943	64.4632	84.5794	112.7969	152.6671	209.3480	290.3359

VII

PRESENT VALUE OF AN ANNUITY OF $1 PER PERIOD

$$\frac{1 - (1 + i)^{-n}}{i}$$

n	$\frac{1}{2}\%$	$\frac{3}{4}\%$	1%	2%	3%	4%	5%	6%
1	0.9950	0.9926	0.9901	0.9804	0.9709	0.9615	0.9524	0.9434
2	1.9851	1.9777	1.9704	1.9416	1.9135	1.8861	1.8594	1.8334
3	2.9702	2.9556	2.9410	2.8839	2.8286	2.7751	2.7232	2.6730
4	3.9505	3.9261	3.9020	3.8077	3.7171	3.6299	3.5460	3.4651
5	4.9259	4.8894	4.8534	4.7135	4.5797	4.4518	4.3295	4.2124
6	5.8964	5.8456	5.7955	5.6014	5.4172	5.2421	5.0757	4.9173
7	6.8621	6.7946	6.7282	6.4720	6.2303	6.0021	5.7864	5.5824
8	7.8230	7.7366	7.6517	7.3255	7.0197	6.7327	6.4632	6.2098
9	8.7791	8.6716	8.5660	8.1622	7.7861	7.4353	7.1078	6.8017
10	9.7304	9.5996	9.4713	8.9826	8.5302	8.1109	7.7217	7.3601
11	10.6770	10.5207	10.3676	9.7868	9.2526	8.7605	8.3064	7.8869
12	11.6189	11.4349	11.2551	10.5753	9.9540	9.3851	8.8633	8.3838
13	12.5562	12.3423	12.1337	11.3484	10.6350	9.9856	9.3936	8.8527
14	13.4887	13.2430	13.0037	12.1062	11.2961	10.5631	9.8986	9.2950
15	14.4166	14.1370	13.8651	12.8493	11.9379	11.1184	10.3797	9.7122
16	15.3399	15.0243	14.7179	13.5777	12.5611	11.6523	10.8378	10.1059
17	16.2586	15.9050	15.5623	14.2919	13.1661	12.1657	11.2741	10.4773
18	17.1728	16.7792	16.3983	14.9920	13.7535	12.6593	11.6896	10.8276
19	18.0824	17.6468	17.2260	15.6785	14.3238	13.1339	12.0853	11.1581
20	18.9874	18.5080	18.0456	16.3514	14.8775	13.5903	12.4622	11.4699
21	19.8880	19.3628	18.8570	17.0112	15.4150	14.0292	12.8212	11.7641
22	20.7841	20.2112	19.6604	17.6580	15.9369	14.4511	13.1630	12.0416
23	21.6757	21.0533	20.4558	18.2922	16.4436	14.8568	13.4886	12.3034
24	22.5629	21.8891	21.2434	18.9139	16.9355	15.2470	13.7986	12.5504
25	23.4456	22.7188	22.0232	19.5235	17.4131	15.6221	14.0939	12.7834
26	24.3240	23.5422	22.7952	20.1210	17.8768	15.9828	14.3752	13.0032
27	25.1980	24.3595	23.5596	20.7069	18.3270	16.3296	14.6430	13.2105
28	26.0677	25.1707	24.3164	21.2813	18.7641	16.6631	14.8981	13.4062
29	26.9330	25.9759	25.0658	21.8444	19.1885	16.9837	15.1411	13.5907
30	27.7941	26.7751	25.8077	22.3965	19.6004	17.2920	15.3725	13.7648
31	28.6508	27.5683	26.5423	22.9377	20.0004	17.5885	15.5928	13.9291
32	29.5033	28.3556	27.2696	23.4683	20.3888	17.8736	15.8027	14.0840
33	30.3515	29.1371	27.9897	23.9886	20.7658	18.1476	16.0025	14.2302
34	31.1956	29.9128	28.7027	24.4986	21.1318	18.4112	16.1929	14.3681
35	32.0354	30.6827	29.4086	24.9986	21.4872	18.6646	16.3742	14.4982
36	32.8710	31.4468	30.1075	25.4888	21.8323	18.9083	16.5469	14.6210
37	33.7025	32.2053	30.7995	25.9695	22.1672	19.1426	16.7113	14.7368
38	34.5299	32.9581	31.4847	26.4406	22.4925	19.3679	16.8679	14.8460
39	35.3531	33.7053	32.1630	26.9026	22.8082	19.5845	17.0170	14.9491
40	36.1722	34.4470	32.8347	27.3555	23.1148	19.7928	17.1591	15.0463
41	36.9873	35.1831	33.4997	27.7995	23.4124	19.9931	17.2944	15.1380
42	37.7983	35.9137	34.1581	28.2348	23.7014	20.1856	17.4232	15.2245
43	38.6053	36.6389	34.8100	28.6616	23.9819	20.3708	17.5459	15.3062
44	39.4082	37.3587	35.4555	29.0800	24.2543	20.5488	17.6628	15.3832
45	40.2072	38.0732	36.0945	29.4902	24.5187	20.7200	17.7741	15.4558
46	41.0022	38.7823	36.7272	29.8923	24.7754	20.8847	17.8801	15.5244
47	41.7932	39.4862	37.3537	30.2866	25.0247	21.0429	17.9810	15.5890
48	42.5803	40.1848	37.9740	30.6731	25.2667	21.1951	18.0772	15.6500
49	43.3635	40.8782	38.5881	31.0521	25.5017	21.3415	18.1687	15.7076
50	44.1428	41.5664	39.1961	31.4236	25.7298	21.4822	18.2559	15.7619

ODD-NUMBERED ANSWERS

Exercise 1.1. [page 7]

1. $3 \in \{2, 3, 4\}$ **3.** $\{2\} \notin \{2, 3, 4\}$ **5.** $5 \notin \{4, 5, 6\}$ **7.** $\emptyset \subset \{4, 5, 6\}$ **9. a.** $\{5, 6, 7\}$
b. $\{5, 6\}, \{5, 7\}, \{6, 7\}$ **c.** $\{5\}, \{6\}, \{7\}$ **d.** $\emptyset$ **11.** $\{x \mid x = 2n - 1, n \text{ a positive integer}\}$
13. $\left\{x \mid x = \dfrac{1}{n}, n \text{ a positive integer}\right\}$ **15.** $\{x \mid x \notin A\}$ **17.** $A' = \{1, 3, 5, 7, 9\} = C$
19. $C' = \{2, 4, 6, 8, 10\} = A$ **21.** $A \cup B = \{1, 2, 3, 4, 5, 6, 8, 10\}$ **23.** $A \cap C = \emptyset$
25. $A' \cup C' = U$ or $\{1, 2, 3, 4, 5, 6, 7, 8, 9, 10\}$ **27.** $A' \cup C = C$ or $\{1, 3, 5, 7, 9\}$

29. 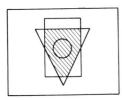   **31.** 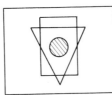   **33.**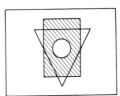

35. 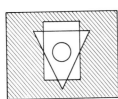   **37.** 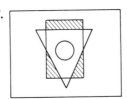   **39.**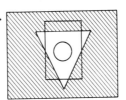

41. $(A')' = A$ **43.** $A \cap U = A$ **45.** $A \cup U = U$ **47.** $A \cup \emptyset = A$ **49.** $\emptyset' \cap \emptyset = \emptyset$ **51.** $\emptyset \cup \emptyset = \emptyset$
53. $A \cup B = \emptyset$ if $A = B = \emptyset$ **55.** $A \cap U = U$ if $A = U$ **57.** $A \cup \emptyset = U$ if $A = U$
59. $A \cap B = A$ if $A \subset B$ **61.** $A \cup B = A \cap B$ if $A = B$ **63.** $A \cap \emptyset = \emptyset$ for any set A
65. 11, 50, 20; 10

Exercise 1.2. [page 12]

1. False **3.** True **5.** True **7.** False **9.** True **11.** False **13.** I **15.** III **17.** II **19.** IV **21.** $7 > 3$ **23.** $-4 < -3$ **25.** $-1 \leq x \leq 1$ **27.** $x > 0$ **29.** $x \geq 0$ **31.** 3 **33.** 7 **35.** -2 **37.** $3x$ if $x \geq 0$, $-3x$ if $x < 0$ **39.** $x + 1$ if $x \geq -1$, $-(x+1)$ if $x < -1$ **41.** $y - 3$ if $y \geq 3$, $-(y-3)$ if $y < 3$ **43.** $|-2| < |-5|$ **45.** $-7 < |-1|$ **47.** $|-3| > 0$ **49.** $2 < 5$ **51.** $7 < 8$ **53.** $|x| < 3$ **55.** $d = 9$ **57.** $d = 29$

Chapter 1 Test [page 15]

1. $\in$ **2.** $\subset$ **3.** $\subset$ **4.** $\in$ **5.** $\{1, 2, 3, 4\}, \{1, 2, 3\}, \{1, 3, 4\}, \{1, 3\}$
6. $\{x \mid x = 5k, \ k \in J\}$ **7.** $\{2, 4, 5, 6, 7, 8, 9\}$ **8.** $\{6, 8\}$ **9.** $\emptyset$ **10.** $\{5, 7, 9\}$
11. $U = \{1, 2, 3, 4, 5, 6, 7, 8, 9\}$ **12.** $\{1, 2, 4\}$ **13.** $\{3, 21\}$ **14.** $\{-8, -1, 0, 3, 21\}$
15. $\{-8, -15/7, -1, 0, 3, 13/2, 21\}$ **16.** $\{-\sqrt{5}, \sqrt{50}\}$ **17.** IV **18.** II **19.** $-6 > -9$
20. $4 \leq y$ **21.** $y \leq 0$ **22.** $y + 2 \geq 0$ **23.** 7 **24.** $x - 5$ if $x - 5 \geq 0$, $5 - x$ if $x - 5 < 0$
25. $x < 4$ **26.** $|x| \geq 6$

Exercise 2.1. [page 19]

1. $-5z^2 + 8z - 2$ **3.** $-y^3 - 6y^2 + y + 3$ **5.** $x^2 + 2y^2$ **7.** 9 **9.** $2x + 1, 4x - 5$
11. $3x^2 - 4x + 2, -x^2 + 4$ **13.** $x^3 + 5x^2 - 4x + 4, x^3 + 3x^2 - 2$
15. $x^5 + 3x^4 - 3x^3 - 2x^2 + 4x + 1, -x^5 + 3x^4 + 3x^3 - 2x^2 - 4x + 1$ **17.** $3x^2 - 5x + 5$
19. $3x^2 - 4x + 4$ **21.** $x^2 - 2x + 6$ **23.** $-1, -21, 1, -2, 20$ **25.** $1, 1, 0, 1, 0$
27. $3, 2h^4 - h^2 + 3, 2h^4 - h^2 + 3$ **29.** $x^2 + 2hx + h^2 + 2x + 2h - 3, x^2 - 2hx + h^2 + 2x - 2h - 3, x^4 + 2x^2 - 3$
31. $1, 1$ **33.** $6, 39$ **35.** n, n

Exercise 2.2. [page 23]

1. $-6x^3y^4$ **3.** $-8x^4y^5$ **5.** x^{3n} **7.** 9 **9.** $a^2bc - ab^2c + 2abc^2$ **11.** $x^2 + 7x + 10$
13. $x^2 - 4xy + 4y^2$ **15.** $10x^2 + 17x + 3$ **17.** $18a^2 - 8b^2$ **19.** $-2ac + bc + 6ad - 3bd$
21. $16a^2 - 24ab + 9b^2$ **23.** $x^3 + 6x^2 + 7x - 4$ **25.** $6x^3 + 7x^2 - 6x + 1$ **27.** $a^4 - ab^3$ **29.** $4a + 4$
31. $20x^2 + 4x$ **33.** $5y^2 - 12y + 5$ **35.** $6z^2 - 24z + 32$
37. $P(x - 1) = x^2 - 5x + 11; P(2 - x) = x^2 - x + 5$ **39.** $[P(3)]^2 = 36; P(3^2) = -78$
41. $(a + b)^2 - (a^2 + b^2) = 2ab$; for $a^2 + b^2 > (a + b)^2$ the product $a \cdot b$ has to be less than 0 ($a > 0$ and $b < 0$ or $a < 0$ and $b > 0$); for $a^2 + b^2 < (a + b)^2$ the product $a \cdot b$ has to be greater than 0 ($a > 0$ and $b > 0$ or $a < 0$ and $b < 0$)

Odd-Numbered Answers

Exercise 2.3. [page 27]

1. $(3x^3y)(3x^2 - x + 2)$ 3. $(x - 4)(x + 1)$ 5. $(x - 6)(x - 2)$ 7. $(y + 2)(y + 2)$ 9. $(x + 5)(x - 5)$
11. $(2x + 3)(x + 1)$ 13. $(3a + 1)(2a + 1)$ 15. $2(3z + 1)(z + 1)$ 17. $x^2y^2(x + 1)(x - 1)$
19. $(x^n + 1)(x^n - 1)$ 21. $x^n(x^2 - x + 2)$ 23. $(y^2 + 2)(y^2 + 1)$ 25. $(2a^2 + 1)(a + 1)(a - 1)$
27. $(-x + y)(3x - y)[x^2 + (y - 2x)^2]$ 29. $(x + y)(x + a)$ 31. $(a - 2b)(a^2 + 2b^2)$
33. $(y - 3x)(y^2 + 3xy + 9x^2)$ 35. $(2x - y)(x^2 - xy + y^2)$ 37. $(a^n - 2)(a^n + 2)$
39. $(x^n - y^n)(x^n + y^n)(x^{2n} + y^{2n})$ 41. $(3x^{2n} - 1)(x^{2n} - 3)$ 43. $2(y^n - 30)(y^n + 24)$
45. $ac - ad + bd - bc = a(c - d) - b(c - d) = (a - b)(c - d)$;
also $ac - ad + bd - bc = bd - ad - bc + ac = (b - a)d - (b - a)c = (b - a)(d - c)$
47. $(x^2 + xy + y^2)(x^2 - xy + y^2)$

Exercise 2.4. [page 31]

1. $4ay^2$ $(a, y \neq 0)$ 3. $2a^2b^2$ $(a, b \neq 0)$ 5. x^n $(x \neq 0)$ 7. x^ny^n $(x, y \neq 0)$ 9. $4a^2 + 2a + 2$
11. $x^2 - 4x - 3$ $(x \neq 0)$ 13. $3x^2 - 2x + 3/5x$ $(x \neq 0)$ 15. $2x - 5$
17. $2y^2 - y + 13/2$, $R = -3/2$ $(y \neq -1/2)$ 19. $x^2 - 2x + 3 + \dfrac{6}{2x - 1}$
21. $2y^3 - y^2 - \dfrac{1}{2}y - \dfrac{5}{4} + \dfrac{-13y/4 + 9/4}{2y^2 + y + 1}$ 23. $x^2 - 2x + 3$

Exercise 2.5. [page 34]

1. a/b $(a, b, c \neq 0)$ 3. 2 $(x + y \neq 0)$ 5. -1 $(b - a \neq 0)$ 7. $-(x + 1)$ $(x \neq 1)$
9. $x^2 - 2x - 3/2$ $(x \neq 0)$ 11. $y + 7$ $(y \neq 2)$ 13. $-(2y + 5)$ $(y \neq 1/2)$ 15. $\dfrac{x + 3}{x - 1}$ $(x \neq 1, -3)$
17. $\dfrac{y - 1}{y + 1}$ $(y \neq 1, -1)$ 19. $\dfrac{n - 5}{n + 4}$ $(n \neq 3, -4)$ 21. $\dfrac{3x + 9}{x - 8}$ $(x \neq 3, 8)$
23. $\dfrac{x^2 + xy + y^2}{x + y}$ $(x \neq y, -y)$ 25. $\dfrac{y^2 + 4}{y^2 + 3}$ $(y \neq 2, -2)$ 27. $9/12$ 29. ab^2/a^2b $(a, b \neq 0)$
31. $\dfrac{3(y - 3)}{y^2 - y - 6}$ $(y \neq -2, 3)$ 33. $\dfrac{3(a^2 - 3a + 9)}{a^3 + 27}$ $(a \neq -3)$ 35. No, $x = 1, 2$
37. For $N > 0$, $a - b < 0$ or $a < b$; for $N < 0$, $a - b > 0$ or $a > b$

Exercise 2.6. [page 38]

1. $\dfrac{2x - 1}{2y}$ 3. 1 5. $\dfrac{2a + 9}{9}$ 7. $\dfrac{5}{2a + 2b}$ 9. $\dfrac{3}{3 - x}$ 11. $\dfrac{-1}{(a + 2)(a + 3)}$ 13. $\dfrac{-3x^2 - 3y^2}{(2x - y)(x - 2y)}$

15. $\dfrac{2y-1}{(y-2)(y+1)(y+1)}$ 17. $\dfrac{n^2-n}{(n+5)(n+3)(n-2)}$ 19. $\dfrac{y(y-11)}{(y-4)(y+4)(y-1)}$ 21. $\dfrac{x^3-2x^2+2x-2}{(x-1)^2}$

23. $\dfrac{4x^3-4x^2+4x+4}{(2x+1)(2x-1)}$ 25. $\dfrac{4}{(y+1)(y+3)}$ 27. $\dfrac{y^2+xz}{(y-x)(y-z)}$ 29. 0

Exercise 2.7. [page 41]

1. $-b^2/a$ 3. $1/ax^2y$ 5. $\dfrac{2x}{x+5}$ 7. $(5ab-4)(4ab+3)$ 9. $x-y$ 11. $\dfrac{x^2-1}{x^2}$ 13. $\dfrac{x+5}{x(x+1)}$

15. $\dfrac{2y^2-6y+3}{3}$ 17. $\dfrac{7}{10a+2}$ 19. $\dfrac{4a^2-3a}{4a+1}$ 21. $\dfrac{-1}{y-3}$ 23. $\dfrac{a-2b}{a+2b}$ 25. $\dfrac{a+6}{a-1}$

27. $\dfrac{x-y-z}{x+y+z}$ 29. $\dfrac{b^2}{3(5b-a)}$

Exercise 2.8. [page 45]

1. $1/5$ 3. $1/9$ 5. 27 7. $1/10$ 9. $5\cdot 3$ 11. $9/5$ 13. $82/9$ 15. $3/16$ 17. x^2y^3

19. x^6y^4 21. $9x^2/y^6$ 23. $4x^4/y$ 25. x^4 27. $1/x^6$ 29. y/x 31. $4x^5y^2$

33. $\dfrac{y^2-x}{xy^2}$ 35. $\dfrac{y^2+x^2}{xy}$ 37. $\dfrac{x}{x-y}$ 39. $\dfrac{x^2+y^2}{xy}$ 41. $y+x$ 43. $\dfrac{xy}{y-x}$ 45. x^{2n-1}

47. 1 49. y^4 51. x^6 53. y/x^{n-1} 55. $1/x^{2n-2}$ 57. 2.54×10^3 59. 6.42×10^5

61. 1.4×10^{-3} 63. 2.30×10^{-5}

Exercise 2.9. [page 50]

1. 4 3. 9 5. $1/4$ 7. $1/8$ 9. $1/4$ 11. $8/27$ 13. x^2 15. $y^{5/4}$ 17. $x^{1/3}$ 19. $y^{1/6}$

21. $1/y^2$ 23. $(x/y)^{5/4}$ 25. $x^n \cdot y^4$ 27. $x^{3n/2}$ 29. $x^{5n/2} \cdot y^{3m/2-1}$ 31. $x^{n-1}y^{m+1}$

33. $x - x^{1/2}$ 35. $y^{5/3} - y$ 37. $x - y$ 39. $(x+y) - (x+y)^{3/2}$ 41. $x(x^{1/2}+1)$

43. $x^{-1/2}(x^{-1}+1)$ 45. $(x+1)^{-1/2}(x)$ 47. 5 49. $2|x|$ 51. $\dfrac{2}{|x|(x+1)^{1/2}}$ $(x > -1)$

Exercise 2.10. [page 56]

1. $\sqrt[3]{2}$ 3. $3\sqrt[3]{x}$ 5. $\sqrt{2y}$ 7. $\sqrt[3]{x^2}$ 9. $x\sqrt[3]{y^2}$ 11. $\sqrt[4]{x^3y}$ 13. $\sqrt{x-y}$ 15. $\dfrac{1}{\sqrt[3]{(x^2+y)^2}}$

17. $y^{2/3}$ 19. $(2xy^2)^{1/4}$ 21. $3(x^3y)^{1/4}$ 23. $(a+b^2)^{1/3}$ 25. 9 27. -3 29. xy^2 31. $2x^2\sqrt{x}$

33. $xy\sqrt[4]{3xy}$ 35. $3y\sqrt{2x}$ 37. 2 39. $b\sqrt[3]{2ab^2}$ 41. $4\sqrt{5}/5$ 43. $\sqrt[3]{2x}/x$ 45. $\sqrt[3]{3y^2}/y$

Odd-Numbered Answers

47. $1/2\sqrt{2}$ **49.** $2(1-\sqrt{3})$ **51.** $\dfrac{x(\sqrt{x}+3)}{x-9}$ **53.** $\dfrac{37-8\sqrt{10}}{27}$ **55.** $\dfrac{-1}{2(1+\sqrt{2})}$

57. $\dfrac{1-x-a}{\sqrt{x+a}+x+a}$ **59.** $-3\sqrt{2}+3$ **61.** $8\sqrt{2}$ **63.** $-\sqrt{5}$ **65.** $8\sqrt{3}$ **67.** $2\sqrt{2}-2$

69. $2\sqrt{3}+\sqrt{6}$ **71.** $1-\sqrt{5}$

Chapter 2 Test [page 59]

1. y^2-1 **2.** $16x-3$ **3.** 17 **4.** 24 **5.** $6x^2+10x-4$ **6.** $20z^2-4z-16$ **7.** $(z-4)(z-3)$
8. $u(u+1)^2(u-1)^2$ **9.** $(2y^n+1)(2y^n-1)$ **10.** $(2t-3)(4t^2+6t+9)$ **11.** $8xy$
12. $3z^2-2z+9$ **13.** $2n+1$ **14.** $r+1$ **15.** $2(2rs-s+3)$ $(r, s \neq 0)$ **16.** $2(x-1)$ $(x \neq -1)$
17. $\dfrac{x-2}{x+2}$ $(x \neq 2, x \neq -2)$ **18.** $\dfrac{4(n+1)}{4-n}$ $(n \neq 4, n \neq -4)$ **19.** $\dfrac{3x+4}{8}$ **20.** $\dfrac{-1}{2x-1}$
21. $\dfrac{10-3x}{2(x-2)}$ **22.** $\dfrac{y+4}{(y-1)(y-2)}$ **23.** $5xy/7ab$ **24.** 1 **25.** $1/12rt$ **26.** x/a **27.** $\dfrac{3(a-b)}{a+b}$
28. -1 **29.** $x^6 y^{12}$ **30.** x^3/y^4 **31.** x^2 **32.** x^2/y^3 **33.** $\dfrac{x^2 y^2+1}{xy}$ **34.** $\dfrac{y-x^2}{x}$ **35.** x^{2n-1}
36. $x^{2n-2} y^4$ **37.** 4.73×10^4 **38.** 4.5×10^{-5} **39.** $x^{11/6}$ **40.** $y^{1/2}/x^{1/3}$ **41.** $1/x^{n/2} y^n$ **42.** xy
43. $y-y^{3/2}$ **44.** y^2-y **45.** $x^{-1/4}(1+x^{3/4})$ **46.** $x^{-2/3}(x^{4/3}-1)$ **47.** $3xy^2\sqrt{xy}$ **48.** $4x\sqrt{y}$
49. $y\sqrt{2x}$ **50.** $3x$ **51.** $\sqrt[3]{xy^2}/y$ **52.** $1/2\sqrt{3y}$ **53.** $2-\sqrt{3}$ **54.** $\dfrac{2x-3\sqrt{x}+1}{x-1}$ **55.** $\dfrac{-7}{2\sqrt{2}-6}$
56. $\dfrac{y-9}{y+4\sqrt{y}+3}$ **57.** $4\sqrt{2}$ **58.** $10\sqrt[3]{3}$ **59.** 3 **60.** $3x+2\sqrt{x}-8$

Exercise 3.1. [page 71]

1. $\{-10\}$ **3.** $\{-2\}$ **5.** $\{0\}$ **7.** $\{1/3\}$ **9.** $\{-7\}$ **11.** $\{15\}$ **13.** $\varnothing$ **15.** $\{-14/5\}$
17. $c = \dfrac{2A-bh}{h}$ $(h \neq 0)$ **19.** $n = \dfrac{1-a+d}{d}$ $(d \neq 0)$ **21.** $y' = \dfrac{3x+1}{x^2-2y^3}$ $(x^2 \neq 2y^3)$
23. $x_1 = \dfrac{x_4}{x_2-2x_3}$ $(x_2 \neq 2x_3)$ **25.** $y = 6(x-x_1)+y_1$ $(x \neq x_1)$ **27.** $-3/5$ **29.** 1.37×10^5 erg/cm^3
31. $P = S(1+r)^{-n}$ **33.** $35; 50$ **35.** 425 **37.** 16 **39.** $15/2$ m/sec^2 **41.** $p_{n-1} = \dfrac{p_n-a}{1-a-b}$

Exercise 3.2. [page 77]

1. $\{0, -2\}$ **3.** $\{2, -7\}$ **5.** $\{-1/2, 3\}$ **7.** $\{4/3, -1/2\}$ **9.** $\{1/2, 1\}$ **11.** $\{1, -10/3\}$
13. $\{2, -2\}$ **15.** $\{2\sqrt{2}, -2\sqrt{2}\}$ **17.** $\{6+\sqrt{5}, 6-\sqrt{5}\}$ **19.** $\{2, -6\}$ **21.** $\{-4, -5\}$
23. $\{1/2, -2\}$ **25.** $\{1, 2\}$ **27.** $\{2, 3/2\}$ **29.** $\{3, -3/2\}$ **31.** $\{\sqrt{5}\}$ **33.** $\{\sqrt{3}, -\sqrt{3}/2\}$
35. $\{2k, -k\}$ **37.** $\left\{\dfrac{1+\sqrt{1-4ac}}{2a}, \dfrac{1-\sqrt{1-4ac}}{2a}\right\}$ **39.** 4   **41.** $\sqrt{66.64}$ m/sec   **43.** $1 - \sqrt{V/A}$
45. 50   **47.** 1, 50   **49.** 1, 5

Exercise 3.3. [page 82]

1. $\{64\}$ **3.** $\{-7\}$ **5.** $\{4\}$ **7.** $\{5/4\}$ **9.** $\{-25\}$ **11.** $\{16\}$ **13.** $\{1\}$ **15.** $\{5\}$ **17.** $\{16\}$
19. $\{4\}$ **21.** $\{1, 3\}$ **23.** $A = \pi r^2$ **25.** $y = 1/x^3$ **27.** $y = \pm\sqrt{a^2 - x^2}$ **29.** $\{1/2, 8\}$
31. $L = 980/\pi^2$ cm

Exercise 3.4. [page 85]

1. $\{25\}$ **3.** $\{\sqrt{2}/2, -\sqrt{2}/2\}$ **5.** $\{2, -7, -3, -2\}$ **7.** $\{64, -8\}$ **9.** $\{1/4, -1/3\}$ **11.** $\{626\}$
13. $\{2, -1, -4\}$ **15.** $\{4, 1\}$ **17.** $\{13\}$ **19.** $\{4, 5, -2, -3\}$

Exercise 3.5. [page 88]

1. $\{x \mid x > 1\}$ **3.** $\{x \mid x > 3\}$ **5.** $\{x \mid x \leq 13/2\}$ **7.** $\{x \mid x < -24\}$ **9.** $\{x \mid -2 \leq x < 4\}$
11. $\{x \mid 2/3 < x < 2\}$ **13.** $\{x \mid -2 < x < 5\}$ **15.** $\{x \mid x \geq 13\}$ **17.** $\{x \mid x \leq 1/3 \text{ or } x > 11\}$
19. $\{x \mid x < 0 \text{ or } x > 2\}$ **21.** $k \leq -2$ **23.** $k < 0$
25. $61.3 \leq P \leq 100$ **27.** $x \geq 575$ **29.** $14 \leq T_f \leq 68$

Exercise 3.6. [page 93]

1. $\{x \mid x < -1 \text{ or } x > 2\}$ **3.** $\{x \mid 0 \leq x \leq 2\}$ **5.** $\{x \mid x < -1 \text{ or } x > 4\}$ **7.** $\{x \mid -\sqrt{5} < x < \sqrt{5}\}$

Odd-Numbered Answers

9. $\{x | x \in R\}$ **11.** $\{x | x < 0 \text{ or } x \geq 1/2\}$ **13.** $\{x | x \leq -4 \text{ or } x \geq 4\}$

15. $\{x | -8/3 < x < -2\}$ **17.** $\{x | x < 0 \text{ or } 2 < x \leq 4\}$ **19.** $\{x | -3 < x < 0 \text{ or } x > 2\}$

21. Since $a > 0$ and $b > 0$, then $2ab > 0$ and $a^2 + 2ab + b^2 > a^2 + b^2$. This is the same as $(a + b)^2 > a^2 + b^2$

23. For all $x \in R$, $(x - 1) \in R$ and $(x - 1)^2 \geq 0$, since the square of all positive or negative real numbers is either positive or zero. Then, since $(x - 1)^2 = x^2 - 2x + 1$, it follows that $x^2 - 2x + 1 \geq 0^2$; adding $2x$ to both sides produces $x^2 + 1 \geq 2x$.

25. $\{x | 2 < x < 8\}$ **27.** $\{t | 2 < t < 6\}$; $\{t | 0 < t < 8\}$

Exercise 3.7. [page 98]

1. $\{6, -6\}$ **3.** $\{-3, 5\}$ **5.** $\{1/3, 1\}$ **7.** $\{-3/2, -7/2\}$ **9.** $\{1/2, 7/2\}$ **11.** $\{22/3, -26/3\}$

13. $\{x | -2 < x < 2\}$ **15.** $\{x | -7 \leq x \leq 1\}$ **17.** $\varnothing$ **19.** $\{x | x < -1 \text{ or } x > 7\}$

21. $\{x | x \leq -3 \text{ or } x \geq 2\}$ **23.** $\{x | -13/2 < x < 11/2\}$ **25.** $|x - 2| < 1$ **27.** $|x + 8| \leq 1$

29. $|4x - 5| \leq 19$

31. (I) If $a \geq 0$, $|-a| = |-(+a)| = |-a| = a$; also $|a| = a$; hence $|-a| = |a|$.
(II) If $a < 0$, $|-a| = |-(-a)| = |a| = a$; also $|a| = |+(-a)| = |-a| = a$; hence $|-a| = |a|$.

33. $|a^2| = a^2$; $|a|^2 = |a| \cdot |a| = a \cdot a = a^2$; hence $|a^2| = |a|^2 = a^2$

35. $x - a < b$ for $x - a \geq 0$ or $-(x - a) < b$ for $x - a < 0$. The union of the two solution sets yields $-b < x - a < b$.

Exercise 3.8. [page 102]

1. 9 nickels, 7 dimes   **3.** 125,000 of the $5 books, 15,000 of the $3.40 books

5. 15 lb of 32% silver alloy must be melted   **7.** $1400 at 3% and $600 at 4%

9. $12,000 at 5% and $36,000 at 3%   **11.** 4, 5, 6   **13.** 88 points   **15.** −40

17. Rate of automobile = 60 mi/hr; rate of plane = 180 mi/hr   **19.** 240 miles

21. 7 and 8 are the two numbers   **23.** 6 and 7; −7 and −6   **25.** $\sqrt{13,000}$ miles

27. 1/2 second to reach 24 feet on way up; $3\frac{1}{2}$ seconds to return to original position   **29.** 1 centimeter

31. 10 feet   **33.** 12 days for the boy alone, 6 days for the father alone   **35.** 94%

Chapter 3 Test [page 106]

1. $\{-7/2\}$ 2. $\{4\}$ 3. $\{-26/21\}$ 4. $\{5/2\}$ 5. $y = \frac{1}{4}x$ 6. $x = 4y$ 7. $\{1/2, 1\}$

8. $\{-1/2, 3\}$ 9. $\{\sqrt{7}, -\sqrt{7}\}$ 10. $\{-4 + \sqrt{3}, -4 - \sqrt{3}\}$ 11. $\left\{-\frac{3}{2} + \frac{\sqrt{13}}{2}, -\frac{3}{2} - \frac{\sqrt{13}}{2}\right\}$

12. $\{1/2, -1\}$ 13. $\left\{\frac{1+\sqrt{17}}{4}, \frac{1-\sqrt{17}}{4}\right\}$ 14. $\left\{\frac{-k+\sqrt{k^2+16}}{2}, \frac{-k-\sqrt{k^2+16}}{2}\right\}$ 15. $k = 9$

16. $k = 4$ or $k = -4$ 17. $\{4, 1\}$ 18. $\{8\}$ 19. $\{2, -2, 1, -1\}$ 20. $\{1/5, -1/4\}$ 21. $x \leq 27$

22. $x > 3/2$ 23. [number line: 0 to 5, open circle at 2, open circle at 8] 24. [number line: -10 to 5, -7 and 2] 25. $-5 < x < 1$

26. $x > 7/2$ or $x < 3$ 27. $\{3, -4\}$ 28. $\{1, -1/3\}$ 29. [number line: -5 to 5, -3 and 7]

30. $|x - 4| \leq 7$ 31. 480,000 units 32. $1.00 33. 2113 and 2263 votes 34. 6 inches 35. $5\frac{1}{3}$ hr

36. 6 mi/hr going; 4 mi/hr returning 37. $12 \leq t \leq 24$

Exercise 4.1. [page 114]

1. $\{(0, 6), (1, 4), (2, 2), (-3, 12), (2/3, 14/3)\}$ 3. $\{(0, 0), (1, -3), (2, 3), (-3, -9/7), (2/3, -9/7)\}$

5. $\{(0, \sqrt{11}), (1, \sqrt{14}), (2, \sqrt{17}), (-3, \sqrt{2}), (2/3, \sqrt{13})\}$

7. a. Domain: $\{2, 5, 7\}$; range: $\{3, 7, 8\}$ b. A function

9. a. Domain: $\{2, 3\}$; range: $\{-1, 4, 6\}$ b. Not a function

11. a. Domain: $\{5, 6, 7\}$; range: $\{5, 6, 7\}$ b. A function 13. Domain: $\{x | x \in R\}$ 15. Domain: $\{x | x \in R\}$

17. Domain: $\{x | x \neq 2\}$ 19. Domain: $\{x | x \geq 3\}$ 21. Domain: $\{x | -2 \leq x \leq 2\}$

23. Domain: $\{x | x \neq 0, 1\}$ 25. Yes 27. Yes 29. No 31. No 33. 2 35. -1 37. 9

39. a^2 41. 1 43. $a - 2$ 45. ± 1 47. ± 3 49. 2 51. 0 53. $2x + h - 1$ 55. Even

57. Even 59. $P = 4\sqrt{A}$ 61. $V = A^{3/2}$ 63. a. 1500 b. 39,999 65. $4400; $75,000

Exercise 4.2. [page 124]

1.
3.
5.
7.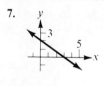
9.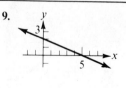

Odd-Numbered Answers

11.   **13.**   **15.** Range: $\{y \mid -1 \leq y \leq 3\}$ **17.** Distance, 5; slope, 4/3

19. Distance, 13; slope 12/5   **21.** Distance, $\sqrt{2}$; slope, 1   **23.** Distance, $3\sqrt{5}$; slope, 1/2

25. Distance, 5; slope, 0   **27.** Distance, 10; slope, undefined   **29.** 7, $\sqrt{68}$, $\sqrt{89}$ **31.** 10, 21, 17

33. From Problem 30, the lengths of the sides of the triangle are 15, $3\sqrt{5}$, and $6\sqrt{5}$; since $15^2 = 225$, $(3\sqrt{5})^2 = 45$, and $(6\sqrt{5})^2 = 180$, it follows that $15^2 = (3\sqrt{5})^2 + (6\sqrt{5})^2$, and the converse of the Pythagorean theorem applies. Hence, the triangle is a right triangle.   **35.** $x - 2y = 8$

37. This follows from the fact that the midpoint of the segment P_1R is $\left(\dfrac{x_1 + x_2}{2}, y_1\right)$ and the midpoint of P_2R is $\left(x_2, \dfrac{y_1 + y_2}{2}\right)$. Then, since triangle P_1OM is similar to triangle P_1P_2R, $P_1O = \dfrac{1}{2}P_1P_2$. Hence, $\left(\dfrac{x_1 + x_2}{2}, \dfrac{y_1 + y_2}{2}\right)$ is the midpoint of P_1P_2.

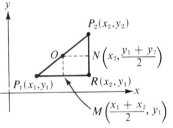

39. 159 long tons   **41.** 61.3 torr   **43.** 12 inches

45. $R(x) = 1500x$ **47.** $C(x) = 50x + 475{,}000$ **49.** 90   **51.** $S = 1000 + 50t$, where t represents years

53. 1.78 trillions of dollars; 2.62 trillions of dollars   **55.** 960.8°F; 300.2°F   **57.** 27.3 cubic meters

Exercise 4.3. [page 131]

1. $4x - y - 7 = 0$ **3.** $x + y - 10 = 0$ **5.** $3x - y = 0$ **7.** $x + 2y + 2 = 0$ **9.** $3x + 4y + 14 = 0$

11. $y - 2 = 0$ **13.** $y = -x + 3$; slope, -1; intercept, 3   **15.** $y = -\dfrac{3}{2}x + \dfrac{1}{2}$; slope, $-3/2$; intercept, 1/2

17. $y = \dfrac{1}{3}x - \dfrac{2}{3}$; slope, 1/3; intercept, $-2/3$ **19.** $y = -\dfrac{2}{3}x - 2$; slope, $-\dfrac{2}{3}$; intercept, -2

21. $3x + 2y - 6 = 0$ **23.** $5x + 2y + 10 = 0$ **25.** $6x - 2y + 3 = 0$ **27.** $3x - y - 5 = 0$

29. $2x - 3y + 19 = 0$ **31.** $x - 2 = 0$

33. Substituting $\dfrac{y_2 - y_1}{x_2 - x_1}$ for m in the formula $y - y_1 = m(x - x_1)$ gives the desired result

35. $f(x) = \dfrac{-x + 11}{3}$ **37.** $C(x) = 50x + 20{,}000$ **39.** $C(x) = 1.5x + 10$; 1.5 cents/year; 40 cents

41. $340,000   **43.** $V = 6000(1 - x/10)$; 4 years

45. Let x be the number of years after its purchase; then $x - 2$ is the number of years after the second year. If V is the value of the car, then

$$V = \begin{cases} 5000 - 1000x & 0 \le x \le 2, \\ 3000 - 500(x - 2) & x > 2; \end{cases}$$

$3000; $1500; 8 years old

47. The slope is $(212 - 32)/(100 - 0) \frac{9}{5}$. From (1), $T_f - 32 = \frac{9}{5}(T_c - 0)$ or $T_f = \frac{9}{5} T_c + 32$.

Exercise 4.4. [page 140]

1. y, 4; x, 1 and 4; minimum, $(5/2, -9/4)$ **3.** y, -7; x, -1 and 7; minimum, $(3, -16)$

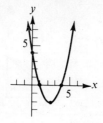
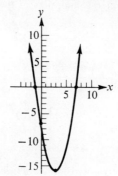

5. y, -4; x, 1 and 4; maximum, $(5/2, 9/4)$ **7.** y, 2; no x; minimum, $(0, 2)$ **9.** y, 0; x, 0 and 3, minimum $\left(\frac{3}{2}, -\frac{9}{4}\right)$

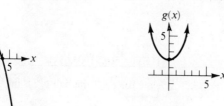
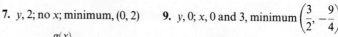
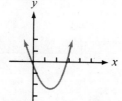

11.

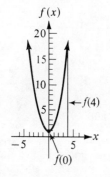

13. Varying k has the effect of translating the graph along the y-axis

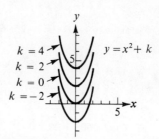

Odd-Numbered Answers

15. a. Parabola **b.** No; there are two values of y associated with each value of x, except $x = 0$
  c. No

17. 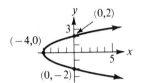   **19.** 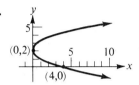   **21.** 4; 4   **23.** 9/2

25. 80 cm/sec; 50 cm/sec
maximum: 90 cm/sec
minimum: 0 cm/sec

27. $30,000 at $x = 100$ **29.** $20,000 at $x = 200$

31. $x =$ width; $y =$ length; $150 = 4x + 2y$; $A(x) = x(75 - 2x)$

33. $x =$ number of fare increases; $P =$ daily profit; $P(x) = (1500 - 50x)(100 + 5x)$ **35.** $C(x) = 8x^2$

37. 224; 224; 0; for $t = 0$ and $t = 5$ the ball is at rooftop level; for $t = 7$ the ball is on the ground

Exercise 4.5. [page 145]

1. 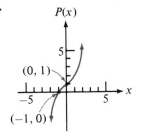   **3.** 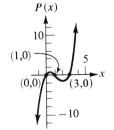   **5.**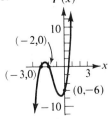

7. 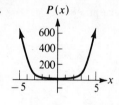   **9.**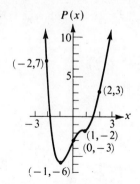

11. $104\pi/3$ cm^3 **13.** $V(x) = x(10 - 2x)^2$

Exercise 4.6. [page 149]

1. 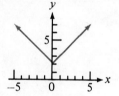   **3.** 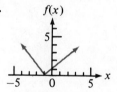   **5.**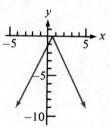

7. 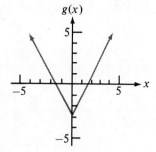   **9.**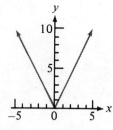

11. 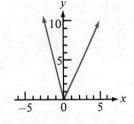   **13.** 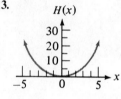   **15.**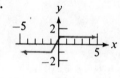

Odd-Numbered Answers

17. 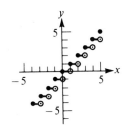 **19.** 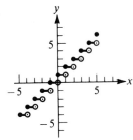 **21.**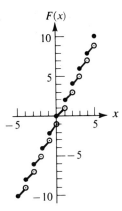

23.  **25.** 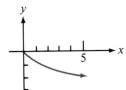 **27.**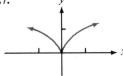

29. 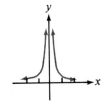 **31.** 16   **33.** 1.5 seconds   **35.** 0.00056; 209 pounds

37. Cost in cents: $C = -c[-x]$, $x > 0$ **39.** 50   **41.** 900   **43.** 4 times   **45.** 24; approximately 9

Exercise 4.7. [page 154]

1. 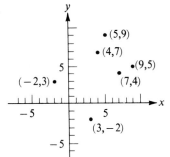 **3.**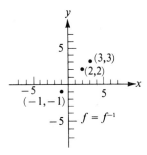

5.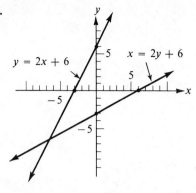

7.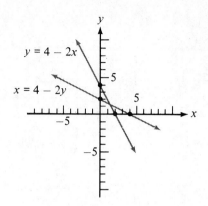

9.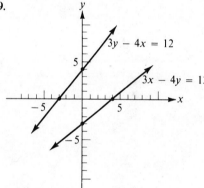

11.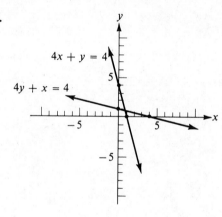

13. $f^{-1}(x) = \sqrt[3]{x}$ **15.** $f^{-1}(x) = \sqrt[3]{\dfrac{x-4}{2}}$ **17.** $f^{-1}(x) = (x-2)^3$

19. $f(x) = x$, $f^{-1}(x) = x$; $f[f^{-1}(x)] = f(x) = x$, $f^{-1}[f(x)] = f^{-1}(x) = x$

21. $f(x) = 4 - 2x$, $f^{-1}(x) = \dfrac{4-x}{2}$; $f[f^{-1}(x)] = f\left[\dfrac{4-x}{2}\right] = 4 - 2\left(\dfrac{4-x}{2}\right) = 4 - 4 + x = x$,

$f^{-1}[f(x)] = f^{-1}[4 - 2x] = \dfrac{4 - (4 - 2x)}{2} = \dfrac{4 - 4 + 2x}{2} = \dfrac{2x}{2} = x$

23. $f(x) = \dfrac{3x - 12}{4}$, $f^{-1}(x) = \dfrac{4x + 12}{3}$; $f[f^{-1}(x)] = f\left[\dfrac{4x + 12}{3}\right] = \dfrac{(4)\left(\dfrac{3x-12}{4}\right) + 12}{3} =$

$\dfrac{3x - 12 + 12}{3} = \dfrac{3x}{3} = x$, $f^{-1}[f(x)] = f^{-1}\left[\dfrac{3x-12}{4}\right] = \dfrac{(3)\left(\dfrac{4x+12}{3}\right) - 12}{4} = \dfrac{4x + 12 - 12}{4} = \dfrac{4x}{4} = x$

Odd-Numbered Answers

25. $f(x) = \sqrt{x+3}$, $f^{-1}(x) = x^2 - 3$, $x \geq 0$; $f(f^{-1}(x)) = f(x^2 - 3) = \sqrt{x^2 - 3 + 3} = \sqrt{x^2} = x$, $x \geq 0$
$f^{-1}(f(x)) = f^{-1}(\sqrt{x+3}) = (\sqrt{x+3})^2 - 3 = (x+3) - 3 = x$

Chapter 4 Test [page 156]

1. Domain: $\{4, 2, 3\}$; range: $\{-1, -4, -5\}$ **2.** Domain: $\{1\}$; range: $\{2, 3, 6\}$
3. Domain: $\{x \mid x \neq -4\}$; range: $\{y \mid y \neq 0\}$ **4.** Domain: $\{x \mid x \geq 6\}$; range: $\{y \mid y \geq 0\}$ **5.** 13 **6.** -5
7. 1 **8.** $x^2 + 2xh + h^2 + 4$ **9.** $\sqrt{41}$; $m = -4/5$ **10.** $\sqrt{5}$; $m = -1/2$ **11.** $4x - y - 15 = 0$
12. $x - 2y + 12 = 0$ **13.** $m = -4$; y-intercept $= 6$; $y = -4x + 6$
14. $m = 3/2$; y-intercept $= -8$; $y = \frac{3}{2}x - 8$ **15.** $2x - 3y + 6 = 0$ **16.** $12x - y - 4 = 0$
17. $C(x) = 15{,}000x + 5{,}000{,}000$ **18.** $50{,}000$; $30{,}000$; $12{,}500$
19. x-intercepts, 3, -2; axis of symmetry, $x = 1/2$; minimum, $(1/2, -25/4)$
20. x-intercepts, 5, 2; axis of symmetry, $x = 7/2$; maximum $(7/2, 9/4)$
21. 5000; $2{,}500{,}000$ **22.** 400 feet; 10 seconds
23. $(-4, -10)$, $(-3, 0)$, $(-2, 0)$, $(-1, -4)$, $(0, -6)$, $(1, 0)$, $(2, 20)$, $(3, 60)$, $(4, 126)$

24.
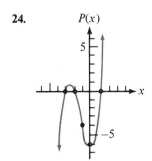

25.
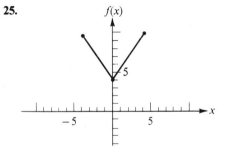

26.
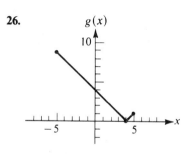

27. 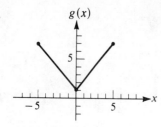 **28.** 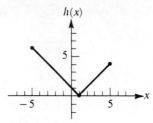 **29.** 125 **30.** 64 posts

31. 45 **32.** 21

33. 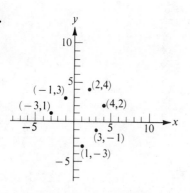 **34.**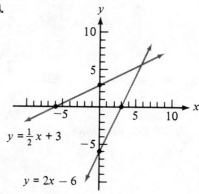

Exercise 5.1. [page 160]

1. (0, 1), (1, 3), (2, 9)   **3.** (−2, −1/25), (0, −1), (2, −25)   **5.** (−3, 8), (0, 1), (3, 1/8)

7. (−2, 1/100), (−1, 1/10), (0, 1)   **9.**   **11.** 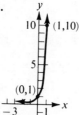   **13.**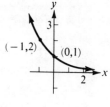

15. 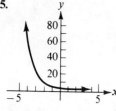   **17.** 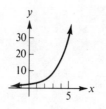   **19.** 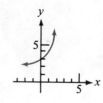   **21.**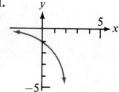

Odd-Numbered Answers

23. 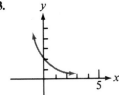 **25.** No; constant function 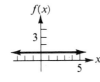 **27.** $\{-2\}$ **29.** $\{3/4\}$ **31.** 3

33. 1000; 1500; 2250; 3375   **35.** 6703

37. 0; 39.3; 63.2; 77.7   **39.** 0.06

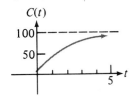

41. 2.7% of I_0 **43.** 12.048 grams

45.

x	-2	-1	-0.5	0	0.5	1	2
$f(x)$	0.02	0.37	0.78	1	0.78	0.37	0.02

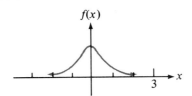

Exercise 5.2. [page 165]

1. $\log_4 16 = 2$ **3.** $\log_3 27 = 3$ **5.** $\log_{1/2} 1/4 = 2$ **7.** $\log_8 1/2 = -1/3$ **9.** $\log_{10} 100 = 2$

11. $\log_{10}(0.1) = -1$ **13.** $2^6 = 64$ **15.** $3^2 = 9$ **17.** $(1/3)^{-2} = 9$ **19.** $10^3 = 1000$

21. $10^{-2} = 0.01$ **23.** 2   **25.** 3   **27.** 1/2   **29.** -1 **31.** 1   **33.** 2   **35.** -1 **37.** $\{2\}$

39. $\{2\}$ **41.** $\{64\}$ **43.** $\{-3\}$ **45.** $\{100\}$ **47.** $\{4\}$

49. By definition, $\log_b 1$ is a number such that $b^{\log_b 1} = 1$. Therefore, $\log_b 1 = 0$.

51. $\log_b b^x = x$ implies $b^x = b^x$, which is true for all real numbers   **53.** $\log_b x + \log_b y$

55. $\log_b x - \log_b y$ **57.** $5 \log_b x$ **59.** $\dfrac{1}{3} \log_b x$ **61.** $\dfrac{1}{2}(\log_b x - \log_b z)$

63. $\frac{1}{3}(\log_{10} x + 2\log_{10} y - \log_{10} z)$ 65. $\log_{10} 2 + \log_{10} \pi + \frac{1}{2}\log_{10} l - \frac{1}{2}\log_{10} g$ 67. $\log_b xy$

69. $\log_b x^2 y^3$ 71. $\log_b \frac{x^3 y}{z^2}$ 73. $\log_{10} \frac{x(x-2)}{z^2}$

75. $\frac{1}{4}\log_{10} 8 + \frac{1}{4}\log_{10} 2 = \frac{1}{4}\log_{10} 2^3 + \frac{1}{4}\log_{10} 2 = \left(\frac{3}{4} + \frac{1}{4}\right)\log_{10} 2 = \log_{10} 2$

77. $\log_{10}[\log_3(\log_5 125)] = \log_{10}[\log_3 (3)] = \log_{10}[1] = 0$

Exercise 5.3. [page 172]

1. 2   3. -3 or $7 - 10$ 5. -4 or $6 - 10$ 7. 4   9. 0.8280   11. $9.9101 - 10$ 13. $8.9031 - 10$
15. 2.3945   17. 4.10   19. 3.67   21. 0.0642   23. 5480   25. 0.000718   27. 0.6246
29. 3.1824   31. 4.5695   33. $9.7095 - 10$ 35. 3.225   37. 10.52   39. 0.05075   41. 0.7495
43. $\log_{10} 3.751$; 0.751 is closer to a tabulated value   45. 9.1   47. 5000   49. 113
51. a. -1.2679 b. -3.5813

Exercise 5.4. [page 178]

1. 7   3. 7.7   5. 6.2   7. 1.0×10^{-3} 9. 2.5×10^{-6} 11. 6.3×10^{-8} 13. 90 dB   15. 60 dB
17. 74.8 dB   19. 2   21. 96 dB   23. 5.9   25. -7.8

Exercise 5.5. [page 182]

1. 4.014   3. 2.299   5. 0.000461   7. 64.34   9. 2.010   11. 3.435×10^{-10} 13. 0.04582
15. 0.2777   17. 9.871   19. 4.746   21. 1.394   23. 3.484   25. 3806   27. 146   29. 2.664
31. 99.5 square centimeters   33. 4.07 square meters   35. $\$1.18 \times 10^{17}$

Exercise 5.6. [page 186]

1. $\left\{\frac{\log_{10} 7}{\log_{10} 2}\right\}$ 3. $\left\{\frac{\log_{10} 8}{\log_{10} 3} - 1\right\}$ 5. $\left\{\frac{1}{2}\left(\frac{\log_{10} 3}{\log_{10} 7} + 1\right)\right\}$ 7. $\left\{\sqrt{\frac{\log_{10} 15}{\log_{10} 4}}, -\sqrt{\frac{\log_{10} 15}{\log_{10} 4}}\right\}$

9. $\left\{\frac{-1}{\log_{10} 3}\right\}$ 11. $\left\{1 - \frac{\log_{10} 15}{\log_{10} 3}\right\}$ 13. $n = \frac{\log_{10} y}{\log_{10} x}$ 15. $t = \frac{\log_{10} y}{k \log_{10} e}$ 17. $\{500\}$ 19. $\{4\}$

21. $\{3\}$ 23. 1.343   25. 5%   27. 2.5%   29. 20 years   31. 12 years   33. 4.9 years

35. $\dfrac{\log_{10} R/(R - CP)}{\log_{10}(1 + i)}$ 37. $L_B = L_A 10^{-0.4(M_B - M_A)}$

Odd-Numbered Answers

Exercise 5.7. [page 191]

1. 3.32 3. 3.41 5. 1.08 7. 0.79 9. 1.0986 11. 2.8332 13. 5.7900 15. 6.1093
17. 1.6487 19. 29.964 21. 1.260 23. 0.8607 25. 0.0821 27. 0.7600 29. $1/3$ 31. 2.10
33. 2.86 35. a. $\log_9 7 = \dfrac{\log_3 7}{\log_3 9} = \dfrac{\log_3 7}{\log_3 3^2} = \dfrac{\log_3 7}{2 \log_3 3} = \dfrac{1}{2} \log_3 7$ b. $\log_{a^2} b = \dfrac{\log_a b}{\log_a a^2} = \dfrac{\log_a b}{2 \log_a a} = \dfrac{1}{2} \log_a b$
37. $(\log_{10} 4 - \log_{10} 2)\log_2 10 = \log_{10}\left(\dfrac{4}{2}\right) \cdot \log_2 10 = \log_{10} 2 \cdot \dfrac{\log_{10} 10}{\log_{10} 2} = \log_{10} 10 = 1$
39. approximately 30 years 41. $-\dfrac{1}{a}\log_e(a-bP)/P$ 43. $4481.70 45. 2.2 centimeters
47. $-\dfrac{L}{R}\log_e(E-iR)/E$

Chapter 5 Test [page 194]

1. [graph]
2. $x = -3$ 3. $\log_3 81 = 4$ 4. $\log_3 \dfrac{1}{9} = -2$ 5. $2^3 = 8$
6. $10^{-4} = 0.0001$ 7. $y = 4$ 9. $\dfrac{1}{3}\log_{10} x + \dfrac{2}{3}\log_{10} y$
10. $\log_{10} 2 + 3 \log_{10} R - \dfrac{1}{2}\log_{10} P - \dfrac{1}{2}\log_{10} Q$ 11. $\log_b \dfrac{x^2}{\sqrt[3]{y}}$ 12. $\log_{10} \dfrac{\sqrt[3]{x^2 y}}{z^3}$ 13. 1.6232
14. $7.4969 - 10$ 15. 2.8340 16. $8.6172 - 10$ 17. 67.4 18. 0.0466 19. 2.655 20. 0.6643
21. 3.5 22. 6.1 23. 2.5×10^{-3} 24. 4×10^{-7} 25. 80 dB 26. 113 dB 27. 50 dB
28. 104 dB 29. 363.2 30. 49.65 31. 1450 32. 0.7711 33. $\dfrac{\log_{10} 7}{\log_{10} 5}$ 34. $\dfrac{\log_{10} 80}{\log_{10} 3} - 1$
35. 5 36. 4 37. 2.18 38. 6.9 39. 4.8676 40. 121.51 41. 0.7866 42. 4.059 grams
43. $3694.55 44. The solution of $2 = e^{kt}$ is $kt = \log_e 2$ or $t = (\log_e 2)/k$ 45. 500 years

Exercise 6.1. [page 201]

1. {(3, 2)} **3.** {(2, 1)} **5.** {(−5, 4)} **7.** {(0, 3/2)} **9.** {(2/3, −1)} **11.** {(1, 2)}

13. {(−19/5, −18/5)} **15.** $\left\{\left(\frac{1}{2}, \frac{1}{4}\right)\right\}$ **17.** {(0, 4)} **19.** $a = 1; b = -1$ **21.** $y = -\frac{10}{3}x + 2$

23. 32 pounds **25.** $7200 at 4%; $8200 at 5% **27.** 4200 tickets at $10, 5800 tickets at $20

29. 10 hours for A, 3 hours for B **31.** 32 and 64 mi/hr

Exercise 6.2. [page 207]

1. {(1, 2, −1)} **3.** {(2, −2, 0)} **5.** {(2, 2, 1)} **7.** {(0, 1, 2)} **9.** Dependent **11.** {(4, −2, 2)}

13. $x^2 + y^2 + 2x + 2y - 23 = 0$ **15.** $a = 1/4, b = 1/2, c = 0$ **17.** $a = 1/3, b = 1/3, c = 1/6$

19. two nonparallel planes intersect in a straight line **21.** 3, 6, 6 **23.** 60 nickels, 20 dimes, 5 quarters

25. 36 ones, 34 fives, 24 tens

27. 40 in A, 20 in B, 40 in C; A receives $400,000, B receives $80,000, C receives $200,000

Exercise 6.3. [page 214]

1. {(−1, −4), (5, 20)} **3.** {(2, 3), (3, 2)} **5.** {(4, −3), (−3, 4)} **7.** {(−1, 3), (−1, −3), (1, 3), (1, −3)}

9. {(−3, $\sqrt{2}$), (−3, −$\sqrt{2}$), (3, $\sqrt{2}$), (3, −$\sqrt{2}$)} **11.** {($\sqrt{3}$, 4), ($\sqrt{3}$, −4), (−$\sqrt{3}$, 4), (−$\sqrt{3}$, −4)}

13. {(1, −2), (−1, 2), (2, −1), (−2, 1)} **15.** {(3, 1), (−3, −1), (−2$\sqrt{7}$, $\sqrt{7}$), (2$\sqrt{7}$, −$\sqrt{7}$)}

17. {(0, 2)} **19.** {(1, 0), (3/2, 1/2)} **21.** {(1, 0)} **23. a.** 1 **b.** 2 **c.** 4

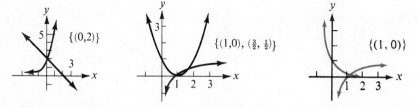

25. 7/2, 5/2 **27.** $4\sqrt{5}$ centimeters, $8\sqrt{5}$ centimeters **29.** 30 feet by 50 feet

31. approximately 17.6 centimeters by 28.4 centimeters

Exercise 6.4. [page 218]

1. (90, 72,000) **3.** (22,760) **5.** (5,275), (50,500) **7.** (7, 11) **9.** (10,800) **11.** (25, 60)

Odd-Numbered Answers

13. $S(x) = \frac{5}{4}x$, $D(x) = -\frac{4}{5}x + 82$ **15. a.** $C(x) = 200x + 50{,}000$, $R(x) = 700x$ **b.** $P(x) = 500x - 50{,}000$
c. $(100, 70{,}000)$ **17.** 522 gallons **19. a.** $R(x) = (200 - 4x)(14 + x)$, $C(x) = 6(200 - 4x)$ **b.** $(50, 0)$
c. $P(x) = (200 - 4x)(14 + x) - 6(200 - 4x)$ **d.** $0 < x < 50$

Chapter 6 Test [page 221]

1. $\{(1/2, 7/2)\}$ **2.** $\{(1, 2)\}$ **3.** $\{(37/13, -10/13)\}$ **4.** $a = 2, b = 5$ **5.** $\{(2, 0, -1)\}$ **6.** $\{(2, 1, -1)\}$
7. $\{(2, -1, 3)\}$ **8.** $a = 2, b = -3, c = 4$ **9.** $\{(1, 2), (4, -13)\}$ **10.** $\{(6/7, -37/7), (2, -3)\}$
11. $\{(2, 3), (2, -3), (-2, 3), (-2, -3)\}$ **12.** $\{(7/8, 17/8)\}$ **13.** 100 milligrams of X, 80 milligrams of Y
14. rounded to a whole number $x = 2$; take the break-even point to be $(2, 100)$ **15.** $(1.5, 4)$

Exercise 7.1. [page 227]

1. 2×2, $\begin{bmatrix} 6 & 2 \\ -1 & 3 \end{bmatrix}$ **3.** 3×2, $\begin{bmatrix} 2 & 3 & 4 \\ -7 & 1 & 0 \end{bmatrix}$ **5.** 3×3, $\begin{bmatrix} 2 & 4 & -2 \\ 3 & 0 & 3 \\ -1 & 1 & 1 \end{bmatrix}$ **7.** 2×4, $\begin{bmatrix} 4 & 2 \\ -3 & 1 \\ -1 & 1 \\ 0 & 6 \end{bmatrix}$

9. $\begin{bmatrix} 3 & 1 \\ 3 & 9 \end{bmatrix}$ **11.** $\begin{bmatrix} 9 & -1 & -1 \\ 2 & 3 & 6 \end{bmatrix}$ **13.** $\begin{bmatrix} -2 & -3 \\ 4 & 0 \end{bmatrix}$ **15.** $\begin{bmatrix} 2 & -9 & -13 \\ 7 & -4 & 1 \\ 9 & 0 & 0 \end{bmatrix}$ **17.** $\begin{bmatrix} 10 \\ 3 \\ -3 \end{bmatrix}$

19. $\begin{bmatrix} 2 & 3 & 4 \\ -1 & 6 & 2 \\ 1 & 0 & 3 \end{bmatrix}$

21. Let X, $X + A$, and B have each entry x_{ij}, $x_{ij} + a_{ij}$, and b_{ij}, respectively. Then since $X + A = B$, $x_{ij} + a_{ij} = b_{ij}$, and since x_{ij}, a_{ij}, and b_{ij} are real numbers, $x_{ij} = b_{ij} - a_{ij}$ (cancellation property of addition); thus, $X = B - A$.

23. $\begin{bmatrix} 2 & 2 \\ -1 & 1 \end{bmatrix}$ **25.** $\begin{bmatrix} 1 & 1 \\ -4 & 3 \end{bmatrix}$

25. Let $A = \begin{bmatrix} a_{11} & a_{12} \\ a_{21} & a_{22} \end{bmatrix}$ and $B = \begin{bmatrix} b_{11} & b_{12} \\ b_{21} & b_{22} \end{bmatrix}$. Then

$[A + B]^t = \begin{bmatrix} a_{11} + b_{11} & a_{21} + b_{21} \\ a_{12} + b_{12} & a_{22} + b_{22} \end{bmatrix} = \begin{bmatrix} a_{11} & a_{21} \\ a_{12} & a_{22} \end{bmatrix} + \begin{bmatrix} b_{11} & b_{21} \\ b_{12} & b_{22} \end{bmatrix} = A^t + B^t$.

Yes

27. Let $A = [b_{ij}]_{m \times n}$, $B = [b_{ij}]_{m \times n}$, and $C = [c_{ij}]_{m \times n}$. Then $(A + B) + C = [(a_{ij} + b_{ij}) + c_{ij}] \, m \times n$
$= [a_{ij} + (b_{ij} + c_{ij})] \, m \times n = A + (B + C)$

31. $\begin{bmatrix} 260 & 44 & 12 & 4 \\ 150 & 30 & 4 & 1 \end{bmatrix}$ **33. a.** 13 **b.** 8 **c.** -2 **d.** 12 **e.** $\begin{bmatrix} 59 & 51 & 26 \\ 33 & 36 & 25 \\ 27 & 22 & 33 \\ 6 & 14 & -8 \end{bmatrix}$

Exercise 7.2. [page 236]

1. $\begin{bmatrix} 0 & -5 & 5 \\ -15 & 5 & -10 \end{bmatrix}$ 3. $[-1]$ 5. $\begin{bmatrix} 1 & -13 \\ 4 & -7 \end{bmatrix}$ 7. $\begin{bmatrix} 30 & -39 \\ 29 & 14 \end{bmatrix}$ 9. $\begin{bmatrix} -5 & -1 \\ 8 & -1 \end{bmatrix}$

11. $\begin{bmatrix} -1 & 0 & -2 \\ 1 & 2 & 8 \\ 0 & 1 & 3 \end{bmatrix}$ 13. $\begin{bmatrix} 1 & 0 & 0 \\ 0 & 1 & 0 \\ 0 & 0 & 1 \end{bmatrix}$ 15. $\begin{bmatrix} 1 & 0 \\ -1 & 2 \end{bmatrix}$ 17. $\begin{bmatrix} 1 & -2 \\ 1 & 2 \end{bmatrix}$ 19. $\begin{bmatrix} -2 & 3 \\ 2 & -4 \end{bmatrix}$

21. $\begin{bmatrix} -1 & 1 \\ -1 & -1 \end{bmatrix}$ 23. $\begin{bmatrix} 1 & -1 \\ 1 & 0 \end{bmatrix}$

25. $A + B = \begin{bmatrix} 0 & 2 \\ -1 & 4 \end{bmatrix}$, $A - B = \begin{bmatrix} -2 & 2 \\ 1 & -1 \end{bmatrix}$, $A^2 = \begin{bmatrix} 1 & 0 \\ 0 & 1 \end{bmatrix}$, $B^2 = \begin{bmatrix} 1 & 0 \\ -3 & 4 \end{bmatrix}$, and $AB = \begin{bmatrix} -3 & 4 \\ -1 & 2 \end{bmatrix}$

a. $(A + B)(A + B) = \begin{bmatrix} 0 & 2 \\ -1 & 4 \end{bmatrix} \cdot \begin{bmatrix} 0 & 2 \\ -1 & 4 \end{bmatrix} = \begin{bmatrix} 2 & 8 \\ -4 & 14 \end{bmatrix}$

and $A^2 + 2AB + B^2 = \begin{bmatrix} 1 & 0 \\ 0 & 1 \end{bmatrix} + 2\begin{bmatrix} -3 & 4 \\ -1 & 2 \end{bmatrix} + \begin{bmatrix} 1 & 0 \\ -3 & 4 \end{bmatrix} = \begin{bmatrix} -4 & 8 \\ -5 & 9 \end{bmatrix}$;

hence, $(A + B)(A + B) \neq A^2 + 2AB + B^2$

b. $(A + B)(A - B) = \begin{bmatrix} 0 & 2 \\ -1 & 4 \end{bmatrix} \cdot \begin{bmatrix} -2 & 2 \\ 1 & -1 \end{bmatrix} = \begin{bmatrix} 2 & -2 \\ 6 & -6 \end{bmatrix}$ and $A^2 - B^2 = \begin{bmatrix} 1 & 0 \\ 0 & 1 \end{bmatrix} - \begin{bmatrix} 1 & 0 \\ -3 & 4 \end{bmatrix} = \begin{bmatrix} 0 & 0 \\ 3 & -3 \end{bmatrix}$;

hence, $(A + B)(A - B) \neq A^2 - B^2$

27. Let $A = \begin{bmatrix} a_{11} & a_{12} \\ a_{21} & a_{22} \end{bmatrix}$, $B = \begin{bmatrix} b_{11} & b_{12} \\ b_{21} & b_{22} \end{bmatrix}$, and $C = \begin{bmatrix} c_{11} & c_{12} \\ c_{21} & c_{22} \end{bmatrix}$

Since $AB = \begin{bmatrix} a_{11}b_{11} + a_{12}b_{21} & a_{11}b_{12} + a_{12}b_{22} \\ a_{21}b_{11} + a_{22}b_{21} & a_{21}b_{12} + a_{22}b_{22} \end{bmatrix}$,

$(AB)C = \begin{bmatrix} (a_{11}b_{11} + a_{12}b_{21})c_{11} + (a_{11}b_{12} + a_{12}b_{22})c_{21} & (a_{11}b_{11} + a_{12}b_{21})c_{12} + (a_{11}b_{12} + a_{12}b_{22})c_{22} \\ (a_{21}b_{11} + a_{22}b_{21})c_{11} + (a_{21}b_{12} + a_{22}b_{22})c_{21} & (a_{21}b_{11} + a_{22}b_{22})c_{12} + (a_{21}b_{12} + a_{22}b_{22})c_{22} \end{bmatrix}$

The element in the first row, first column is given by $a_{11}b_{11}c_{11} + a_{12}b_{21}c_{11} + a_{11}b_{12}c_{21} + a_{12}b_{22}c_{21} = (a_{11}b_{11}c_{11} + a_{11}b_{12}c_{21}) + (a_{12}b_{21}c_{11} + a_{12}b_{22}c_{21}) = a_{11}(b_{11}c_{11} + b_{12}c_{21}) + a_{12}(b_{21}c_{11} + b_{22}c_{21})$. When the remaining elements are treated in a similar manner

$(AB)C = \begin{bmatrix} a_{11}(b_{11}c_{11} + b_{12}c_{21}) + a_{12}(b_{21}c_{11} + b_{22}c_{21}) & a_{11}(b_{11}c_{12} + b_{12}c_{22}) + a_{12}(b_{21}c_{12} + b_{22}c_{22}) \\ a_{21}(b_{11}c_{11} + b_{12}c_{21}) + a_{22}(b_{21}c_{11} + b_{22}c_{21}) & a_{21}(b_{11}c_{12} + b_{12}c_{22}) + a_{22}(b_{21}c_{12} + b_{22}c_{22}) \end{bmatrix}$

$= \begin{bmatrix} a_{11} & a_{12} \\ a_{21} & a_{22} \end{bmatrix} \cdot \begin{bmatrix} b_{11}c_{11} + b_{12}c_{21} & b_{11}c_{12} + b_{12}c_{22} \\ b_{21}c_{11} + b_{22}c_{21} & b_{21}c_{12} + b_{22}c_{22} \end{bmatrix}$

$= \begin{bmatrix} a_{11} & a_{12} \\ a_{21} & a_{22} \end{bmatrix} \cdot \left(\begin{bmatrix} b_{11} & b_{12} \\ b_{21} & b_{22} \end{bmatrix} \cdot \begin{bmatrix} c_{11} & c_{12} \\ c_{21} & c_{22} \end{bmatrix} \right)$

$= A(BC).$

Odd-Numbered Answers

29. Let $A = \begin{bmatrix} a_{11} & a_{12} \\ a_{21} & a_{22} \end{bmatrix}$, $B = \begin{bmatrix} b_{11} & b_{12} \\ b_{21} & b_{22} \end{bmatrix}$, and $C = \begin{bmatrix} c_{11} & c_{12} \\ c_{21} & c_{22} \end{bmatrix}$. Then $B + C = \begin{bmatrix} b_{11}+c_{11} & b_{12}+c_{12} \\ b_{21}+c_{21} & b_{22}+c_{22} \end{bmatrix}$;

hence $(B+C)A = \begin{bmatrix} (b_{11}+c_{11})a_{11}+(b_{12}+c_{12})a_{21} & (b_{11}+c_{11})a_{12}+(b_{12}+c_{12})a_{22} \\ (b_{21}+c_{22})a_{11}+(b_{22}+c_{22})a_{21} & (b_{21}+c_{22})a_{12}+(b_{22}+c_{22})a_{22} \end{bmatrix}$. The element in the first row, first column is given by $b_{11}a_{11} + c_{11}a_{21} + b_{12}a_{21} + c_{12}a_{21} = (b_{11}a_{11} = (b_{12}a_{21}) + (c_{11}a_{21} + c_{12}a_{21})$. When the remaining elements are treated in a similar manner,

$$(B+C)A = \begin{bmatrix} (b_{11}a_{11}+b_{12}a_{21})+(c_{11}a_{11}+c_{12}a_{21}) & (b_{11}a_{12}+b_{12}a_{22})+(c_{11}a_{12}+c_{12}a_{22}) \\ (b_{21}a_{11}+b_{22}a_{21})+(c_{21}a_{11}+c_{22}a_{21}) & (b_{21}a_{12}+b_{22}a_{22})+(c_{21}a_{12}+c_{22}a_{22}) \end{bmatrix}$$

$$= \begin{bmatrix} b_{11}a_{11}+b_{12}a_{21} & b_{11}a_{12}+b_{12}a_{22} \\ b_{21}a_{11}+b_{22}a_{21} & b_{21}a_{12}+b_{22}a_{22} \end{bmatrix} + \begin{bmatrix} c_{11}a_{11}+c_{12}a_{21} & c_{11}a_{12}+c_{12}a_{22} \\ c_{21}a_{11}+c_{22}a_{21} & c_{21}a_{12}+c_{22}a_{22} \end{bmatrix}$$

$$= \begin{bmatrix} b_{11} & b_{12} \\ b_{21} & b_{22} \end{bmatrix} \cdot \begin{bmatrix} a_{11} & a_{12} \\ a_{21} & a_{22} \end{bmatrix} + \begin{bmatrix} c_{11} & c_{12} \\ c_{21} & c_{22} \end{bmatrix} \cdot \begin{bmatrix} a_{11} & a_{12} \\ a_{21} & a_{22} \end{bmatrix}$$

$$= BC + CA.$$

31. $(A_{2 \times 2} \cdot B_{2 \times 2}) = \begin{bmatrix} a_{11}b_{11}+a_{12}b_{21} & a_{11}b_{12}+a_{12}b_{22} \\ a_{21}b_{11}+a_{22}b_{21} & a_{21}b_{22}+a_{22}b_{22} \end{bmatrix}$

Hence, $(A_{2 \times 2} \cdot B_{2 \times 2})^t = \begin{bmatrix} a_{11}b_{11}+a_{12}b_{21} & a_{21}b_{11}+a_{22}b_{21} \\ a_{11}b_{12}+a_{12}b_{22} & a_{21}b_{12}+a_{22}b_{22} \end{bmatrix}$. By the commutative property of multiplication for real numbers,

$$(A_{2 \times 2} \cdot B_{2 \times 2})^t = \begin{bmatrix} b_{11}a_{11}+b_{21}a_{12} & b_{11}a_{21}+b_{21}a_{22} \\ b_{12}a_{11}+b_{22}a_{12} & b_{12}a_{21}+b_{22}a_{22} \end{bmatrix} = \begin{bmatrix} b_{11} & b_{21} \\ b_{12} & b_{22} \end{bmatrix} \cdot \begin{bmatrix} a_{11} & a_{21} \\ a_{12} & a_{22} \end{bmatrix} = B^t_{2 \times 2} \cdot A^t_{2 \times 2}$$

33. $-1/2$ 35. $11/16$

37. $AB = \begin{bmatrix} 1 & 3 & 1 & 2 \\ 1 & 1 & 1 & 2 \end{bmatrix}$; An entry of 1 means any of P_5, P_6, P_7 or P_8 has had only one secondary contact with the infected X or Y; an entry of 2 means two secondary contacts with X or Y, and so on.

39. $[500 \ 1000 \ 1500] + [25 \ 60 \ 105] \begin{bmatrix} 10 & 0 & 0 \\ 0 & 20 & 0 \\ 0 & 0 & 30 \end{bmatrix} = [750 \ 2200 \ 4650]$

	State sales tax	City sales tax
41. All amplifiers	$120	$20
All tuners	$144	$24
All speakers	$288	$48

43. $\begin{bmatrix} 950 & 475 & 380 \\ 760 & 190 & 475 \\ 1900 & 570 & 190 \\ 950 & 50 & 950 \end{bmatrix}$ 45. 385

Exercise 7.3. [page 245]

1. $\{(2, -1)\}$ 3. $\{(5, 1)\}$ 5. $\{(8, 1)\}$ 7. $\{(-2, 2, 0)\}$ 9. $\{(5/4, 5/2, -1/2)\}$

11. $\{(-77/27, -8/27, 29/27)\}$ 13. $\begin{bmatrix} k & 0 \\ 0 & 1 \end{bmatrix}\begin{bmatrix} a & b \\ c & d \end{bmatrix} = \begin{bmatrix} k\cdot a + 0\cdot c & k\cdot b + 0\cdot d \\ 0\cdot a + 1\cdot c & 0\cdot b + 1\cdot d \end{bmatrix} = \begin{bmatrix} ka & kb \\ c & d \end{bmatrix}$

15. $\begin{bmatrix} 0 & 1 \\ 1 & 0 \end{bmatrix}\begin{bmatrix} a & b \\ c & d \end{bmatrix} = \begin{bmatrix} 0\cdot a + 1\cdot c & 0\cdot b + 1\cdot d \\ 1\cdot a + 0\cdot c & 1\cdot b + 0\cdot d \end{bmatrix} = \begin{bmatrix} c & d \\ a & b \end{bmatrix}$

17. $\begin{bmatrix} 1 & 0 \\ k & 1 \end{bmatrix}\begin{bmatrix} a & b \\ c & d \end{bmatrix} = \begin{bmatrix} 1\cdot a + 0\cdot c & 1\cdot b + 0\cdot d \\ k\cdot a + 1\cdot c & k\cdot b + 1\cdot d \end{bmatrix} = \begin{bmatrix} a & b \\ ka+c & kb+d \end{bmatrix}$

19. 5 in A, 20 in B, 75 in C

Exercise 7.4. [page 249]

1. $M_{11} = \begin{vmatrix} 0 & 3 & -1 \\ 1 & 2 & 2 \\ -1 & 3 & 1 \end{vmatrix}, \quad A_{11} = \begin{vmatrix} 0 & 3 & -1 \\ 1 & 2 & 2 \\ -1 & 3 & 1 \end{vmatrix}$

3. $M_{23} = \begin{vmatrix} 2 & 1 & 0 \\ -2 & 1 & 2 \\ 1 & -1 & 1 \end{vmatrix}, \quad A_{23} = -\begin{vmatrix} 2 & 1 & 0 \\ -2 & 1 & 2 \\ 1 & -1 & 1 \end{vmatrix}$

5. $M_{31} = \begin{vmatrix} 1 & -2 & 0 \\ 0 & 3 & -1 \\ -1 & 3 & 1 \end{vmatrix}, \quad A_{31} = \begin{vmatrix} 1 & -2 & 0 \\ 0 & 3 & -1 \\ -1 & 3 & 1 \end{vmatrix}$

7. $M_{44} = \begin{vmatrix} 2 & 1 & -2 \\ 1 & 0 & 3 \\ -2 & 1 & 2 \end{vmatrix}, \quad A_{44} = \begin{vmatrix} 2 & 1 & -2 \\ 1 & 0 & 3 \\ -2 & 1 & 2 \end{vmatrix}$

9. 1   11. 0   13. -30 15. $3x+1$ 17. 3   19. 0   21. -1 23. 0   25. x^3
27. $x = 5$ 29. $x = 3$

31. Expanding by elements in the first row yields $\begin{vmatrix} 0 & 1 & 0 & 0 \\ 1 & 0 & 3 & 2 \\ 5 & -1 & 2 & 1 \\ 1 & 0 & 1 & 1 \end{vmatrix} = -\begin{vmatrix} 1 & 3 & 2 \\ 5 & 2 & 1 \\ 1 & 1 & 1 \end{vmatrix}$;

expanding by elements in the third row yields $-\begin{vmatrix} 3 & 2 \\ 2 & 1 \end{vmatrix} + \begin{vmatrix} 1 & 2 \\ 5 & 1 \end{vmatrix} - \begin{vmatrix} 1 & 3 \\ 5 & 2 \end{vmatrix} = 1 - 9 + 13 = 5.$

33. 2, 6, 24

35. Let $A = \begin{bmatrix} a_{11} & a_{12} \\ a_{21} & a_{22} \end{bmatrix}$ and $\delta(A) = a_{11}a_{22} - a_{12}a_{21}$. It follows that

$aA = \begin{bmatrix} aa_{11} & aa_{12} \\ aa_{21} & aa_{22} \end{bmatrix}$ and $\delta(aA) = a^2 a_{11}a_{22} - a^2 a_{12}a_{21} = a^2(a_{11}a_{22} - a_{12}a_{21}) = a^2 \delta(A)$.

37. Let $A = \begin{bmatrix} a_{11} & a_{12} \\ a_{21} & a_{22} \end{bmatrix}$ and $B = \begin{bmatrix} b_{11} & b_{12} \\ b_{21} & b_{22} \end{bmatrix}$; then $\delta(A) = a_{11}a_{22} - a_{12}a_{21}$

Odd-Numbered Answers

and $\delta(B) = b_{11}b_{22} - b_{12}b_{21}$, $AB = \begin{bmatrix} a_{11}b_{11} + a_{12}b_{21} & a_{11}b_{12} + a_{12}b_{22} \\ a_{21}b_{11} + a_{22}b_{21} & a_{21}b_{12} + a_{22}b_{22} \end{bmatrix}$

and $\delta(AB) = (a_{11}b_{11} + a_{12}b_{21})(a_{21}b_{12} + a_{22}b_{22}) - (a_{11}b_{12} + a_{12}b_{22})(a_{21}b_{11} + a_{22}b_{21})$, which simplifies to $a_{11}b_{11}a_{22}b_{22}$
$- a_{11}b_{12}a_{22}b_{21} - a_{12}b_{22}a_{21}b_{11} + a_{12}b_{21}a_{21}b_{12}$

$$= a_{11}a_{22}(b_{11}b_{22} - b_{12}b_{21}) - a_{12}a_{21}(b_{11}b_{22} - b_{12}b_{21}) = (a_{11}a_{22} - a_{12}a_{21})(b_{11}b_{22} - b_{12}b_{21})$$
$$= \delta(A) \cdot \delta(B).$$

Exercise 7.5. [page 257]

1. $\begin{bmatrix} 3 & -2 \\ -1 & 1 \end{bmatrix}$ **3.** $\frac{1}{5}\begin{bmatrix} 1 & 3 \\ -1 & 2 \end{bmatrix}$ **5.** $\delta(A) = 0$; no inverse

7. $\frac{1}{6}\begin{bmatrix} 2 & 2 & -5 \\ -4 & 2 & 1 \\ 0 & 0 & 3 \end{bmatrix}$ **9.** $\frac{1}{3}\begin{bmatrix} -2 & -3 & -1 \\ -1 & 0 & 1 \\ 6 & -6 & 3 \end{bmatrix}$ **11.** $\delta(A) = 0$; no inverse **13.** $-1\begin{bmatrix} 4 & -7 \\ -3 & 5 \end{bmatrix}$

15. $\frac{1}{2}\begin{bmatrix} -2 & -4 \\ 4 & 7 \end{bmatrix}$ **17.** No inverse **19.** No inverse **21.** $\begin{bmatrix} 2 & -3 & 11 \\ -2 & 4 & -13 \\ 1 & -2 & 7 \end{bmatrix}$

23. $-1\begin{bmatrix} 0 & 0 & -1 \\ 0 & -1 & 0 \\ -1 & 0 & 0 \end{bmatrix}$

25. Let $A \cdot B = \begin{bmatrix} 2 & 3 \\ 1 & -1 \end{bmatrix} \cdot \begin{bmatrix} 0 & 1 \\ 3 & 1 \end{bmatrix}$; $A \cdot B = \begin{bmatrix} 9 & 5 \\ -3 & 0 \end{bmatrix}$ and $\delta(AB) = 15$,

so $(A \cdot B)^{-1} = \frac{1}{15}\begin{bmatrix} 0 & -5 \\ 3 & 9 \end{bmatrix}$; also, since $\delta(A) = -5$ and $\delta(B) = -3$, $B^{-1} = -\frac{1}{3}\begin{bmatrix} 1 & -1 \\ -3 & 0 \end{bmatrix}$

and $A^{-1} = \frac{1}{5}\begin{bmatrix} -1 & -3 \\ -1 & 2 \end{bmatrix}$; $B^{-1} \cdot A^{-1} = \frac{1}{15}\begin{bmatrix} 0 & -5 \\ 3 & 9 \end{bmatrix}$; hence $(A \cdot B)^{-1} = B^{-1} \cdot A^{-1}$.

27. Let $A = \begin{bmatrix} 3 & 0 & 1 \\ 2 & 1 & 0 \\ 0 & 1 & 2 \end{bmatrix}$; then $\delta(A) = 8$ and $A^{-1} = \frac{1}{8}\begin{bmatrix} 2 & 1 & -1 \\ -4 & 6 & 2 \\ 2 & -3 & 3 \end{bmatrix}$.

Let $B = \begin{bmatrix} 2 & 1 & 0 \\ 1 & 1 & 2 \\ 0 & 1 & 0 \end{bmatrix}$; then $\delta(B) = -1/4$ and $B^{-1} = -\frac{1}{4}\begin{bmatrix} -2 & 0 & 2 \\ 0 & 0 & -4 \\ 1 & -2 & 1 \end{bmatrix}$;

$A \cdot B = \begin{bmatrix} 6 & 4 & 0 \\ 5 & 3 & 2 \\ 1 & 3 & 2 \end{bmatrix}$, $\delta(A \cdot B) = -32$, and $(A \cdot B)^{-1} = -\frac{1}{32}\begin{bmatrix} 0 & -8 & 8 \\ -8 & 12 & -12 \\ 12 & -14 & -2 \end{bmatrix}$;

$B^{-1} \cdot A^{-1} = -\frac{1}{4}\begin{bmatrix} -2 & 0 & 2 \\ 0 & 0 & -4 \\ 1 & -2 & 1 \end{bmatrix} \cdot \frac{1}{8}\begin{bmatrix} 2 & 1 & -1 \\ -4 & 6 & 2 \\ 2 & -3 & 3 \end{bmatrix} = -\frac{1}{32}\begin{bmatrix} 0 & -8 & 8 \\ -8 & 12 & -12 \\ 12 & -14 & -2 \end{bmatrix}$;

hence $(A \cdot B)^{-1} = B^{-1} \cdot A^{-1}$.

29. Let $A = \begin{bmatrix} a_{11} & a_{12} \\ a_{21} & a_{22} \end{bmatrix}$; then $\delta(A) = (a_{11}a_{22} - a_{12}a_{21}) \neq 0$, since A is nonsingular, and hence $\dfrac{1}{\delta(A)}$ is defined and A^{-1} exists

$$A^{-1} = \frac{1}{\delta(A)} \begin{bmatrix} a_{22} & -a_{12} \\ -a_{21} & a_{11} \end{bmatrix} = \begin{bmatrix} \dfrac{a_{22}}{\delta(A)} & \dfrac{-a_{12}}{\delta(A)} \\ \dfrac{-a_{21}}{\delta(A)} & \dfrac{a_{11}}{\delta(A)} \end{bmatrix}$$

$$\delta(A^{-1}) = \frac{a_{22}a_{11}}{[\delta(A)]^2} - \frac{a_{12}a_{21}}{[\delta(A)]^2} = \frac{a_{22}a_{11} - a_{12}a_{21}}{[\delta(A)]^2} = \frac{\delta(A)}{[\delta(A)]^2} = \frac{1}{\delta(A)}$$

Exercise 7.6. [page 263]

1. $\{(1, 1)\}$ **3.** $\{(2, 2)\}$ **5.** $\{(6, 4)\}$ **7.** $\{(1, 1, 1)\}$ **9.** $\{(1, 1, 0)\}$ **11.** $\{(1, -2, 3)\}$

13. $\{(3, -1, -2)\}$ **15.** $X = \begin{bmatrix} 40 \\ 80 \end{bmatrix}$ **17.** $X = \begin{bmatrix} 72 \\ 104 \\ 40 \end{bmatrix}$ **19.** $X = \begin{bmatrix} 5 \\ 10 \\ 4 \end{bmatrix}$

Exercise 7.7. [page 269]

1. $\{(13/5, 3/5)\}$ **3.** $\{(22/7, 20/7)\}$ **5.** $\{(6, 4)\}$ **7.** Inconsistent **9.** $\{(4, 1)\}$

11. $\left\{\left(\dfrac{1}{a+b}, \dfrac{1}{a+b}\right)\right\}$ $(a \neq -b)$ **13.** $\{(1, 1, 0)\}$ **15.** $\{(1, -2, 3)\}$ **17.** $\{(3, -1, -2)\}$

19. $\{(-1/3, -25/24, -5/8)\}$ **21.** $\{(1, -1/3, 1/2)\}$ **23.** $\{(w, x, y, z)\} = \{(2, -1, 1, 0)\}$

25. $A = \begin{vmatrix} a_1 & b_1 \\ a_2 & b_2 \end{vmatrix}$ and $\delta(A) = a_1 b_2 - a_2 b_1$; $A_y = \begin{vmatrix} a_1 & c_1 \\ a_2 & c_2 \end{vmatrix}$ and $\delta(A_y) = a_1 c_2 - a_2 c_1 = 0$, so $a_1 c_2 = a_2 c_1$;

$A_x = \begin{vmatrix} c_1 & b_1 \\ c_2 & b_2 \end{vmatrix}$ and $\delta(A_x) = b_2 c_1 - b_1 c_2 = 0$, so $b_1 c_2 = b_2 c_1$;

hence, $a_1 c_2 / b_1 c_2 = a_2 c_1 / b_2 c_1$; $a_1/b_1 = a_2/b_2$ and $a_1 b_2 = a_2 b_1$, so $a_1 b_2 - a_2 b_1 = 0$; therefore $\delta(A) = 0$

27. 5 ounces of X, 3 ounces of Y, 10 ounces of Z

Chapter 7 Test [page 271]

1. $\begin{bmatrix} 1 & -1 \\ 1 & 1 \end{bmatrix}$ **2.** $\begin{bmatrix} 2 & 5 & -2 \\ 14 & -1 & 12 \end{bmatrix}$ **3.** $\begin{bmatrix} -2 & -3 & 1 \\ 3 & -7 & 6 \\ -5 & 2 & 5 \end{bmatrix}$ **4.** $\begin{bmatrix} -8 & -3 & -6 \\ 6 & -3 & 0 \\ -9 & 5 & 5 \end{bmatrix}$ **5.** $\begin{bmatrix} -21 & 7 \\ -14 & 0 \\ -7 & -7 \end{bmatrix}$

6. $[13]$ **7.** $\begin{bmatrix} -13 & 3 \\ -19 & 27 \end{bmatrix}$ **8.** $\begin{bmatrix} 18 & 7 & 25 \\ 8 & -1 & 11 \\ 3 & 0 & 3 \end{bmatrix}$ **9.** $\{(2, -1)\}$ **10.** $\{(2, 1, 1)\}$ **11.** -3

Odd-Numbered Answers

12. 7 **13.** −3 **14.** 14 **15.** $\dfrac{1}{34}\begin{bmatrix} -3 & 2 \\ 11 & 4 \end{bmatrix}$ **16.** $\dfrac{1}{6}\begin{bmatrix} 1 & 3 & -2 \\ -3 & -3 & 6 \\ 1 & -3 & 4 \end{bmatrix}$

17. $\{(-2, 1)\}$ **18.** $\{(3, -1, 1)\}$ **19.** $X = \begin{bmatrix} 125 \\ 275 \end{bmatrix}$ **20.** $X = \begin{bmatrix} 58 \\ 44 \\ 55 \end{bmatrix}$

21. $\{(-16/7, -13/7)\}$ **22.** $\{(2, -1, 0)\}$

Exercise 8.1. [page 277]

1.   **3.** 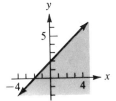 **5.** 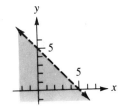 **7.**

9. 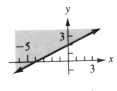 **11.** 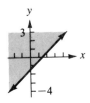 **13.**   **15.**

17.   **19.** 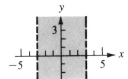 **21.**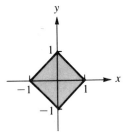

Exercise 8.2. [page 280]

1. 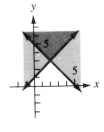 **3.** 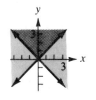 **5.** 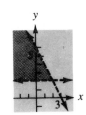 **7.**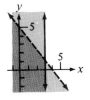

9.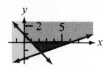

11.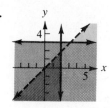

13.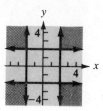

15. ∅

17.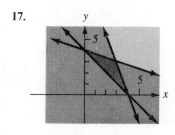

19. $x \geq 0, y \geq 0$
$x \geq 20$
$y \geq 30$
$x + y \leq 100$
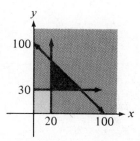

21. $x \geq 0, y \geq 0$
$y \geq 2x$
$x + y \leq 50$
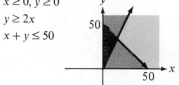

23. $x \geq 0, y \geq 0$
$\frac{3}{4}x + \frac{1}{4}y \leq 120$
$\frac{1}{2}x + \frac{1}{2}y \leq 140$
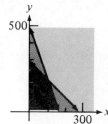

Exercise 8.3 [page 285]

1. Vertices: (0, 2), (2, 0), (2, 4), (4, 2)

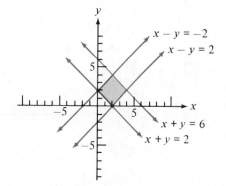

3. Vertices: (0,0), (0, 4), (2, 4), (10/3, 10/3), (5, 0)

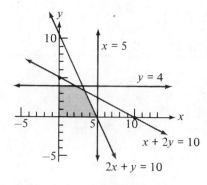

Odd-Numbered Answers

5. Vertices: (0, 4), (3, 6), (5, 4), (5, 2), (0, −3)   **7.** No   **9.** Yes

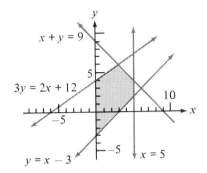

11. The half-planes $x > 1$ and $x < 0$ are convex, but their union is not

Exercise 8.4. [page 289]

1. Maximum, 14; minimum, 2   **3.** Maximum, 80/3; minimum, 0   **5.** Maximum, 17; minimum, −16

7. Maximum, 36 at (7, 3); minimum, 0 along the line segment joining (1, 3) and (2, 6)

9. Minimum, 20 at (20/7, 12/7)   **11.** Maximum profit, $830   **13.** Maximum profit, $790

15. 200 standard models and 400 deluxe models

Exercise 8.5. [page 296]

1. $F = 70$ at (10, 10)   **3.** $F = 525$ at (2.5, 10)   **5.** $F = 5200$ at (6, 4)   **7.** $F = 13$ at (2, 3, 0)

9. Maximum number $= 96$; 64 of X, 32 of Y, 0 of Z

11. Maximum profit $P = \$140$; 6 of tool A and 2 of tool B

13. Maximum number of passengers $= 5700$, for 6 707's, 9 747's, and 3 DC-10's

Exercise 8.6. [page 301]

1. Maximize $F = 18x_1 + 10x_2$
subject to $9x_1 + x_2 \leq 9$
$x_1 + 2x_2 \leq 27$

3. Minimize $G = 75y_1 + 45y_2 + 90y_3$
subject to $y_1 + y_2 + 3y_3 \geq 10$
$y_1 + 3y_2 + 2y_3 > 10$
$3y_1 + y_2 + y_3 \geq 12$

5. Minimum of $G = 31$ at $y_1 = 1/2$, $y_2 = 7/2$

7. a. Minimize $G = 12y_1 + 45y_2 + 24y_3$
subject to $y_1 + 2y_2 + 3y_3 \geq 22$
$y_1 + 5y_2 + y_3 \geq 40$
$y_1 \geq 0, y_2 \geq 0$

b. Maximize $F = 22x_1 + 40x_2$
subject to
$x_1 + x_2 \leq 12$
$2x_1 + 5x_2 \leq 45$
$3x_1 + x_2 \leq 24$
$x_1 \geq 0, x_2 \geq 0$

c. Minimum of $G = 390$

9. Minimum cost = $3500 when factory A runs 5/3 hours and factory B runs 20/3 hours

Chapter 8 Test [page 304]

1.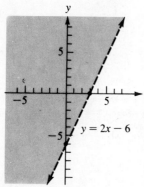

2.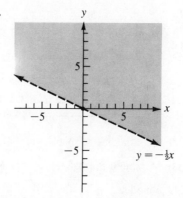

3.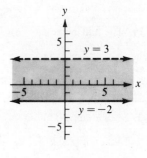

4.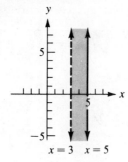

5.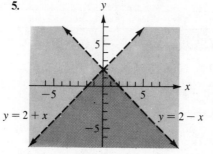

6.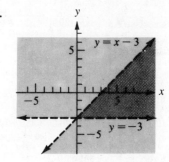

7.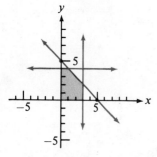

8.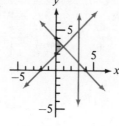

Odd-Numbered Answers

9. Minimum, 0; maximum, 8   **10.** Minimum, 0; maximum, 11
11. Maximum profit, $1100   **12.** Maximum profit, $210   **13.** $F = 200$ at $(4, 4)$
14. Maximize $F = 12x_1 + 6x_2$ **15.** Minimum of $G = 70$ at $y_1 = 4, y_2 = 2$
subject to $3x_1 + x_2 \le 1$
$4x_1 + 3x_2 \le 1$
$x_1 \ge 0, x_2 \ge 0$

Exercise 9.1. [page 311]

1. $-4, -3, -2, -1$ **3.** $-1/2, 1, 7/2, 7$ **5.** $2, 3/2, 4/3, 5/4$ **7.** $0, 1, 3, 6$ **9.** $-1, 1, -1, 1$
11. $1, 0, -1/3, 1/2$ **13.** $11, 15, 19$ **15.** $x + 2, x + 3, x + 4$ **17.** $2x + 7, 2x + 10, 2x + 13$
19. $32, 128, 512, 2048$ **21.** $8/3, 16/3, 32/3, 64/3$ **23.** $x/a, -x^2/a^2, x^3/a^3, -x^4/a^4$ **25.** $4n + 3, 31$
27. $-5n + 8, -92$ **29.** $48(2)^{n-1}, 1536$ **31.** $-\frac{1}{3}(-3)^{n-1}, -243$ **33.** $2, 3, 41$
35. Twenty-eighth   **37.** 3   **39.** $20{,}000\,(1.3)^n$ **41.** 32   **43.** $P_n = -500 + (n-1)50$
45. $34.50; $n \le 30$ **47.** $25{,}000\,(0.8)^n$ **49.** 1/2 mile

Exercise 9.2. [page 318]

1. $1 + 4 + 9 + 16$ **3.** $-\frac{1}{2} + \frac{1}{4} - \frac{1}{8}$ **5.** $1 + \frac{1}{2} + \frac{1}{4} + \cdots$ **7.** $\sum_{j=1}^{4} x^{2j-1}$ **9.** $\sum_{j=1}^{5} j^2$
11. $\sum_{j=1}^{\infty} j(j+1)$ **13.** $\sum_{j=1}^{\infty} \frac{j+1}{j}$ **15.** 63   **17.** 806   **19.** -6 **21.** 1092   **23.** 31/32
25. 364/729   **27.** 168   **29.** $3/4, 7/8, 15/16, 31/32; 1$ **31.** $p = 4; q = -3$ **33.** 196 bricks   **35.** 7260
37. 42.8 milligrams   **39.** $615

Exercise 9.3. [page 325]

1. $\lim_{n \to \infty} s_n = 0$ **3.** $\lim_{n \to \infty} s_n = 1$ **5.** $\lim_{n \to \infty} s_n$ is undefined   **7.** $\lim_{n \to \infty} s_n = 0$
9. Convergent, $\lim_{n \to \infty} \left| 0 - \frac{1}{2^n} \right| = 0$ **11.** Divergent, $\lim_{n \to \infty} n$ is undefined
13. Convergent, $\lim_{n \to \infty} \left| 0 - (-1)^{n-1} \frac{1}{2^{n-1}} \right| = 0$ **15.** 24   **17.** No sum   **19.** 2   **21.** 31/99
23. 2408/999   **25.** 29/225   **27.** 31.25 milligrams   **29.** $800,000   **31.** 30 feet

Exercise 9.4. [page 331]

1. $8 \cdot 7 \cdot 6 \cdot 5 \cdot 4 \cdot 3 \cdot 2 \cdot 1$ **3.** $6 \cdot 5 \cdot 4 \cdot 3 \cdot 2 \cdot 1$ **5.** $5 \cdot 4 \cdot 3 \cdot 2 \cdot 1 = 120$ **7.** $\dfrac{9 \cdot 8 \cdot 7!}{7!} = 72$

9. $\dfrac{5 \cdot 4 \cdot 3 \cdot 2 \cdot 1 \cdot 7}{8 \cdot 7!} = 15$ **11.** $\dfrac{8 \cdot 7 \cdot 6!}{2 \cdot 1 \cdot 6!} = 28$ **13.** $3!$ **15.** $\dfrac{6!}{2!}$ **17.** $\dfrac{8!}{5!}$ **19.** $\dfrac{6!}{5!\,1!} = 6$

21. $\dfrac{3!}{3!\,0!} = 1$ **23.** $\dfrac{7!}{0!\,7!} = 1$ **25.** $\dfrac{5!}{2!\,3!} = 10$ **27.** $(n)(n-1)(n-2)\cdots 3 \cdot 2 \cdot 1$

29. $(3n)(3n-1)(3n-2)\cdots 3 \cdot 2 \cdot 1$ **31.** $(n-2)(n-3)(n-4)\cdots 3 \cdot 2 \cdot 1$ **33.** $(n+2)(n+1)$

35. $\dfrac{n+1}{n+3}$ **37.** $\dfrac{2n-1}{2n-2}$ **39.** $x^5 + 15x^4 + 90x^3 + 270x^2 + 405x + 243$

41. $x^4 - 12x^3 + 54x^2 - 108x + 81$ **43.** $8x^3 - 6x^2y + \dfrac{3}{2}xy^2 - \dfrac{1}{8}y^3$

45. $\dfrac{1}{64}x^6 + \dfrac{3}{8}x^5 + \dfrac{15}{4}x^4 + 20x^3 + 60x^2 + 96x + 64$

47. $x^{20} + 20x^{19}y + \dfrac{20 \cdot 19}{2!}x^{18}y^2 + \dfrac{20 \cdot 19 \cdot 18}{3!}x^{17}y^3 + \cdots$ or

$\dbinom{20}{0}x^{20} + \dbinom{20}{1}x^{19}y + \dbinom{20}{2}x^{18}y^2 + \dbinom{20}{3}x^{17}y^3 + \cdots$

49. $a^{12} + 12a^{11}(-2b) + \dfrac{12 \cdot 11}{2!}a^{10}(-2b)^2 + \dfrac{12 \cdot 11 \cdot 10}{3!}a^9(-2b)^3 + \cdots$ or

$\dbinom{12}{0}a^{12} + \dbinom{12}{1}a^{11}(-2b) + \dbinom{12}{2}a^{10}(-2b)^2 + \dbinom{12}{3}a^9(-2b)^3 + \cdots$

51. $x^{10} + 10x^9(-\sqrt{2}) + \dfrac{10 \cdot 9}{2!}x^8(-\sqrt{2})^2 + \dfrac{10 \cdot 9 \cdot 8}{3!}x^7(-\sqrt{2})^3 + \cdots$ or

$\dbinom{10}{0}x^{10} + \dbinom{10}{1}x^9(-\sqrt{2}) + \dbinom{10}{2}x^8(-\sqrt{2})^2 + \dbinom{10}{3}x^7(-\sqrt{2})^3 + \cdots$

53. 1.22 **55.** $-3003a^{10}b^5$ **57.** $3360x^6y^4$ **59. a.** $1 - x + x^2 - x^3 + \cdots$ **b.** $1 - x + x^2 - x^3 + \cdots$

61. $1, 5, 10, 10, 5, 1$ **63.** $\$1480$

Chapter 9 Test [page 336]

1. $13, 16, 19$ **2.** $a-4, a-6, a-8$ **3.** $-18, 54, -162$ **4.** $3/2, 9/4, 27/8$

5. $s_n = 5n - 8;\ s_7 = 27$ **6.** $s_n = (-2)(-1/3)^{n-1};\ s_5 = -2/81$ **7.** 25 **8.** Sixth term

9. $2 + 6 + 12 + 20$ **10.** $\displaystyle\sum_{k=1}^{\infty} x^{k+1}$ **11.** 119 **12.** $121/243$ **13.** 3 **14.** $8/3$ **15.** $1/2$

Odd-Numbered Answers

16. 4/9   **17.** $5 \cdot (2 \cdot 1)$ **18.** 48   **19.** 21   **20.** $1/n(n+1)!$ **21.** $x^{10} - 20x^9 y + 180x^8 y^2 - 960x^7 y^3$
22. $-15,360 x^3 y^7$ **23.** gift c.

Exercise 10.1. [page 342]

1. 1; 5; 8   **3.** 2; 6; 16   **5.** 2; 2; 4   **7.** 4   **9.** 24   **11.** 16   **13.** 64   **15.** 24   **17.** 216
19. 375   **21.** 30   **23.** 10   **25.** 48   **27.** $\dfrac{5!}{2!}$ or 60   **29.** $\dfrac{8!}{3!}$ or 6720
31. $P_{5,3} = \dfrac{5!}{2!} = \dfrac{5 \cdot 4!}{2!} = 5\left(\dfrac{4!}{2!}\right) = 5\,(P_{4,2})$ **33.** $P_{n,3} = \dfrac{n!}{(n-3)!} = \dfrac{n(n-1)!}{(n-3)!} = n\left[\dfrac{(n-1)!}{(n-3)!}\right] = n(P_{n-1,2})$ **35.** 9
37. 24   **39.** 48   **41.** 400   **43.** 120   **45.** 1,860,480   **47.** 17,576,000

Exercise 10.2. [page 346]

1. 7   **3.** 15   **5.** $C_{52,5}$ **7.** $C_{13,5} \cdot C_{13,5} \cdot C_{13,3}$ **9.** $4\,C_{13,5}$ **11.** 164   **13.** 10   **15.** 210
17. 12   **19.** 3003   **21.** 5400   **23.** 840

Exercise 10.3. [page 350]

1. $\{1, 2, 3, 4, 5, 6\}$; $\{3, 4, 5, 6\}$; 2/3   **3.** $\{(H, H), (H, T), (T, H), (T, T)\}$; $\{(H, H),(T, T)\}$; 1/2   **5.** 1/6
7. 1/18   **9.** 5/9   **11.** 1/52   **13.** 3/26   **15.** 1/17   **17.** 11/221   **19.** 1/190   **21.** 3/38
23. 10/21   **25.** $\{$BBB, GGG, BGG, GBG, GGB, BBG, BGB, GBB$\}$ **27.** 15/28   **29.** 3/10
31. 1/250   **33.** 1/6

Exercise 10.4. [page 356]

1. 1/6   **3.** 7/8   **5.** 13/18   **7.** 5/33   **9.** 1/11   **11.** 5/22   **13.** 14/33   **15.** 15/22   **17.** 3/13
19. $2.14; less   **21.** 5.3 cents   **23.** $3.29   **25.** 11/50   **27.** 3/5   **29.** 0.86   **31.** 5/8   **33.** 39/50
35. 4/5

Exercise 10.5. [page 361]

1. 20/91; no   **3. a.** 2/45   **b.** 28/75   **c.** 4/225   **d.** 1/9   **5. a.** 11/36   **b.** 5/36   **c.** No
7. a. 1/2   **b.** 1/2   **c.** 1/4   **d.** 1/4   **e.** 1/8   **9.** 1/4   **11. a.** 15/77   **b.** 16/77   **c.** 46/77
13. 1/16   **15.** 1/36   **17.** 1/8   **19.** 0.054   **21.** 19/27   **23.** 0.9025   **25. a.** 71/72   **b.** 5/9
c. 5/36   **d.** 61/72   **27. a.** 1/210   **b.** 29/210   **c.** 29/70

Exercise 10.6. [page 367]

1. 500/1296   **3.** 150/1296   **5.** 171/1296   **7.** 15/64   **9.** 7/64   **11.** 5/16   **13.** 98,415/1,000,000
15. 32/81   **17. a.** 4096/1,000,000   **b.** 27,648/100,000   **c.** 45,568/100,000   **d.** 46,656/1,000,000
19. 6196/1,048,576

Exercise 10.7. [page 370]

1. 2/11   **3.** 29/30   **5.** 8/15   **7.** 4/5   **9.** 33/2550; 6/51   **11.** 1/6   **13.** 5/9   **15.** 63/65
17. a. 1/10   **b.** 9/10   **c.** 3/5   **d.** 3/8   **e.** 1/40   **f.** 1/6   **g.** 3/4   **h.** 21/22   **19.** 0.48, 0.24
21. 0.75

Chapter 10 Test [page 373]

1. 16   **2.** 64   **3.** 128   **4.** 360   **5.** 792   **6.** 1,033,885,600   **7.** 200   **8.** 1/17   **9.** 1/221
10. 10/17   **11.** 25/102   **12.** 26/51   **13.** 80/221   **14.** 41/663   **15.** 8/7   **16.** 51/2704
17. 69/676   **18.** 20/221   **19.** 295/663   **20.** 0.015   **21.** 1/4   **22.** 27/128; 3/8

Exercise 11.1. [page 379]

1. $5920   **3.** $1800   **5.** $6404   **7.** $3581.60   **9.** $12,205.40   **11.** $2364.09   **13.** 5 years
15. 12%   **17.** $157,920   **19.** 6.14%   **21.** $4008.89   **23.** $15,283.50
25. $80,000 at 6%; $90,000 at 8%

Exercise 11.2. [page 383]

1. $2097.90   **3.** $25,304.85   **5.** $183,928   **7.** $164,593.50   **9.** $174,469.11   **11.** 42 years
13. $126   **15.** $489.44   **17.** $291.19

Exercise 11.3. [page 387]

1. $38,608.50   **3.** $75,948   **5.** $84,579.40; $84,579.76   **7.** $2237.77   **9.** 21   **11.** $41.70
13. 17 years   **15.** $1000

Exercise 11.4. [page 392]

1. $1491.80   **3.** $7510.50   **5.** $23,620   **7.** $1642   **9.** $4257.30   **11.** approximately $18\frac{1}{3}$ years
13. approximately 7%   **15.** $1136.60   **17.** $27,183; $26,916   **19.** $18,771

Odd-Numbered Answers

21.

Future Value S
631.24
633.39
634.49
635.24
635.54
635.61
635.62
635.62

Chapter 11 Test [page 394]

1. $5970.50; $5978 **2.** 10 years **3. a.** $21,600 **b.** $33,600 **c.** $186.67 **4.** $5637.10; $5975.30
5. $52,239.30 **6.** $109,378.65 **7.** $242.76 **8.** $3309.38 **9.** $3980.24 **10.** 25

Exercise 12.1. [page 403]

1. 4 **3.** −8 **5.** 6 **7.** 1 **9.** −6 **11.** 12 **13.** −1 **15.** $y = -6x - 9$ **17.** $y = 7x - 1$
19. $10x$ **21.** 60 **23.** 80 **25.** 16; 32

Exercise 12.2. [page 407]

1. $6x$ **3.** $2x$ **5.** $2x - 6$ **7.** $4x + 1$ **9.** $-2x + 10$ **11.** $3x^2$ **13.** $3x^2 + 1$ **15.** $-1/x^2$
17. $-2/x^3$ **19.** $1/(x+1)^2$ **21.** $1/4$ **23.** $y = 2x + 2$ **25.** $y = 11x - 12$ **27.** 5 **29.** −8

Exercise 12.3. [page 412]

1. 0 **3.** $4x^3$ **5.** $\frac{1}{3}x^{-2/3}$ **7.** $\frac{3}{2}x^{1/2}$ **9.** $-2/x^3$ **11.** $-2x^{-3/2}$ **13.** $0.52x^{-0.48}$ **15.** 2
17. $2x$ **19.** $4x + 5$ **21.** $2x - 3$ **23.** $-0.2x + 0.5$ **25.** $12x^2 - 8x$ **27.** $x^2 - 10x + 11$
29. $-1/x^2$ **31.** $3x^{1/2} - 4x^{-3/2}$ **33.** $18x + 6$ **35.** $y = 2x - 1$ **37.** $y = 11x - 16$ **39.** $y = \frac{1}{2}x + 9$
41. $(3, -9)$ **43.** $(0, 0), (2, -4)$ **45.** $\left(1, \frac{5}{6}\right), \left(2, \frac{2}{3}\right)$ **47.** $(3, 12)$

Exercise 12.4. [page 417]

1. $(5x + 1)7 + (7x - 2)5$ **3.** $x^{1/3}(6x - 4) + (3x^2 - 4x + 6)x^{-2/3}/3$ **5.** $(3x^2 + 1)4x + (2x^2 - 4)6x = 24x^3 - 20x$

7. $(4x+10)(2x+11)+(x^2+11x-7)4$ 9. $(\sqrt{x}+6)x^{-1/2}/2+(\sqrt{x}-4)x^{-1/2}/2 = 1 - x^{-1/2}$
11. $(8x^3-2x^2)(-6x-6)+(-3x^2-6x+3)(24x^2-4x)$
13. $(2x^4+2x^{1/2}+x)(4x^3-2x^{-1/2})+(x^4-4x^{1/2}+1)(8x^3+x^{-1/2}+1)$ 15. $2(x^2+5x)(2x+5)$
17. $-2/(2x+1)^2$ 19. $1/(x+1)^2$ 21. $(-2x^2+4x)/4x^4 = (-x+2)/2x^3$ 23. $-2x/(5x^2-1)^2$
25. $[(x^2-x+1)(2x+8)-(x^2+8x-3)(2x-1)]/(x^2-x+1)^2$ 27. $1/2\sqrt{x}(\sqrt{x}+1)^2$
29. $-30x^2/(9x^3-1)^2$ 31. $-1/(x+2)^2$ 33. $(2x+1)(24x+7)+2(3x+1)(4x+1)$
35. $(2x^2+4)[(x^3-1)(32x^3+9)+(8x^4+9x)3x^2]+(x^3-1)(8x^4+9x)4x$
37. $[(2x+3)(12x+4)-2(x+1)(6x-2)]/(2x+3)^2$
39. $[2x(x^2+1)(x^3-4)-x^2(5x^4+3x^2-8x)]/(x^2+1)^2(x^3-4)^2$ 41. $y=16x-10$ 43. $y=-12x+8$

Exercise 12.5. [page 420]

1. $2(x+1)$ 3. $2(4x-4)^{-1/2}$ 5. $x(x^2+1)^{-1/2}$ 7. $3(7x^2-x)^2(14x-1)$ 9. $2(-x+2)^{-3}$
11. $0.021(0.1x+9)^{-0.79}$ 13. $10(-3x^3+x^2+2x)^9(-9x^2+2x+2)$ 15. $2(6x-1)^{-2/3}$
17. $\frac{8}{3}(4x+1)^{-1/3}$ 19. $-12x^2(x^3-1)^{-5}$ 21. $-12x(x^2-1)^{-5/3}$ 23. $\frac{1}{2}x(x+1)^{-1/2}+(x+1)^{1/2}$
25. $\frac{1}{2}[(5x+1)(2x+3)]^{-1/2}(20x+17)$ 27. $x(2x+1)(x^2-2)^{-1/2}+2\sqrt{x^2-2}$ 29. $(2+x)/2(x+1)^{3/2}$
31. $[x(x^2-1)(x^2+1)^{-1/2}-2x\sqrt{x^2-1}]/(x^2-1)^2$ 33. $\frac{1}{2}\left[\frac{x}{x+1}\right]^{-1/2}\frac{1}{(x+1)^2}$
35. $\sqrt{2x+1}(3x+1)^{-2/3}+\sqrt[3]{3x+1}(2x+1)^{-1/2}$ 37. $y=8x-7$ 39. $y=\frac{3}{2}x+1$

Exercise 12.6. [page 424]

1. $-4x$ 3. $1/y$ 5. $-x/y$ 7. $3x/4y$ 9. $1/(3y^2+2y+1)$ 11. $y/(y^2-x)$
13. $(6-3x^2-2xy^2)/2x^2y$ 15. $-y^{1/2}/x^{1/2}$ 17. $x/2y(1+y^2)$ 19. $-1/y(x-1)^2$
21. $y=\frac{2}{5}x+\frac{21}{5}, y=-\frac{2}{5}x-\frac{21}{5}$ 23. $y=-\frac{4}{3}x+\frac{25}{3}, y=\frac{4}{3}x-\frac{25}{3}$ 25. $y=-4x+8$
27. $y=-\frac{1}{3}x+3$ 29. $y=-4x+1$

Exercise 12.7. [page 427]

1. $12x^2$ 3. $-\frac{3}{2}x^{-3/2}$ 5. 10   7. $12x$ 9. $4/x^3$ 11. $-2x^{-5/3}-x^{-3/2}$ 13. $2/(x+1)^3$
15. $108(3x+7)^2$ 17. $4(1-x)^{-3}$ 19. $-\frac{9}{4}x^4(x^3+1)^{-3/2}+3x(x^3+1)^{-1/2}$ 21. 0   23. $\frac{40}{3}x^{-8/3}$

Odd-Numbered Answers

25. 24 **27.** $6(1-x)^{-4}$ **29.** $6/(x+1)^4$ **31.** $-(x^2+2y^2)/4y^3 = -1/8y^3$
33. $-[y^2+(1-x)^2]/y^3 = -1/y^3$

Exercise 12.8. [page 432]

1. $1/x$ **3.** $2/(2x+1)$ **5.** $(2x+1)/(x^2+x)$ **7.** $3/x$ **9.** $2/x$
11. $[1+x(x^2+1)^{-1/2}]/(x+\sqrt{x^2+1}) = 1/\sqrt{x^2+1}$ **13.** $3(\ln x)^2/x$ **15.** $2/x + 6/(2x-1)$ **17.** $-1/x$
19. $x + 2x \ln x$ **21.** $(1-\ln x)/x^2$ **23.** $-2xy/(4y-1)$ **25.** $-1/x^2$ **27.** $(2-2x^2)/(x^2+1)^2$
29. $1/x \ln x$ **31.** $y = 5x$ **33.** $y = \dfrac{2}{e}x$ **35.** $x^x[1+\ln x]$ **37.** $\dfrac{(x+1)^{3/2}x^{4/3}}{(x^2+1)^5}\left[\dfrac{3/2}{x+1}+\dfrac{4/3}{x}-\dfrac{10x}{x^2+1}\right]$
39. $(x^4+x^2)^{2x}\left[\dfrac{8x^2+4}{x^2+1}+2\ln(x^4+x^2)\right]$

Exercise 12.9. [page 436]

1. $4e^{4x}$ **3.** $-3e^{-3x}$ **5.** $2e^{2x+6}$ **7.** $-2xe^{-x^2}$ **9.** $\dfrac{1}{2}x^{-1/2}e^{\sqrt{x}}$ **11.** $\dfrac{-e^{1/x}}{x^2}$ **13.** $e^{x^3-x}(3x^2-1)$
15. $e^{(2x+1)^2}(4x+2)$ **17.** $2(e^{4x}+e^{-4x})$ **19.** $(18x+54)e^{9x}+2e^{9x}$ **21.** $2e^{-x}/(2x+4) - e^{-x}\ln(2x+4)$
23. $(1-x)/e^x$ **25.** $e^x/(e^x+1)^2$ **27.** $6(x+1)^3e^{6x} + 3(x+1)^2e^{6x}$ **29.** $2(x^2+4e^{x^3})(2x+12x^2e^{x^3})$
31. $dy/dx = -1/(1+4e^{4y})$ **33.** e^{-x} **35.** $4x^2e^{x^2}+2e^{x^2}$ **37.** $8e^x/(e^x+8)^2$ **39.** $y = -2x+2$
41. $y = 3ex - 2e$ **43.** $b^x/\log_b e$ or $b^x \ln b$ **45.** $3^x/\log_3 e$ or $3^x \ln 3$

Chapter 12 Test [page 438]

1. $6x-10$ **2.** $2x+4$ **3.** 4 **4.** -5 **5.** $18x^2+1/x^2$ **6.** $4x^{-1/2}-12x^3$
7. $\dfrac{1}{2}(2x^2+x)^{-1/2}(4x+1)$ **8.** $4(x^3+7x)^3(3x^2+7)$ **9.** $5(x^3+4x^2+1)+(5x+2)(3x^2+8x)$
10. $(6x^2-1)(5x^4-2x)+12x(x^5-x^2)$ **11.** $-4/x^2 - 2/x^3$ **12.** $(1-x)/2\sqrt{x}(x+1)^2$
13. $14x/(7x^2+4)$ **14.** $6x/(x^2+1)$ **15.** $e^x/(e^x+1)$ **16.** $2xe^{x^2}/(e^{x^2}+1)$ **17.** $10e^{10x}$ **18.** $-3e^{-3x}$
19. $e^{\sqrt{2x+8}}(2x+8)^{-1/2}$ **20.** $-e^{-(x^2+1)^2}4x(x^2+1)$ **21.** $e^x + e^x \ln x + xe^x \ln x$
22. $8[e^x + e^x \ln x + xe^x \ln x]$ **23.** $-(1+y^2)/(2xy-1)$ **24.** $1/(2y+4)$ **25.** $48x^2(x^2+6)^2 + 8(x^2+6)^3$
26. $24(-x+1)^{-5}$ **27.** $3 + 2\ln x$ **28.** $-2(2x^2+y^2)/y^3 = -30/y^3$ **29.** $y = 10x - 15$ **30.** $y = 6x - 6$

Exercise 13.1. [page 443]

1. increasing on $x_1 \leq x \leq x_3$, $x_5 \leq x \leq x_7$
   decreasing on $x_3 \leq x \leq x_5$, $x_7 \leq x \leq x_9$

3. increasing on the x-axis

5. increasing on $-\infty < x \leq 2$
   decreasing on $2 \leq x < \infty$

7. increasing on $1 \leq x < \infty$
   decreasing on $-\infty < x \leq 1$

9. decreasing on the x-axis

11. increasing on $-\infty < x \leq -1$
    decreasing on $-1 \leq x \leq 1$
    increasing on $1 \leq x < \infty$

13. increasing on the x-axis

15. increasing on $-\infty < x \leq -1$
    decreasing on $-1 \leq x \leq 0$
    increasing on $0 \leq x < \infty$

17. decreasing on $-\infty < x \leq -3$
    increasing on $-3 \leq x \leq 2$
    decreasing on $2 \leq x < \infty$

19. decreasing on $-\infty < x \leq -1$
    increasing on $-1 \leq x \leq 0$
    decreasing on $0 \leq x \leq 1$
    increasing on $1 \leq x < \infty$

21. concave upward on $0 < x < \infty$
    concave downward on $-\infty < x < 0$
    point of inflection: $(0, 0)$

23. concave upward on $3 < x < \infty$
    concave downward on $-\infty < x < 3$
    point of inflection: $(3, -54)$

25. concave upward on $-2 < x < \infty$
    concave downward on $-\infty < x < -2$
    point of inflection: $(-2, 14)$

27. concave upward on the x-axis

29. concave upward on $-\infty < x < 0$
    concave downward on $0 < x < 1$
    concave upward on $1 < x < \infty$
    points of inflection: $(0, 0), (1, -1)$

31. increasing on $0 \leq x \leq 2500$

Exercise 13.2. [page 449]

1.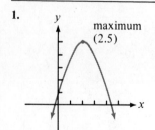
maximum (2.5)

3.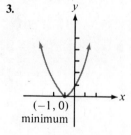
$(-1, 0)$ minimum

5.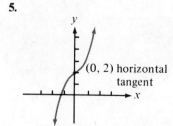
$(0, 2)$ horizontal tangent

Odd-Numbered Answers

7.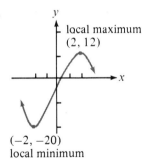

9.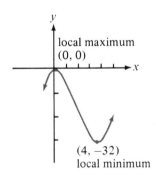

11.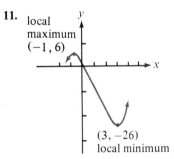

13.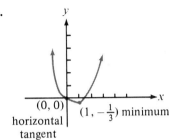

15.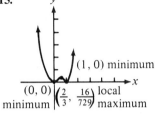

17.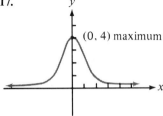

19.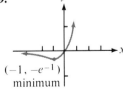

21.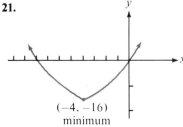

23.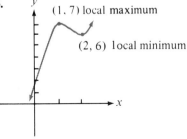

25.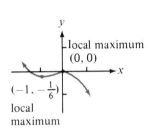

27.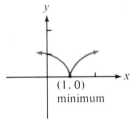

29. ce^{-1}/k **31.** $\sqrt{2bq/a}$ **33.** 700   **35.** $115   **37.** 1024 feet

Exercise 13.3. [page 454]

1. 40, 40 3. $\frac{1}{2}, \frac{1}{2}$ 5. $-\frac{1}{2}$ 7. 250 feet by 250 feet

9. 3200 feet by 400 feet, using 400 feet of wire along the stream 11. 325

13. $r = \sqrt[3]{\frac{16}{\pi}}$ inches, $h = 2\sqrt[3]{\frac{16}{\pi}}$ inches 15. 30 inches by 30 inches by 15 inches

17. $\frac{100}{9}$ inches by $\frac{200}{9}$ inches by $\frac{100}{3}$ inches 19. $x = \frac{1}{4}$ mile

21. Do not cut the wire. The maximum area is enclosed by bending the wire into a circle of radius $5/\pi$ feet. This is an example of an absolute maximum occurring at an endpoint of an interval

Exercise 13.4. [page 459]

1. $2\pi r$ 3. $\frac{\sqrt{3}}{2} x$ 5. re^{-kt} mg/min 7. $abce^{-ct-be^{-ct}}$ 9. -300 units per dollar

11. 532π cm^3/min 13. $-16k$ watts/cm^3 15. $v(2) = 36, a(2) = 50$ 17. $v(1) = 2.5, a(1) = 2.5$

19. 6 seconds; $v(6) = -96$ ft/sec

21. $v(2) = 60$ cm/min, $a(2) = 24$ cm/min^2; increasing on $0 \le t \le 6$, decreasing on $6 \le t < \infty$; 108 cm/min

23. $(ca - b)/(x + c)^2$

Exercise 13.5. [page 463]

1. $7/\pi$ ft/min 3. $1/9\pi$ cm/sec 5. 1 in/sec 7. $2\sqrt{3}/3$ ft/sec 9. $5/2\pi$ ft/min

11. $(8\pi P \times 10^{-9})/vl$ cm^3/min^2

Exercise 13.6. [page 469]

1. 5/3 3. 8/21 5. 3/2 7. 9/8; 9% 9. $0 \le x < 125$ 11. 28; 27

13. $MC(10) = 40$, 15. 398 17. increasing on 19. -30
    $MC(11) = 44$, $30 \le x < \infty$
    exact cost is 42

Chapter 13 Test [page 471]

1. decreasing on $-\infty < x \le -3$
   increasing on $-3 \le x \le 4$
   decreasing on $4 \le x < \infty$

2. decreasing on $-\infty < x \le 0$
   increasing on $0 \le x < \infty$

Odd-Numbered Answers

3. concave upward on $2 < x < \infty$
   concave downward on $-\infty < x < 2$
   point of inflection: $(2, -4)$

4. concave downward on $-\infty < x < -2$
   concave upward on $-2 < x < 0$
   concave downward on $0 < x < 2$
   points of inflection: $(-2, 224)$, $(2, -224)$, $(0, 0)$

5.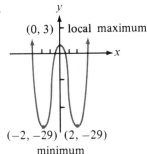

6.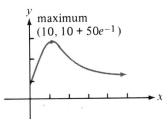

7. Travel $15 - 8/\sqrt{3}$ miles up the road from camp and $16/\sqrt{3}$ miles diagonally across the beach

8. 100 calculators 9. -0.84 deg/min 10. $100\sqrt{\pi}$ in²/sec 11. 11; $MP(14) = 12$ 12. 1

Exercise 14.1. [page 477]

1. $5x + c$ 3. $\dfrac{x^5}{5} + c$ 5. $\dfrac{2}{5}x^{5/2} + c$ 7. $3x^{1/3} + c$ 9. $\dfrac{3}{2}x^4 + c$ 11. $2x^{1/2} + c$

13. $-\dfrac{40}{3}x^{-3/2} + c$ 15. $x^3 + x + c$ 17. $\dfrac{x^3}{3} - \dfrac{3}{2}x^2 + x + c$ 19. $6x^{3/2} - 2x^4 + c$

21. $\dfrac{x^3}{3} + 2x^2 + 4x + c$ 23. $\dfrac{x^4}{4} + 3x^2 + c$ 25. $\dfrac{6}{5}x^{5/6} + 4x^{1/2} + c$ 27. $x^2 + x + c$

29. $x^4 + 4x^2 - 3x + c$ 31. $y = 2x^2 + 3$ 33. $y = x^3 + x + 14$ 35. $y = 6x^{3/2} - 48$

37. $P(x) = 300x - 0.01x^2 - 29{,}000$ 39. $Q(x) = 0.01x + 800/x - 4$ 41. $s(t) = t^2 + 4t + 5$

43. $s(t) = -t^3 + 5$ 45. $s(t) = 3t^2 + 2t - 4$

Exercise 14.2. [page 484]

1. $8 \ln x + c$ 3. $x^2 - 4 \ln x + c$ 5. $\dfrac{x^3}{3} + 2 \ln x - \dfrac{x^{-3}}{3} + c$ 7. $\dfrac{1}{11}(x + 1)^{11} + c$ 9. $\dfrac{1}{18}(3x + 1)^6 + c$

11. $\dfrac{2}{3}(x + 1)^{3/2} + c$ 13. $\dfrac{2}{25}(5x + 3)^{5/2} + c$ 15. $\dfrac{3}{5}(x + 1)^{5/3} + c$ 17. $\dfrac{1}{4}(x^2 - 4)^4 + c$

19. $\dfrac{1}{30}(2x^3 + 1)^5 + c$ 21. $-\dfrac{1}{10}(5x^2 + 2)^{-1} + c$ 23. $\dfrac{1}{12}(4x^2 + 4x + 1)^3 + c$ 25. $(x^2 + 2x + 2)^{1/2} + c$

27. $\ln(x-3)+c$ **29.** $\frac{1}{4}\ln(4x+9)+c$ **31.** $\frac{1}{2}\ln(x^2+1)+c$ **33.** $\frac{1}{2}\ln(x^2+2x+8)+c$

35. $\frac{1}{3}\ln(x^3+5)+c$ **37.** $\frac{1}{2}e^{2x}+c$ **39.** $-e^{-x}+c$ **41.** $\frac{e^x-e^{-x}}{2}+c$ **43.** $\frac{1}{2}e^{x^2}+c$

45. $-\frac{1}{3}e^{-x^3}+c$ **47.** $e^{x^2+x}+c$ **49.** $\frac{1}{10}(e^{2x}+1)^5+c$ **51.** $\ln(e^x+1)+c$ **53.** $\frac{1}{2}(\ln x)^2+c$

Exercise 14.3. [page 489]

1. $-xe^{-x}-e^{-x}+c$ **3.** $\frac{1}{5}xe^{5x}-\frac{1}{25}e^{5x}+c$ **5.** $\frac{1}{2}x^2e^{2x}-\frac{1}{2}xe^{2x}+\frac{1}{4}e^{2x}+c$ **7.** xe^x+c

9. $x\ln 2x - x + c$ **11.** $2x\ln x - 2x + c$ **13.** $\frac{1}{2}x^2\ln x - \frac{x^2}{4}+c$ **15.** $\frac{2}{3}x^{3/2}\ln x - \frac{4}{9}x^{3/2}+c$

17. $x(\ln x)^2 - 2x\ln x + 2x + c$ **19.** $(x+1)\ln(x+1) - x + c$ **21.** $\frac{1}{2}(x+1)^2\ln(x+1) - \frac{1}{4}(x+1)^2 + c$

23. $\frac{1}{6}x(x+2)^6 - \frac{1}{42}(x+2)^7 + c$ **25.** $\frac{1}{6}x(4x+1)^{3/2} - \frac{1}{60}(4x+1)^{5/2}+c$

27. $\frac{1}{5}(x+1)(x+6)^5 - \frac{1}{30}(x+6)^6 + c$ **29.** $\frac{1}{2}e^{x^2}(x^2-1)+c$

31. $\frac{1}{n+1}x^{n+1}\ln x - \frac{1}{(n+1)^2}x^{n+1} + c, n \neq -1$

Exercise 14.4. [page 493]

1. 15 **3.** 3/2 **5.** 4/3 **7.** 7/4 **9.** 0 **11.** 9/2 **13.** 0 **15.** 156 **17.** $1+\ln 2$ **19.** 26/3

21. $-81/4$ **23.** 15/8 **25.** $e-e^{-1}$ **27.** $\frac{1}{2}(e^4-1)$ **29.** $\frac{1}{2}(e-1)$ **31.** $\ln 8 - \ln 2 = \ln 4$

33. $\frac{1}{2}(\ln 10 - \ln 2) = \frac{1}{2}\ln 5$ **35.** 49/20 **37.** e^2 **39.** $8\ln 4 - 15/4$ **41.** 1/3 **43.** 0

45. $\ln(e+1) - \ln 2$ **47.** $\int_1^4 3x^2 dx = 63$ $\int_1^2 3x^2 dx + \int_2^4 3x^2 dx = 7 + 56 = 63$

Chapter 14 Test [page 495]

1. $\frac{2}{5}x^{2.5}+c$ **2.** $-50x^{-0.02}+c$ **3.** x^2-8x+c **4.** $6x^3-3x^2+5x+c$ **5.** $x^{1/2}+c$

6. $\frac{1}{2}\ln x + c$ **7.** $\frac{1}{6}(2x+1)^3+c$ **8.** $\frac{2}{5}(x+1)^{5/2}+c$ **9.** $\frac{1}{3}(3x^2-4)^{1/2}+c$ **10.** $2(2x^3-4)^{1/2}+c$

Odd-Numbered Answers

11. $\frac{1}{3}\ln(x^3 - 6x^2 + 1) + c$ **12.** $\frac{1}{2}(x^3 - 6x^2 + 1)^{2/3} + c$ **13.** $-10e^{-0.1x} + c$ **14.** $2x^2 - 7x + \frac{1}{9}e^{9x} + c$

15. $\frac{1}{2}xe^{2x} - \frac{1}{4}e^{2x} + c$ **16.** $\frac{1}{4}e^{2x^2} + c$ **17.** $2e^{\sqrt{x}} + c$ **18.** $-e^{1/x} + c$ **19.** $\frac{1}{10}(\ln x)^{10} + c$

20. $\frac{1}{10}x^{10}\ln x - \frac{1}{100}x^{10} + c$ **21.** 2 **22.** 1/5 **23.** 4/9 **24.** $\ln 3$ **25.** $3\ln 2 - 1$

26. $\frac{7}{2}\ln 2 - \frac{3}{4}$

Exercise 15.1. [page 502]

1. 16 **3.** 2 **5.** 1 **7.** 20 **9.** 9/2 **11.** 8 **13.** 9 **15.** 6 **17.** 18 **19.** 4/3
21. 23/3 **23.** 8/3 **25.** 27/4 **27.** 14/3 **29.** 82/3 **31.** 3/2 **33.** 1 **35.** $1 + \ln 2$
37. $e - e^{-3}$ **39.** $e + e^{-2} - 1$ **41.** $6\ln 6 - 5$ **43.** 1

Exercise 15.2. [page 506]

1. 8 **3.** 9 **5.** 3 **7.** 4/3 **9.** 4/3 **11.** 9 **13.** 4/3 **15.** 81/4 **17.** 36 **19.** $\frac{3}{4} - \ln 2$
21. 32/3 **23.** 5 **25.** 1 **27.** $e^3 + e^{-3} - 2$ **29.** $e + e^{-1} - 2$

Exercise 15.3. [page 510]

1. 3 **3.** 4 **5.** 11 **7.** 0 **9.** 13/6 **11.** $\frac{1}{6}(e^3 - e^{-3})$ **13.** $\frac{1}{2}\ln 5$ **15.** $\frac{1}{7}(8\ln 8 - 7)$

17. $\frac{PR^2}{6vl}$ **19.** 22,500 **21.** 625/6

Exercise 15.4. [page 516]

1. $6\frac{2}{3}$ l/min **3.** $\pi PR^4/8vl$ **5.** 44 units of dollars **7.** $CS = 1800, PS = 1200$ **9.** $CS = 6, PS = 4\frac{2}{3}$

11. $CS = 288, PS = 125\frac{1}{3}$ **13.** $30,000 **15.** $1500 **17.** $34,418.75 **19.** $\frac{e^{-2} - e^{-1/2}}{e^{-2} - e^{-1/4}} \approx 0.73$

Chapter 15 Test [page 518]

1. a. 16 **b.** 20 **2.** 36 **3. a.** 8/3 **b.** 59/6 **4.** 32/3 **5.** 45 **6.** 64 **7.** 249 units of dollars

8. $CS = 32, PS = 6\frac{2}{3}$ **9.** $30,000 **10.** $75,852

Exercise 16.1. [page 523]

1. 12; -16 **3.** 19; 7   **5.** 2; 4   **7.** 38; 9   **9.** 1/8; 1   **11.** 72.4 cal/hr   **13.** $V(x, y) = \dfrac{xy(150 - xy)}{x + y}$

15. $P(x, y) = 40x + 125y$ **17.** $7091   **19.**

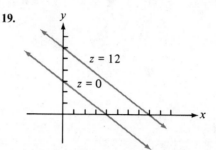

Exercise 16.2. [page 527]

1. $\partial z/\partial x = 5$
    $\partial z/\partial y = 3$

3. $\partial z/\partial x = 2x$
    $\partial z/\partial y = 2y$

5. $\partial z/\partial x = 2x + y^2$
    $\partial z/\partial y = 2xy + 12y^3$

7. $\partial z/\partial x = 39x^2 + 8xy^{1/2} - 1$
    $\partial z/\partial y = 2x^2 y^{-1/2}$

9. $\partial z/\partial x = 6x + 4$
    $\partial z/\partial y = 0$

11. $\partial z/\partial x = 6(3x + 4y)$
    $\partial z/\partial y = 8(3x + 4y)$

13. $\partial z/\partial x = 3(x^3 - 2x + y^2)^2(3x^2 - 2)$
    $\partial z/\partial y = 3(x^3 - 2x + y^2)^2(2y)$

15. $\partial z/\partial x = y^2 e^{xy^2}$
    $\partial z/\partial y = 2xy e^{xy^2}$

17. $\partial z/\partial x = 6/(6x + y^2)$
    $\partial z/\partial y = 2y/(6x + y^2)$

19. $\partial z/\partial x = e^{x^2 y}(2x^2 y + 1)$
    $\partial z/\partial y = x^3 e^{x^2 y}$

21. $\partial z/\partial x = 4x^4 y/(x^4 + 2y^2) + y \ln(x^4 + 2y^2)$
    $\partial z/\partial y = 4xy^2/(x^4 + 2y^2) + x \ln(x^4 + 2y^2)$

23. $\partial z/\partial x = 1/2\sqrt{x}(2y - 6)$
    $\partial z/\partial y = -2\sqrt{x}/(2y - 6)^2$

25. $\partial z/\partial x = -7y/(x - 3y)^2$
    $\partial z/\partial y = 7x/(x - 3y)^2$

27. $\left.\dfrac{\partial z}{\partial x}\right|_{(2, 1)} = 8$

29. $\left.\dfrac{\partial z}{\partial y}\right|_{(1, 2)} = 5$

31. $\partial V/\partial r = 2\pi rh$
    $\partial V/\partial h = \pi r^2$

33. $C = 0.15xy + 1.2/x + 1.2/y$
    $\partial C/\partial x = 0.15y - 1.2/x^2$
    $\partial C/\partial y = 0.15x - 1.2/y^2$

35. $\partial z/\partial x = -5$, $\partial z/\partial y = 4$, $\partial w/\partial x = 2$, $\partial w/\partial y = -3$

37. $-1/25$; $-9/50$

39. $-3/10$; $-3/10$

Odd-Numbered Answers

Exercise 16.3. [page 531]

1. $\partial^2 z/\partial x^2 = 8$ 3. $\partial^2 z/\partial x^2 = 4y$ 5. $\partial^2 z/\partial x^2 = 6xy^2 + 2$ 7. $\partial^2 z/\partial x^2 = y^2 e^{xy}$
9. $\partial^2 z/\partial x^2 = 24x^2(x^2 + y^2) + 6(x^2 + y^2)^2$ 11. $\partial^2 z/\partial x^2 = 4y/(4x + y)^2 + 4/(4x + y)$ 13. $\partial^2 z/\partial y^2 = -6y^2$
15. $\partial^2 z/\partial y^2 = 6x^2 y - 3xy^{-3/2}$ 17. $\partial^2 z/\partial y^2 = 2/y^3$ 19. $\partial^2 z/\partial y^2 = 10/(1 + 6x)$ 21. $\partial^2 z/\partial y^2 = 40x^2 e^{-2xy}$
23. $\partial^2 z/\partial y^2 = (12x^4 y^2 - 4y^6)/(x^4 + y^4)^2$ 25. $\partial^2 x/\partial x \partial y = 15x^2 y^4$ 27. $\partial^2 z/\partial x \partial y = 20xy - 2$
29. $\partial^2 z/\partial x \partial y = 2x^3 y e^{x^2 y} + 2x e^{x^2 y}$ 31. $\partial^2 z/\partial x \partial y = 1/x$ 33. $\partial^2 z/\partial x \partial y = 24x^3 y^2 - 2x = \partial^2 z/\partial y \partial x$

35. For example, $f_x = A\alpha x^{\alpha - 1} y^\beta = \dfrac{\alpha}{x}(Ax^\alpha y^\beta) = \dfrac{\alpha z}{x}$

$f_{xx} = A\alpha(\alpha - 1)x^{\alpha - 2} y^\beta = \dfrac{\alpha(\alpha - 1)}{x^2}(Ax^\alpha y^\beta)$

$= \dfrac{\alpha(\alpha - 1)z}{x^2}$

Exercise 16.4. [page 536]

1. local minimum value: $f(0, 0) = 4$ 3. local maximum value: $f(3, 4) = 25$
5. local minimum value: $f(1, -2) = -55$ 7. local minimum value: $f(1, 3) = -15$
9. local minimum value: $f(1, 1) = -10$, local maximum value: $f(-1, -1) = 10$
11. local maximum value: $f(1, 1) = 6$ 13. no extrema
15. $x = 6$, $y = 6$, $z = 6$ will give a maximum product of 216
17. maximum profit occurs at $x = 100$, $y = 175$ 19. No, $P(x, y)$ has no critical points.
21. Selling prices: $x = 60$ cents, $y = 40$ cents. The restaurant should sell 50 of the better burgers and 20 of the cheaper kind

Exercise 16.5. [page 543]

1. constrained maximum value: $f(1, 1) = 2$
   constrained minimum value: $f(-1, -1) = -2$
3. constrained minimum value: $f(2, 1) = 5$
5. constrained minimum value: $f(2, 4) = 71$
7. constrained minimum value: $f(0, 2, 0) = 4, f(0, -2, 0) = 4$
9. constrained maximum value: $f(0, 3, 0) = 9, f(0, -3, 0) = 9$
   constrained minimum value: $f(1, 0, 0) = 1, f(-1, 0, 0) = 1$
11. constrained maximum volume: $512\sqrt{3}/9$; dimensions: $\dfrac{8}{\sqrt{3}} \times \dfrac{8}{\sqrt{3}} \times \dfrac{8}{\sqrt{3}}$

Chapter 16 Test [page 544]

1. 42; -26

2. -2; 80

3. $\partial z/\partial x = 3x^2 - 6xy^3$
   $\partial z/\partial y = -9x^2 y^2 + 24y^3$

4. $\partial z/\partial x = 2x + \dfrac{1}{2} x^{-1/2} y^{1/2}$
   $\partial z/\partial y = \dfrac{1}{2} x^{1/2} y^{-1/2}$

5. $\partial z/\partial x = -y e^{-(x+5y^2)}$
   $\partial z/\partial y = -10 y^2 e^{-(x+5y^2)} + e^{-(x+5y^2)}$

6. $\partial z/\partial x = 2x/(x^2 + 8y^4)$
   $\partial z/\partial y = 32 y^3/(x^2 + 8y^4)$

7. $\partial^2 z/\partial x^2 = 8 - 24 x^2 y^5$
   $\partial^2 z/\partial y^2 = -40 x^4 y^3 + 2$

8. $\partial^2 z/\partial x^2 = 4 x^{-3} y^4 + 12 x^{-5}$
   $\partial^2 z/\partial y^2 = -6y + 24 x^{-1} y^2$

9. $\partial^2 z/\partial x\, \partial y = 72(4x + 3y)$

10. $\partial^2 z/\partial x \partial y = 21 x^2 y^6 e^{x^3 + y^7}$

11. local minimum value: $f(-1, 3) = 0$

12. local maximum value: $f(-2, -2) = 9$

13. constrained maximum value: $f(3.6, 1.2) = 21.6$

14. constrained maximum value: $f(4, 4) = 26, f(-4, -4) = 26$;
    constrained minimum value: $f(4, -4) = -6, f(-4, 4) = -6$

15. Minimum cost of 108 cents occurs when the box has dimensions of 6 inches by 6 inches by 3 inches

INDEX

Abscissa, 110
Absolute extrema, 451
Absolute inequality, 86
Absolute value
   definition of, 12
   equations involving, 95
   functions involving, 145
   inequalities involving, 96
Acceleration, 457
Addition of matrices, 224
Additive inverse
   of a matrix, 225
   of a real number, 545
Algebraic expression, 16
Amortization, 386
Annuities, 380ff
   future value of, 381
   present value of, 385
Antiderivative, 473ff
Antilogarithm, 170
Appreciation, 133
Area
   between two graphs, 503ff
   under a graph, 496ff
Arithmetic progression
   definition of, 309
   nth term of an, 309
   sum of n terms of an, 315
Associative law
   of addition, 226, 545
   of multiplication, 231, 546
Augmented matrix, 243
Average rate of change, 401
Average value of a function, 507ff
Average velocity, 402

Base
   of a logarithm, 157
   of a power, 16, 158
Bernoulli trials, 363
Binomial, 17
Binomial expansion
   coefficients in a, 330
   rth term of a, 331
Binomial probability, 363ff
Binomial theorem, 330
Bougher-Lambert law, 161
Bracket function, 147
Break-even point, 216

Cardiac output, 515
Cartesian coordinate system
   in the plane, 109
   in three dimensions, 206, 520
Cartesian product, 109
Celsius scale, 123
Chain rule, 420
Change in profit, 511
Characteristic of a logarithm, 169
Circular permutation, 344
Coefficient matrix, 243
Cofactor, 247
Column matrix, 223
Column vector, 223
Combinations, 345
Common difference, 309
Common logarithm, 168
Common ratio, 310
Commutative law
   of addition, 546
   of multiplication, 546

Complement of a set, 7
Complementary events, 348
Completing the square in a quadratic equation, 75
Complex fractions, 39
Compound interest, 184, 376, 390
Concavity, 441ff
Conditional inequality
   definition of, 86
   solution by sign graph, 90
Conditional probability, 368ff
Conformable matrices, 231
Conjugate of a binomial, 54
Consistent equations, 198
Constant function, 119
Constrained extrema, 538ff
Consumer's surplus, 512
Continuous compounding of interest, 388ff
Continuous function, 414, 439
Convergent sequence, 322
Convex polygon, 283
Convex set, 282
Counting properties, 338
Critical point, 533
Critical value, 445
Cramer's rule, 265ff
Cube root, 47

Decreasing function, 159, 440
Definite integral, 490ff
Degree of a polynomial, 17
Deleted Cartesian set, 359
Demand function, 218, 512
Dependent events, 359

605

Depreciation, 130
Derivative
  applications of, 439ff
  chain rule, 420
  of a constant, 410
  definition of, 404
  of exponential functions, 434
  higher-order, 425, 529
  implicit, 422
  logarithmic, 433
  of logarithmic functions, 429
  partial, 524ff
  of powers, 409, 419
  of a product, 413
  of a quotient, 414
  as a rate of change, 456ff
  of a sum, 410
Determinant(s)
  cofactor of an element of a, 247
  in Cramer's rule, 265ff
  definition of. 246
  expansion of a, 248
  in inverse of a matrix, 252ff
  of a matrix, 246
  minor of an element of a, 247
  order of a, 246
  solution of system of equations by, 266
  value of a, 247
Diagonal matrix, 233
Diagonal of a matrix, 232
Difference
  of matrices, 225
  of real numbers, 546
Differential, 480
Dimension of a matrix, 223
Direct variation, 149
Directed distance, 120
Discontinuous function, 439
Discriminant, 76
Disjoint sets, 7
Distance between two points, 120
Distributive law, 231, 546
Division
  definition for real numbers, 546
  of polynomials, 28ff
Domain
  of an exponential function, 158
  of a function, 111
  of a logarithmic function, 163
  of a relation, 111
Dual problem, 298ff

e
  decimal approximation for, 159, 389
  defined as a limit of a sequence, 388
  logarithms to the base, 188
  table for e^x, 552
  table for $\log_e$, 553
Elasticity of demand, 465
Element of a set, 3
Elementary transformation
  of an equation, 67
  of an inequality, 86
  of matrices, 241
Empty set, 3
Equality of matrices, 224
Equation(s)
  consistent, 198
  dependent, 198, 200
  equivalent, 68
  exponential, 184ff
  first-degree, 69
  inconsistent, 198
  involving absolute value, 95
  involving radicals, 80ff
  linear, 69
  logarithmic, 184ff
  matrix form for linear system of, 243, 259
  of a parabola, 134
  quadratic, 73ff
  second-degree, 73ff
  of a straight line, 117ff
  systems of linear, in three variables, 264ff
  systems of linear, in two variables, 197ff
  for word problems, 99ff
Equilibrium point, 217
Equivalent equations, 68
Equivalent expressions, 16
Equivalent fractions, 33ff
Equivalent inequalities, 86
Equivalent systems of equations, 198
Events, 348
Expansion of a binomial, 330
Expansion of a determinant by cofactors, 248
Exponent(s)
  definition of, 17
  laws of, 22, 29, 44
  as a logarithm, 164
  rational, 47ff
Exponential equations, 184
Exponential functions, 158ff
Extraneous solutions, 80

Factorial notation, 328
Factoring polynomials, 25ff
Fahrenheit scale, 123
First-degree equations, 69
First-degree function, 117
First-degree relation, 275
First-derivative test for a local maximum or minimum, 446
Fraction(s), 33ff, 547
Function(s)
  bracket, 147
  constant, 119
  continuous, 439
  decreasing, 159, 440
  definition of, 111
  domain of a, 111
  exponential, 158ff
  even, 117
  graph of a, 112, 117ff
  increasing, 159, 440
  inverse of a, 152
  involving absolute value, 145
  linear, 117ff
  logarithmic, 162ff
  marginal, 467
  notation, 113
  objective, 286
  odd, 117
  one-to-one, 151
  polynomial, 142ff
  power, 149
  quadratic, 134
  range of a, 111
  several variables, 519ff
  step, 147
  zeros of a, 137
Future value, 377, 390

Geometric progression
  definition of, 307
  nth term of a, 310
  sum of n terms of a, 316
Gompertz curve, 460
Graph(s)
  of an absolute-value function, 145
  of a bracket function, 147
  of a constant function, 119
  of an exponential function, 158ff
  of inequalities, 88, 90
  intercepts of a, 118
  of inverse relations, 151
  of a linear equation in three variables, 206
  of a linear function, 117
  of a linear inequality, 88, 276
  of a logarithmic function, 163
  of an ordered pair, 110
  of a polynomial function, 142
  of a quadratic function, 134ff
  of a real number, 11
  of a relation, 110ff
  of a second-degree equation, 136
  of systems of equations, 200, 206, 213

Index

of systems of linear inequalities, 278
turning point of a, 143
Greater than, 11

Half-plane, 276
Half-space, 284
Hematocrit level, 450
Higher-order derivatives, 425, 529
Horizontal line, 118

Identity element
  for addition, 226, 545
  for multiplication, 233, 546
Implicit differentiation, 422ff
Inconsistent equations, 198, 206
Increasing function, 159, 440
Indefinite integral, 473
Independent events, 359
Index of a radical, 52
Index of summation, 318
Inequalities
  absolute, 86
  conditional, 86
  involving absolute value, 96ff
  linear, 87
  quadratic, 90ff
  solution of, 86ff
  systems of, 278ff
  transformations of, 86
Instantaneous rate of change, 402
Integers, set of, 10
Integral
  definite, 490ff
  indefinite, 473
Integral exponents, 43
Integration, 473ff
Integration by parts, 485ff
Intensity level, 175
Intercept
  form for a linear equation, 128
  of a line, 118
Interest, 377, 378
Interpolation, 170
Intersection of sets, 5
Inverse
  additive, for matrices, 226
  of an exponential function, 162ff
  of a function, 152
  multiplicative for a square matrix, 251ff
Inverse relations, 151
Inverse variation, 149
Irrational number, 10

Kelvin scale, 127

Lagrange multipliers, 538ff
Laminar flow, 139

Learning curve, 159
Least common denominator, 36
Least common multiple, 36
Length of a line segment, 119
Leontief model, 261
Less than, 11
Line graph, 88
Linear combination, 199
Linear depreciation, 130
Linear equation(s)
  definition, 69
  intercept form of a, 129
  point-slope form of a, 127
  slope-intercept form of a, 128
  standard form of a, 117
  systems of, 197
  two-point form of a, 132
Linear function, 118, 285
Linear inequality, 87
Linear interpolation, 170
Linear programming, 285ff
Linear system(s)
  coefficient matrix of a, 243, 259
  solution of, by determinants, 266
    by linear combination, 185, 190
    by matrices, 243, 259
  in three variables, 203ff
  in two variables, 197ff
Linearly dependent equations, 198
Linearly independent equations, 198
Local extrema, 141, 444
Logarithm(s)
  to base e, 167, 188ff
  to base 10, 167ff
  changing base of a, 188
  characteristic of a, 169
  common, 167
  computations with, 179ff
  as an exponent, 163
  laws of, 164
  mantissa of a, 169
  natural, 188
  table of $\log_e$, 553
  table of $\log_{10}$, 550–551
Logarithmic differentiation, 433
Logarithmic equations, 184ff
Logarithmic functions, 162ff
Logistic function, 192, 459

Malthus, law of, 196
Mantissa of a logarithm, 169
Mapping, 111
Marginal functions, 467
Mathematical expectation, 356
Matrix (matrices), 222ff
  augmented, 243
  coefficient, 243

column, 223
definition of, 223
determinant of a square, 246ff
diagonal, 233
difference of two, 225
elementary transformation of a, 241
entries of a, 223
equality of, 224
identity, for addition, 226
  for multiplication, 233
inverse of a square, 251ff
negative of a, 225
nonsingular, 242, 254
order of a, 223
principal diagonal of a, 232
product of two, 230
product with a real number, 229
rank of a, 268
row, 223
row-equivalent, 241
singular, 254
square, 232
sum of two, 224
transpose of a, 224
zero, 225
Maximum point
  local, 143
  of an objective function, 286
  of a parabola, 136
Member of a set, 3
Method of substitution, 200
Minimum point
  local, 143
  of an objective function, 286
  of a parabola, 136
Minor, 247
Mixed partial derivatives, 529
Monomial, 17
Multiplication
  identity matrix for, 233
  of a matrix and a real number, 229
Multiplicative identity, 233, 546
Multiplicative inverse, 18, 250, 546
Multiplicity of solutions, 74
Mutually exclusive events, 355

Natural logarithm, 188
Natural number, 3, 10
Negative
  of a matrix, 225
  of a number, 11
  real numbers, 11
Nonnegative real numbers, 11
Nonsingular matrix, 242, 254
Normal distribution, 162
Notation
  scientific, 47

Notation (*continued*)
   summation, 317
nth root, 47
nth term
   of an arithmetic progression, 309
   of a geometric progression, 310
Null set, 3
Number line, 10

Objective function, 286
Odds, 353
One-to-one function, 151
Open sentence, 67
Operations on sets, 5ff
Order
   of a determinant, 246
   of a matrix, 223
   of a radical, 52
   symbols for, 11
Ordered pair, 109
Ordinate, 110
Origin, 11
Outcome, 348

Parabola, 134
   axis of symmetry, 135
   vertex of, 135
Partial differentiation, 524ff
Partial sums, 323
Permutation(s)
   circular, 344
   definition of, 340
   distinguishable, 341
   of n things at a time, 340
   of n things r at a time, 341
pH, 175
Pivotal
   column, 293
   entry, 293
   row, 293
Plane, equation of, 206
Point of inflection, 442
Point-slope form of a linear equation, 127
Poiseuille's law, 516
Polynomial(s)
   completely factored, 25
   definition of a, 17
   degree of a, 17
   prime, 25
   products of, 21ff
   quotient of, 28ff
   sums of, 16
Polynomial function
   definition of, 142
   graph of a, 143ff
   turning points of graph of a, 143
Positive real number, 11

Power(s)
   definition of, 16
   with integral exponents, 43
   with rational exponents, 47
Power function, 148
Power rule, 409, 419
Present value, 377, 384ff
Prime polynomial, 25
Primal, 299
Probability
   a posteriori, 348, 349
   a priori, 348
   binomial, 363ff
   conditional, 368ff
   function, 348
   meaning of, 348
   of mutually exclusive events, 355
Producer's surplus, 512
Product(s)
   involving fractions, 39ff
   of matrices, 230
   of a matrix and a real number, 229
   of polynomials, 21ff
   of rational expressions, 39ff
Product rule, 413
Profit, 138

Quadratic equation(s), 73ff
   discriminant of a, 76
   graph of a, 134ff
   solution of, 73ff
   standard form of a, 73
   systems involving, 209ff
Quadratic formula, 75
Quadratic function, 134ff
Quadratic inequalities, 90ff
Quotient(s)
   of polynomials, 28ff
   of rational expressions, 39
Quotient rule, 414

Radical expression, 52
Radicand, 52
Range
   of exponential function, 159
   of a function, 111
   of a logarithmic function, 163
   of a relation, 111
   of summation, 318
Rank of a matrix, 268
Rate of change of a function
   average, 401
   instantaneous, 402
Rational exponents, 47
Rational expression, 33
Rational number(s), 10
   decimal numeral for a, 10
   set of, 10

Rationalizing
   denominators, 53
   numerators, 54
Real numbers, set of, 10
Reciprocal, 18
Rectangular coordinate system, 109
Related rates, 461ff
Relation(s)
   definition of, 110
   domain of a, 111
   inverse, 151
   range of a, 111
Repeating decimals, 10
Revenue, 138
Reynold's number, 450
Richter scale, 179
Rise of a line segment, 121
Root(s)
   cube, 47
   of an equation, 67
   nth, 47
   square, 47
Row-equivalent matrices, 241
Row matrix, 223
Row vector, 223
Rules of differentiation, 408ff
Run of a line segment, 121

Sample point, 348
Sample space, 348
Scientific notation, 47
Second-degree equation, 73
Second-derivative test for a local maximum or minimum, 448
Sequence(s)
   alternating, 323
   arithmetic, 308
   convergent, 322
   definition of, 307
   divergent, 323
   general term of a, 308
   geometric, 309
   limit of a, 322
   strictly increasing, 321
Series
   arithmetic, 315
   definition of, 314
   geometric, 316
   infinite geometric, 324
Set(s)
   complement of a, 7
   designation of a, 3
   disjoint, 7
   element of a, 3
   empty, 3
   of integers, 10
   intersection of, 5
   or irrational numbers, 10
   of real numbers, 10

Index

union of, 5
universal, 4
Set-builder notation, 4
Set function, probability, 348
Sigma notation, 317
Sign graph, 90
Simple interest, 122, 375
Simplex method, 291ff
Singular matrix, 254
Slack variables, 292
Slope of a line segment, 121
Slope of a tangent line, 398
Slope-intercept form of a linear equation, 128
Solution(s)
  of an equation, 67
  of equations by substitution, 84
  of exponential equations, 184ff
  of a linear equation, 69
  of a linear inequality, 87ff
  of a linear system using matrices, 259ff
  of a linear system using determinants, 265
  of quadratic equations, 73ff
  of quadratic inequalities, 90ff
  of a system by substitution, 200
  of systems in three variables, 203ff
  of systems in two variables, 197ff
  of systems of inequalities, 278ff
  of systems of nonlinear equations, 209ff
Solution set(s)
  of an equation, 68
  of an inequality, 86
  of a system, 198, 203
Square matrix, 232
Square root, 47
Standard form
  of a linear equation, 117
  of a quadratic equation, 73

Step function, 147
Subset
  of a combination, 345
  of an event, 348
  definition, 3
Substitution, solutions of equations by, 84
Subtraction
  of matrices, 225
  of real numbers, 546
Sum(s)
  of an infinite geometric progression, 324
  of an infinite series, 318, 324
  involving fractions, 36ff
  of matrices, 224
  of polynomials, 16ff
Summation
  index of, 318
  range of, 318
Summation notation, 317
Supply function, 218, 512
System(s)
  equivalent, 198, 204
  of linear equations in three variables, 203ff
  of linear equations in two variables, 197ff
  of linear inequalities, 278ff
  of nonlinear equations, 209ff

Table(s)
  amount of an annuity of $1 per period, 556
  compound amount of $1, 554
  of e^x, 552
  interpolation in, 170
  of $\log_e$, 553
  of $\log_{10}$, 550–551
  present value of an annuity of $1 per period, 557
  present value of $1, 555

Tangent to a graph, 397
Terminating decimals, 10
Tests for a local maximum or minimum
  for functions of two variables, 534
  using the first derivative, 446
  using the second derivative, 448
Total revenue, 513
Transformations, elementary
  of equations, 68
  of inequalities, 86
  of matrices, 241
Transpose of a matrix, 224
Trinomial, 17
Turning point of a graph, 143
Two-point form of a linear equation, 132

Unconditional inequality, 86
Union of sets, 5
Universal set, 4

Value
  of a derivative, 405
  of a determinant, 247
  of a function, 113
Variable, 4
Variation
  direct, 149
  inverse, 149
Vector
  column, 223
  row, 223
Velocity, 402, 457
Venn diagrams, 4
Vertical line, 119

Word problems, 99ff

Zeros of a function, 137
Zero matrix, 225